Essentials of Ecology

Robert VanTuyl

Essentials of Ecology

FIFTH EDITION

G. Tyler Miller, Jr.

Scott E. Spoolman

BROOKS/COLE
CENGAGE Learning™

Australia • Brazil • Japan • Korea • Mexico • Singapore • Spain • United Kingdom • United States

BROOKS/COLE
CENGAGE Learning™

Essentials of Ecology, 5e
G. Tyler Miller, Jr. and Scott E. Spoolman

Vice President, Editor-in-Chief: Michelle Julet

Publisher: Yolanda Cossio

Development Editor: Christopher Delgado

Assistant Editor: Lauren Oliveira

Editorial Assistant: Samantha Arvin

Media Editor: Kristina Razmara

Marketing Manager: Amanda Jellerichs

Marketing Assistant: Katherine Malatesta

Marketing Communications Manager:
Linda Yip

Project Manager, Editorial Production:
Andy Marinkovich

Creative Director: Rob Hugel

Art Director: John Walker

Print Buyer: Karen Hunt

Permissions Editor: John Hill

Production Service/Compositor:
Thompson-Steele, Inc.

Text Designer: Carolyn Deacy

Photo Researcher: Abigail Reip

Copy Editor: Andrea Fincke

Illustrator: Patrick Lane, ScEYEence Studios;
Rachel Ciemma

Cover Image: © JUPITERIMAGES/Comstock
Images/Alamy

For product information and technology assistance, contact us at
Cengage Learning Customer & Sales Support, 1-800-354-9706

For permission to use material from this text or product,
submit all requests online at **www.cengage.com/permissions**
Further permissions questions can be e-mailed to
permissionrequest@cengage.com

Library of Congress Control Number: 2008933001

ISBN-13: 978-0-495-55795-1

ISBN-10: 0-495-55795-1

Brooks/Cole
10 Davis Drive
Belmont, CA 94002-3098
USA

Cengage Learning is a leading provider of customized learning solutions with office locations around the globe, including Singapore, the United Kingdom, Australia, Mexico, Brazil, and Japan. Locate your local office at **international.cengage.com/region**

Cengage Learning products are represented in Canada by Nelson Education, Ltd.

For your course and learning solutions, visit **academic.cengage.com**.

Purchase any of our products at your local college store or at our preferred online store **www.ichapters.com**.

Printed in China by China Translation & Printing Services Limited
2 3 4 5 6 7 12 11 10 09

Brief Contents

About the Cover Photo

Scarlet Macaw   This strikingly beautiful parrot species lives in the subtropical forests in Central and South America, including Costa Rica, southern Panama, and the Amazon Basin in Brazil and Peru. They have a lifespan of 30 to 50 years and eat mostly seeds and fruits. The squawks and screams of these noisy birds can be heard for long distances throughout the forests. The scarlet macaws are threatened by their popularity as pets, which is due to their beautiful plumage and affectionate ways with humans. Under an international agreement, it is illegal to remove them from the wild without special permits. However, a number of these rare parrots are illegally captured, smuggled from their native habitats to the United States and Canada, and sold on the black market for thousands of dollars a piece. During their trip north many of the smuggled birds die from stress and poor care. An even worse threat for the scarlet macaw is the clear-cutting and fragmentation of much of its forest habitat, which is taking place at a rapid and increasing rate. For these reasons, scarlet macaws and a number of other tropical bird species are threatened with extinction.

Detailed Contents

Photo 1 The endangered brown pelican was protected in the first U.S. wildlife refuge in Florida.

SuperStock

Photo 2 Homeless people in Calcutta India

Photo 3 Endangered ring-tailed lemur in Madagascar

Photo 4 Temperate deciduous forest, winter, Rhode Island (USA)

Photo 5 Sea star species helps to control mussel populations in intertidal zone communities in the U.S. Pacific northwest.

Nanosys

Photo 6 Flexible solar cells manufactured with use of nanotechnology

International Development Enterprises

Photo 8 Treadle pump used to supply irrigation water in parts of Bangladesh and India

Francois Suchel/Peter Arnold, Inc.

Photo 7 Cypress swamp, an inland wetland in U.S. state of Tennessee

SUSTAINING BIODIVERSITY

Photo 9 Energy efficient *straw bale house* in Crested Butte, Colorado (USA) during construction

Photo 10 Completed energy efficient *straw bale house* in Crested Butte, Colorado (USA)

Photo 11 Roof garden in Wales, Machynlleth (UK)

Photo 12 Photochemical smog in Mexico City, Mexico

SUPPLEMENTS

Pierre A. Pittet/UN Food and Agriculture Organization

Photo 13 Cow dung is collected and burned as a fuel for cooking and heating in India.

For Instructors

What's New

In this edition, we build on proven strengths of past editions with the following major new features:

- New concept-centered approach
- Quantitative Data Analysis or Ecological Footprint Analysis exercise at the end of each chapter and additional Data Analysis exercises in the Supplements
- New design along with many new pieces of art and photographs
- Comprehensive review section at the end of each chapter with review questions that include all chapter key terms in boldface

This edition also introduces a new coauthor, **Scott Spoolman,** who worked as a contributing editor on this and other environmental science textbooks by Tyler Miller for more than 4 years. (See About the Authors, p. xxiii.)

New Concept-Centered Approach

Each major chapter section is built around one to three **key concepts**—a major new feature of this edition. These concepts state the most important take-away messages of each chapter. They are listed at the front of each chapter (see Chapter 9, p. 184), and each chapter section begins with a key question and concepts (see Chapter 9, pp. 189, 193, and 206), which are highlighted and referenced throughout each chapter.

A logo ⟵CONCEPT LINKS in the margin links the material in each chapter to appropriate key concepts in foregoing chapters (see pp. 101, 145, and 219).

New Design

The concepts approach is well-served by our new design, which showcases the concepts, core case studies, and other new features as well as proven strengths of this textbook. The new design (see Chapter 1, pp. 5–27), which enhances visual learning, also incorporates a thoroughly updated art program with 134 new or upgraded diagrams and 44 new photos—amounting to half of the book's 337 figures.

Sustainability Remains as the Integrating Theme of This Book

Sustainability, a watchword in the 21st Century for those concerned about the environment, is the overarching theme of this introductory ecological textbook. You can see the sustainability emphasis by looking at the Brief Contents (p. iii).

Four **scientific principles of sustainability** play a major role in carrying out this book's sustainability theme. These principles are introduced in Chapter 1, depicted in Figure 1-17 (p. 23 and the back cover of the student edition), and used throughout the book, with each reference is marked in the margin by 🌿. (See Chapter 3, pp. 59, 60, 65, 74, and 75.)

Core Case Studies and the Sustainability Theme

Each chapter opens with a **Core Case Study** (see Chapter 5, p. 100), which is applied throughout the chapter. These connections to the **Core Case Study** are indicated in the book's margin by ⟵CORE CASE STUDY. (See Chapter 5, pp. 102, 103, 104, 108, 110, 111, 119, and 120.)

Each chapter ends with a *Revisiting* box (see Chapter 5, p. 119), which connects the **Core Case Study** and other material in the chapter to the four **scientific principles of sustainability. Thinking About** exercises placed throughout each chapter (see Chapter 7, pp. 144, 145, 146, 148, 152, 157, and 159) challenge students to make these and other connections for themselves.

Five Subthemes Guide the Way toward Sustainability

In the previous edition of this book, we used five major subthemes, which are carried on in this new edition: *natural capital, natural capital degradation, solutions, trade-offs,* and *individuals matter* (see diagram on back cover of student edition).

- *Natural capital.* Sustainability focuses on the natural resources and natural services that support all life and economies. Examples of diagrams that illustrate this subtheme are Figures 1-3 (p. 8), 8-4 (p. 165), and 10-4 (p. 217).

- **Natural capital degradation.** We describe how human activities can degrade natural capital. Examples of diagrams that illustrate this subtheme are Figures 1-7 (p. 12), 6-A (p. 124), and 10-15 (p. 225).

- **Solutions.** Next comes the search for *solutions* to natural capital degradation and other environmental problems. We present proposed solutions in a balanced manner and challenge students to use critical thinking to evaluate them. A number of figures and chapter sections and subsections present proven and possible solutions to various environmental problems. Examples are Section 9-4 (pp. 206–211), Figure 10-17 (p. 227), and Figure 10-19 (p. 231).

- **Trade-Offs.** The search for solutions involves *trade-offs,* because any solution requires weighing advantages against disadvantages. (See p. 9 and Figure 10-9, p. 220.)

- **Individuals Matter.** Throughout the book *Individuals Matter* boxes describe what various concerned citizens and scientists have done to help us work toward sustainability. (See pp. 205, 230, and 261.) Also, several *What Can You Do?* boxes describe how readers can deal with the problems we face. Examples are Figures 9-18 (p. 201), 9-24 (p. 210), and 10-29 (p. 245).

Case Studies

In addition to the 11 Core Case Studies described above, 31 additional **Case Studies** (see pp. 93–95, 177–178, and 257–259) appear throughout the book. (See items in BOLD type in the Detailed Contents, pp. v–xiv.) The total of 42 case studies provides an in-depth look at specific environmental problems and their possible solutions.

Critical Thinking

The introduction on *Learning Skills* describes critical thinking skills (pp. 2–4). Specific critical thinking exercises are used throughout the book in several ways:

- As 66 **Thinking About** exercises. This *interactive approach to learning* reinforces textual and graphic information and concepts by asking students to analyze material immediately after it is presented rather than waiting until the end of the chapter (see pp. 56, 62, and 87).

- In all Science Focus boxes (see pp. 54, 188, and 195).

- In the captions of many of the book's figures (see Figures 5-17, p. 117; 7-12, p. 151; and 8-5, p. 166).

- As 10 *How Would You Vote?* exercises (see pp. 10, 114, and 223).

- As end-of-chapter questions (see pp. 120 and 212).

Visual Learning

This book's 233 diagrams—90 of them new to this edition—are designed to present complex ideas in understandable ways relating to the real world. (See Figures 3-18, p. 68; 4-2, p. 79; and 7-11, p. 149.) We have also carefully selected 104 photographs—34 of them new to this edition—to illustrate key ideas. (See Figures 3-4, p. 53; 4-10, p. 89; and 10-11, p. 222.) We have avoided the common practice of including numerous "filler" photographs that are not very effective or that show the obvious.

And to enhance visual learning, nearly 53 *CengageNOW* animations, many referenced in figures (see Figures 8-15, p. 175 and 10-26, p. 241), are available online. *CengageNOW* provides students with a more complete learning experience that takes what students read on the page and places it into a more interactive environment.

Major Changes in This Edition: A Closer Look

Major changes in this new edition include the following:

- New co-author (see p. xxiii)

- Concept-centered approach with each chapter section built around one to three **Key Concepts** that provide the most important messages of each chapter. Each chapter also links material to related key concepts from previous chapters. All of the Key Concepts, listed by chapter, can be found in Supplement 11, page S61.

- New design serving the concept-centered approach and integration of Core Case Studies, with 134 new or upgraded figures and 34 new photographs.

- Expansion of the sustainability theme built around the four **scientific principles of sustainability** (Figure 1-17, p. 23 and the back cover of the student edition)

- Reduced the number of chapters from 12 to 11 by rearranging and combining some material to improve flow.

- 2 new chapter opening **Core Case Studies** (pp. 28 and 50)

- 26 **Science Focus** boxes that provide greater depth on scientific concepts and on the work of environmental scientists (see pp. 197, 235, and 253).

- Connections to *The Habitable Planet,* a set of 13 videos produced by Annenberg Media. Each half-hour video describes research that two different scientists are doing on a particular environmental problem (see pp. 72, 218, and 254).

- *Review* section at the end of each chapter with comprehensive review questions that include all key terms in boldface. (See pp. 74, 75, and 180.)

- A Data Analysis or Ecological Footprint Analysis exercise at the end of each chapter (see pp. 26, 76, 98–99, and 274) and additional exercises analyzing graphs or maps in the book's Supplements (see pp. S7, S14, and S27).

- **Research Frontier** boxes list key areas of cutting-edge research, with links to such research provided on the website for this book (see pp. 71, 96, and 172).

- **Green Career** items in the text list various green careers with further information found on the website for this book (see pp. 72, 73, and 244).

- Student projects listed by chapter are found in Supplement 10, pp. S59–S60. Some instructors may find these useful for getting students more deeply involved in key environmental issues.

- *Active Graphing* exercises in *CengageNOW* for many chapters that involve students in the graphing and evaluation of data.

- Improved flow and content based on input from 47 new reviewers (identified by an asterisk in the List of Reviewers on pp. xx–xxii).

- More than 2,000 updates based on information and data published in 2005, 2006, 2007, and 2008.

- Integration of material on the growing ecological and economic impacts of China. (See Index citations for China.)

- Many new or expanded topics including expanded treatment of ecological footprints (Figures 1-9, p. 14, and 1-10, p. 15, and ecological footprint calculations at the end of a number of chapters); additional maps of global economic, population, hunger, health, and waste production data (Supplement 3, pp. S10–S19); revisiting Easter Island (p. 31); tipping points (p. 46); tropical forest losses (p. 50); hurricanes and New Orleans (pp. 177–178); tropical forest fragmentation (p. 195); vultures and rabies (p. 197); disappearing honeybees (pp. 202–203); threatened polar bears (p. 203); Jane Goodall (p. 205); effects of gray wolves on the Yellowstone ecosystem (p. 235); Blackfoot reconciliation ecology (pp. 244–245); restoring mangroves (p. 255); and endangered marine turtles (pp. 259–260).

In-Text Study Aids

Each chapter begins with a list of *key questions and concepts* showing how the chapter is organized and what students will be learning. When a new term is introduced and defined, it is printed in boldface type, and all such terms are summarized in the glossary at the end of the book and highlighted in review questions at the end of each chapter.

Sixty-six *Thinking About* exercises reinforce learning by asking students to think critically about the implications of various environmental issues and solutions immediately after they are discussed in the text. The captions of many figures contain questions that involve students in thinking about and evaluating their content.

Each chapter ends with a *Review* section containing a detailed set of review questions that include all chapter key terms in boldface (p. 75), followed by a set of *Critical Thinking* (p. 180) questions to encourage students to think critically and apply what they have learned to their lives.

Supplements for Students

A multitude of electronic supplements available to students take the learning experience beyond the textbook:

- *CengageNOW* is an online learning tool that helps students access their unique study needs. Students take a pre-test and a personalized study plan provides them with specific resources for review. A post-test then identifies content that might require further study. *How Do I Prepare* tutorials, another feature of *CengageNOW,* walk students through basic math, chemistry, and study skills to help them brush up quickly and be ready to succeed in their course.

- *WebTutor* on WebCT or Blackboard provides qualified adopters of this textbook with access to a full array of study tools, including flash cards, practice quizzes, animations, exercises, and web links.

- *Audio Study Tools.* Students can download these useful study aids, which contain valuable information such as reviews of important concepts, key terms, questions, clarifications of common misconceptions, and study tips.

- Access to *InfoTrac® College Edition* for teachers and students using *CengageNOW* and *WebTutor* on WebCT or Blackboard. This fully searchable online library gives users access to complete environmental articles from several hundred current periodicals and others dating back over 20 years.

The following materials for this textbook are available on the companion website at

academic.cengage.com/biology/miller

- *Chapter Summaries* help guide student reading and study of each chapter.

- *Flash Cards* and *Glossary* allow students to test their mastery of each chapter's Key Terms.

- *Chapter Tests* provide multiple-choice practice quizzes.

- Information on a variety of *Green Careers.*

- *Readings* list major books and articles consulted in writing each chapter and include suggestions for articles, books, and websites that provide additional information.

- *What Can You Do?* offers students resources for what they can do to effect individual change on key environmental issues.

- *Weblinks* and *Research Frontier Links* offer an extensive list of websites with news and research related to each chapter.

Other student learning tools include:

- *Essential Study Skills for Science Students* by Daniel D. Chiras. This book includes chapters on developing good study habits, sharpening memory, getting the most out of lectures, labs, and reading assignments, improving test-taking abilities, and becoming a critical thinker. Available for students on instructor request.

- *Lab Manual.* New to this edition, this lab manual includes both hands-on and data analysis labs to help your students develop a range of skills. Create a custom version of this Lab Manual by adding labs you have written or ones from our collection with Cengage Custom Publishing. An Instructor's Manual for the labs will be available to adopters.

- *What Can You Do?* This guide presents students with a variety of ways that they can affect the environment, and shows them how to track the effect their actions have on their ecological footprint. Available for students on instructor request.

Supplements for Instructors

- *PowerLecture.* This DVD, available to adopters, allows you to create custom lectures in Microsoft® PowerPoint using lecture outlines, all of the figures and photos from the text, bonus photos, and animations from *CengageNOW.* PowerPoint's editing tools allow use of slides from other lectures, modification or removal of figure labels and leaders, insertion of your own slides, saving slides as JPEG images, and preparation of lectures for use on the Web.

- *Instructor's Manual.* Available to adopters. Updated and reorganized, the Instructor's Manual has been thoughtfully revised to make creating your lectures even easier. Some of the features new to this edition include the integration of the case studies and feature boxes into the lecture outline, a new section on teaching tips, and a revised video reference list with web resources. Also available on Power-Lecture.

- *Test Bank.* Available to Adopters. The test bank contains thousands of questions and answers in a variety of formats. New to this edition are short essay questions to further challenge your students' understanding of the topics. Also available on PowerLecture.

- *Transparencies.* Featuring all the illustrations from the chapters, this set contains 250 printed Transparencies of key figures, and 250 electronic Masters. These electronic Masters will allow you to print, in color, only those additional figures you need.

- *ABC Videos for Environmental Science.* The 45 informative and short video clips cover current news stories on environmental issues from around the world. These clips are a great way to start a lecture or spark a discussion. Available on DVD with a workbook, on the PowerLecture, and in *Cengage-NOW* with additional internet activities.

- *ExamView.* This full-featured program helps you create and deliver customized tests (both print and online) in minutes, using its complete word processing capabilities.

Other Textbook Options

Instructors wanting a book with a different length and emphasis can use one of our three other books that we have written for various types of environmental science courses: *Living in the Environment,* 16th edition (674 pages, Brooks/Cole 2009), *Environmental Science,* 12th edition (430 pages, Brooks/Cole 2008), and *Sustaining the Earth: An Integrated Approach,* 9th edition (339 pages, Brooks/Cole, 2009).

Help Us Improve This Book

Let us know how you think this book can be improved. If you find any errors, bias, or confusing explanations, please e-mail us about them at

mtg89@hotmail.com

spoolman@tds.net

Most errors can be corrected in subsequent printings of this edition, as well as in future editions.

Acknowledgments

We wish to thank the many students and teachers who have responded so favorably to the 4 previous editions of *Essentials of Ecology,* 15 previous editions of *Living in the Environment,* the 12 editions of *Environmental Science,* and the 8 editions of *Sustaining the Earth,* and who have corrected errors and offered many helpful suggestions for improvement. We are also deeply indebted to the more than 295 reviewers, who pointed out errors and suggested many important improvements in the various editions of these four books. We especially want to thank the reviewers of the latest edition of this book, who are identified by an asterisk in the master list of reviewers on pp. xx–xxii.

It takes a village to produce a textbook, and the members of the talented production team, listed on the copyright page, have made vital contributions as well. Our special thanks go to developmental editor Christopher Delgado, production editors Andy Marinkovich and Nicole Barone, copy editor Andrea Fincke, layout expert Bonnie Van Slyke, photo researcher Abigail Reip, artist Patrick Lane, media editor Kristina Razmara, assistant editor Lauren Oliveira, editorial assistant Samantha Arvin, and Brooks/Cole's hard-working sales staff.

We also thank Ed Wells and the dedicated team who developed the *Laboratory Manual* to accompany this book, and the people who have translated this book into eight languages for use throughout much of the world.

We also deeply appreciate having had the opportunity to work with Jack Carey, former biology publisher at Brooks/Cole, for 40 years before his recent retirement. We now are fortunate and excited to be working with Yolanda Cossio, the biology publisher at Brooks/Cole.

G. Tyler Miller, Jr.
Scott Spoolman

Guest Essayists

Guest essays by the following authors are available on *CengageNOW:* **M. Kat Anderson,** ethnoecologist with the National Plant Center of the USDA's Natural Resource Conservation Center; **Lester R. Brown,** president, Earth Policy Institute; **Michael Cain,** ecologist and adjunct professor at Bowdoin College; **Herman E. Daly,** senior research scholar at the School of Public Affairs, University of Maryland; **Garrett Hardin,** professor emeritus (now deceased) of human ecology, University of California, Santa Barbara; **Paul G. Hawken,** environmental author and business leader; **Jane Heinze-Fry,** environmental educator; **Amory B. Lovins,** energy policy consultant and director of research, Rocky Mountain Institute; **Bobbi S. Low,** professor of resource ecology, University of Michigan; **Lester W. Milbrath,** former director of the research program in environment and society, State University of New York, Buffalo; **Peter Montague,** director, Environmental Research Foundation; **Norman Myers,** tropical ecologist and consultant in environment and development; **David W. Orr,** professor of environmental studies, Oberlin College; **Vandana Shiva,** physicist, educator, environmental consultant; **Nancy Wicks,** ecopioneer and director of Round Mountain Organics; **Donald Worster,** environmental historian and professor of American history, University of Kansas.

Quantitative Exercise Contributors

Dr. Dean Goodwin and his colleagues, Berry Cobb, Deborah Stevens, Jeannette Adkins, Jim Lehner, Judy Treharne, Lonnie Miller, and Tom Mowbray, provided excellent contributions to the Data Analysis and Ecological Footprint Analysis exercises.

Cumulative Reviewers (Reviewers of the 5th edition are indicated by an asterisk.)

Barbara J. Abraham, Hampton College; Donald D. Adams, State University of New York at Plattsburgh; Larry G. Allen, California State University, Northridge; Susan Allen-Gil, Ithaca College; James R. Anderson, U.S. Geological Survey; Mark W. Anderson, University of Maine; Kenneth B. Armitage, University of Kansas; Samuel Arthur, Bowling Green State University; Gary J. Atchison, Iowa State University; *Thomas W. H. Backman, Lewis Clark State University; Marvin W. Baker, Jr., University of Oklahoma; Virgil R. Baker, Arizona State University; *Stephen W. Banks, Louisiana State University in Shreveport; Ian G. Barbour, Carleton College; Albert J. Beck, California State University, Chico; *Eugene C. Beckham, Northwood University; *Diane B. Beechinor, Northeast Lakeview College; W. Behan, Northern Arizona University; *David Belt, Johnson County Community College; Keith L. Bildstein, Winthrop College; *Andrea Bixler, Clarke College; Jeff Bland, University of Puget Sound; Roger G. Bland, Central Michigan University; Grady Blount II, Texas A&M University, Corpus Christi; *Lisa K. Bonneau, University of Missouri-Kansas City; Georg Borgstrom, Michigan State University; Arthur C. Borror, University of New Hampshire; John H. Bounds, Sam Houston State University; Leon F. Bouvier, Population Reference Bureau; Daniel J. Bovin, Universitè Laval; *Jan Boyle, University of Great Falls; *James A. Brenneman, University of Evansville; Michael F. Brewer, Resources for the Future, Inc.; Mark M. Brinson, East Carolina University; Dale Brown, University of Hartford; Patrick E. Brunelle, Contra Costa College; Terrence J. Burgess, Saddleback College North; David Byman, Pennsylvania State University, Worthington–Scranton; Michael L. Cain, Bowdoin College, Lynton K. Caldwell, Indiana University; Faith Thompson Campbell, Natural Resources Defense Council, Inc.; *John S. Campbell, Northwest College; Ray Canterbery, Florida State University; Ted J. Case, University of San Diego; Ann Causey, Auburn University; Richard A. Cellarius, Evergreen State University; William U. Chandler, Worldwatch Institute; F. Christman, University of North Carolina, Chapel Hill; Lu Anne Clark, Lansing Community College; Preston Cloud, University of California, Santa Barbara; Bernard C. Cohen, University of Pittsburgh; Richard A. Cooley, University of California, Santa Cruz; Dennis J. Corrigan; George Cox, San Diego State University; John D. Cunningham, Keene State College; Herman E. Daly, University of Maryland; Raymond F. Dasmann, University of California, Santa Cruz; Kingsley Davis, Hoover Institution; Edward E. DeMartini, University of California, Santa Barbara; *James Demastes, University of Northern Iowa; Charles E. DePoe, Northeast Louisiana University; Thomas R. Detwyler, University of Wisconsin; *Bruce DeVantier, Southern Illinois University Carbondale; Peter H. Diage, University of California, Riverside; *Stephanie Dockstader, Monroe Community College; Lon D. Drake, University of Iowa; *Michael Draney, University of Wisconsin - Green Bay; David DuBose, Shasta College; Dietrich Earnhart, University of Kansas; *Robert East, Washington & Jefferson College; T. Edmonson, University of Washington; Thomas Eisner, Cornell University; Michael Esler, Southern Illinois University; David E. Fairbrothers, Rutgers University; Paul P. Feeny, Cornell University; Richard S. Feldman, Marist College; *Vicki Fella-Pleier, La Salle University; Nancy Field, Bellevue Community College; Allan Fitzsimmons, University of Kentucky; Andrew J. Friedland, Dartmouth College; Kenneth O. Fulgham, Humboldt State University; Lowell L. Getz, University of Illinois at Urbana–Champaign; Frederick F. Gilbert, Washington State University; Jay Glassman, Los Angeles Valley College; Harold Goetz, North Dakota State University; *Srikanth Gogineni, Axia College of University of Phoenix; Jeffery J. Gordon, Bowling Green State University; Eville Gorham, University of Minnesota; Michael Gough, Resources for the Future; Ernest M. Gould, Jr., Harvard University; Peter Green, Golden West College; Katharine B. Gregg, West Virginia Wesleyan College; Paul K. Grogger, University of Colorado at Colorado Springs; L. Guernsey, Indiana State University; Ralph Guzman, University of California, Santa Cruz; Raymond Hames, University of Nebraska, Lincoln; *Robert Hamilton IV, Kent State University, Stark Campus; Raymond E. Hampton, Central Michigan University; Ted L. Hanes, California State University, Fullerton; William S. Hardenbergh, Southern Illinois University at Carbondale; John P. Harley, Eastern Kentucky University; Neil A. Harriman, University of Wisconsin, Oshkosh; Grant A. Harris, Washington State University; Harry S. Hass, San Jose City College; Arthur N. Haupt, Population Reference Bureau; Denis A. Hayes, environmental consultant; Stephen Heard, University of Iowa; Gene Heinze-Fry, Department of Utilities, Commonwealth of Massachusetts; Jane Heinze-Fry, environmental educator; John G. Hewston, Humboldt State University; David L. Hicks, Whitworth College; Kenneth M. Hinkel, University of Cincinnati; Eric Hirst, Oak Ridge National Laboratory; Doug Hix, University of Hartford; S. Holling, University of British Columbia; Sue Holt, Cabrillo College; Donald Holtgrieve, California State University, Hayward; *Michelle Homan, Gannon University; Michael H. Horn, California State University, Fullerton; Mark A. Hornberger, Bloomsburg University; Marilyn Houck, Pennsylvania State University; Richard D. Houk, Winthrop College; Robert J. Huggett, College of William and Mary; Donald Huisingh, North Carolina State University; *Catherine Hurlbut, Florida Community College at Jacksonville; Marlene K. Hutt, IBM; David R. Inglis, University of Massachusetts; Robert Janiskee, University of South Carolina; Hugo H. John, University of Connecticut; Brian A. Johnson, University of Pennsylvania, Bloomsburg; David I. Johnson, Michigan State University; Mark Jonasson, Crafton Hills College; *Zoghlul Kabir, Rutgers/New Brunswick; Agnes

Kadar, Nassau Community College; Thomas L. Keefe, Eastern Kentucky University; *David Kelley, University of St. Thomas; William E. Kelso, Louisiana State University; Nathan Keyfitz, Harvard University; David Kidd, University of New Mexico; Pamela S. Kimbrough; Jesse Klingebiel, Kent School; Edward J. Kormondy, University of Hawaii–Hilo/West Oahu College; John V. Krutilla, Resources for the Future, Inc.; Judith Kunofsky, Sierra Club; E. Kurtz; Theodore Kury, State University of New York at Buffalo; Steve Ladochy, University of Winnipeg; *Troy A. Ladine, East Texas Baptist University; *Anna J. Lang, Weber State University; Mark B. Lapping, Kansas State University; *Michael L. Larsen, Campbell University; *Linda Lee, University of Connecticut; Tom Leege, Idaho Department of Fish and Game; *Maureen Leupold, Genesee Community College; William S. Lindsay, Monterey Peninsula College; E. S. Lindstrom, Pennsylvania State University; M. Lippiman, New York University Medical Center; Valerie A. Liston, University of Minnesota; Dennis Livingston, Rensselaer Polytechnic Institute; James P. Lodge, air pollution consultant; Raymond C. Loehr, University of Texas at Austin; Ruth Logan, Santa Monica City College; Robert D. Loring, DePauw University; Paul F. Love, Angelo State University; Thomas Lovering, University of California, Santa Barbara; Amory B. Lovins, Rocky Mountain Institute; Hunter Lovins, Rocky Mountain Institute; Gene A. Lucas, Drake University; Claudia Luke; David Lynn; Timothy F. Lyon, Ball State University; Stephen Malcolm, Western Michigan University; Melvin G. Marcus, Arizona State University; Gordon E. Matzke, Oregon State University; Parker Mauldin, Rockefeller Foundation; Marie McClune, The Agnes Irwin School (Rosemont, Pennsylvania); Theodore R. McDowell, California State University; Vincent E. McKelvey, U.S. Geological Survey; Robert T. McMaster, Smith College; John G. Merriam, Bowling Green State University; A. Steven Messenger, Northern Illinois University; John Meyers, Middlesex Community College; Raymond W. Miller, Utah State University; Arthur B. Millman, University of Massachusetts, Boston; *Sheila Miracle, Southeast Kentucky Community & Technical College; Fred Montague, University of Utah; Rolf Monteen, California Polytechnic State University; *Debbie Moore, Troy University Dothan Campus; *Michael K. Moore, Mercer University; Ralph Morris, Brock University, St. Catherine's, Ontario, Canada; Angela Morrow, Auburn University; William W. Murdoch, University of California, Santa Barbara; Norman Myers, environmental consultant; Brian C. Myres, Cypress College; A. Neale, Illinois State University; Duane Nellis, Kansas State University; Jan Newhouse, University of Hawaii, Manoa; Jim Norwine, Texas A&M University, Kingsville; John E. Oliver, Indiana State University; *Mark Olsen, University of Notre Dame; Carol Page, copyeditor; Eric Pallant, Allegheny College; *Bill Paletski, Penn State University; Charles F. Park, Stanford University; Richard J. Pedersen, U.S. Department of Agriculture, Forest Service; David Pelliam, Bureau of Land Management, U.S. Department of Interior; *Murray Paton Pendarvis, Southeastern Louisiana University; *Dave Perault, Lynchburg College; Rodney Peterson, Colorado State University; Julie Phillips, De Anza College; John Pichtel, Ball State University; William S. Pierce, Case Western Reserve University; David Pimentel, Cornell University; Peter Pizor, Northwest Community College; Mark D. Plunkett, Bellevue Community College; Grace L. Powell, University of Akron; James H. Price, Oklahoma College; Marian E. Reeve, Merritt College; Carl H. Reidel, University of Vermont; Charles C. Reith, Tulane University; Roger Revelle, California State University, San Diego; L. Reynolds, University of Central Arkansas; Ronald R. Rhein, Kutztown University of Pennsylvania; Charles Rhyne, Jackson State University; Robert A. Richardson, University of Wisconsin; Benjamin F. Richason III, St. Cloud State University; Jennifer Rivers, Northeastern University; Ronald Robberecht, University of Idaho; William Van B. Robertson, School of Medicine, Stanford University; C. Lee Rockett, Bowling Green State University; Terry D. Roelofs, Humboldt State University; *Daniel Ropek, Columbia George Community College; Christopher Rose, California Polytechnic State University; Richard G. Rose, West Valley College; Stephen T. Ross, University of Southern Mississippi; Robert E. Roth, Ohio State University; *Dorna Sakurai, Santa Monica College; Arthur N. Samel, Bowling Green State University; *Shamili Sandiford, College of DuPage; Floyd Sanford, Coe College; David Satterthwaite, I.E.E.D., London; Stephen W. Sawyer, University of Maryland; Arnold Schecter, State University of New York; Frank Schiavo, San Jose State University; William H. Schlesinger, Ecological Society of America; Stephen H. Schneider, National Center for Atmospheric Research; Clarence A. Schoenfeld, University of Wisconsin, Madison; *Madeline Schreiber, Virginia Polytechnic Institute; Henry A. Schroeder, Dartmouth Medical School; Lauren A. Schroeder, Youngstown State University; Norman B. Schwartz, University of Delaware; George Sessions, Sierra College; David J. Severn, Clement Associates; *Don Sheets, Gardner-Webb University; Paul Shepard, Pitzer College and Claremont Graduate School; Michael P. Shields, Southern Illinois University at Carbondale; Kenneth Shiovitz; F. Siewert, Ball State University; E. K. Silbergold, Environmental Defense Fund; Joseph L. Simon, University of South Florida; William E. Sloey, University of Wisconsin, Oshkosh; Robert L. Smith, West Virginia University; Val Smith, University of Kansas; Howard M. Smolkin, U.S. Environmental Protection Agency; Patricia M. Sparks, Glassboro State College; John E. Stanley, University of Virginia; Mel Stanley, California State Polytechnic University, Pomona; *Richard Stevens, Monroe Community College; Norman R. Stewart, University of Wisconsin, Milwaukee; Frank E. Studnicka, University of Wisconsin, Platteville; Chris Tarp, Contra Costa College; Roger E. Thibault, Bowling Green State University; William L. Thomas, California State University, Hayward; Shari Turney, copyeditor; John D. Usis, Youngstown State University; Tinco E. A. van Hylckama, Texas Tech University; Robert R. Van Kirk, Humboldt State University; Donald

E. Van Meter, Ball State University; *Rick Van Schoik, San Diego State University; Gary Varner, Texas A&M University; John D. Vitek, Oklahoma State University; Harry A. Wagner, Victoria College; Lee B. Waian, Saddleback College; Warren C. Walker, Stephen F. Austin State University; Thomas D. Warner, South Dakota State University; Kenneth E. F. Watt, University of California, Davis; Alvin M. Weinberg, Institute of Energy Analysis, Oak Ridge Associated Universities; Brian Weiss; Margery Weitkamp, James Monroe High School (Granada Hills, California); Anthony Weston, State University of New York at Stony Brook; Raymond White, San Francisco City College; Douglas Wickum, University of Wisconsin, Stout; Charles G. Wilber, Colorado State University; Nancy Lee Wilkinson, San Francisco State University; John C. Williams, College of San Mateo; Ray Williams, Rio Hondo College; Roberta Williams, University of Nevada, Las Vegas; Samuel J. Williamson, New York University; *Dwina Willis, Freed-Hardeman University; Ted L. Willrich, Oregon State University; James Winsor, Pennsylvania State University; Fred Witzig, University of Minnesota at Duluth; *Martha Wolfe, Elizabethtown Community and Technical College; George M. Woodwell, Woods Hole Research Center; *Todd Yetter, University of the Cumberlands; Robert Yoerg, Belmont Hills Hospital; Hideo Yonenaka, San Francisco State University; *Brenda Young, Daemen College; *Anita Zvodsk, Barry University; Malcolm J. Zwolinski, University of Arizona.

About the Authors

G. Tyler Miller, Jr.

G. Tyler Miller, Jr., has written 58 textbooks for introductory courses in environmental science, basic ecology, energy, and environmental chemistry. Since 1975, Miller's books have been the most widely used textbooks for environmental science in the United States and throughout the world. They have been used by almost 3 million students and have been translated into eight languages.

Miller has a Ph.D. from the University of Virginia and has received two honorary doctorate degrees for his contributions to environmental education. He taught college for 20 years and developed an innovative interdisciplinary undergraduate science program before deciding to write environmental science textbooks full time since 1975. Currently, he is the President of Earth Education and Research, devoted to improving environmental education.

He describes his hopes for the future as follows:

If I had to pick a time to be alive, it would be the next 75 years. Why? First, there is overwhelming scientific evidence that we are in the process of seriously degrading our own life support system. In other words, we are living unsustainably. Second, within your lifetime we have the opportunity to learn how to live more sustainably by working with the rest of nature, as described in this book.

I am fortunate to have three smart, talented, and wonderful sons—Greg, David, and Bill. I am especially privileged to have Kathleen as my wife, best friend, and research associate. It is inspiring to have a brilliant, beautiful (inside and out), and strong woman who cares deeply about nature as a lifemate. She is my hero. I dedicate this book to her and to the earth.

Scott E. Spoolman

Scott Spoolman is a writer and textbook editor with over 25 years of experience in educational publishing. He has worked with Tyler Miller since 2003 as a contributing editor on earlier editions of *Essentials of Ecology, Living in the Environment, Environmental Science,* and *Sustaining the Earth.*

Spoolman holds a master's degree in science journalism from the University of Minnesota. He has authored numerous articles in the fields of science, environmental engineering, politics, and business. He worked as an acquisitions editor on a series of college forestry textbooks. He has also worked as a consulting editor in the development of over 70 college and high school textbooks in fields of the natural and social sciences.

In his free time, he enjoys exploring the forests and waters of his native Wisconsin along with his family— his wife, environmental educator Gail Martinelli, and his children, Will and Katie.

Spoolman has the following to say about his collaboration with Tyler Miller:

I am honored to be joining with Tyler Miller as a coauthor to continue the Miller tradition of thorough, clear, and engaging writing about the vast and complex field of environmental science. This is the greatest and most rewarding challenge I have ever faced. I share Tyler Miller's passion for ensuring that these textbooks and their multimedia supplements will be valuable tools for students and instructors. To that end, we strive to introduce this interdisciplinary field in ways that will be informative and sobering, but also tantalizing and motivational.

If the flip side of any problem is indeed an opportunity, then this truly is one of the most exciting times in history for students to start an environmental career. Environmental problems are numerous, serious, and daunting, but their possible solutions generate exciting new career opportunities. We place high priorities on inspiring students with these possibilities, challenging them to maintain a scientific focus, pointing them toward rewarding and fulfilling careers, and in doing so, working to help sustain life on earth.

Learning Skills

Students who can begin early in their lives to think of things as connected, even if they revise their views every year, have begun the life of learning.

MARK VAN DOREN

Why Is It Important to Study Environmental Science?

Welcome to **environmental science**—an *interdisciplinary* study of how the earth works, how we interact with the earth, and how we can deal with the environmental problems we face. Because environmental issues affect every part of your life, the concepts, information, and issues discussed in this book and the course you are taking will be useful to you now and throughout your life.

Understandably, we are biased, but *we strongly believe that environmental science is the single most important course in your education.* What could be more important than learning how the earth works, how we are affecting its life support system, and how we can reduce our environmental impact?

We live in an incredibly challenging era. We are becoming increasingly aware that during this century we need to make a new cultural transition in which we learn how to live more sustainably by sharply reducing the degradation of our life-support system. We hope this book will inspire you to become involved in this change in the way we view and treat the earth, which sustains us and our economies and all other living things.

You Can Improve Your Study and Learning Skills

Maximizing your ability to learn should be one of your most important lifetime educational goals. It involves continually trying to *improve your study and learning skills.* Here are some suggestions for doing so:

Develop a passion for learning. As the famous physicist and philosopher Albert Einstein put it, "I have no special talent. I am only passionately curious."

Get organized. Becoming more efficient at studying gives you more time for other interests.

Make daily to-do lists in writing. Put items in order of importance, focus on the most important tasks, and assign a time to work on these items. Because life is full of uncertainties, you might be lucky to accomplish half of the items on your daily list. Shift your schedule as needed to accomplish the most important items.

Set up a study routine in a distraction-free environment. Develop a written daily study schedule and stick to it. Study in a quiet, well-lighted space. Work while sitting at a desk or table—not lying down on a couch or bed. Take breaks every hour or so. During each break, take several deep breaths and move around; this will help you to stay more alert and focused.

Avoid procrastination—putting work off until another time. Do not fall behind on your reading and other assignments. Set aside a particular time for studying each day and make it a part of your daily routine.

Do not eat dessert first. Otherwise, you may never get to the main meal (studying). When you have accomplished your study goals, reward yourself with dessert (play or leisure).

Make hills out of mountains. It is psychologically difficult to climb a mountain, which is what reading an entire book, reading a chapter in a book, writing a paper, or cramming to study for a test can feel like. Instead, break these large tasks (mountains) down into a series of small tasks (hills). Each day, read a few pages of a book or chapter, write a few paragraphs of a paper, and review what you have studied and learned. As American automobile designer and builder Henry Ford put it, "Nothing is particularly hard if you divide it into small jobs."

Look at the big picture first. Get an overview of an assigned reading in this book by looking at the *Key Questions and Concepts* box at the beginning of each chapter. It lists key questions explored in the chapter sections and the corresponding key concepts, which are the critical lessons to be learned in the chapter. Use this list as a chapter roadmap. When you finish a chapter you can also use it to review.

Ask and answer questions as you read. For example, "What is the main point of a particular subsection or paragraph?" Relate your own questions to the key questions and key concepts being addressed in each major chapter section. In this way, you can flesh out a chapter outline to help you understand the chapter material. You may even want to do such an outline in writing.

Focus on key terms. Use the glossary in this textbook to look up the meanings of terms or words you do not understand. This book shows all key terms in **boldface** type and lesser, but still important, terms in *italicized* type. The review questions at the end of each chapter also include the chapter's key terms in boldface. Flash cards for testing your mastery of key terms for each chapter are available on the website for this book, or you can make your own by putting a term on one side of an index card or piece of paper and its meaning on the other side.

Interact with what you read. We suggest that you mark key sentences and paragraphs with a highlighter or pen. Consider putting an asterisk in the margin next to material you think is important and double asterisks next to material you think is especially important. Write comments in the margins, such as *beautiful, confusing, misleading,* or *wrong*. You might fold down the top corners of pages on which you highlighted passages and the top and bottom corners of especially important pages. This way, you can flip through a chapter or book and quickly review the key ideas.

Review to reinforce learning. Before each class session, review the material you learned in the previous session and read the assigned material.

Become a good note taker. Do not try to take down everything your instructor says. Instead, write down main points and key facts using your own shorthand system. Review, fill in, and organize your notes as soon as possible after each class.

Write out answers to questions to focus and reinforce learning. Answer the critical thinking questions found in *Thinking About* boxes throughout chapters, in many figure captions, and at the end of each chapter. These questions are designed to inspire you to think critically about key ideas and connect them to other ideas and to your own life. Also answer the review questions found at the end of each chapter. The website for each chapter has an additional detailed list of review questions. Writing out your answers to the critical thinking and review questions can reinforce your learning. Save your answers for review and preparation for tests.

Use the buddy system. Study with a friend or become a member of a study group to compare notes, review material, and prepare for tests. Explaining something to someone else is a great way to focus your thoughts and reinforce your learning. Attend any review sessions offered by instructors or teaching assistants.

Learn your instructor's test style. Does your instructor emphasize multiple-choice, fill-in-the-blank, true-or-false, factual, or essay questions? How much of the test will come from the textbook and how much from lecture material? Adapt your learning and studying methods to your instructor's style. It may not exactly match your own, but the reality is that your instructor is in charge.

Become a good test taker. Avoid cramming. Eat well and get plenty of sleep before a test. Arrive on time or early. Calm yourself and increase your oxygen intake by taking several deep breaths. (Do this also about every 10–15 minutes while taking the test.) Look over the test and answer the questions you know well first.

Then work on the harder ones. Use the process of elimination to narrow down the choices for multiple-choice questions. Paring them down to two choices gives you a 50% chance of guessing the right answer. For essay questions, organize your thoughts before you start writing. If you have no idea what a question means, make an educated guess. You might get some partial credit and avoid getting a zero. Another strategy for getting some credit is to show your knowledge and reasoning by writing something like this: "If this question means so and so, then my answer is _____."

Develop an optimistic but realistic outlook. Try to be a "glass is half-full" rather than a "glass is half-empty" person. Pessimism, fear, anxiety, and excessive worrying (especially over things you cannot control) are destructive and lead to inaction. Try to keep your energizing feelings of realistic optimism slightly ahead of any immobilizing feelings of pessimism. Then you will always be moving forward.

Take time to enjoy life. Every day, take time to laugh and enjoy nature, beauty, and friendship. You can do this without falling behind in your work and living under a cloud of guilt and anxiety if you become an effective and efficient learner.

You Can Improve Your Critical Thinking Skills: Becoming a Good Baloney Detector

Critical thinking involves developing skills to analyze information and ideas, judge their validity, and make decisions. Critical thinking helps you to distinguish between facts and opinions, evaluate evidence and arguments, take and defend informed positions on issues, integrate information, see relationships, and apply your knowledge to dealing with new and different problems and to your own lifestyle choices. Here are some basic skills for learning how to think more critically.

Question everything and everybody. Be skeptical, as any good scientist is. Do not believe everything you hear and read, including the content of this textbook, without evaluating the information you receive. Seek other sources and opinions. As Albert Einstein put it, "The important thing is not to stop questioning."

Identify and evaluate your personal biases and beliefs. Each of us has biases and beliefs taught to us by our parents, teachers, friends, role models, and experience. What are your basic beliefs, values, and biases? Where did they come from? What assumptions are they based on? How sure are you that your beliefs, values, and assumptions are right and why? According to the American psychologist and philosopher William James, "A great many people think they are thinking when they are merely rearranging their prejudices."

Be open-minded and flexible. Be open to considering different points of view. Suspend judgment until you gather more evidence, and be capable of changing your mind. Recognize that there may be a number of useful and acceptable solutions to a problem and that very few issues are black or white. There are trade-offs involved in dealing with any environmental issue, as you will learn in this book. One way to evaluate divergent views is to try to take the viewpoints of other people. How do they see the world? What are their basic assumptions and beliefs? Are their positions logically consistent with their assumptions and beliefs?

Be humble about what you know. Some people are so confident in what they know that they stop thinking and questioning. To paraphrase American writer Mark Twain, "It's not what we don't know that's so bad. It's what we know is true, but just ain't so, that hurts us."

Evaluate how the information related to an issue was obtained. Are the statements you heard or read based on firsthand knowledge and research or on hearsay? Are unnamed sources used? Is the information based on reproducible and widely accepted scientific studies (*reliable science,* p. 33) or on preliminary scientific results that may be valid but need further testing (*tentative* or *frontier science,* p. 33)? Is the information based on a few isolated stories or experiences (*anecdotal information*) or on carefully controlled studies with the results reviewed by experts in the field involved (*peer review*)? Is it based on unsubstantiated and dubious scientific information or beliefs (*unreliable science,* p. 34)?

Question the evidence and conclusions presented. What are the conclusions or claims? What evidence is presented to support them? Does the evidence support them? Is there a need to gather more evidence to test the conclusions? Are there other, more reasonable conclusions?

Try to uncover differences in basic beliefs and assumptions. On the surface most arguments or disagreements involve differences in opinions about the validity or meaning of certain facts or conclusions. Scratch a little deeper and you will find that most disagreements are usually based on different (and often hidden) basic assumptions concerning how we look at and interpret the world around us. Uncovering these basic differences can allow the parties involved to understand where each is "coming from" and to agree to disagree about their basic assumptions, beliefs, or principles.

Try to identify and assess any motives on the part of those presenting evidence and drawing conclusions. What is their expertise in this area? Do they have any unstated assumptions, beliefs, biases, or values? Do they have a personal agenda? Can they benefit financially or politically from acceptance of their evidence and conclusions? Would investigators with different basic assumptions or beliefs take the same data and come to different conclusions?

Expect and tolerate uncertainty. Recognize that science is an ever-changing adventure that provides only a degree of certainty. Scientists can disprove things but they cannot establish absolute proof or certainty. However, the widely accepted results of reliable science have a high degree of certainty.

Do the arguments used involve logical fallacies or debating tricks? Here are six of many examples. *First,* attack the presenter of an argument rather than the argument itself. *Second,* appeal to emotion rather than facts and logic. *Third,* claim that if one piece of evidence or one conclusion is false, then all other related pieces of evidence and conclusions are false. *Fourth,* say that a conclusion is false because it has not been scientifically proven. (Scientists never prove anything absolutely, but they can often establish high degrees of certainty, as discussed on pp. 33–34.) *Fifth,* inject irrelevant or misleading information to divert attention from important points. *Sixth,* present only either/or alternatives when there may be a number of options.

Do not believe everything you read on the Internet. The Internet is a wonderful and easily accessible source of information, providing alternative explanations and opinions on almost any subject or issue—much of it not available in the mainstream media and scholarly articles. Web logs, or blogs, have become a major source of information, even more important than standard news media for some people. However, because the Internet is so open, anyone can post anything they want to a blog or other website with no editorial control or review by experts. As a result, evaluating information on the Internet is one of the best ways to put into practice the principles of critical thinking discussed here. Use and enjoy the Internet, but think critically and proceed with caution.

Develop principles or rules for evaluating evidence. Develop a written list of principles to serve as guidelines for evaluating evidence and claims. Continually evaluate and modify this list on the basis of your experience.

Become a seeker of wisdom, not a vessel of information. Many people believe that the main goal of education is to learn as much as you can by gathering more and more information. We believe that the primary goal is to learn how to sift through mountains of facts and ideas to find the few *nuggets of wisdom* that are the most useful for understanding the world and for making decisions. This book is full of facts and numbers, but they are useful only to the extent that they lead to an understanding of key ideas, scientific laws, theories, concepts, and connections. The major goals of the study of environmental science are to find out how nature works and sustains itself (*environmental wisdom*) and to use *principles of environmental wisdom* to help make human societies and economies more sustainable, more just, and more beneficial and enjoyable for all. As writer Sandra Carey put it, "Never mistake knowledge for wisdom. One helps you make a living; the other helps you make a life." Or as American writer Walker Percy suggested "some individuals with a high intelligence but lacking wisdom can get all A's and flunk life."

To help you practice critical thinking, we have supplied questions throughout this book—at the end of each chapter, and throughout each chapter in brief boxes labeled *Thinking About* and in the captions of many figures. There are no right or wrong answers to many of these questions. A good way to improve your critical thinking skills is to compare your answers with those of your classmates and to discuss how you arrived at your answers.

Know Your Own Learning Style

People have different ways of learning and it can be helpful to know your own learning style. *Visual learners* learn best from reading and viewing illustrations and diagrams. They can benefit from using flash cards (available on the website for this book) to memorize key terms and ideas. This is a highly visual book with many carefully selected photographs and diagrams designed to illustrate important ideas, concepts, and processes.

Auditory learners learn best by listening and discussing. They might benefit from reading aloud while studying and using a tape recorder in lectures for study and review. *Logical learners* learn best by using concepts and logic to uncover and understand a subject rather than relying mostly on memory.

Part of what determines your learning style is how your brain works. According to the *split-brain hypothesis,* the left hemisphere of your brain is good at logic, analysis, and evaluation, and the right half of the brain is good at visualizing, synthesizing, and creating. Our goal is to provide material that stimulates both sides of your brain.

The study and critical thinking skills encouraged in this book and in most courses largely involve the left brain. However, you can improve these skills by giving your left brain a break and letting your creative side loose. You can do this by brainstorming ideas with classmates with the rule that no left-brain criticism is allowed until the session is over.

When you are trying to solve a problem, rest, meditate, take a walk, exercise, or do something to shut down your controlling left-brain activity, and allow the right side of your brain to work on the problem in a less controlled and more creative manner.

This Book Presents a Positive and Realistic Environmental Vision of the Future

There are always *trade-offs* involved in making and implementing environmental decisions. Our challenge is to give a fair and balanced presentation of different viewpoints, advantages and disadvantages of various technologies and proposed solutions to environmental problems, and good and bad news about environmental problems without injecting personal bias.

Studying a subject as important as environmental science and ending up with no conclusions, opinions, and beliefs means that both teacher and student have failed. However, any conclusions one does reach must result from a process of thinking critically to evaluate different ideas and understand the trade-offs involved. Our goal is to present a positive vision of our environmental future based on realistic optimism.

Help Us Improve This Book

Researching and writing a book that covers and connects ideas in a wide variety of disciplines is a challenging and exciting task. Almost every day, we learn about some new connection in nature.

In a book this complex, there are bound to be some errors—some typographical mistakes that slip through and some statements that you might question, based on your knowledge and research. We invite you to contact us and point out any bias, correct any errors you find, and suggest ways to improve this book. Please e-mail your suggestions to Tyler Miller at **mtg@hotmail.com** or Scott Spoolman at **spoolman@tds.net**.

Now start your journey into this fascinating and important study of how the earth works and how we can leave the planet in a condition at least as good as what we found. Have fun.

Study nature, love nature, stay close to nature. It will never fail you.
FRANK LLOYD WRIGHT

Environmental Problems, Their Causes, and Sustainability

Living in an Exponential Age

Two ancient kings enjoyed playing chess. The winner claimed a prize from the loser. After one match, the winning king asked the losing king to pay him by placing one grain of wheat on the first square of the chessboard, two grains on the second square, four on the third, and so on, with the number doubling on each square until all 64 squares were filled.

The losing king, thinking he was getting off easy, agreed with delight. It was the biggest mistake he ever made. He bankrupted his kingdom because the number of grains of wheat he had promised was probably more than all the wheat that has ever been harvested!

This fictional story illustrates the concept of **exponential growth,** by which a quantity increases at a *fixed percentage* per unit of time, such as 2% per year. Exponential growth is deceptive. It starts off slowly, but after only a few doublings, it grows to enormous numbers because each doubling is more than the total of all earlier growth.

Here is another example. Fold a piece of paper in half to double its thickness. If you could continue doubling the thickness of the paper 42 times, the stack would reach from the earth to the moon—386,400 kilometers (240,000 miles) away. If you could double it 50 times, the folded paper would almost reach the sun—149 million kilometers (93 million miles) away!

Because of exponential growth in the human population (Figure 1-1), in 2008 there were 6.7 billion people on the planet. Collectively, these people consume vast amounts of food, water, raw materials, and energy and in the process produce huge amounts of pollution and wastes. Unless death rates rise sharply, there will probably be 9.3 billion of us by 2050 and perhaps as many as 10 billion by the end of this century.

The exponential rate of global population growth has declined since 1963. Even so, each day we add an average of 225,000 more people to the earth's population. This is roughly equivalent to adding a new U.S. city of Los Angeles, California, every 2 months, a new France every 9 months, and a new United States—the world's third most populous country—about every 4 years.

No one knows how many people the earth can support, and at what level of resource consumption or affluence, without seriously degrading the ability of the planet to support us and other forms of life and our economies. But there are some disturbing warning signs. Biologists estimate that, by the end of this century, our exponentially increasing population and resource consumption could cause the irreversible loss of one-third to one-half of the world's known different types of plants and animals.

There is also growing evidence and concern that continued exponential growth in human activities such as burning *fossil fuels* (carbon-based fuels such as coal, natural gas, and gasoline) and clearing forests will change the earth's climate during this century. This could ruin some areas for farming, shift water supplies, eliminate many of the earth's unique forms of life, and disrupt economies in various parts of the world.

Great news: We have solutions to these problems that we could implement within a few decades, as you will learn in this book.

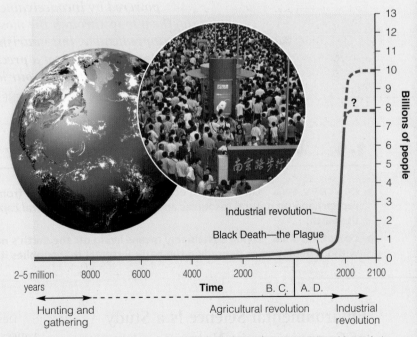

Figure 1-1 *Exponential growth:* the J-shaped curve of past exponential world population growth, with projections to 2100 showing possible population stabilization with the J-shaped curve of growth changing to an S-shaped curve. (This figure is not to scale.) (Data from the World Bank and United Nations; photo L. Yong/UNEP/Peter Arnold, Inc)

1-1 What is an environmentally sustainable society?

CONCEPT 1-1A Our lives and economies depend on energy from the sun (*solar capital*) and on natural resources and natural services (*natural capital*) provided by the earth.

CONCEPT 1-1B Living sustainably means living off the earth's natural income without depleting or degrading the natural capital that supplies it.

1-2 How can environmentally sustainable societies grow economically?

CONCEPT 1-2 Societies can become more environmentally sustainable through economic development dedicated to improving the quality of life for everyone without degrading the earth's life support systems.

1-3 How are our ecological footprints affecting the earth?

CONCEPT 1-3 As our ecological footprints grow, we are depleting and degrading more of the earth's natural capital.

1-4 What is pollution, and what can we do about it?

CONCEPT 1-4 Preventing pollution is more effective and less costly than cleaning up pollution.

1-5 Why do we have environmental problems?

CONCEPT 1-5A Major causes of environmental problems are population growth, wasteful and unsustainable resource use, poverty, exclusion of environmental costs of resource use from the market prices of goods and services, and attempts to manage nature with insufficient knowledge.

CONCEPT 1-5B People with different environmental worldviews often disagree about the seriousness of environmental problems and what we should do about them.

1-6 What are four scientific principles of sustainability?

CONCEPT 1-6 Nature has sustained itself for billions of years by using solar energy, biodiversity, population control, and nutrient cycling—lessons from nature that we can apply to our lifestyles and economies.

---

*This is a *concept-centered* book, with each major chapter section built around one to three key concepts derived from the natural or social sciences. Key questions and concepts are summarized at the beginning of each chapter. You can use this list as a preview and as a review of the key ideas in each chapter.

Note: Supplements 2 (p. S4), 3 (p. S10), 4 (p. S20), 5 (p. S31), and 6 (p. S39) can be used with this chapter.

> *Alone in space, alone in its life-supporting systems,*
> *powered by inconceivable energies,*
> *mediating them to us through the most delicate adjustments,*
> *wayward, unlikely, unpredictable, but nourishing, enlivening, and enriching*
> *in the largest degree—is this not a precious home for all of us?*
> *Is it not worth our love?*
>
> BARBARA WARD AND RENÉ DUBOS

1-1 What Is an Environmentally Sustainable Society?

▶ **CONCEPT 1-1A** Our lives and economies depend on energy from the sun (*solar capital*) and on natural resources and natural services (*natural capital*) provided by the earth.

▶ **CONCEPT 1-1B** Living sustainably means living off the earth's natural income without depleting or degrading the natural capital that supplies it.

Environmental Science Is a Study of Connections in Nature

The **environment** is everything around us. It includes all of the living and the nonliving things with which we interact. And it includes a complex web of relationships that connect us with one another and with the world we live in.

Despite our many scientific and technological advances, we are utterly dependent on the environment for air, water, food, shelter, energy, and everything else we need to stay alive and healthy. As a result, we are part of, and not apart from, the rest of nature.

This textbook is an introduction to **environmental science,** an *interdisciplinary* study of how humans interact with the environment of living and nonliving

Links: refers to the Core Case Study. refers to the book's sustainability theme. indicates links to key concepts in earlier chapters.

Table 1-1

Major Fields of Study Related to Environmental Science

Major Fields	Subfields
Biology: study of living things (organisms)	**Ecology:** study of how organisms interact with one another and with their nonliving environment
	Botany: study of plants
	Zoology: study of animals
Chemistry: study of chemicals and their interactions	**Biochemistry:** study of the chemistry of living things
Earth science: study of the planet as a whole and its nonliving systems	**Climatology:** study of the earth's atmosphere and climate
	Geology: study of the earth's origin, history, surface, and interior processes
	Hydrology: study of the earth's water resources
	Paleontology: study of fossils and ancient life
Social sciences: studies of human society	**Anthropology:** study of human cultures
	Demography: study of the characteristics of human populations
	Geography: study of the relationships between human populations and the earth's surface features
	Economics: study of the production, distribution, and consumption of goods and services
	Political Science: study of the principles, processes, and structure of government and political institutions
Humanities: study of the aspects of the human condition not covered by the physical and social sciences	**History:** study of information and ideas about humanity's past
	Ethics: study of moral values and concepts concerning right and wrong human behavior and responsibilities
	Philosophy: study of knowledge and wisdom about the nature of reality, values, and human conduct

things. It integrates information and ideas from the *natural sciences,* such as biology, chemistry, and geology, the *social sciences,* such as geography, economics, political science, and demography (the study of populations), and the *humanities,* including philosophy and ethics (Table 1-1 and Figure 1-2). The goals of environmental science are to learn *how nature works, how the environment affects us, how we affect the environment, and how to deal with environmental problems and live more sustainably.*

A key subfield of environmental science is **ecology,** the biological science that studies how **organisms,** or living things, interact with their environment and with each other. Every organism is a member of a certain **species:** a group of organisms with distinctive traits and, for sexually reproducing organisms, can mate and produce fertile offspring. For example, all humans are members of a species that biologists have named *Homo sapiens sapiens.* A major focus of ecology is the study of ecosystems. An **ecosystem** is a set of

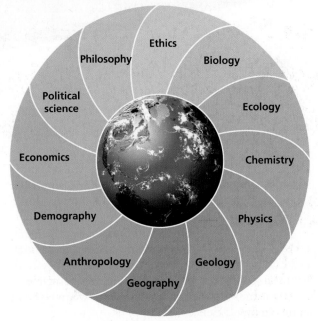

Figure 1-2
Environmental science is an interdisciplinary study of connections between the earth's life-support system and human activities.

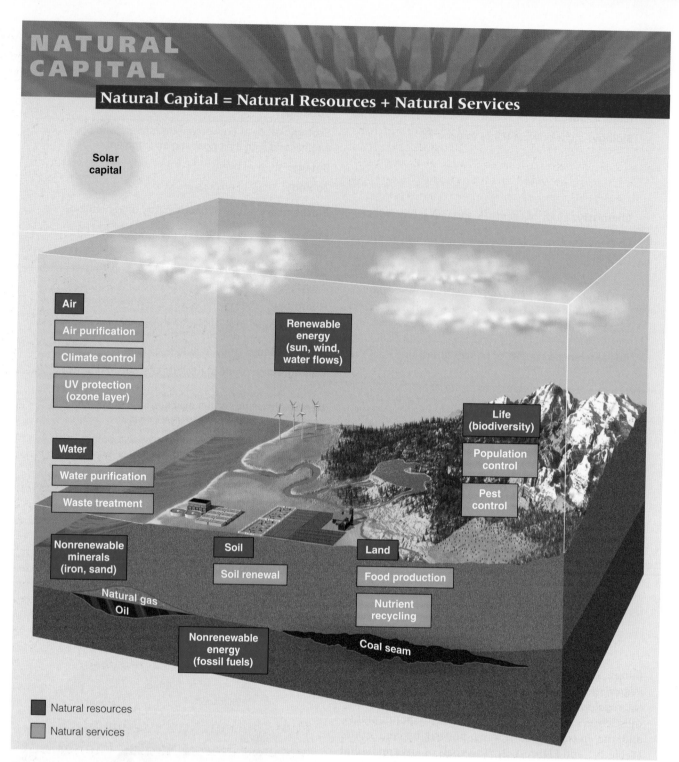

Natural Capital = Natural Resources + Natural Services

Solar capital

Air
- Air purification
- Climate control
- UV protection (ozone layer)

Renewable energy (sun, wind, water flows)

Life (biodiversity)

Population control

Pest control

Water
- Water purification
- Waste treatment

Nonrenewable minerals (iron, sand)

Soil
- Soil renewal

Land
- Food production
- Nutrient recycling

Natural gas
Oil

Nonrenewable energy (fossil fuels)

Coal seam

■ Natural resources

□ Natural services

Figure 1-3 Key *natural resources* (blue) and *natural services* (orange) that support and sustain the earth's life and economies (**Concept 1-1A**).

organisms interacting with one another and with their environment of nonliving matter and energy within a defined area or volume.

We should not confuse environmental science and ecology with **environmentalism,** a social movement dedicated to protecting the earth's life-support systems for us and all other forms of life. Environmentalism is practiced more in the political and ethical arenas than in the realm of science.

Sustainability Is the Central Theme of This Book

Sustainability is the ability of the earth's various natural systems and human cultural systems and economies to survive and adapt to changing environmental conditions indefinitely. It is the central theme of this book, and its components provide the subthemes of this book.

A critical component of sustainability is **natural capital**—the natural resources and natural services that keep us and other forms of life alive and support our economies (Figure 1-3). **Natural resources** are materials and energy in nature that are essential or useful to humans. These resources are often classified as *renewable* (such as air, water, soil, plants, and wind) or *nonrenewable* (such as copper, oil, and coal). **Natural services** are functions of nature, such as purification of air and water, which support life and human economies. Ecosystems provide us with these essential services at no cost.

One vital natural service is **nutrient cycling,** the circulation of chemicals necessary for life, from the environment (mostly from soil and water) through organisms and back to the environment (Figure 1-4). For example, *topsoil,* the upper layer of the earth's crust, provides the nutrients that support the plants, animals, and microorganisms that live on land; when they die and decay, they resupply the soil with these nutrients. Without this service, life as we know it could not exist.

Natural capital is supported by **solar capital:** energy from the sun (Figure 1-3). Take away solar energy, and all natural capital would collapse. Solar energy warms the planet and supports *photosynthesis*—a complex chemical process that plants use to provide food for themselves and for us and most other animals. This direct input of solar energy also produces indirect forms of renewable solar energy such as wind, flowing water, and biofuels made from plants and plant residues. Thus, our lives and economies depend on energy from the sun (*solar capital*) and natural resources and natural services (*natural capital*) provided by the earth (**Concept 1-1A**).

A second component of sustainability—and another sub-theme of this text—is to recognize that many human activities can *degrade natural capital* by using normally renewable resources faster than nature can renew them. For example, in parts of the world, we are clearing mature forests much faster than nature can replenish them. We are also harvesting many species of ocean fish faster than they can replenish themselves.

This leads us to a third component of sustainability. Environmental scientists search for *solutions* to problems such as the degradation of natural capital. However, their work is limited to finding the scientific solutions, while the political solutions are left to political processes. For example, scientific solutions might be to stop chopping down biologically diverse, mature forests, and to harvest fish no faster than they can replenish themselves. But implementing such solutions could require government laws and regulations.

The search for solutions often involves conflicts. When scientists argue for protecting a diverse natural forest to help prevent the premature extinction of various life forms, for example, the timber company that had planned to harvest trees in that forest might protest. Dealing with such conflicts often involves making *trade-offs,* or compromises—a fourth component of sustainability. In the case of the timber company, it might be persuaded to plant a tree farm in an area that had

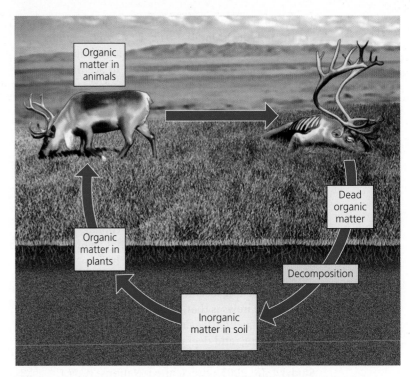

Figure 1-4 *Nutrient cycling:* an important natural service that recycles chemicals needed by organisms from the environment (mostly from soil and water) through organisms and back to the environment.

already been cleared or degraded, in exchange for preserving the natural forest.

Any shift toward environmental sustainability should be based on scientific concepts and results that are widely accepted by experts in a particular field, as discussed in more detail in Chapter 2. In making such a shift, *individuals matter*—another subtheme of this book. Some people are good at thinking of new ideas and inventing innovative technologies or solutions. Others are good at putting political pressure on government officials and business leaders, acting either alone or in groups to implement those solutions. In any case, a shift toward sustainability for a society ultimately depends on the actions of individuals within that society.

Environmentally Sustainable Societies Protect Natural Capital and Live Off Its Income

The ultimate goal is an **environmentally sustainable society**—one that meets the current and future basic resource needs of its people in a just and equitable manner without compromising the ability of future generations to meet their basic needs.

Imagine you win $1 million in a lottery. If you invest this money and earn 10% interest per year, you will have a sustainable income of $100,000 a year that you can live off of indefinitely, while allowing interest to accumulate on what is left after each withdrawal, without depleting your capital. However, if you spend

$200,000 per year, even while allowing interest to accumulate, your capital of $1 million will be gone early in the seventh year. Even if you spend only $110,000 per year and still allow the interest to accumulate, you will be bankrupt early in the eighteenth year.

The lesson here is an old one: *Protect your capital and live off the income it provides.* Deplete or waste your capital, and you will move from a sustainable to an unsustainable lifestyle.

The same lesson applies to our use of the earth's natural capital—the global trust fund that nature provides for us. *Living sustainably* means living off **natural income,** the renewable resources such as plants, animals, and soil provided by natural capital. This means preserving the earth's natural capital, which supplies this income, while providing the human population with adequate and equitable access to this natural income for the foreseeable future (**Concept 1-1B**).

The bad news is that, according to a growing body of scientific evidence, we are living unsustainably by wasting, depleting, and degrading the earth's natural capital at an exponentially accelerating rate (**Core Case Study**).* In 2005, the United Nations (U.N.) released its *Millennium Ecosystem Assessment.*

CORE CASE STUDY

*The opening Core Case Study is used as a theme to connect and integrate much of the material in each chapter. The logo indicates these connections

According to this 4-year study by 1,360 experts from 95 countries, human activities are degrading or overusing about 62% of the earth's natural services (Figure 1-3). In its summary statement, the report warned that "human activity is putting such a strain on the natural functions of Earth that the ability of the planet's ecosystems to sustain future generations can no longer be taken for granted." The good news is that the report suggests we have the knowledge and tools to conserve the planet's natural capital, and it describes commonsense strategies for doing this.

— RESEARCH FRONTIER* —

A crash program to gain better and more comprehensive information about the health of the world's life-support systems. See **academic.cengage.com/biology/miller**.

— HOW WOULD YOU VOTE?** ☑ —

Do you believe that the society you live in is on an unsustainable path? Cast your vote online at **academic .cengage.com/biology/miller**.

*Environmental science is a developing field with many exciting research frontiers that are identified throughout this book.

**To cast your vote, go the website for this book and then to the appropriate chapter (in this case, Chapter 1). In most cases, you will be able to compare how you voted with others using this book.

1-2　How Can Environmentally Sustainable Societies Grow Economically?

▶ **CONCEPT 1-2** Societies can become more environmentally sustainable through economic development dedicated to improving the quality of life for everyone without degrading the earth's life support systems.

There Is a Wide Economic Gap between Rich and Poor Countries

Economic growth is an increase in a nation's output of goods and services. It is usually measured by the percentage of change in a country's **gross domestic product (GDP):** the annual market value of all goods and services produced by all firms and organizations, foreign and domestic, operating within a country. Changes in a country's economic growth per person are measured by **per capita GDP:** the GDP divided by the total population at midyear.

The value of any country's currency changes when it is used in other countries. Because of such differences, a basic unit of currency in one country can buy more of a particular thing than the basic unit of currency of another country can buy. Consumers in the

first country are said to have more *purchasing power* than consumers in the second country have. To help compare countries, economists use a tool called *purchasing power parity (PPP)*. By combining per capita GDP and PPP, for any given country, they arrive at a **per capita GDP PPP**—a measure of the amount of goods and services that a country's average citizen could buy in the United States.

While economic growth provides people with more goods and services, **economic development** has the goal of using economic growth to improve living standards. The United Nations classifies the world's countries as economically developed or developing based primarily on their degree of industrialization and their per capita GDP PPP. The **developed countries** (with 1.2 billion people) include the United States, Canada, Japan, Australia, New Zealand, and most countries of

Europe. Most are highly industrialized and have a high per capita GDP PPP.

All other nations (with 5.5 billion people) are classified as **developing countries,** most of them in Africa, Asia, and Latin America. Some are *middle-income, moderately developed countries* such as China, India, Brazil, Turkey, Thailand, and Mexico. Others are *low-income, least developed countries* where per capita GDP PPP is steadily declining. These 49 countries with 11% of the world's population include Angola, Congo, Belarus, Nigeria, Nicaragua, and Jordan. Figure 2 on p. S10 in Supplement 3 is a map of high-, upper middle-, lower middle-, and low-income countries.

Figure 1-5 compares some key characteristics of developed and developing countries. About 97% of the projected increase in the world's population between 2008 and 2050 is expected to take place in developing countries, which are least equipped to handle such large population increases.

We live in a world of haves and have-nots. Despite a 40-fold increase in economic growth since 1900, *more than half of the people in the world live in extreme poverty and try to survive on a daily income of less than $2. And one of every six people, classified as desperately poor, struggle to survive on less than $1 a day.* (All dollar figures are in U.S. dollars.) (Figure 1-6)

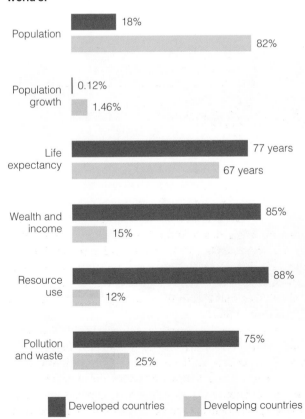

Percentage of World's:

Population — Developed: 18%, Developing: 82%

Population growth — Developed: 0.12%, Developing: 1.46%

Life expectancy — Developed: 77 years, Developing: 67 years

Wealth and income — Developed: 85%, Developing: 15%

Resource use — Developed: 88%, Developing: 12%

Pollution and waste — Developed: 75%, Developing: 25%

■ Developed countries   ■ Developing countries

Figure 1-5 *Global outlook:* comparison of developed and developing countries, 2008. (Data from the United Nations and the World Bank)

Sean Sprague/Peter Arnold, Inc.

Figure 1-6 *Extreme poverty:* boy searching for items to sell in an open dump in Rio de Janeiro, Brazil. Many children of poor families who live in makeshift shanty-towns in or near such dumps often scavenge all day for food and other items to help their families survive. This means that they cannot go to school.

Some economists call for continuing conventional economic growth, which has helped to increase food supplies, allowed people to live longer, and stimulated mass production of an array of useful goods and services for many people. They also see such growth as a cure for poverty, maintaining that some of the resulting increase in wealth trickles down to countries and people near the bottom of the economic ladder.

Other economists call for us to put much greater emphasis on **environmentally sustainable economic development.** This involves using political and economic systems to *discourage* environmentally harmful and unsustainable forms of economic growth that degrade natural capital, and to *encourage* environmentally beneficial and sustainable forms of economic development that help sustain natural capital (**Concept 1-2**).

┌─ **THINKING ABOUT**
Economic Growth and Sustainability

Is exponential economic growth incompatible with environmental sustainability? What are three types of goods whose exponential growth would promote environmental sustainability?

1-3 How Are Our Ecological Footprints Affecting the Earth?

▶ **CONCEPT 1-3** As our ecological footprints grow, we are depleting and degrading more of the earth's natural capital.

Some Resources Are Renewable

From a human standpoint, a **resource** is anything obtained from the environment to meet our needs and wants. **Conservation** is the management of natural resources with the goal of minimizing resource waste and sustaining resource supplies for current and future generations.

Some resources, such as solar energy, fresh air, wind, fresh surface water, fertile soil, and wild edible plants, are directly available for use. Other resources such as petroleum, iron, water found underground, and cultivated crops, are not directly available. They become useful to us only with some effort and technological ingenuity. For example, petroleum was a mysterious fluid until we learned how to find, extract, and convert (refine) it into gasoline, heating oil, and other products that could be sold.

Solar energy is called a **perpetual resource** because it is renewed continuously and is expected to last at least 6 billion years as the sun completes its life cycle.

On a human time scale, a **renewable resource** can be replenished fairly quickly (from hours to hundreds of years) through natural processes as long as it is not used up faster than it is renewed. Examples include forests, grasslands, fisheries, freshwater, fresh air, and fertile soil.

The highest rate at which a renewable resource can be used *indefinitely* without reducing its available supply is called its **sustainable yield.** When we exceed a renewable resource's natural replacement rate, the available supply begins to shrink, a process known as **environmental degradation,** as shown in Figure 1-7.

We Can Overexploit Commonly Shared Renewable Resources: The Tragedy of the Commons

There are three types of property or resource rights. One is *private property* where individuals or firms own

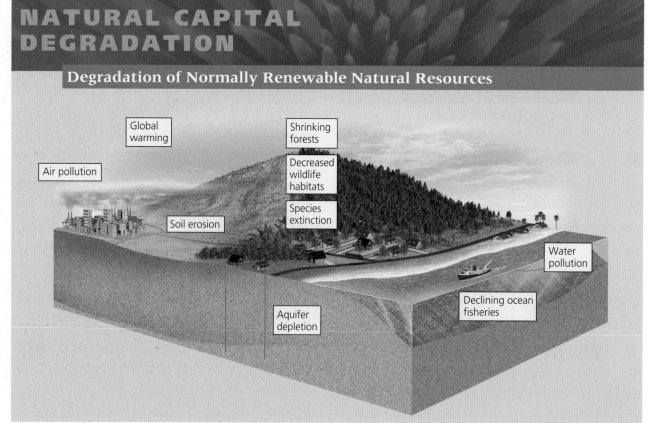

NATURAL CAPITAL DEGRADATION

Degradation of Normally Renewable Natural Resources

Global warming

Shrinking forests

Air pollution

Decreased wildlife habitats

Species extinction

Soil erosion

Water pollution

Aquifer depletion

Declining ocean fisheries

Figure 1-7
Degradation of normally renewable natural resources and services in parts of the world, mostly as a result of rising population and resource use per person.

the rights to land, minerals, or other resources. Another is *common property* where the rights to certain resources are held by large groups of individuals. For example, roughly one-third of the land in the United States is owned jointly by all U.S. citizens and held and managed for them by the government. Another example is land that belongs to a whole village and can be used by anyone for activities such as grazing cows or sheep.

A third category consists of *open access renewable resources,* owned by no one and available for use by anyone at little or no charge. Examples of such shared renewable resources include clean air, underground water supplies, and the open ocean and its fish.

Many common property and open access renewable resources have been degraded. In 1968, biologist Garrett Hardin (1915–2003) called such degradation the *tragedy of the commons.* It occurs because each user of a shared common resource or open-access resource reasons, "If I do not use this resource, someone else will. The little bit that I use or pollute is not enough to matter, and anyway, it's a renewable resource."

When the number of users is small, this logic works. Eventually, however, the cumulative effect of many people trying to exploit a shared resource can exhaust or ruin it. Then no one can benefit from it. Such resource degradation results from the push to satisfy the short-term needs and wants of a growing number of people. It threatens our ability to ensure the long-term economic and environmental sustainability of open-access resources such as clean air or an open-ocean fishery.

One solution is to *use shared resources at rates well below their estimated sustainable yields* by reducing use of the resources, regulating access to the resources, or doing both. For example, the most common approach is for governments to establish laws and regulations limiting the annual harvests of various types of ocean fish that are being harvested at unsustainable levels in their coastal waters. Another approach is for nations to enter into agreements that regulate access to open-access renewable resources such as the fish in the open ocean.

Another solution is to *convert open-access resources to private ownership.* The reasoning is that if you own something, you are more likely to protect your investment. That sounds good, but this approach is not practical for global open-access resources—such as the atmosphere, the open ocean, and most wildlife species—that cannot be divided up and converted to private property.

Some Resources Are Not Renewable

Nonrenewable resources exist in a fixed quantity, or *stock,* in the earth's crust. On a time scale of millions to billions of years, geological processes can renew such resources. But on the much shorter human time scale of hundreds to thousands of years, these resources can be depleted much faster than they are formed. Such exhaustible resources include *energy resources* (such as coal and oil), *metallic mineral resources* (such as copper and aluminum), and *nonmetallic mineral resources* (such as salt and sand).

As such resources are depleted, human ingenuity can often find substitutes. For example, during this century, a mix of renewable energy resources such as wind, the sun, flowing water, and the heat in the earth's interior could reduce our dependence on nonrenewable fossil fuels such as oil and coal. Also, various types of plastics and composite materials can replace certain metals. But sometimes there is no acceptable or affordable substitute.

Some nonrenewable resources, such as copper and aluminum, can be recycled or reused to extend supplies. **Reuse** is using a resource over and over in the same form. For example, glass bottles can be collected, washed, and refilled many times (Figure 1-8). **Recycling** involves collecting waste materials and processing them into new materials. For example, discarded aluminum cans can be crushed and melted to make new

Mark Edwards/Peter Arnold, Inc.

Figure 1-8 *Reuse:* This child and his family in Katmandu, Nepal, collect beer bottles and sell them for cash to a brewery where they will be reused.

aluminum cans or other aluminum products. But energy resources such as oil and coal cannot be recycled. Once burned, their energy is no longer available to us.

Recycling nonrenewable metallic resources takes much less energy, water, and other resources and produces much less pollution and environmental degradation than exploiting virgin metallic resources. Reusing such resources takes even less energy and other resources and produces less pollution and environmental degradation than recycling does.

Figure 1-9 *Consumption of natural resources.* The top photo shows a family of five subsistence farmers with all their possessions. They live in the village of Shingkhey, Bhutan, in the Himalaya Mountains, which are sandwiched between China and India in South Asia. The bottom photo shows a typical U.S. family of four living in Pearland, Texas, with their possessions .

Both photos by Peter Menzel

Our Ecological Footprints Are Growing

Many people in developing countries struggle to survive. Their individual use of resources and the resulting environmental impact is low and is devoted mostly to meeting their basic needs (Figure 1-9, top). By contrast, many individuals in more affluent nations consume large amounts of resources way beyond their basic needs (Figure 1-9, bottom).

Supplying people with resources and dealing with the resulting wastes and pollution can have a large environmental impact. We can think of it as an **ecological footprint**—the amount of biologically productive land and water needed to supply the people in a particular country or area with resources and to absorb and recycle the wastes and pollution produced by such resource use. The **per capita ecological footprint** is the average ecological footprint of an individual in a given country or area.

If a country's, or the world's, total ecological footprint is larger than its *biological capacity* to replenish its renewable resources and absorb the resulting waste products and pollution, it is said to have an *ecological deficit.* The World Wildlife Fund (WWF) and the Global Footprint Network estimated that in 2003 (the latest data available) humanity's global ecological footprint exceeded the earth's *biological capacity* by about 25% (Figure 1-10, right). That figure was about 88% in the world's high-income countries, with the United States having the world's largest total ecological footprint. If the current exponential growth in the use of renewable resources continues, the Global Footprint Network estimates that by 2050 humanity will be trying to use twice as many renewable resources as the planet can supply (Figure 1-10, bottom) (**Concept 1-3**). See Figure 3 on p. S24 and Figure 5 on pp. S27 in Supplement 4 for maps of the human ecological footprints for the world and the United States, and Figure 4 on p. S26 for a map of countries that are ecological debtors and those that are ecological creditors.

The per capita ecological footprint is an estimate of how much of the earth's renewable resources an individual consumes. After the oil-rich United Arab Emirates, the United States has the world's second largest per capita ecological footprint. In 2003 (the latest data available), its per capita ecological footprint was about 4.5 times the average global footprint per person, 6 times larger than China's per capita footprint, and 12 times the average per capita footprint in the world's low-income countries.

According to William Rees and Mathis Wackernagel, the developers of the ecological footprint concept, it would take the land area of about *five more planet earths* for the rest of the world to reach current U.S. levels of consumption with existing technology. Put another way, if everyone consumed as much as the average American does today, the earth's natural capital could support only about 1.3 billion people—not

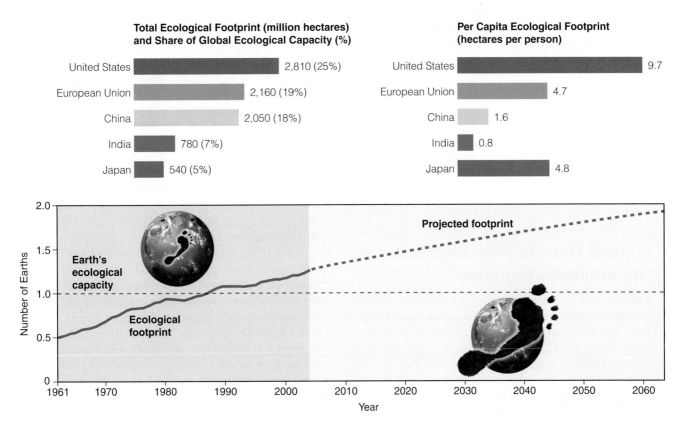

Total Ecological Footprint (million hectares) and Share of Global Ecological Capacity (%)

United States	2,810 (25%)
European Union	2,160 (19%)
China	2,050 (18%)
India	780 (7%)
Japan	540 (5%)

Per Capita Ecological Footprint (hectares per person)

United States	9.7
European Union	4.7
China	1.6
India	0.8
Japan	4.8

Figure 1-10 Natural capital use and degradation: total and per capita ecological footprints of selected countries (top). In 2003, humanity's total or global ecological footprint was about 25% higher than the earth's ecological capacity (bottom) and is projected to be twice the planet's ecological capacity by 2050. **Question:** If we are living beyond the earth's biological capacity, why do you think the human population and per capita resource consumption are still growing exponentially? (Data from Worldwide Fund for Nature, Global Footprint Network)

today's 6.7 billion. In other words, we are living unsustainably by depleting and degrading some of the earth's irreplaceable natural capital and the natural renewable income it provides as our ecological footprints grow and spread across the earth's surface (**Concept 1-3**). For more on this subject, see the Guest Essay by Michael Cain at CengageNOW™. See the Case Study that follows about the growing ecological footprint of China.

THINKING ABOUT
Your Ecological Footprint

Estimate your own ecological footprint by visiting the website **www.myfootprint.org/**. What are three things you could do to reduce your ecological footprint?

■ CASE STUDY
China's New Affluent Consumers

More than a billion super-affluent consumers in developed countries are putting immense pressure on the earth's natural capital. Another billion consumers are attaining middle-class, affluent lifestyles in rapidly developing countries such as China, India, Brazil, South Korea, and Mexico. The 700 million middle-class consumers in China and India number more than twice the size of the entire U.S. population, and the number is growing rapidly. In 2006, the World Bank projected that by 2030 the number of middle-class consumers

living in today's developing nations will reach 1.2 billion—about four times the current U.S. population.

China is now the world's leading consumer of wheat, rice, meat, coal, fertilizers, steel, and cement, and it is the second largest consumer of oil after the United States. China leads the world in consumption of goods such as television sets, cell phones, refrigerators, and soon, personal computers. On the other hand, after 20 years of industrialization, two-thirds of the world's most polluted cities are in China; this pollution threatens the health of urban dwellers. By 2020, China is projected to be the world's largest producer and consumer of cars and to have the world's leading economy in terms of GDP PPP.

Suppose that China's economy continues growing exponentially at a rapid rate and its projected population size reaches 1.5 billion by 2033. Then China will need two-thirds of the world's current grain harvest, twice the world's current paper consumption, and more than the current global production of oil.

According to environmental policy expert Lester R. Brown:

The western economic model—the fossil fuel–based, automobile-centered, throwaway economy—is not going to work for China. Nor will it work for India, which by 2033 is projected to have a population even larger than China's, or for the other 3 billion people in developing countries who are also dreaming the "American dream."

For more details on the growing ecological footprint of China, see the Guest Essay by Norman Myers for this chapter at CengageNOW.

Cultural Changes Have Increased Our Ecological Footprints

Culture is the whole of a society's knowledge, beliefs, technology, and practices, and human cultural changes have had profound effects on the earth.

Evidence of organisms from the past and studies of ancient cultures suggest that the current form of our species, *Homo sapiens sapiens,* has walked the earth for perhaps 90,000–195,000 years—less than an eye-blink in the 3.56 billion years of life on the earth. Until about 12,000 years ago, we were mostly *hunter–gatherers* who obtained food by hunting wild animals or scavenging their remains and gathering wild plants. Early hunter–gathers lived in small groups and moved as needed to find enough food for survival.

Since then, three major cultural changes have occurred. *First* was the *agricultural revolution,* which began 10,000–12,000 years ago when humans learned how to grow and breed plants and animals for food, clothing, and other purposes. *Second* was the *industrial–medical revolution,* beginning about 275 years ago when people invented machines for the large-scale production of goods in factories. This involved learning how to get energy from fossil fuels, such as coal and oil, and how to grow large quantities of food in an efficient manner. *Finally,* the *information–globalization revolution* began about 50 years ago, when we developed new technologies for gaining rapid access to much more information and resources on a global scale.

Each of these cultural changes gave us more energy and new technologies with which to alter and control more of the planet to meet our basic needs and increasing wants. They also allowed expansion of the human population, mostly because of increased food supplies and longer life spans. In addition, they each resulted in greater resource use, pollution, and environmental degradation as our ecological footprints expanded (Figure 1-10) and allowed us to dominate the planet.

Many environmental scientists and other analysts call for us to bring about a new **environmental,** or **sustainability, revolution** during this century. It would involve learning how to reduce our ecological footprints and live more sustainability.

For more background and details on environmental history, see Supplement 5 (p. S31).

1-4   What Is Pollution and What Can We Do about It?

▶ **CONCEPT 1-4** Preventing pollution is more effective and less costly than cleaning up pollution.

Pollution Comes from a Number of Sources

Pollution is any in the environment that is harmful to the health, survival, or activities of humans or other organisms. Pollutants can enter the environment naturally, such as from volcanic eruptions, or through human activities, such as burning coal and gasoline and discharging chemicals into rivers and the ocean.

The pollutants we produce come from two types of sources. **Point sources** are single, identifiable sources. Examples are the smokestack of a coal-burning power or industrial plant (Figure 1-11), the drainpipe of a factory, and the exhaust pipe of an automobile. **Nonpoint sources** are dispersed and often difficult to identify. Examples are pesticides blown from the land into the air and the runoff of fertilizers and

pesticides from farmlands, lawns, gardens, and golf courses into streams and lakes. It is much easier and cheaper to identify and control or prevent pollution from point sources than from widely dispersed nonpoint sources.

There are two main types of pollutants. **Biodegradable pollutants** are harmful materials that can be broken down by natural processes. Examples are human sewage and newspapers. **Nondegradable pollutants** are harmful materials that natural processes cannot break down. Examples are toxic chemical elements such as lead, mercury, and arsenic (see Supplement 6, p. S39, for an introduction to basic chemistry).

Pollutants can have three types of unwanted effects. *First,* they can disrupt or degrade life-support systems for humans and other species. *Second,* they can damage wildlife, human health, and property. *Third,* they can create nuisances such as noise and unpleasant smells, tastes, and sights.

We Can Clean Up Pollution or Prevent It

Consider the smoke produced by a steel mill. We can try to deal with this problem by asking two entirely different questions. One question is "how can we clean up the smoke?" The other is "how can we avoid producing the smoke in the first place?"

The answers to these questions involve two different ways of dealing with pollution. One is **pollution cleanup,** or **output pollution control,** which involves cleaning up or diluting pollutants after they have been produced. The other is **pollution prevention,** or **input pollution control,** which reduces or eliminates the production of pollutants.

Environmental scientists have identified three problems with relying primarily on pollution cleanup. *First,* it is only a temporary bandage as long as population and consumption levels grow without corresponding improvements in pollution control technology. For example, adding catalytic converters to car exhaust systems has reduced some forms of air pollution. At the same time, increases in the number of cars and the total distance each car travels have reduced the effectiveness of this cleanup approach.

Second, cleanup often removes a pollutant from one part of the environment only to cause pollution in another. For example, we can collect garbage, but the garbage is then *burned* (perhaps causing air pollution and leaving toxic ash that must be put somewhere), *dumped*

Ray Pfortner/Peter Arnold, Inc.

Figure 1-11 *Point-source air pollution* from a pulp mill in New York State (USA).

on the land (perhaps causing water pollution through runoff or seepage into groundwater), or *buried* (perhaps causing soil and groundwater pollution).

Third, once pollutants become dispersed into the environment at harmful levels, it usually costs too much or is impossible to reduce them to acceptable levels.

Pollution prevention (front-of-the-pipe) and pollution cleanup (end-of-the-pipe) solutions are both needed. But environmental scientists, some economists, and some major companies urge us to put more emphasis on prevention because it works better and in the long run is cheaper than cleanup (**Concept 1-4**).

1-5   Why Do We Have Environmental Problems?

▶ **CONCEPT 1-5A** Major causes of environmental problems are population growth, wasteful and unsustainable resource use, poverty, exclusion of environmental costs of resource use from the market prices of goods and services, and attempts to manage nature with insufficient knowledge.

▶ **CONCEPT 1-5B** People with different environmental worldviews often disagree about the seriousness of environmental problems and what we should do about them.

Experts Have Identified Five Basic Causes of Environmental Problems

As we run more and more of the earth's natural resources through the global economy, in many parts of the world, forests are shrinking, deserts are expanding, soils are eroding, and agricultural lands are deteriorat-

ing. In addition, the lower atmosphere is warming, glaciers are melting, sea levels are rising, and storms are becoming more destructive. And in many areas, water tables are falling, rivers are running dry, fisheries are collapsing, coral reefs are disappearing, and various species are becoming extinct.

According to a number of environmental and social scientists, the major causes of these and other

Causes of Environmental Problems

| Population growth | Unsustainable resource use | Poverty | Excluding environmental costs from market prices | Trying to manage nature without knowing enough about it |

Figure 1-12 Environmental and social scientists have identified five basic causes of the environmental problems we face (**Concept 1-5A**). **Question:** What are three ways in which your lifestyle contributes to these causes?

environmental problems are population growth, wasteful and unsustainable resource use, poverty, failure to include the harmful environmental costs of goods and services in their market prices, and insufficient knowledge of how nature works (Figure 1-12 and **Concept 1-5A**).

We have discussed the exponential growth of the human population (**Core Case Study**), and here we will examine other major causes of environmental problems in more detail.

CORE CASE STUDY

Poverty Has Harmful Environmental and Health Effects

Poverty occurs when people are unable to meet their basic needs for adequate food, water, shelter, health, and education. Poverty has a number of harmful environmental and health effects (Figure 1-13). The daily lives of half of the world's people, who are trying to live on the equivalent of less than $2 a day, are focused on getting enough food, water, and cooking and heating fuel to survive. Desperate for short-term survival, some of these people deplete and degrade forests, soil, grasslands, fisheries, and wildlife, at an ever-increasing rate. They do not have the luxury of worrying about long-term environmental quality or sustainability.

Poverty affects population growth. To many poor people, having more children is a matter of survival. Their children help them gather fuel (mostly wood and animal dung), haul drinking water, and tend crops and livestock. Their children also help to care for them in their old age (which is their 40s or 50s in the poorest countries) because they do not have social security, health care, and retirement funds.

While poverty can increase some types of environmental degradation, the reverse is also true. Pollution and environmental degradation have a severe impact on the poor and can increase poverty. Consequently, many of the world's desperately poor people die prematurely from several preventable health problems.

One such problem is *malnutrition* from a lack of protein and other nutrients needed for good health

(Figure 1-14). The resulting weakened condition can increase the chances of death from normally nonfatal illnesses, such as diarrhea and measles. A second problem is limited access to adequate sanitation facilities and clean drinking water. More than 2.6 billion people (38% of the world's population) have no decent bathroom facilities. They are forced to use fields, backyards, ditches, and streams. As a result, more than 1 billion people—one of every seven—get water for drinking, washing, and cooking from sources polluted by human and animal feces. A third problem is severe respiratory disease and premature death from inhaling indoor air pollutants produced by burning wood or coal in open fires or in poorly vented stoves for heat and cooking.

According to the World Health Organization, these factors cause premature death for at least 7 million people each year. *This amounts to about 19,200 premature deaths per day, equivalent to 96 fully loaded 200-passenger airliners crashing every day with no survivors!* Two-thirds of those dying are children younger than age 5. The news media rarely cover this ongoing human tragedy.

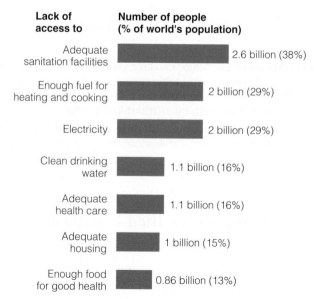

Lack of access to	Number of people (% of world's population)
Adequate sanitation facilities	2.6 billion (38%)
Enough fuel for heating and cooking	2 billion (29%)
Electricity	2 billion (29%)
Clean drinking water	1.1 billion (16%)
Adequate health care	1.1 billion (16%)
Adequate housing	1 billion (15%)
Enough food for good health	0.86 billion (13%)

Figure 1-13 Some harmful results of poverty. **Question:** Which two of these effects do you think are the most harmful? Why? (Data from United Nations, World Bank, and World Health Organization)

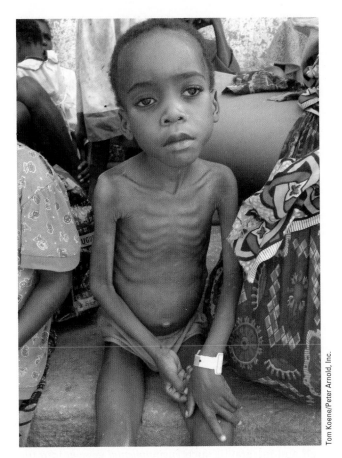

Figure 1-14 *Global Outlook:* in developing countries, one of every three children under age 5, such as this child in Lunda, Angola, suffers from severe malnutrition caused by a lack of calories and protein. According to the World Health Organization, each day at least 13,700 children under age 5 die prematurely from malnutrition and infectious diseases, most from drinking contaminated water and being weakened by malnutrition.

The *great news* is that we have the means to solve the environmental, health, and social problems resulting from poverty within 20–30 years if we can find the political and ethical will to act.

Affluence Has Harmful and Beneficial Environmental Effects

The harmful environmental effects of poverty are serious, but those of affluence are much worse (Figure 1-10, top). The lifestyles of many affluent consumers in developed countries and in rapidly developing countries such as India and China (p. 15) are built upon high levels of consumption and unnecessary waste of resources. Such affluence is based mostly on the assumption—fueled by mass advertising—that buying more and more things will bring happiness.

This type of affluence has an enormous harmful environmental impact. It takes about 27 tractor-trailer loads of resources per year to support one American, or 7.9 billion truckloads per year to support the entire U.S. population. Stretched end-to-end, each year these trucks would reach beyond the sun!

While the United States has far fewer people than India, the average American consumes about 30 times as much as the average citizen of India and 100 times as much as the average person in the world's poorest countries. As a result, the average environmental impact, or ecological footprint per person, in the United States is much larger than the average impact per person in developing countries (Figure 1-10, top).

On the other hand, affluence can lead people to become more concerned about environmental quality. It also provides money for developing technologies to reduce pollution, environmental degradation, and resource waste.

In the United States and most other affluent countries, the air is cleaner, drinking water is purer, and most rivers and lakes are cleaner than they were in the 1970s. In addition, the food supply is more abundant and safer, the incidence of life-threatening infectious diseases has been greatly reduced, lifespans are longer, and some endangered species are being rescued from premature extinction.

Affluence financed these improvements in environmental quality, based on greatly increased scientific research and technological advances. And education spurred citizens insist that businesses and elected officials improve environmental quality. Affluence and education have also helped to reduce population growth in most developed countries. However, a downside to wealth is that it allows the affluent to obtain the resources they need from almost anywhere in the world without seeing the harmful environmental impacts of their high-consumption life styles.

> **THINKING ABOUT**
> **The Poor, the Affluent, and Exponentially Increasing Population Growth**
>
> CORE CASE STUDY
>
> Some see rapid population growth of the poor in developing countries as the primary cause of our environmental problems. Others say that the much higher resource use per person in developed countries is a more important factor. Which factor do you think is more important? Why?

Prices Do Not Include the Value of Natural Capital

When companies use resources to create goods and services for consumers, they are generally not required to pay the environmental costs of such resource use. For example, fishing companies pay the costs of catching fish but do not pay for the depletion of fish stocks. Timber companies pay for clear-cutting forests but not for the resulting environmental degradation and loss of wildlife habitat. The primary goal of these companies is to maximize their profits, so they do not voluntarily pay these harmful environmental costs or even try to assess them, unless required to do so by government laws or regulations.

As a result, the prices of goods and services do not include their harmful environmental costs. Thus, consumers are generally not aware of them and have no effective way to evaluate the resulting harmful effects on the earth's life-support systems and on their own health.

Another problem is that governments give companies tax breaks and payments called *subsidies* to assist them in using resources to run their businesses. This helps to create jobs and stimulate economies, but it can also result in degradation of natural capital, again because the value of the natural capital is not included in the market prices of goods and services. We explore this problem and some possible solutions in later chapters.

People Have Different Views about Environmental Problems and Their Solutions

Differing views about the seriousness of our environmental problems and what we should do about them arise mostly out of differing environmental worldviews. Your **environmental worldview** is a set of assumptions and values reflecting how you think the world works and what you think your role in the world should be. This involves **environmental ethics,** which are our beliefs about what is right and wrong with how we treat the environment. Here are some important *ethical questions* relating to the environment:

- Why should we care about the environment?

- Are we the most important beings on the planet or are we just one of the earth's millions of different forms of life?

- Do we have an obligation to see that our activities do not cause the premature extinction of other species? Should we try to protect all species or only some? How do we decide which species to protect?

- Do we have an ethical obligation to pass on to future generations the extraordinary natural world in a condition at least as good as what we inherited?

- Should every person be entitled to equal protection from environmental hazards regardless of race, gender, age, national origin, income, social class, or any other factor?

THINKING ABOUT
Our Responsibilities

How would you answer each of the questions above? Compare your answers with those of your classmates. Record your answers and, at the end of this course, return to these questions to see if your answers have changed.

People with widely differing environmental worldviews can take the same data, be logically consistent, and arrive at quite different conclusions because they start with different assumptions and moral, ethical, or religious beliefs (**Concept 1-5B**). Environmental worldviews are discussed in detail in Chapter 25, but here is a brief introduction.

The **planetary management worldview** holds that we are separate from nature, that nature exists mainly to meet our needs and increasing wants, and that we can use our ingenuity and technology to manage the earth's life-support systems, mostly for our benefit, indefinitely.

The **stewardship worldview** holds that we can and should manage the earth for our benefit, but that we have an ethical responsibility to be caring and responsible managers, or *stewards,* of the earth. It says we should encourage environmentally beneficial forms of economic growth and development and discourage environmentally harmful forms.

The **environmental wisdom worldview** holds that we are part of, and totally dependent on, nature and that nature exists for all species, not just for us. It also calls for encouraging earth-sustaining forms of economic growth and development and discouraging earth-degrading forms. According to this view, our success depends on learning how life on earth sustains itself and integrating such *environmental wisdom* into the ways we think and act.

Many of the ideas for the stewardship and environmental wisdom worldviews are derived from the writings of Aldo Leopold (Individuals Matter, p. 22).

We Can Learn to Make Informed Environmental Decisions

The first step for dealing with an environmental problem is to carry out scientific research on the nature of the problem and to evaluate possible solutions to the problem. Once this is done, other factors involving the social sciences and the humanities (Table 1-1) must be used to evaluate each proposed solution. This involves considering various *human values.* What are its projected short-term and long-term beneficial and harmful environmental, economic, and health effects? How much will it cost? Is it ethical? Figure 1-15 shows the major steps involved in making an environmental decision.

We Can Work Together to Solve Environmental Problems

Making the shift to more sustainable societies and economies involves building what sociologists call **social capital.** This involves getting people with different views and values to talk and listen to one another, find common ground based on understanding and trust, and work together to solve environmental and other problems. This means nurturing openness, communication, cooperation, and hope and discouraging close-mindedness, polarization, confrontation, and fear.

Solutions to environmental problems are not black and white, but rather all shades of gray because proponents of all sides of these issues have some legitimate and useful insights. In addition, any proposed solution has short- and long-term advantages and disadvantages that must be evaluated (Figure 1-15). This means that citizens who strive to build social capital also search for *trade-off solutions* to environmental problems—an im-

portant theme of this book. They can also try to agree on shared visions of the future and work together to develop strategies for implementing such visions beginning at the local level, as citizens of Chattanooga, Tennessee (USA), have done.

■ CASE STUDY

The Environmental Transformation of Chattanooga, Tennessee

Local officials, business leaders, and citizens have worked together to transform Chattanooga, Tennessee (USA), from a highly polluted city to one of the most sustainable and livable cities in the United States (Figure 1-16).

During the 1960s, U.S. government officials rated Chattanooga as having the dirtiest air in the United States. Its air was so polluted by smoke from its coke ovens and steel mills that people sometimes had to turn on their vehicle headlights in the middle of the day. The Tennessee River, flowing through the city's industrial center, bubbled with toxic waste. People and industries fled the downtown area and left a wasteland of abandoned and polluting factories, boarded-up buildings, high unemployment, and crime.

In 1984, the city decided to get serious about improving its environmental quality. Civic leaders started a *Vision 2000* process with a 20-week series of community meetings in which more than 1,700 citizens from all walks of life gathered to build a consensus about what the city could be at the turn of the century. Citizens identified the city's main problems, set goals, and brainstormed thousands of ideas for solutions.

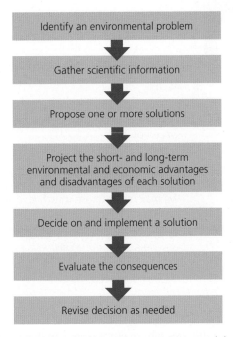

Figure 1-15 Steps involved in making an environmental decision.

Chattanooga Area Convention and Visitors Bureau

Figure 1-16 Since 1984, citizens have worked together to make the city of Chattanooga, Tennessee, one of the most sustainable and best places to live in the United States.

INDIVIDUALS MATTER

Aldo Leopold's Environmental Ethics

According to *Aldo Leopold* (Figure 1-A), the role of the human species should be to protect nature, not conquer it.

In 1933, Leopold became a professor at the University of Wisconsin and in 1935, he was one of the founders of the U.S. Wilderness Society. Through his writings and teachings, he became one of the leaders of the *conservation* and *environmental movements* of the 20th century. In doing this, he laid important groundwork for the field of environmental ethics.

Leopold's weekends of planting, hiking, and observing nature at his farm in Wisconsin provided material for his most famous book, *A Sand County Almanac,* published after his death in 1949. Since then, more than 2 million copies of this environmental classic have been sold.

The following quotations from his writings reflect Leopold's *land ethic,* and they form the basis for many of the beliefs of the modern stewardship and environmental wisdom worldviews:

- All ethics so far evolved rest upon a single premise: that the individual is

Courtesy of the University of Wisconsin—Madison Archives

Figure 1-A Individuals Matter: *Aldo Leopold* (1887–1948) was a forester, writer, and conservationist. His book *A Sand County Almanac* (published after his death) is considered an environmental classic that inspired the modern environmental and conservation movement.

a member of a community of interdependent parts.

- To keep every cog and wheel is the first precaution of intelligent tinkering.
- That land is a community is the basic concept of ecology, but that land is to be loved and respected is an extension of ethics.
- The land ethic changes the role of Homo sapiens from conqueror of the

land-community to plain member and citizen of it.

- We abuse land because we regard it as a commodity belonging to us. When we see land as a community to which we belong, we may begin to use it with love and respect.
- Anything is right when it tends to preserve the integrity, stability, and beauty of the biotic community. It is wrong when it tends otherwise.

By 1995, Chattanooga had met most of its original goals. The city had encouraged zero-emission industries to locate there and replaced its diesel buses with a fleet of quiet, zero-emission electric buses, made by a new local firm.

The city also launched an innovative recycling program after environmentally concerned citizens blocked construction of a garbage incinerator that would have emitted harmful air pollutants. These efforts paid off. Since 1989, the levels of the seven major air pollutants in Chattanooga have been lower than those required by federal standards.

Another project involved renovating much of the city's low-income housing and building new low-income rental units. Chattanooga also built the nation's largest freshwater aquarium, which became the centerpiece for downtown renewal. The city developed a riverfront park along both banks of the Tennessee River running through downtown. The park draws more than 1 million visitors per year. As property values and living conditions have improved, people and businesses have moved back downtown.

In 1993, the community began the process again in *Revision 2000.* Goals included transforming an abandoned and blighted area in South Chattanooga into a mixed community of residences, retail stores, and zero-

emission industries where employees can live near their workplaces. Most of these goals have been implemented.

Chattanooga's environmental success story, enacted by people working together to produce a more livable and sustainable city, is a shining example of what other cities can do by building their social capital.

Individuals Matter

Chattanooga's story shows that a key to finding solutions to environmental problems is to recognize that most social change results from individual actions and individuals acting together (using *social capital*) to bring about change through *bottom-up* grassroots action. In other words, *individuals matter*—another important theme of this book. Here are two pieces of good news. First, research by social scientists suggests that it takes only 5–10% of the population of a community, a country, or the world to bring about major social change. Second, such research also shows that significant social change can occur much more quickly than most people think.

Anthropologist Margaret Mead summarized our potential for social change: "Never doubt that a small group of thoughtful, committed citizens can change the world. Indeed, it is the only thing that ever has."

1-6 What Are Four Scientific Principles of Sustainability?

▶ **CONCEPT 1-6** Nature has sustained itself for billions of years by using solar energy, biodiversity, population control, and nutrient cycling—lessons from nature that we can apply to our lifestyles and economies.

Studying Nature Reveals Four Scientific Principles of Sustainability

How can we live more sustainably? According to environmental scientists, we should study how life on the earth has survived and adapted to major changes in environmental conditions for billions of years. We could make the transition to more sustainable societies by applying these *lessons from nature* to our lifestyles and economies, as summarized below and in Figure 1-17 (**Concept 1-6**).

- *Reliance on Solar Energy:* the sun (solar capital) warms the planet and supports photosynthesis used by plants to provide food for themselves and for us and most other animals.

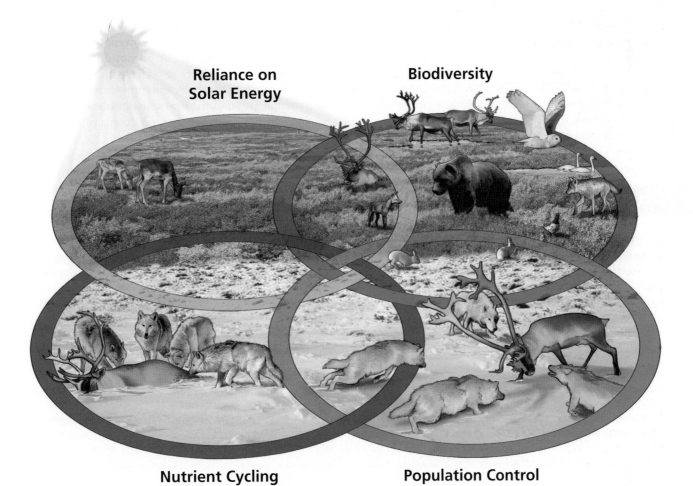

Reliance on Solar Energy

Biodiversity

Nutrient Cycling

Population Control

Figure 1-17 Four scientific principles of sustainability: These four interconnected principles of sustainability are derived from learning how nature has sustained a variety of life forms on the earth for about 3.56 billion years. The top left oval shows sunlight stimulating the production of vegetation in the arctic tundra during its brief summer (*solar energy*) and the top right oval shows some of the diversity of species found there during the summer (*biodiversity*). The bottom right oval shows arctic gray wolves stalking a caribou during the long cold winter (*population control*). The bottom left oval shows arctic gray wolves feeding on their kill. This, plus huge numbers of tiny decomposers that convert dead matter to soil nutrients, recycle all materials needed to support the plant growth shown in the top left and right ovals (*nutrient cycling*).

Current Emphasis	Sustainability Emphasis
Pollution cleanup	Pollution prevention
Waste disposal (bury or burn)	Waste prevention
Protecting species	Protecting habitat
Environmental degradation	Environmental restoration
Increasing resource use	Less resource waste
Population growth	Population stabilization
Depleting and degrading natural capital	Protecting natural capital

Figure 1-18 Solutions: some shifts involved in bringing about the *environmental* or *sustainability revolution.* **Question:** Which three of these shifts do you think are most important? Why?

- *Biodiversity* (short for *biological diversity*): the astounding variety of different organisms, the genes they contain, the ecosystems in which they exist, and the natural services they provide have yielded countless ways for life to adapt to changing environmental conditions throughout the earth's history.

- *Population Control:* competition for limited resources among different species places a limit on how much their populations can grow.

- *Nutrient Cycling:* natural processes recycle chemicals that plants and animals need to stay alive and reproduce (Figure 1-4). There is little or no waste in natural systems.

Using the four **scientific principles of sustainability** to guide our lifestyles and economies could help us bring about an *environmental* or *sustainability revolution* during your lifetime (see the Guest Essay by Lester R. Brown at CengageNOW). Figure 1-18 lists some of the shifts involved in bringing about this new cultural change by learning how to live more sustainably.

Scientific evidence indicates that we have perhaps 50 years and no more than 100 years to make such crucial cultural changes. If this is correct, sometime during this century we could come to a critical fork in the road, at which point we will choose a path toward sustainability or continue on our current unsustainable course. Everything you do, or do not do, will play a role in our collective choice of which path we will take. One of the goals of this book is to provide a realistic environmental vision of the future that, instead of immobilizing you with fear, gloom, and doom, will energize you by inspiring realistic hope.

REVISITING   Exponential Growth and Sustainability

We face an array of serious environmental problems. This book is about *solutions* to these problems. Making the transition to more sustainable societies and economies challenges us to devise ways to slow down the harmful effects of exponential growth (**Core Case Study**) and to use the same power of exponential growth to implement more sustainable lifestyles and economies.

The key is to apply the four **scientific principles of sustainability** (Figure 1-17 and **Concept 1-6**) to the design of our economic and social systems and to our individual lifestyles. We can use such information to help slow human population growth, sharply reduce poverty, curb the unsustainable forms of resource use that are eating away at the earth's natural capital, build social capital, and create a better world for ourselves, our children, and future generations.

Exponential growth is a double-edged sword. It can cause environmental harm. But we can also use it positively to amplify beneficial changes in our lifestyles and economies by applying the four **scientific principles of sustainability**. Through our individual and collective actions or inactions, we choose which side of that sword to use.

We are rapidly altering the planet that is our only home. If we make the right choices during this century, we can create an extraordinary and sustainable future on our planetary home. If we get it wrong, we face irreversible ecological disruption that could set humanity back for centuries and wipe out as many as half of the world's species.

You have the good fortune to be a member of the 21st century *transition generation,* which will decide what path humanity takes. What a challenging and exciting time to be alive!

What's the use of a house
if you don't have a decent planet to put it on?

HENRY DAVID THOREAU

1. Review the Key Questions and Concepts for this chapter on p. 6. What is **exponential growth?** Why is living in an exponential age a cause for concern for everyone living on the planet?

2. Define **environment.** Distinguish among **environmental science, ecology,** and **environmentalism.** Distinguish between an **organism** and a **species.** What is an **ecosystem?** What is **sustainability?** Explain the terms **natural capital, natural resources, natural services, solar capital,** and **natural capital degradation.** What is **nutrient cycling** and why is it important? Describe the ultimate goal of an **environmentally sustainable society.** What is **natural income?**

3. What is the difference between **economic growth** and **economic development?** Distinguish among **gross domestic product (GDP), per capita GDP,** and **per capita GDP PPP.** Distinguish between **developed countries** and **developing countries** and describe their key characteristics. What is **environmentally sustainable economic development?**

4. What is a **resource?** What is **conservation?** Distinguish among a **renewable resource, nonrenewable resource,** and **perpetual resource** and give an example of each. What is **sustainable yield?** Define and give three examples of **environmental degradation.** What is the tragedy of the commons? Distinguish between **recycling** and **reuse** and give an example of each. What is an **ecological footprint?** What is a **per capita ecological footprint?** Compare the total and per capita ecological footprints of the United States and China.

5. What is **culture?** Describe three major cultural changes that have occurred since humans arrived on the earth.

Why has each change led to more environmental degradation? What is the **environmental** or **sustainability revolution?**

6. Define **pollution.** Distinguish between **point sources** and **nonpoint sources** of pollution. Distinguish between **biodegradable pollutants** and **nondegradable pollutants** and give an example of each. Distinguish between **pollution cleanup** and **pollution prevention** and give an example of each. Describe three problems with solutions that rely mostly on pollution cleanup.

7. Identify five basic causes of the environmental problems that we face today. What is **poverty?** In what ways do poverty and affluence affect the environment? Explain the problems we face by not including the harmful environmental costs in the prices of goods and services.

8. What is an **environmental worldview?** What is **environmental ethics?** Distinguish among the **planetary management, stewardship,** and **environmental wisdom worldviews.** Describe Aldo Leopold's environmental ethics. What major steps are involved in making an environmental decision? What is **social capital?**

9. Discuss the lessons we can learn from the environmental transformation of Chattanooga, Tennessee (USA). Explain why individuals matter in dealing with the environmental problems we face.

10. What are four **scientific principles of sustainability?** Explain how exponential growth (**Core Case Study**) affects them.

Note: Key Terms are in **bold** type.

1. List three ways in which you could apply **Concepts 1-5A** and **1-6** to making your lifestyle more environmentally sustainable.

2. Describe two environmentally beneficial forms of exponential growth (**Core Case Study**).

3. Explain why you agree or disagree with the following propositions:
   a. Stabilizing population is not desirable because, without more consumers, economic growth would stop.
   b. The world will never run out of resources because we can use technology to find substitutes and to help us reduce resource waste.

4. Suppose the world's population stopped growing today. What environmental problems might this help solve? What environmental problems would remain? What economic problems might population stabilization make worse?

5. When you read that at least 19,200 people die prematurely each day (13 per minute) from preventable malnutrition and infectious disease, do you **(a)** doubt that it is true, **(b)** not want to think about it, **(c)** feel hopeless, **(d)** feel sad, **(e)** feel guilty, or **(f)** want to do something about this problem?

6. What do you think when you read that **(a)** the average American consumes 30 times more resources than

the average citizen of India, and **(b)** human activities are projected to make the earth's climate warmer? Are you skeptical, indifferent, sad, helpless, guilty, concerned, or outraged? Which of these feelings help perpetuate such problems, and which can help solve them?

7. For each of the following actions, state one or more of the four **scientific principles of sustainability** (Figure 1-17) that are involved: **(a)** recycling soda cans; **(b)** using a rake instead of leaf blower; **(c)** choosing to have no more than one child; **(d)** walking to class instead of driving; **(e)** taking your own reusable bags to the grocery store to carry things home in; **(f)** volunteering to help restore a prairie ; and **(g)** lobbying elected officials to require that 20% of your country's electricity be produced by renewable wind power by 2020.

8. Explain why you agree or disagree with each of the following statements: **(a)** humans are superior to other forms of life, **(b)** humans are in charge of the earth, **(c)** all economic growth is good, **(d)** the value of other forms of life depends only on whether they are useful to us, **(e)** because all forms of life eventually become extinct we should not worry about whether our activities cause their premature extinction, **(f)** all forms of life have an

inherent right to exist, **(g)** nature has an almost unlimited storehouse of resources for human use, **(h)** technology can solve our environmental problems, **(i)** I do not believe I have any obligation to future generations, and **(j)** I do not believe I have any obligation to other forms of life.

9. What are the basic beliefs of your environmental worldview (p. 20)? Record your answer. Then at the end of this course, return to your answer to see if your environmental worldview has changed. Are the beliefs included in your environmental worldview consistent with your answers to question 8? Are your environmental actions consistent with your environmental worldview?

10. List two questions that you would like to have answered as a result of reading this chapter.

Note: See Supplement 13 (p. S78) for a list of Projects related to this chapter.

ECOLOGICAL FOOTPRINT ANALYSIS

If a country's or the world's *ecological footprint per person* (Figure 1-10, p. 15) is larger than its *biological capacity per person* to replenish its renewable resources and to absorb the resulting waste products and pollution, it is said to have an *ecological*

deficit. If the reverse is true, it has an *ecological credit* or *reserve.* Use the data below to calculate the ecological deficit or credit for various countries. (For a map of ecological creditors and debtors, see Figure 4 on p. S26 in Supplement 4.)

Place	Per Capita Ecological Footprint (hectares per person)*	Per Capita Biocapacity (hectares per person)	Ecological Credit (+) or Debit (−) (hectares per person)
World	2.2	1.8	− 0.4
United States	9.8	4.7	
China	1.6	0.8	
India	0.8	0.4	
Russia	4.4	0.9	
Japan	4.4	0.7	
Brazil	2.1	9.9	
Germany	4.5	1.7	
United Kingdom	5.6	1.6	
Mexico	2.6	1.7	
Canada	7.6	14.5	

Source: Date from WWF, *Living Planet Report 2006.*

*1 hectare = 2.47 acres

1. Which two countries have the largest ecological deficits?

2. Which two countries have an ecological credit?

3. Rank the countries in order from the largest to the smallest per capita footprint.

LEARNING ONLINE

Log on to the Student Companion Site for this book at **academic.cengage.com/biology/miller**, and choose Chapter 1 for many study aids and ideas for further reading and research. These include flash cards, practice quizzing, Weblinks, information on Green Careers, and InfoTrac® College Edition articles.

Science, Matter, Energy, and Systems

Carrying Out a Controlled Scientific Experiment

One way in which scientists learn about how nature works is to conduct a *controlled experiment.* To begin, scientists isolate *variables,* or factors that can change within a system or situation being studied. An experiment involving *single-variable analysis* is designed to isolate and study the effects of one variable at a time.

To do such an experiment, scientists set up two groups. One is the *experimental group* in which a chosen variable is changed in a known way, and the other is the *control group* in which the chosen variable is not changed. If the experiment is designed and run properly, differences between the two groups should result from the variable that was changed in the experimental group.

In 1963, botanist F. Herbert Bormann, forest ecologist Gene Likens, and their colleagues began carrying out a classic controlled experiment. The goal was to compare the loss of water and nutrients from an uncut forest ecosystem (the *control site*) with one that was stripped of its trees (the *experimental site*).

They built V-shaped concrete dams across the creeks at the bottoms of several forested valleys in the Hubbard Brook Experimental Forest in New Hampshire (Figure 2-1). The dams were anchored on impenetrable bedrock, so that all surface water

leaving each forested valley had to flow across a dam where scientists could measure its volume and dissolved nutrient content.

In the first experiment, the investigators measured the amounts of water and dissolved plant nutrients that entered and left an undisturbed forested area (the control site) (Figure 2-1, left). These measurements showed that an undisturbed mature forest is very efficient at storing water and retaining chemical nutrients in its soils.

The next experiment involved setting up an experimental forested area. One winter, the investigators cut down all trees and shrubs in one valley (the experimental site), left them where they fell, and sprayed the area with herbicides to prevent the regrowth of vegetation. Then they compared the inflow and outflow of water and nutrients in this experimental site (Figure 2-1, right) with those in the control site (Figure 2-1, left) for 3 years.

With no plants to help absorb and retain water, the amount of water flowing out of the deforested valley increased by 30–40%. As this excess water ran rapidly over the ground, it eroded soil and carried dissolved nutrients out of the deforested site. Overall, the loss of key nutrients from the experimental forest was six to eight times that in the nearby control forest.

Figure 2-1 Controlled field experiment to measure the effects of deforestation on the loss of water and soil nutrients from a forest. V–notched dams were built into the impenetrable bedrock at the bottoms of several forested valleys (left) so that all water and nutrients flowing from each valley could be collected and measured for volume and mineral content. These measurements were recorded for the forested valley (left), which acted as the control site. Then all the trees in another valley (the experimental site) were cut (right) and the flows of water and soil nutrients from this experimental valley were measured for 3 years.

2-1 What is science?

CONCEPT 2-1 Scientists collect data and develop theories, models, and laws about how nature works.

2-2 What is matter?

CONCEPT 2-2 Matter consists of elements and compounds, which are in turn made up of atoms, ions, or molecules.

2-3 How can matter change?

CONCEPT 2-3 When matter undergoes a physical or chemical change, no atoms are created or destroyed (the law of conservation of matter).

2-4 What is energy and how can it be changed?

CONCEPT 2-4A When energy is converted from one form to another in a physical or chemical change, no energy is created or destroyed (first law of thermodynamics).

CONCEPT 2-4B Whenever energy is changed from one form to another, we end up with lower-quality or less usable energy than we started with (second law of thermodynamics).

2-5 What are systems and how do they respond to change?

CONCEPT 2-5A Systems have inputs, flows, and outputs of matter and energy, and their behavior can be affected by feedback.

CONCEPT 2-5B Life, human systems, and the earth's life-support systems must conform to the law of conservation of matter and the two laws of thermodynamics.

Note: Supplements 1 (p. S2), 2 (p. S4), 5 (p. S31), and 6 (p. S39) can be used with this chapter.

Science is an adventure of the human spirit.
It is essentially an artistic enterprise, stimulated largely by curiosity,
served largely by disciplined imagination,
and based largely on faith in the reasonableness, order,
and beauty of the universe.

WARREN WEAVER

2-1 What Is Science?

▶ **CONCEPT 2-1** Scientists collect data and develop theories, models, and laws about how nature works.

Science Is a Search for Order in Nature

Have you ever seen an area in a forest where all the trees were cut down? If so, you might wonder about the effects of cutting down all those trees. You might wonder how it affected the animals and people living in that area and how it affected the land itself. That is what scientists Bormann and Likens (**Core Case Study**) thought about when they designed their experiment.

Such curiosity is what motivates scientists. **Science** is an endeavor to discover how nature works and to use that knowledge to make predictions about what is likely to happen in nature. It is based on the assumption that events in the natural world follow or-derly cause-and-effect patterns that can be understood through careful observation, measurements, experimentation, and modeling. Figure 2-2 (p. 30) summarizes the scientific process.

There is nothing mysterious about this process. You use it all the time in making decisions. Here is an example of applying the scientific process to an everyday situation:

Observation: You try to switch on your flashlight and nothing happens.

Question: Why didn't the light come on?

Hypothesis: Maybe the batteries are dead.

Test the hypothesis: Put in new batteries and try to switch on the flashlight.

Figure 2-2 *What scientists do.* The essence of science is this process for testing ideas about how nature works. Scientists do not necessarily follow the exact order of steps shown here. For example, sometimes a scientist might start by formulating a hypothesis to answer the initial question and then run experiments to test the hypothesis.

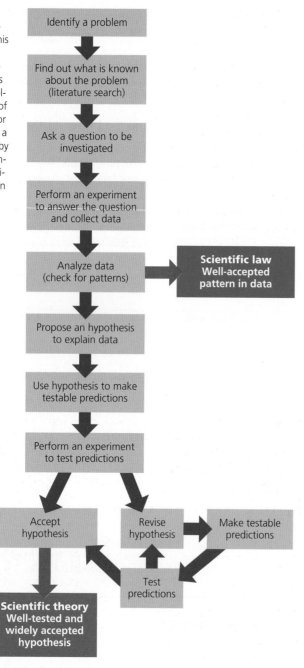

Identify a problem

↓

Find out what is known about the problem (literature search)

↓

Ask a question to be investigated

↓

Perform an experiment to answer the question and collect data

↓

Analyze data (check for patterns) → **Scientific law** Well-accepted pattern in data

↓

Propose an hypothesis to explain data

↓

Use hypothesis to make testable predictions

↓

Perform an experiment to test predictions

↓

Accept hypothesis   Revise hypothesis → Make testable predictions

Test predictions

↓

Scientific theory Well-tested and widely accepted hypothesis

Result: Flashlight still does not work.

New hypothesis: Maybe the bulb is burned out.

Experiment: Replace bulb with a new bulb.

Result: Flashlight works when switched on.

Conclusion: Second hypothesis is verified.

Here is a more formal outline of steps scientists often take in trying to understand nature, although not always in the order listed:

- *Identify a problem.* Bormann and Likens (**Core Case Study**) identified the loss of water and soil nutrients from cutover forests as a problem worth studying. CORE CASE STUDY

- *Find out what is known about the problem.* Bormann and Likens searched the scientific literature to find out what was known about retention and loss of water and soil nutrients in forests.

- *Ask a question to be investigated.* The scientists asked: "How does clearing forested land affect its ability to store water and retain soil nutrients?

- *Collect data to answer the question.* To collect **data**—information needed to answer their questions—scientists make observations of the subject area they are studying. Scientific observations involve gathering information by using human senses of sight, smell, hearing, and touch and extending those senses by using tools such as rulers, microscopes, and satellites. Often scientists conduct **experiments,** or procedures carried out under controlled conditions to gather information and test ideas. Bormann and Likens collected and analyzed data on the water and soil nutrients flowing from a patch of an undisturbed forest (Figure 2-1, left) and from a nearby patch of forest where they had cleared the trees for their experiment (Figure 2-1, right).

- *Propose a hypothesis to explain the data.* Scientists suggest a **scientific hypothesis,** a possible and testable explanation of what they observe in nature or in the results of their experiments. The data collected by Bormann and Likens show a decrease in the ability of a cleared forest to store water and retain soil nutrients such as nitrogen. They came up with the following hypothesis to explain their data: When a forest is cleared, it retains less water and loses large quantities of its soil nutrients when water from rain and melting snow flows across its exposed soil.

- *Make testable predictions.* Scientists use a hypothesis to make testable or logical predictions about what should happen if the hypothesis is valid. They often do this by making "If . . . then" predictions. Bormann and Likens predicted that *if* their original hypothesis was valid for nitrogen, *then* a cleared forest should also lose other soil nutrients such as phosphorus.

- *Test the predictions with further experiments, models, or observations.* To test their prediction, Bormann and Likens repeated their controlled experiment and measured the phosphorus content of the soil. Another way to test predictions is to develop a **model,** an approximate representation or simulation of a system being studied. Since Bormann and Likens performed their experiments, scientists have developed increasingly sophisticated mathematical and computer models of how forest systems work. Data from Bormann and Likens's research and that of other scientists can be fed into such models and used to predict the loss of phosphorus and other types of soil nutrients. These predictions can be compared with the actual measured losses to test the validity of the models.

- *Accept or reject the hypothesis.* If their new data do not support their hypotheses, scientists come up with other testable explanations. This process continues until there is general agreement among scientists in the field being studied that a particular hypothesis is the best explanation of the data. After Bormann and Likens confirmed that the soil in a cleared forest also loses phosphorus, they measured losses of other soil nutrients, which also supported their hypothesis. A well-tested and widely accepted scientific hypothesis or a group of related hypotheses is called a **scientific theory.** Thus, Bormann and Likens and their colleagues developed a theory that trees and other plants hold soil in place and help it

to retain water and nutrients needed by the plants for their growth.

Important features of the scientific process are *curiosity, skepticism, peer review, reproducibility,* and *openness to new ideas.* Good scientists are extremely curious about how nature works. But they tend to be highly skeptical of new data, hypotheses, and models until they can be tested and verified. **Peer review** happens when scientists report details of the methods and models they used, the results of their experiments, and the reasoning behind their hypotheses for other scientists working in the same field (their peers) to examine and criticize. Ideally, other scientists repeat and analyze the work to see if the data can be reproduced and whether the proposed hypothesis is reasonable and useful (Science Focus, below).

For example, Bormann and Likens (**Core Case Study**) submitted the results of their for- CORE CASE STUDY

SCIENCE FOCUS

Easter Island: Some Revisions to a Popular Environmental Story

For years, the story of Easter Island has been used in textbooks as an example of how humans can seriously degrade their own life-support system. It concerns a civilization that once thrived and then largely disappeared from a small, isolated island in the great expanse of the South Pacific, located about 3,600 kilometers (2,200 miles) off the coast of Chile.

Scientists used anthropological evidence and scientific measurements to estimate the ages of certain artifacts found on Easter Island (also called Rapa Nui). They hypothesized that about 2,900 years ago, Polynesians used double-hulled, seagoing canoes to colonize the island. The settlers probably found a paradise with fertile soil that supported dense and diverse forests and lush grasses. According to this hypothesis, the islanders thrived, and their population increased to as many as 15,000 people.

Measurements made by scientists seemed to indicate that over time, the Polynesians began living unsustainably by using the island's forest and soil resources faster than they could be renewed. When they used up the large trees, the islanders could no longer build their traditional seagoing canoes for fishing in deeper offshore waters, and no one could escape the island by boat.

Without the once-great forests to absorb and slowly release water, springs and streams dried up, exposed soils were

eroded, crop yields plummeted, and famine struck. There was no firewood for cooking or keeping warm. According to the original hypothesis, the population and the civilization collapsed as rival clans fought one another for dwindling food supplies, and the island's population dropped sharply. By the late 1870s, only about 100 native islanders were left.

In 2006, anthropologist Terry L. Hunt, Director of the University of Hawaii Rapa Nui Archeological Field School, evaluated the accuracy of past measurements and other evidence and carried out new measurements to estimate the ages of various artifacts. He used these data to formulate an alternative hypothesis describing the human tragedy on Easter Island.

Hunt came to several new conclusions. *First,* the Polynesians arrived on the island about 800 years ago, not 2,900 years ago. *Second,* their population size probably never exceeded 3,000, contrary to the earlier estimate of up to 15,000. *Third,* the Polynesians did use the island's trees and other vegetation in an unsustainable manner, and by 1722, visitors reported that most of the island's trees were gone.

But one question not answered by the earlier hypothesis was, why did the trees never grow back? Recent evidence and Hunt's new hypothesis suggest that rats (which either came along with the original settlers as stowaways or were brought along

as a source of protein for the long voyage) played a key role in the island's permanent deforestation. Over the years, the rats multiplied rapidly into the millions and devoured the seeds that would have regenerated the forests.

Another of Hunt's conclusions was that after 1722, the population of Polynesians on the island dropped to about 100, mostly from contact with European visitors and invaders. Hunt hypothesized that these newcomers introduced fatal diseases, killed off some of the islanders, and took large numbers of them away to be sold as slaves.

This story is an excellent example of how science works. The gathering of new scientific data and reevaluation of older data led to a revised hypothesis that challenges our thinking about the decline of civilization on Easter Island. As a result, the tragedy may not be as clear an example of human-caused ecological collapse as was once thought. However, there is evidence that other earlier civilizations did suffer ecological collapse largely from unsustainable use of soil, water, and other resources, as described in Supplement 5 on p. S31.

Critical Thinking

Does the new doubt about the original Easter Island hypothesis mean that we should not be concerned about using resources unsustainably on the island in space we call Earth? Explain.

est experiments to a respected scientific journal. Before publishing this report, the journal editors had it reviewed by other soil and forest experts. Other scientists have repeated the measurements of soil content in undisturbed and cleared forests of the same type and also in different types of forests. Their results have also been subjected to peer review. In addition, computer models of forest systems have been used to evaluate this problem, with the results subjected to peer review.

Scientific knowledge advances in this way, with scientists continually questioning measurements, making new measurements, and sometimes coming up with new and better hypotheses (Science Focus, p. 31). As a result, good scientists are *open to new ideas* that have survived the rigors of the scientific process.

Scientists Use Reasoning, Imagination, and Creativity to Learn How Nature Works

Scientists arrive at conclusions, with varying degrees of certainty, by using two major types of reasoning. **Inductive reasoning** involves using specific observations and measurements to arrive at a general conclusion or hypothesis. It is a form of "bottom-up" reasoning that goes from the specific to the general. For example, suppose we observe that a variety of different objects fall to the ground when we drop them from various heights. We can then use inductive reasoning to propose that *all objects fall to the earth's surface when dropped.*

Depending on the number of observations made, there may be a high degree of certainty in this conclusion. However, what we are really saying is "All objects that we or other observers have dropped from various heights have fallen to the earth's surface." Although it is extremely unlikely, we cannot be *absolutely sure* that no one will ever drop an object that does not fall to the earth's surface.

Deductive reasoning involves using logic to arrive at a specific conclusion based on a generalization or premise. It is a form of "top-down" reasoning that goes from the general to the specific. For example,

Generalization or premise: All birds have feathers.

Example: Eagles are birds.

Deductive conclusion: Eagles have feathers.

THINKING ABOUT

The Hubbard Brook Experiment and Scientific Reasoning

 CORE CASE STUDY

In carrying out and interpreting their experiment, did Bormann and Likens rely primarily on inductive or deductive reasoning?

Deductive and inductive reasoning and critical thinking skills (pp. 2–3) are important scientific tools. But scientists also use intuition, imagination, and creativity to explain some of their observations in nature. Often such ideas defy conventional logic and current scientific knowledge. According to physicist Albert Einstein, "There is no completely logical way to a new scientific idea." Intuition, imagination, and creativity are as important in science as they are in poetry, art, music, and other great adventures of the human spirit, as reflected by scientist Warren Weaver's quotation found at the opening of this chapter.

Scientific Theories and Laws Are the Most Important Results of Science

If an overwhelming body of observations and measurements supports a scientific hypothesis, it becomes a scientific theory. *Scientific theories are not to be taken lightly.* They have been tested widely, are supported by extensive evidence, and are accepted by most scientists in a particular field or related fields of study.

Nonscientists often use the word *theory* incorrectly when they actually mean *scientific hypothesis,* a tentative explanation that needs further evaluation. The statement, "Oh, that's just a theory," made in everyday conversation, implies that the theory was stated without proper investigation and careful testing—the opposite of the scientific meaning of the word.

Another important and reliable outcome of science is a **scientific law,** or **law of nature:** a well-tested and widely accepted description of what we find happening over and over again in the same way in nature. An example is the *law of gravity,* based on countless observations and measurements of objects falling from different heights. According to this law, all objects fall to the earth's surface at predictable speeds.

A scientific law is no better than the accuracy of the observations or measurements upon which it is based (see Figure 1 in Supplement 1 on p. S3). But if the data are accurate, a scientific law cannot be broken, unless and until we get contradictory new data.

Scientific theories and laws have a high probability of being valid, but they are not infallible. Occasionally, new discoveries and new ideas can overthrow a well-accepted scientific theory or law in what is called a **paradigm shift.** It occurs when the majority of scientists in a field or related fields accept a new *paradigm,* or framework for theories and laws in a particular field.

A good way to summarize the most important outcomes of science is to say that scientists collect data and develop theories, models, and laws that describe and explain how nature works (**Concept 2-1**). Scientists use reasoning and critical thinking skills. But the best scientists also use intuition, imagination, and creativity in asking important questions, developing hypotheses, and designing ways to test them.

For a superb look at how science works and what scientists do, see the Annenberg video series, *The Habitable Planet: A Systems Approach to Environmental Science* (see the website at **www.learner.org/resources/series209 .html**). Each of the 13 videos describes how scientists working on two different problems related to a certain subject are learning about how nature works. Also see Video 2, *Thinking Like Scientists,* in another Annenberg series, *Teaching High School Science* (see the website at **www.learner.org/resources/series126.html**).

The Results of Science Can Be Tentative, Reliable, or Unreliable

A fundamental part of science is *testing*. Scientists insist on testing their hypotheses, models, methods, and results over and over again to establish the reliability of these scientific tools and the resulting conclusions.

Media news reports often focus on disputes among scientists over the validity of data, hypotheses, models, methods, or results (see Science Focus, below). This helps to reveal differences in the reliability of various scientific tools and results. Simply put, some science is more reliable than other science, depending on how carefully it has been done and on how thoroughly the hypotheses, models, methods, and results have been tested.

Sometimes, preliminary results that capture news headlines are controversial because they have not been widely tested and accepted by peer review. They are not yet considered reliable, and can be thought of as **tentative science** or **frontier science.** Some of these results will be validated and classified as reliable and some will be discredited and classified as unreliable. At the frontier stage, it is normal for scientists to disagree about the meaning and accuracy of data and the validity of hypotheses and results. This is how scientific knowledge advances.

By contrast, **reliable science** consists of data, hypotheses, theories, and laws that are widely accepted by scientists who are considered experts in the field under study. The results of reliable science are based on

SCIENCE FOCUS

The Scientific Consensus over Global Warming

Based on measurements and models, it is clear that carbon dioxide and other gases in the atmosphere play a major role in determining the temperature of the atmosphere through a natural warming process called the *natural greenhouse effect.* Without the presence of these *greenhouse gases* in the atmosphere, the earth would be too cold for most life as we know it to exist, and you would not be reading these words. The earth's natural greenhouse effect is one of the most widely accepted theories in the atmospheric sciences and is an example of *reliable science.*

Since 1980, many climate scientists have been focusing their studies on three major questions:

- How much has the earth's atmosphere warmed during the past 50 years?

- How much of the warming is the result of human activities such as burning oil, gas, and coal and clearing forests, which add carbon dioxide and other greenhouse gases to the atmosphere?

- How much is the atmosphere likely to warm in the future and how might this affect the climate of different parts of the world?

To help clarify these issues, in 1988, the United Nations and the World Meteorological Organization established the Intergovernmental Panel on Climate Change (IPCC) to study how the climate system works, document past climate changes, and project future climate changes. The IPCC network includes more than 2,500 climate experts from 70 nations.

Since 1990, the IPCC has published four major reports summarizing the scientific consensus among these climate experts. In its 2007 report, the IPCC came to three major conclusions:

- It is *very likely* (a 90–99% probability) that the lower atmosphere is getting warmer and has warmed by about 0.74 C° (1.3 F°) between 1906 and 2005.

- Based on analysis of past climate data and use of 19 climate models, it is *very likely* (a 90–99% probability) that human activities, led by emissions of carbon dioxide from burning fossil fuels, have been the main cause of the observed atmospheric warming during the past 50 years.

- It is *very likely* that the earth's mean surface temperature will increase by about 3 C° (5.4 F°) between 2005 and 2100, unless we make drastic cuts in greenhouse gas emissions from power plants, factories, and cars that burn fossil fuels.

This scientific consensus among most of the world's climate experts is currently considered the most *reliable science* we have on this subject.

As always, there are individual scientists who disagree with the scientific consensus view. Typically, they question the reliability of certain data, say we don't have enough data to come to reliable conclusions, or question some of the hypotheses or models involved. However, in the case of global warming, they are in a distinct and declining minority.

Media reports are sometimes confusing or misleading because they present reliable science along with a quote from a scientist in the field who disagrees with the consensus view, or from someone who is not an expert in the field. This can cause public distrust of well-established reliable science, such as that reported by the IPCC, and may sometimes lead to a belief in ideas that are not widely accepted by the scientific community. (See the Guest Essay on environmental reporting by Andrew C. Revkin at CengageNOW.)

Critical Thinking

Find a newspaper article or other media report that presents the scientific consensus view on global warming and then attempts to balance it with a quote from a scientist who disagrees with the consensus view. Try to determine: **(a)** whether the dissenting scientist is considered an expert in climate science, **(b)** whether the scientist has published any peer reviewed papers on the subject, and **(c)** what organizations or industries are supporting the dissenting scientist.

the self-correcting process of testing, open peer review, reproducibility, and debate. New evidence and better hypotheses (Science Focus, p. 31) may discredit or alter tried and accepted views and even result in paradigm shifts. But unless that happens, those views are considered to be the results of reliable science.

Scientific hypotheses and results that are presented as reliable without having undergone the rigors of peer review, or that have been discarded as a result of peer review, are considered to be **unreliable science.** Here are some critical thinking questions you can use to uncover unreliable science:

- Was the experiment well designed? Did it involve enough testing? Did it involve a control group? **(Core Case Study)** ↰ CORE CASE STUDY

- Have the data supporting the proposed hypotheses been verified? Have the results been reproduced by other scientists?

- Do the conclusions and hypotheses follow logically from the data?

- Are the investigators unbiased in their interpretations of the results? Are they free of a hidden agenda? Were they funded by an unbiased source?

- Have the conclusions been verified by impartial peer review?

- Are the conclusions of the research widely accepted by other experts in this field?

If the answer to each of these questions is "yes," then the results can be classified as reliable science. Otherwise, the results may represent tentative science that needs further testing and evaluation, or they can be classified as unreliable science.

Environmental Science Has Some Limitations

Before continuing our study of environmental science, we need to recognize some of its limitations, as well as those of science in general. *First,* scientists can disprove things but they cannot prove anything absolutely, because there is always some degree of uncertainty in scientific measurements, observations, and models.

SCIENCE FOCUS

Statistics and Probability

Statistics consists of mathematical tools used to collect, organize, and interpret numerical data. For example, suppose we weigh each individual in a population of 15 rabbits. We can use statistics to calculate the *average* weight of the population. To do this, we add up the weights of the 15 rabbits and divide the total by 15. Similarly, Bormann and Likens **(Core Case Study)** made ↰ CORE CASE STUDY many measurements of nitrate levels in the water flowing from their undisturbed and cut patches of forests (Figure 2-1) and then averaged the results to get the most reliable value.

Scientists also use the statistical concept of probability to evaluate their results. **Probability** is the chance that something will happen. For example, if you toss a nickel, what is the probability or chance that it will come up heads? If your answer is 50%, you are correct. The chance of the nickel coming up heads is ½, which can also be expressed as 50% or 0.5. Probability is often expressed as a number between 0 and 1 written as a decimal (such as 0.5).

Now suppose you toss the coin 10 times and it comes up heads 6 times. Does this mean that the probability of it coming up heads is 0.6 or 60%? The answer is no because the *sample size*—the number of objects or events studied—was too small to yield a statistically accurate result. If you increase your sample size to 1,000 by tossing the coin 1,000 times, you are almost certain to get heads 50% of the time and tails 50% of the time.

It is important when doing scientific research to take samples in different places, in order to get a comprehensive evaluation of the variable being studied. It is also critical to have a large enough sample size to give an accurate estimate of the overall probability of an event happening.

For example, if you wanted to study the effects of a certain air pollutant on the needles of pine trees, you would need to locate different stands of the same type of pine tree that are all exposed to the pollutant over a certain period of time. At each location, you would need to measure the levels of the pollutant in the atmosphere at different times and average the results. You would also need to make measurements of the damage (such as needle loss) to a large enough sample of trees in each location over a certain time period. Then you would average the results in each location and compare the results from all locations.

If the average results were consistent in different locations, you could then say that there is a certain probability, say 60% (or 0.6), that this type of pine tree suffered a certain percentage loss of its needles when exposed to a specified average level of the pollutant over a given time. You would also need to run other experiments to determine that natural needle loss, extreme temperatures, insects, plant diseases, drought, or other factors did not cause the needle losses you observed. As you can see, getting reliable scientific results is not a simple process.

Critical Thinking

What does it mean when an international body of the world's climate experts says that there is a 90–99% chance (probability of 0.9–0.99) that human activities, led by emissions of carbon dioxide from burning fossil fuels, have been the main cause of the observed atmospheric warming during the past 50 years? Why would the probability never be 100%?

Instead scientists try to establish that a particular hypothesis, theory, or law has a very high *probability* (90–99%) of being true and thus is classified as reliable science. Most scientists rarely say something like, "Cigarettes cause lung cancer." Rather, they might say, "Overwhelming evidence from thousands of studies indicates that people who smoke have an increased risk of developing lung cancer."

THINKING ABOUT
Scientific Proof

Does the fact that science can never prove anything absolutely mean that its results are not valid or useful? Explain.

Second, scientists are human and cannot be expected to be totally free of bias about their results and hypotheses. However, bias can be minimized and often uncovered by the high standards of evidence required through peer review, although some scientists are bypassing traditional peer review by publishing their results online.

A *third* limitation involves use of statistical tools. There is no way to measure accurately how much soil is eroded annually worldwide, for example. Instead, scientists use statistical sampling and methods to estimate such numbers (Science Focus, at left). Such results should not be dismissed as "only estimates" because they can indicate important trends.

A *fourth* problem is that many environmental phenomena involve a huge number of interacting variables and complex interactions, which makes it too costly to test one variable at a time in controlled experiments such as the one described in the **Core Case Study** that opens this chapter. To help deal with this problem, scientists develop mathematical models that include the interactions of many variables. Running such models on computers can sometimes overcome this limitation and save both time and money. In addition, computer models can be used to simulate global experiments on phenomena like climate change, which are impossible to do in a controlled physical experiment.

Finally, the scientific process is limited to understanding the natural world. It cannot be applied to moral or ethical questions, because such questions are about matters for which we cannot collect data from the natural world. For example, we can use the scientific process to understand the effects of removing trees from an ecosystem, but this process does not tell us whether it is right or wrong to remove the trees.

Much progress has been made, but we still know too little about how the earth works, its current state of environmental health, and the environmental impacts of our activities. These knowledge gaps point to important *research frontiers,* several of which are highlighted throughout this text.

2-2   What Is Matter?

▶ **CONCEPT 2-2** Matter consists of elements and compounds, which are in turn made up of atoms, ions, or molecules.

Matter Consists of Elements and Compounds

To begin our study of environmental science, we start at the most basic level, looking at matter—the stuff that makes up life and its environment. **Matter** is anything that has mass and takes up space. It is made up of **elements,** each of which is a fundamental substance that has a unique set of properties and cannot be broken down into simpler substances by chemical means. For example, gold is an element; it cannot be broken down chemically into any other substance.

Some matter is composed of one element, such as gold or silver, but most matter consists of **compounds:** combinations of two or more different elements held together in fixed proportions. For example, water is a compound made of the elements hydrogen and oxygen, which have chemically combined with one another. (See Supplement 6 on p. S39 for an expanded discussion of basic chemistry.)

To simplify things, chemists represent each element by a one- or two-letter symbol. Table 2-1 (p. 36), lists the elements and their symbols that you need to know to understand the material in this book. Just four elements—oxygen, carbon, hydrogen, and nitrogen—make up about 96% of your body weight and that of most other living things.

Atoms, Ions, and Molecules Are the Building Blocks of Matter

The most basic building block of matter is an **atom:** the smallest unit of matter into which an element can be divided and still retain its chemical properties. The idea that all elements are made up of atoms is called the **atomic theory** and is the most widely accepted scientific theory in chemistry.

Table 2-1

Elements Important to the Study of Environmental Science

Element	Symbol	Element	Symbol
Hydrogen	H	Bromine	Br
Carbon	C	Sodium	Na
Oxygen	O	Calcium	Ca
Nitrogen	N	Lead	Pb
Phosphorus	P	Mercury	Hg
Sulfur	S	Arsenic	As
Chlorine	Cl	Uranium	U
Fluorine	F		

Atoms are incredibly small. In fact, more than 3 million hydrogen atoms could sit side by side on the period at the end of this sentence. If you could view them with a supermicroscope, you would find that each different type of atom contains a certain number of three different types of *subatomic particles:* positively charged **protons (p), neutrons (n)** with no electrical charge, and negatively charged **electrons (e).**

Each atom consists of an extremely small and dense center called its **nucleus**—which contains one or more protons and, in most cases, one or more neutrons— and one or more electrons moving rapidly somewhere around the nucleus in what is called an *electron probability cloud* (Figure 2-3). Each atom (except for *ions,* expained at right) has equal numbers of positively

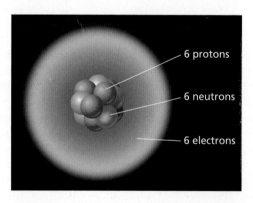

6 protons

6 neutrons

6 electrons

Figure 2-3 Greatly simplified model of a carbon-12 atom. It consists of a nucleus containing six positively charge protons and six neutral neutrons. There are six negatively charged electrons found outside its nucleus. We cannot determine the exact locations of the electrons. Instead, we can estimate the *probability* that they will be found at various locations outside the nucleus—sometimes called an *electron probability cloud.* This is somewhat like saying that there are six airplanes flying around inside a cloud. We don't know their exact location, but the cloud represents an area where we can probably find them.

charged protons and negatively charged electrons. Because these electrical charges cancel one another, *atoms as a whole have no net electrical charge.*

Each element has a unique **atomic number,** equal to the number of protons in the nucleus of its atom. Carbon (C), with 6 protons in its nucleus (Figure 2-3), has an atomic number of 6, whereas uranium (U), a much larger atom, has 92 protons in its nucleus and an atomic number of 92.

Because electrons have so little mass compared to protons and neutrons, *most of an atom's mass is concentrated in its nucleus.* The mass of an atom is described by its **mass number:** the total number of neutrons and protons in its nucleus. For example, a carbon atom with 6 protons and 6 neutrons in its nucleus has a mass number of 12, and a uranium atom with 92 protons and 143 neutrons in its nucleus has a mass number of 235 (92 + 143 = 235).

Each atom of a particular element has the same number of protons in its nucleus. But the nuclei of atoms of a particular element can vary in the number of neutrons they contain, and therefore, in their mass numbers. Forms of an element having the same atomic number but different mass numbers are called **isotopes** of that element. Scientists identify isotopes by attaching their mass numbers to the name or symbol of the element. For example, the three most common isotopes of carbon are carbon-12 (Figure 2-3, with six protons and six neutrons), carbon-13 (with six protons and seven neutrons), and carbon-14 (with six protons and eight neutrons). Carbon-12 makes up about 98.9% of all naturally occurring carbon.

A second building block of matter is an **ion**—an atom or groups of atoms with one or more net positive or negative electrical charges. An ion forms when an atom gains or loses one or more electrons. An atom that loses one or more of its electrons becomes an ion with one or more positive electrical charges, because the number of positively charged protons in its nucleus is now greater than the number of negatively charged electrons outside its nucleus. Similarly, when an atom gains one or more electrons, it becomes an ion with one or more negative electrical charges, because the number of negatively charged electrons is greater than the number of positively charged protons in its nucleus.

Ions containing atoms of more than one element are the basic units found in some compounds (called *ionic compounds*). For more details on how ions form see p. S39 in Supplement 6.

The number of positive or negative charges carried by an ion is shown as a superscript after the symbol for an atom or a group of atoms. Examples encountered in this book include a *positive* hydrogen ion (H^+), with one positive charge, an aluminum ion (Al^{3+}) with three positive charges, and a *negative* chloride ion (Cl^-) with one negative charge. These and other ions listed in Table 2-2 are used in other chapters in this book.

One example of the importance of ions in our study of environmental science is the nitrate ion (NO_3^-), a nutrient essential for plant growth. Figure 2-4 shows measurements of the loss of nitrate ions from the deforested area (Figure 2-1, right) in the controlled experiment run by Bormann and Likens (**Core Case Study**). Numerous chemical analyses of

CORE CASE STUDY

the water flowing through the dams of the cleared forest area showed an average 60-fold rise in the concentration of NO_3^- compared to water running off of the uncleared forest area. The stream below this valley became covered with algae whose populations soared as a result of an excess of nitrate plant nutrients. After a few years, however, vegetation began growing back on the cleared valley and nitrate levels in its runoff returned to normal levels.

Ions are also important for measuring a substance's **acidity** in a water solution, a chemical characteristic that helps determine how a substance dissolved in water will interact with and affect its environment. Scientists use **pH** as a measure of acidity, based on the amount of hydrogen ions (H^+) and hydroxide ions (OH^-) contained in a particular volume of a solution. Pure water (not tap water or rainwater) has an equal number of H^+ and OH^- ions. It is called a *neutral solution* and has a pH of 7. An *acidic solution* has more hydrogen ions than hydroxide ions and has a pH less than 7. A *basic solution* has more hydroxide ions than hydrogen ions and has a pH greater than 7. (See Figure 5 on p. S41 in Supplement 6 for more details.)

The third building block of matter is a **molecule:** a combination of two or more atoms of the same or different elements held together by forces called *chemical bonds.* Molecules are the basic units of some compounds

Table 2-2

Ions Important to the Study of Environmental Science

Positive Ion	Symbol	Negative Ion	Symbol
hydrogen ion	H^+	chloride ion	Cl^-
sodium ion	Na^+	hydroxide ion	OH^-
calcium ion	Ca^{2+}	nitrate ion	NO_3^-
aluminum ion	Al^{3+}	sulfate ion	SO_4^{2-}
ammonium ion	NH_4^+	phosphate ion	PO_4^{3-}

Table 2-3

Compounds Important to the Study of Environmental Science

Compound	Formula	Compound	Formula
sodium chloride	$NaCl$	methane	CH_4
carbon monoxide	CO	glucose	$C_6H_{12}O_6$
carbon dioxide	CO_2	water	H_2O
nitric oxide	NO	hydrogen sulfide	H_2S
nitrogen dioxide	NO_2	sulfur dioxide	SO_2
nitrous oxide	N_2O	sulfuric acid	H_2SO_4
nitric acid	HNO_3	ammonia	NH_3

(called *molecular compounds*). Examples are shown in Figure 4 on p. S41 in Supplement 6.

Chemists use a **chemical formula** to show the number of each type of atom or ion in a compound. This shorthand contains the symbol for each element present and uses subscripts to represent the number of atoms or ions of each element in the compound's basic structural unit. Examples of compounds and their formulas encountered in this book are sodium chloride ($NaCl$) and water (H_2O, read as "H-two-O"). These and other compounds important to our study of environmental science are listed in Table 2-3.

You may wish to mark the pages containing Tables 2-1 through 2-3, as they could be useful references for understanding material in other chapters.

CENGAGE**NOW**™ Examine atoms—their parts, how they work, and how they bond together to form molecules—at CengageNOW™.

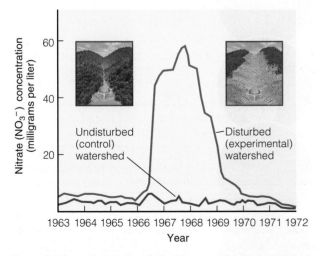

Figure 2-4 Loss of nitrate ions (NO_3^-) from a deforested watershed in the Hubbard Brook Experimental Forest in New Hampshire (Figure 2-1, right). The average concentration of nitrate ions in runoff from the deforested experimental watershed was 60 times greater than in a nearby unlogged watershed used as a control (Figure 2-1, left). (Data from F. H. Bormann and Gene Likens)

Organic Compounds Are the Chemicals of Life

Table sugar, vitamins, plastics, aspirin, penicillin, and most of the chemicals in your body are **organic compounds,** which contain at least two carbon atoms combined with atoms of one or more other elements. All other compounds are called **inorganic compounds.** One exception, methane (CH_4), has only one carbon atom but is considered an organic compound.

The millions of known organic (carbon-based) compounds include the following:

- *Hydrocarbons:* compounds of carbon and hydrogen atoms. One example is methane (CH_4), the main component of natural gas, and the simplest organic compound. Another is octane (C_8H_{18}), a major component of gasoline.

- *Chlorinated hydrocarbons:* compounds of carbon, hydrogen, and chlorine atoms. An example is the insecticide DDT ($C_{14}H_9Cl_5$).

- *Simple carbohydrates* (simple sugars): certain types of compounds of carbon, hydrogen, and oxygen atoms. An example is glucose ($C_6H_{12}O_6$), which most plants and animals break down in their cells to obtain energy. (For more details see Figure 8 on p. S42 in Supplement 6.)

Larger and more complex organic compounds, essential to life, are composed of *macromolecules.* Some of these molecules, called *polymers,* are formed when a number of simple organic molecules (*monomers*) are linked together by chemical bonds, somewhat like rail cars linked in a freight train. The three major types of organic polymers are

- *complex carbohydrates* such as cellulose and starch, which consist of two or more monomers of simple sugars such as glucose (see Figure 8 on p. S42 in Supplement 6),

- *proteins* formed by monomers called *amino acids* (see Figure 9 on p. S42 in Supplement 6), and

- *nucleic acids* (DNA and RNA) formed by monomers called *nucleotides* (see Figures 10 and 11 on p. S43 in Supplement 6).

Lipids, which include fats and waxes, are a fourth type of macromolecule essential for life (see Figure 12 on p. S43 in Supplement 6).

Matter Comes to Life through Genes, Chromosomes, and Cells

The story of matter, starting with the hydrogen atom, becomes more complex as molecules grow in complexity. This is no less true when we examine the fundamental components of life. The bridge between nonliving and living matter lies somewhere between macromolecules and **cells**—the fundamental structural units of life, which we explore in more detail in the next chapter.

Above, we mentioned nucleotides in DNA (see Figures 10 and 11 on p. S43 in Supplement 6). Within some DNA molecules are certain sequences of nucleotides called **genes.** Each of these distinct pieces of DNA contains instructions, called *genetic information,* for making specific proteins. Each of these coded units of genetic information concerns a specific **trait,** or characteristic passed on from parents to offspring during reproduction in an animal or plant.

Thousands of genes, in turn, make up a single **chromosome,** a special DNA molecule together with a number of proteins. Genetic information coded in your chromosomal DNA is what makes you different from an oak leaf, an alligator, or a flea, and from your parents. In other words, it makes you human, but it also makes you unique. The relationships of genetic material to cells are depicted in Figure 2-5.

A human body contains trillions of cells, each with an identical set of genes.

Each human cell (except for red blood cells) contains a nucleus.

Each cell nucleus has an identical set of chromosomes, which are found in pairs.

A specific pair of chromosomes contains one chromosome from each parent.

Each chromosome contains a long DNA molecule in the form of a coiled double helix.

Genes are segments of DNA on chromosomes that contain instructions to make proteins—the building blocks of life.

Figure 2-5 Relationships among cells, nuclei, chromosomes, DNA, and genes.

Matter Occurs in Various Physical Forms

The atoms, ions, and molecules that make up matter are found in three *physical states:* solid, liquid, and gas. For example, water exists as ice, liquid water, or water vapor depending on its temperature and the surrounding air pressure. The three physical states of any sample of matter differ in the spacing and orderliness of its atoms, ions, or molecules. A solid has the most compact and orderly arrangement, and a gas the least compact and orderly arrangement. Liquids are somewhere in between.

Some Forms of Matter Are More Useful than Others

Matter quality is a measure of how useful a form of matter is to humans as a resource, based on its availability and *concentration,* or amount of it that is contained in a given area or volume. **High-quality matter** is highly concentrated, is typically found near the earth's surface, and has great potential for use as a resource. Low-quality matter is not highly concentrated, is often located deep underground or dispersed in the ocean or atmosphere, and usually has little potential for use as a resource. See Figure 2-6 for examples illustrating differences in matter quality.

High Quality   **Low Quality**

Solid   Gas

Salt   Solution of salt in water

Coal   Coal-fired power plant emissions

Gasoline   Automobile emissions

Aluminum can   Aluminum ore

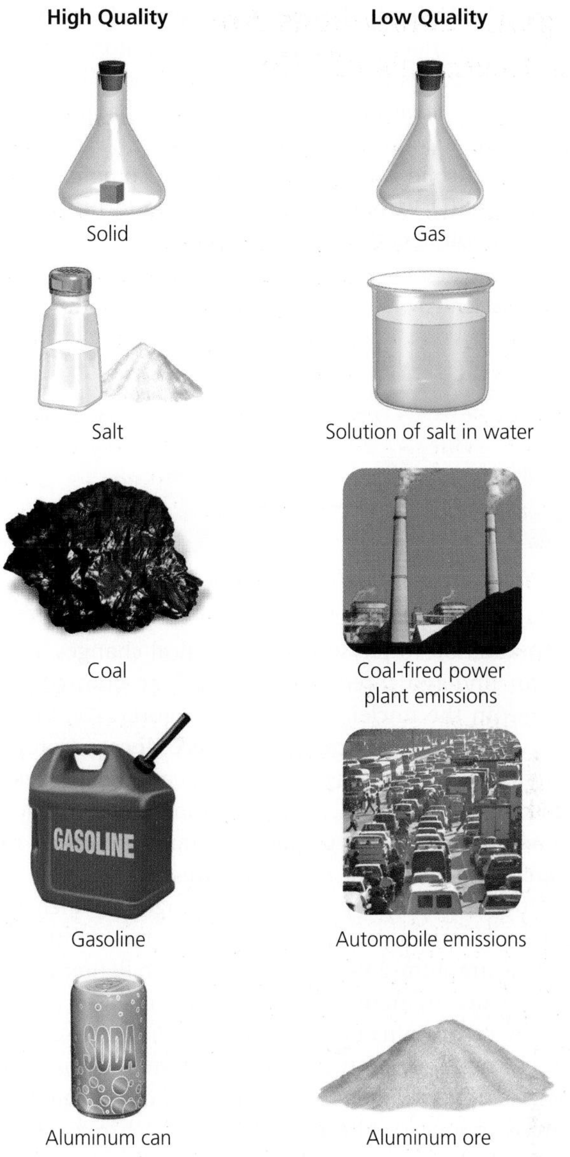

Figure 2-6 Examples of differences in matter quality. *High-quality matter* (left column) is fairly easy to extract and is highly concentrated; *low-quality matter* (right column) is not highly concentrated and is more difficult to extract than high-quality matter.

2-3 How Can Matter Change?

▶ **CONCEPT 2-3** When matter undergoes a physical or chemical change, no atoms are created or destroyed (the law of conservation of matter).

Matter Undergoes Physical, Chemical, and Nuclear Changes

When a sample of matter undergoes a **physical change,** its *chemical composition,* or the arrangement of its atoms or ions within molecules does not change. A piece of aluminum foil cut into small pieces is still aluminum foil. When solid water (ice) melts or liquid water boils, none of the H_2O molecules are changed.

The molecules are simply arranged in different spatial (physical) patterns.

— THINKING ABOUT —
Controlled Experiments and Physical Changes

CORE CASE STUDY

How would you set up a controlled experiment (**Core Case Study**) to verify that when water changes from one physical state to another, its chemical composition does not change?

In a **chemical change,** or **chemical reaction,** there is a change in the arrangement of atoms or ions within molecules of the substances involved. Chemists use *chemical equations* to represent what happens in a chemical reaction. For example, when coal burns completely, the solid carbon (C) in the coal combines with oxygen gas (O_2) from the atmosphere to form the gaseous compound carbon dioxide (CO_2).

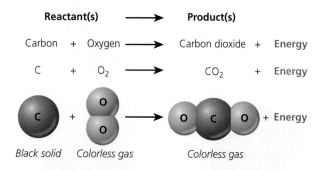

In addition to physical and chemical changes, matter can undergo three types of **nuclear changes,** or changes in the nuclei of its atoms (Figure 2-7). In the first type, called **natural radioactive decay,** isotopes spontaneously emit fast-moving subatomic particles, high-energy radiation such as gamma rays, or both (Figure 2-7, top). The unstable isotopes are called **radioactive isotopes** or **radioisotopes.**

Nuclear fission is a nuclear change in which the nuclei of certain isotopes with large mass numbers (such as uranium-235) are split apart into lighter nuclei when struck by neutrons; each fission releases two or three neutrons plus energy (Figure 2-7, middle). Each of these neutrons, in turn, can trigger an additional fission reaction. Multiple fissions within a certain amount of mass produce a **chain reaction,** which releases an enormous amount of energy.

Nuclear fusion is a nuclear change in which two isotopes of light elements, such as hydrogen, are forced together at extremely high temperatures until they fuse to form a heavier nucleus (Figure 2-7, bottom). A tremendous amount of energy is released in this process. Fusion of hydrogen nuclei to form helium nuclei is the source of energy in the sun and other stars.

We Cannot Create or Destroy Matter

We can change elements and compounds from one physical, chemical, or nuclear form to another, but we can never create or destroy any of the atoms involved in any physical or chemical change. All we can do is rearrange the atoms, ions, or molecules into different spatial patterns (physical changes) or combinations (chemical changes). These statements, based on many thousands of measurements, describe a scientific law known as the **law of conservation of matter:** when a physical or chemical change occurs, no atoms are created or destroyed (**Concept 2-3**).

This law means there is no "away" as in "to throw away." *Everything we think we have thrown away remains here with us in some form.* We can reuse or recycle some materials and chemicals, but the law of conservation of matter means we will always face the problem of what to do with some quantity of the wastes and pollutants we produce.

We talk about consuming matter as if matter is being used up or destroyed, but the law of conservation of matter says that this is impossible. What is meant by *matter consumption,* is not destruction of matter, but rather conversion of matter from one form to another.

2-4   What Is Energy and How Can It Be Changed?

▶ **CONCEPT 2-4A** When energy is converted from one form to another in a physical or chemical change, no energy is created or destroyed (first law of thermodynamics).

▶ **CONCEPT 2-4B** Whenever energy is changed from one form to another, we end up with lower-quality or less usable energy than we started with (second law of thermodynamics).

Energy Comes in Many Forms

Energy is the capacity to do work or transfer heat. Work is done when something is moved. The amount of work done is the product of the force applied to an object to move it a certain distance (work = force × distance).

For example, it takes a certain amount of muscular force to lift this book to a certain height.

There are two major types of energy: *moving energy* (called *kinetic energy*) and *stored energy* (called *potential energy*). Moving matter has **kinetic energy** because it has mass and velocity. Examples are wind (a mov-

Radiactive decay

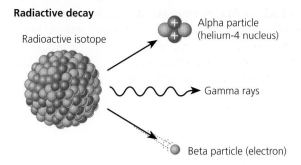

Radioactive isotope

Alpha particle
(helium-4 nucleus)

Gamma rays

Beta particle (electron)

Radioactive decay occurs when nuclei of unstable isotopes spontaneously emit fast-moving chunks of matter (alpha particles or beta particles), high-energy radiation (gamma rays), or both at a fixed rate. A particular radioactive isotope may emit any one or a combination of the three items shown in the diagram.

Nuclear fission

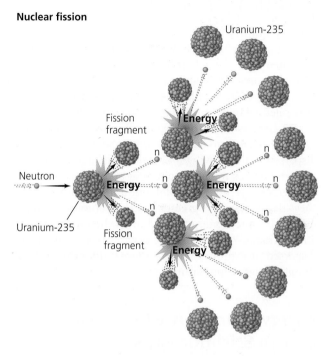

Uranium-235

Fission fragment

Neutron

Energy

Uranium-235

Fission fragment

Energy

Nuclear fission occurs when the nuclei of certain isotopes with large mass numbers (such as uranium-235) are split apart into lighter nuclei when struck by a neutron and release energy plus two or three more neutrons. Each neutron can trigger an additional fission reaction and lead to a *chain reaction*, which releases an enormous amount of energy.

Nuclear fusion

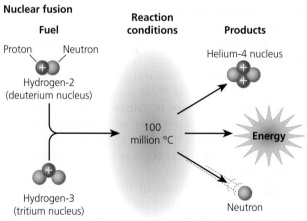

Fuel

Reaction conditions

Products

Proton   Neutron

Hydrogen-2
(deuterium nucleus)

Helium-4 nucleus

100 million °C

Energy

Hydrogen-3
(tritium nucleus)

Neutron

Nuclear fusion occurs when two isotopes of light elements, such as hydrogen, are forced together at extremely high temperatures until they fuse to form a heavier nucleus and release a tremendous amount of energy.

Figure 2-7 Types of nuclear changes: natural radioactive decay (top), nuclear fission (middle), and nuclear fusion (bottom).

ing mass of air), flowing water, and electricity (flowing electrons).

Another form of kinetic energy is **heat:** the total kinetic energy of all moving atoms, ions, or molecules within a given substance. When two objects at differ-

ent temperatures contact one another, heat flows from the warmer object to the cooler object.

Heat can be transferred from one place to another by three different methods: *radiation* (the emission of electromagnetic energy), *conduction* (the transfer of

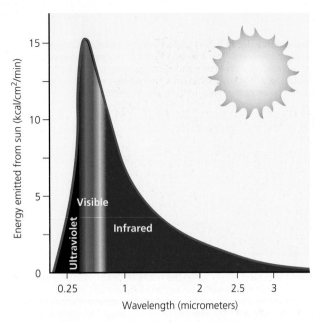

Energy emitted from sun (kcal/cm²/min)

Visible
Ultraviolet
Infrared

Wavelength (micrometers)

CENGAGENOW **Active Figure 2-8** *Solar capital:* the spectrum of electromagnetic radiation released by the sun consists mostly of visible light. *See an animation based on this figure at CengageNOW.*

kinetic energy between substances in contact with one another), and *convection* (the movement of heat within liquids and gases from warmer to cooler portions).

In **electromagnetic radiation,** another form of kinetic energy, energy travels in the form of a *wave* as a result of changes in electric and magnetic fields. There are many different forms of electromagnetic radiation, each having a different *wavelength* (distance between successive peaks or troughs in the wave) and *energy content*. Forms of electromagnetic radiation with short wavelengths, such as gamma rays, X rays, and ultraviolet (UV) radiation, have a higher energy content than do forms with longer wavelengths, such as visible light and infrared (IR) radiation (Figure 2-8). Visible light makes up most of the spectrum of electromagnetic radiation emitted by the sun (Figure 2-8).

CENGAGENOW Find out how color, wavelengths, and energy intensities of visible light are related at CengageNOW.

The other major type of energy is **potential energy,** which is stored and potentially available for use. Examples of potential energy include a rock held in your hand, an unlit match, the chemical energy stored in gasoline molecules, and the nuclear energy stored in the nuclei of atoms.

Potential energy can be changed to kinetic energy. Hold this book up, and it has potential energy; drop it on your foot, and its potential energy changes to kinetic energy. When a car engine burns gasoline, the potential energy stored in the chemical bonds of gasoline molecules changes into mechanical (kinetic) energy, which propels the car, and heat. Potential energy stored in the

molecules of carbohydrates you eat becomes kinetic energy when your body uses it to move and do other forms of work.

CENGAGENOW Witness how a Martian might use kinetic and potential energy at CengageNOW.

Some Types of Energy Are More Useful Than Others

Energy quality is a measure of an energy source's capacity to do useful work. **High-quality energy** is concentrated and has a high capacity to do useful work. Examples are very high-temperature heat, nuclear fission, concentrated sunlight, high-velocity wind, and energy released by burning natural gas, gasoline, or coal.

By contrast, **low-quality energy** is dispersed and has little capacity to do useful work. An example is heat dispersed in the moving molecules of a large amount of matter (such as the atmosphere or an ocean) so that its temperature is low. The total amount of heat stored in the Atlantic Ocean is greater than the amount of high-quality chemical energy stored in all the oil deposits of Saudi Arabia. Yet because the ocean's heat is so widely dispersed, it cannot be used to move things or to heat things to high temperatures.

Energy Changes Are Governed by Two Scientific Laws

Thermodynamics is the study of energy transformations. Scientists have observed energy being changed from one form to another in millions of physical and chemical changes. But they have never been able to detect the creation or destruction of any energy in such changes. The results of these experiments have been summarized in the **law of conservation of energy,** also known as the **first law of thermodynamics:** When energy is converted from one form to another in a physical or chemical change, no energy is created or destroyed (**Concept 2-4A**).

This scientific law tells us that when one form of energy is converted to another form in any physical or chemical change, *energy input always equals energy output.* No matter how hard we try or how clever we are, we cannot get more energy out of a system than we put in. This is one of nature's basic rules.

People talk about consuming energy but the first law says that it is impossible to use up energy. *Energy consumption,* then, means converting energy from one form to another with no energy being destroyed or created in the process.

Because the first law of thermodynamics states that energy cannot be created or destroyed, only converted from one form to another, you may be tempted to think

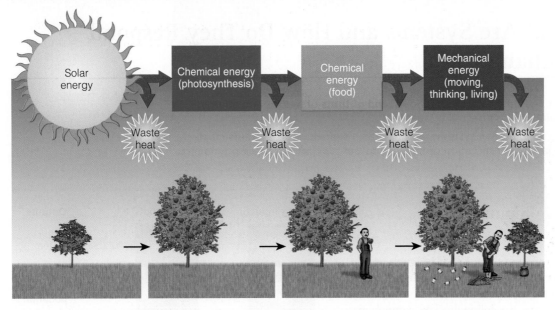

CENGAGENOW **Active Figure 2-9** The second law of thermodynamics in action in living systems. Each time energy changes from one form to another, some of the initial input of high-quality energy is degraded, usually to low-quality heat that is dispersed into the environment. *See an animation based on this figure at* CengageNOW.
Question: What are three things that you did during the past hour that degraded high-quality energy?

there will always be enough energy. Yet if you fill a car's tank with gasoline and drive around or use a flashlight battery until it is dead, something has been lost. But what is it? The answer is *energy quality,* the amount of energy available that can perform useful work.

Countless experiments have shown that whenever energy changes from one form to another, we always end up with less usable energy than we started with. These results have been summarized in the **second law of thermodynamics:** When energy changes from one form to another, we always end up with lower-quality or less usable energy than we started with (**Concept 2-4B**). This lower-quality energy usually takes the form of heat given off at a low temperature to the environment. There it is dispersed by the random motion of air or water molecules and becomes even less useful as a resource.

In other words, *energy always goes from a more useful to a less useful form when it is changed from one form to another.* No one has ever found a violation of this fundamental scientific law. It is another one of nature's basic rules.

Consider three examples of the second law of thermodynamics in action. *First,* when you drive a car, only about 6% of the high-quality energy available in its gasoline fuel actually moves the car, according to energy expert Amory Lovins. (See his Guest Essay at CengageNOW.) The remaining 94% is degraded to low-quality heat that is released into the environment. Thus, 94% of the money you spend for gasoline is not used to transport you anywhere.

Second, when electrical energy in the form of moving electrons flows through filament wires in an incandescent lightbulb, about 5% of it changes into useful light, and 95% flows into the environment as low-quality heat. In other words, the *incandescent lightbulb* is really an energy-wasting *heat bulb.*

Third, in living systems, solar energy is converted into chemical energy (food molecules) and then into mechanical energy (used for moving, thinking, and living). During each conversion, high-quality energy is degraded and flows into the environment as low-quality heat. Trace the flows and energy conversions in Figure 2-9 to see how this happens.

The second law of thermodynamics also means *we can never recycle or reuse high-quality energy to perform useful work.* Once the concentrated energy in a serving of food, a liter of gasoline, or a chunk of uranium is released, it is degraded to low-quality heat that is dispersed into the environment.

Energy efficiency, or **energy productivity,** is a measure of how much useful work is accomplished by a particular input of energy into a system. There is plenty of room for improving energy efficiency. Scientists estimate that only 16% of the energy used in the United States ends up performing useful work. The remaining 84% is either unavoidably wasted because of the second law of thermodynamics (41%) or unnecessarily wasted (43%). Thus, thermodynamics teaches us an important lesson: the cheapest and quickest way to get more energy is to stop wasting almost half the energy we use. We explore energy waste and energy efficiency in depth in Chapters 15 and 16.

CENGAGENOW See examples of how the first and second laws of thermodynamics apply in our world at CengageNOW.

2-5 What Are Systems and How Do They Respond to Change?

> ▶ **CONCEPT 2-5A** Systems have inputs, flows, and outputs of matter and energy, and their behavior can be affected by feedback.
>
> ▶ **CONCEPT 2-5B** Life, human systems, and the earth's life-support systems must conform to the law of conservation of matter and the two laws of thermodynamics.

Systems Have Inputs, Flows, and Outputs

A **system** is a set of components that function and interact in some regular way. The human body, a river, an economy, and the earth are all systems.

Most systems have the following key components: **inputs** from the environment, **flows** or **throughputs** of matter and energy within the system at certain rates, and **outputs** to the environment (Figure 2-10) (**Concept 2-5A**). One of the most powerful tools used by environmental scientists to study how these components of systems interact is computer modeling. (Science Focus, below)

Systems Respond to Change through Feedback Loops

When people ask you for feedback, they are usually seeking your response to something they said or did. They might feed this information back into their mental processes to help them decide whether and how to change what they are saying or doing.

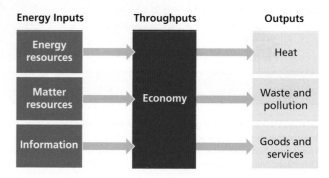

Figure 2-10 Inputs, throughput, and outputs of an economic system. Such systems depend on inputs of matter and energy resources and outputs of waste and heat to the environment. Such a system can become unsustainable if the throughput of matter and energy resources exceeds the ability of the earth's natural capital to provide the required resource inputs or the ability of the environment to assimilate or dilute the resulting heat, pollution, and environmental degradation.

Similarly, most systems are affected one way or another by **feedback,** any process that increases (positive feedback) or decreases (negative feedback) a change to a system (**Concept 2-5A**). Such a process, called a **feedback loop,** occurs when an output of matter, energy,

SCIENCE FOCUS

The Usefulness of Models

Scientists use *models,* or simulations, to learn how systems work. Some of our most powerful and useful technologies are mathematical and computer models.

Making a mathematical model usually requires going through three steps many times. *First,* scientists make guesses about systems they are modeling and write down equations to express these estimates. *Second,* they compute the likely behavior of a system implied by such equations. *Third,* they compare the system's projected behavior with observations of its actual behavior, also considering existing experimental data.

Mathematical models are particularly useful when there are many interacting variables, when the time frame of events being modeled is long, and when controlled experiments are impossible or too expensive to conduct.

After building and testing a mathematical model, scientists use it to predict what is *likely* to happen under a variety of conditions. In effect, they use mathematical models to answer *if–then* questions: "*If* we do such and such, *then* what is likely to happen now and in the future?" This process can give us a variety of projections or scenarios of possible futures or outcomes based on different assumptions. Mathematical models (like all other models) are no better than the assumptions on which they are built and the data fed into them.

Using data collected by Bormann and Likens in their Hubbard Brook experiment (**Core Case Study**), scientists created mathematical models to describe a forest and evaluate what happens to soil nutrients or other variables if the forest is disturbed or cut down.

Other areas of environmental science where computer modeling is becoming increasingly important include the studies of climate change, deforestation, biodiversity loss, and ocean systems.

Critical Thinking

What are two limitations of computer models? Do their limitations mean that we should not rely on such models? Explain.

or information is fed back into the system as an input and leads to changes in that system.

A **positive feedback loop** causes a system to change further in the same direction (Figure 2-11). In the Hubbard Brook experiments, for example (**Core Case Study**), researchers found that when vegetation was removed from a stream valley, flowing water from precipitation caused erosion and loss of nutrients, which caused more vegetation to die. With even less vegetation to hold soil in place, flowing water caused even more erosion and nutrient loss, which caused even more plants to die.

Such accelerating positive feedback loops are of great concern in several areas of environmental science. One of the most alarming is the melting of polar ice, which has occurred as the temperature of the atmosphere has risen during the past few decades. As that ice melts, there is less of it to reflect sunlight, and more water is exposed to sunlight. Because water is darker, it absorbs more solar energy, making the area warmer and causing the ice to melt faster, thus exposing more water. The melting of polar ice thus accelerates, causing a number of serious problems that we explore further in Chapter 19.

A **negative, or corrective, feedback loop** causes a system to change in the opposite direction from which is it moving. A simple example is a thermostat, a device that controls how often, and how long a heating or cooling system runs (Figure 2-12). When the furnace in a house is turned on and begins heating the house, the thermostat can be set to turn the furnace off when the temperature in the house reaches the set number. The house then stops getting warmer and starts to cool.

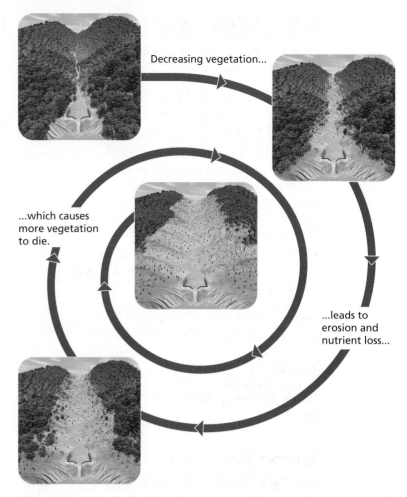

Figure 2-11 *Positive feedback loop.* Decreasing vegetation in a valley causes increasing erosion and nutrient losses, which in turn causes more vegetation to die, which allows for more erosion and nutrient losses. The system receives feedback that continues the process of deforestation.

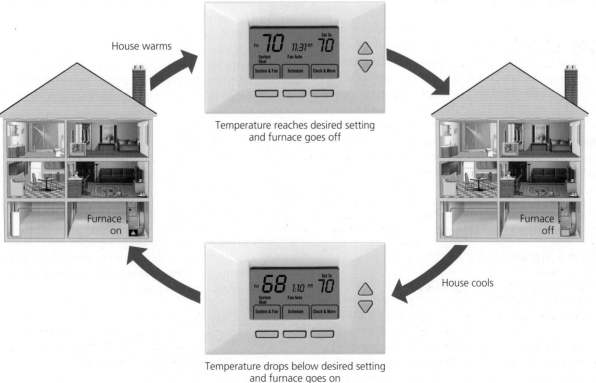

Figure 2-12 *Negative feedback loop.* When a house being heated by a furnace gets to a certain temperature, its thermostat is set to turn off the furnace, and the house begins to cool instead of continuing to get warmer. When the house temperature drops below the set point, this information is fed back, and the furnace is turned on and runs until the desired temperature is reached. The system receives feedback that reverses the process of heating or cooling.

THINKING ABOUT
Hubbard Brook and Feedback Loops

CORE
CASE
STUDY

How might experimenters have employed a negative feedback loop to stop, or correct, the positive
feedback loop that resulted in increasing erosion and nutrient losses in the Hubbard Brook experimental forest?

An important case of a negative feedback loop is the recycling and reuse of some resources such as aluminum, copper, and glass. For example, an aluminum can is one output of a mining and manufacturing system. When that output becomes an input, as the can is recycled and used in place of raw aluminum to make a new product, that much less aluminum is mined and the environmental impact of the mining-manufacturing system is lessened. Such a negative feedback loop therefore can promote sustainability and reduce the environmental impact of human activities by reducing the use of matter and energy resources and the amount of pollution and solid waste produced by use of such material.

Time Delays Can Allow a System to Reach a Tipping Point

Complex systems often show **time delays** between the input of a feedback stimulus and the response to it. For example, scientists could plant trees in a degraded area such as the Hubbard Brook experimental forest to slow erosion and nutrient losses (**Core Case Study**), but it would take years for the trees and other vegetation to grow enough to accomplish this purpose.

CORE
CASE
STUDY

Time delays can also allow an environmental problem to build slowly until it reaches a *threshold level,* or **tipping point,** causing a fundamental shift in the behavior of a system. Prolonged delays dampen the negative feedback mechanisms that might slow, prevent, or halt environmental problems. In the Hubbard Brook example, if erosion and nutrient losses reached a certain point where the land could not support vegetation, then an irreversible tipping point would have been reached, and it would be futile to plant trees to try to restore the system. Other environmental problems that can reach tipping point levels are population growth, leaks from toxic waste dumps, global climate change, and degradation of forests from prolonged exposure to air pollutants.

System Effects Can Be Amplified through Synergy

A **synergistic interaction,** or **synergy,** occurs when two or more processes interact so that the combined effect is greater than the sum of their separate effects. Scientific studies reveal such an interaction between smoking and inhaling asbestos particles. Lifetime smokers have ten times the risk that nonsmokers have of getting lung cancer. And individuals exposed to asbes-

tos particles for long periods increase their risk of getting lung cancer fivefold. But people who smoke and are exposed to asbestos have 50 times the risk that nonsmokers have of getting lung cancer.

Similar dangers can result from combinations of certain air pollutants that, when combined, are more hazardous to human health than they would be acting independently. We examine such hazards further in Chapter 17.

On the other hand, synergy can be helpful. Suppose we want to persuade an elected official to vote for a certain environmental law. You could write, e-mail, or visit the official. But you may have more success if you can get a group of potential voters to do such things. In other words, the combined or synergistic efforts of people working together can be more effective than the efforts of each person acting alone.

RESEARCH FRONTIER

Identifying environmentally harmful and beneficial synergistic interactions. See **academic.cengage.com/biology/miller**.

Human Activities Can Have Unintended Harmful Results

One of the lessons we can derive from the four **scientific principles of sustainability** (see back cover) is that *everything we do affects someone or something in the environment in some way.* In other words, any action in a complex system has multiple and often unintended, unpredictable effects. As a result, most of the environmental problems we face today are unintended results of activities designed to increase the quality of human life.

For example, clearing trees from the land to plant crops can increase food production and feed more people. But it can also lead to soil erosion, flooding, and a loss of biodiversity, as Easter Islanders and other civilizations learned the hard way (Science Focus, p. 31, and Supplement 5 on p. S31).

One factor that can lead to an environmental surprise is a *discontinuity* or abrupt change in a previously stable system when some *environmental threshold* or *tipping point* is crossed. Scientific evidence indicates that we are now reaching an increasing number of such tipping points. For example, we have depleted fish stocks in some parts of the world to the point where it is not profitable to harvest them. Other examples, such as deforested areas turning to desert, coral reefs dying, species disappearing, glaciers melting, and sea levels rising, will be discussed in later chapters.

RESEARCH FRONTIER

Tipping points for various environmental systems such as fisheries, forests coral reefs, and the earth's climate system. See **academic.cengage.com/biology/miller.**

Life, economic and other human systems, and the earth's life support systems depend on matter and energy, and therefore they must obey the law of conservation of matter and the two laws of thermodynamics (**Concept 2-5B**). Without these laws, economic growth based on using matter and energy resources to produce goods and services (Figure 2-10) could be expanded indefinitely and cause even more serious environmental problems. But these scientific laws place limits on what we can do with matter and energy resources.

A Look Ahead

In the next six chapters, we apply the three basic laws of matter and thermodynamics and the four **scientific principles of sustainability** (see

back cover) to living systems. Chapter 3 shows how the sustainability principles related to solar energy and nutrient cycling apply in ecosystems. Chapter 4 focuses on using the biodiversity principle to understand the relationships between species diversity and evolution. Chapter 5 examines how the biodiversity and population control principles relate to interactions among species and how such interactions regulate population size. In Chapter 6, we apply the principles of biodiversity and population control to the growth of the human population. In Chapter 7, we look more closely at terrestrial biodiversity in different types of deserts, grasslands, and forests. Chapter 8 examines aquatic biodiversity in aquatic systems such as oceans, lakes, wetlands, and rivers.

REVISITING The Hubbard Brook Experimental Forest and Sustainability

The controlled experiment discussed in the **Core Case Study** that opened this chapter revealed that clearing a mature forest degrades some of its natural capital (Figure 1-7, p. 12). Specifically, the loss of trees and vegetation altered the ability of the forest to retain and recycle water and other critical plant nutrients—a crucial ecological function based on one of the four **scientific principles of sustainability** (see back cover). In other words, the uncleared forest was a more sustainable system than a similar area of cleared forest (Figures 2-1 and 2-4).

This loss of vegetation also violated the other three scientific principles of sustainability. For example, the cleared forest had fewer plants that could use solar energy to produce food for

animals. And the loss of plants and animals reduced the life-sustaining biodiversity of the cleared forest. This in turn reduced some of the interactions between different types of plants and animals that help control their populations.

Humans clear forests to grow food and build cities. The key question is, how far can we go in expanding our ecological footprints (Figure 1-10, p. 15) without threatening the quality of life for our own species and the other species that keep us alive and support our economies? To live sustainably, we need to find and maintain a balance between preserving undisturbed natural systems and modifying other natural systems for our use.

The second law of thermodynamics holds, I think, the supreme position among laws of nature. . . . If your theory is found to be against the second law of thermodynamics, I can give you no hope.

ARTHUR S. EDDINGTON

REVIEW

1. Review the Key Questions and Concepts for this chapter on p. 29. Describe the controlled scientific experiment carried out at the Hubbard Brook Experimental Forest. What is **science?** Describe the steps involved in the scientific process. What is **data?** What is an **experiment?** What is a **model?** Distinguish among a **scientific hypothesis, scientific theory,** and **scientific law (law of nature)**. What is **peer review** and why is it important? Explain why scientific theories are not to be taken lightly and why people often use the term "theory" incorrectly.

2. Distinguish between **inductive reasoning** and **deductive reasoning** and give an example of each. Explain

why scientific theories and laws are the most important results of science.

3. What is a **paradigm shift?** Distinguish among **tentative science (frontier science), reliable science,** and **unreliable science.** Describe the scientific consensus concerning global warming. What is **statistics?** What is **probability** and what is its role in scientific conclusions? What are five limitations of science and environmental science?

4. What is **matter?** Distinguish between an **element** and a **compound** and give an example of each. Distinguish among **atoms, ions,** and **molecules** and give an

example of each. What is the **atomic theory?** Distinguish among **protons, neutrons,** and **electrons.** What is the **nucleus** of an atom? Distinguish between the **atomic number** and the **mass number** of an element. What is an **isotope?** What is **acidity?** What is **pH?**

5. What is a **chemical formula?** Distinguish between **organic compounds** and **inorganic compounds** and give an example of each. Distinguish among complex carbohydrates, proteins, nucleic acids, and lipids. What is a **cell?** Distinguish among **genes, traits,** and **chromosomes.** What is **matter quality?** Distinguish between **high-quality matter** and **low-quality matter** and give an example of each.

6. Distinguish between a **physical change** and a **chemical change (chemical reaction)** and give an example of each. What is a **nuclear change?** Explain the differences among **natural radioactive decay, nuclear fission,** and **nuclear fusion.** What is a **radioactive isotope (radioisotope)?** What is a **chain reaction?** What is the **law of conservation of matter** and why is it important?

7. What is **energy?** Distinguish between **kinetic energy** and **potential energy** and give an example of each. What is **heat?** Define and give two examples of **electromagnetic radiation.** What is **energy quality?** Distinguish between **high-quality energy** and **low-quality energy** and give an example of each.

8. What is the **law of conservation of energy (first law of thermodynamics)** and why is it important? What is the **second law of thermodynamics** and why is it important? Explain why this law means that we can never recycle or reuse high-quality energy. What is **energy efficiency (energy productivity)** and why is it important?

9. Define and give an example of a **system?** Distinguish among the **input, flow (throughput),** and **output** of a system. Why are scientific models useful? What is **feedback?** What is a **feedback loop?** Distinguish between a **positive feedback loop** and a **negative (corrective) feedback loop** in a system, and give an example of each. Distinguish between a **time delay** and a **synergistic interaction (synergy)** in a system and give an example of each. What is a **tipping point?**

10. Explain how human activities can have unintended harmful environmental results. Relate the four **scientific principles of sustainability** to the Hubbard Brook Experimental Forest controlled experiment (**Core Case Study**).

Note: Key Terms are in **bold** type.

CRITICAL THINKING

1. What ecological lesson can we learn from the controlled experiment on the clearing of forests described in the **Core Case Study** that opened this chapter?

2. Think of an area you have seen where some significant change has occurred to a natural system. What is a question you might ask in order to start a scientific process to evaluate the effects of this change, similar to the process described in the **Core Case Study**?

3. Describe a way in which you have applied the scientific process described in this chapter (Figure 2-2) in your own life, and state the conclusion you drew from this process. Describe a new problem that you would like to solve using this process.

4. Respond to the following statements:
   a. Scientists have not absolutely proven that anyone has ever died from smoking cigarettes.
   b. The natural greenhouse theory—that certain gases (such as water vapor and carbon dioxide) warm the lower atmosphere—is not a reliable idea because it is just a scientific theory.

5. A tree grows and increases its mass. Explain why this phenomenon is not a violation of the law of conservation of matter.

6. If there is no "away" where organisms can get rid of their wastes, why is the world not filled with waste matter?

7. Someone wants you to invest money in an automobile engine, claiming that it will produce more energy than the energy in the fuel used to run it. What is your response? Explain.

8. Use the second law of thermodynamics to explain why a barrel of oil can be used only once as a fuel, or in other words, why we cannot recycle high-quality energy.

9. a. Imagine you have the power to revoke the law of conservation of matter for one day. What are three things you would do with this power?
   b. Imagine you have the power to violate the first law of thermodynamics for one day. What are three things you would do with this power?

10. List two questions that you would like to have answered as a result of reading this chapter.

Note: See Supplement 13 (p. S78) for a list of Projects related to this chapter.

Marine scientists from the U.S. state of Maryland have produced the following two graphs as part of a report on the current health of the Chesapeake Bay. They are pleased with the recovery of the striped bass population but are concerned about the decline of the blue crab population, because blue crabs are consumed by mature striped bass. Their hypothesis is that as the population of striped bass increases, the population of blue crab decreases.

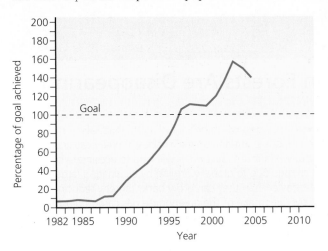

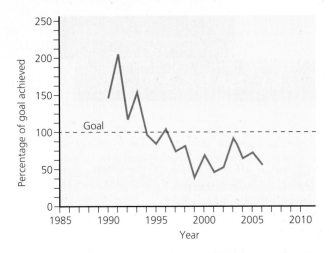

Using the data in the above graphs, answer the following questions:

1. Which years confirm their hypothesis?

2. Which years do not support their hypothesis?

3. If the crab population reaches 100% of the goal figure, what would you predict the striped bass goal figure would be?

LEARNING ONLINE

Log on to the Student Companion Site for this book at **academic.cengage.com/biology/miller**, and choose Chapter 2 for many study aids and ideas for further reading and research. These include flash cards, practice quizzing, Weblinks, information on Green Careers, and InfoTrac® College Edition articles.

Ecosystems: What Are They and How Do They Work?

Tropical Rain Forests Are Disappearing

Tropical rain forests are found near the earth's equator and contain an incredible variety of life. These lush forests are warm year round and have high humidity and heavy rainfall almost daily. Although they cover only about 2% of the earth's land surface, studies indicate that they contain up to half of the world's known terrestrial plant and animal species. For these reasons, they make an excellent natural laboratory for the study of *ecosystems*—communities of organisms interacting with one another and with the physical environment of matter and energy in which they live.

So far, at least half of these forests have been destroyed or disturbed by humans cutting down trees, growing crops, grazing cattle, and building settlements (Figure 3-1), and the degradation of these centers of life (biodiversity) is increasing. Ecologists warn that without strong conservation measures, most of these forests will probably be gone or severely degraded within your lifetime.

Scientists project that disrupting these ecosystems will have three major harmful effects. *First,* it will reduce the earth's vital biodiversity by destroying or degrading the habitats of many of their unique plant and animal species, thereby causing their premature extinction. *Second,* it will help to accelerate climate change due to global warming by eliminating large areas of trees faster than they can grow back, thereby reducing the trees' overall uptake of the greenhouse gas carbon dioxide.

Third, it will change regional weather patterns in ways that will prevent the return of diverse tropical rain forests in cleared or degraded areas. Once this tipping point is reached, tropical rain forest in such areas will become less diverse tropical grassland.

Ecosystems recycle materials and provide humans and other organisms with essential natural services (Figure 1-3, p. 8) and natural resources such as nutrients (Figure 1-4, p. 9). In this chapter, we look more closely at how ecosystems work and how human activities, such as stripping a large area of its trees, can disrupt the cycling of nutrients within ecosystems and the flow of energy through them.

Figure 3-1 Natural capital degradation: satellite image of the loss of tropical rain forest, cleared for farming, cattle grazing, and settlements, near the Bolivian city of Santa Cruz between June 1975 (left) and May 2003 (right).

3-1 What is ecology?

CONCEPT 3-1 Ecology is the study of how organisms interact with one another and with their physical environment of matter and energy.

3-2 What keeps us and other organisms alive?

CONCEPT 3-2 Life is sustained by the flow of energy from the sun through the biosphere, the cycling of nutrients within the biosphere, and gravity.

3-3 What are the major components of an ecosystem?

CONCEPT 3-3A Ecosystems contain living (biotic) and nonliving (abiotic) components.

CONCEPT 3-3B Some organisms produce the nutrients they need, others get their nutrients by consuming other organisms, and some recycle nutrients back to producers by decomposing the wastes and remains of organisms.

3-4 What happens to energy in an ecosystem?

CONCEPT 3-4A Energy flows through ecosystems in food chains and webs.

CONCEPT 3-4B As energy flows through ecosystems in food chains and webs, the amount of chemical energy available to organisms at each succeeding feeding level decreases.

3-5 What happens to matter in an ecosystem?

CONCEPT 3-5 Matter, in the form of nutrients, cycles within and among ecosystems and the biosphere, and human activities are altering these chemical cycles.

3-6 How do scientists study ecosystems?

CONCEPT 3-6 Scientists use field research, laboratory research, and mathematical and other models to learn about ecosystems.

Note: Supplements 2 (p. S4), 4 (p. S20), 6 (p. S39), 7 (p. S46), and 13 (p. S78) can be used with this chapter.

The earth's thin film of living matter is sustained by grand-scale cycles of chemical elements.

G. EVELYN HUTCHINSON

3-1   What Is Ecology?

▶ **CONCEPT 3-1** Ecology is the study of how organisms interact with one another and with their physical environment of matter and energy.

Cells Are the Basic Units of Life

All organisms (living things) are composed of **cells:** the smallest and most fundamental structural and functional units of life. They are minute compartments covered with a thin membrane and within which the processes of life occur. The idea that all living things are composed of cells is called the **cell theory** and it is the most widely accepted scientific theory in biology. Organisms may consist of a single cell (bacteria, for instance) or huge numbers of cells, as is the case for most plants and animals.

On the basis of their cell structure, organisms can be classified as either *eukaryotic* or *prokaryotic.* A **eukaryotic cell** is surrounded by a membrane and has a distinct *nucleus* (a membrane-bounded structure containing genetic material in the form of DNA) and

several other internal parts called *organelles,* which are also surrounded by membranes (Figure 3-2a, p. 52). Most organisms consist of eukaryotic cells. A **prokaryotic cell** is also surrounded by a membrane, but it has no distinct nucleus and no other internal parts surrounded by membranes (Figure 3-2b, p. 52). All bacteria consist of a single prokaryotic cell. The relationships among cells and their genetic material were shown in Figure 2-5 (p. 38).

Species Make Up the Encyclopedia of Life

For a group of sexually reproducing organisms, a **species** is a set of individuals that can mate and produce fertile offspring. Every organism is a member of a certain species with certain traits. Scientists have developed a

---

Figure 3-2 Natural capital: (a) generalized structure of a *eukaryotic cell* and **(b)** *prokaryotic cell*. Note that a prokaryotic cell lacks a distinct nucleus and generalized structure of a *eukaryotic cell*.

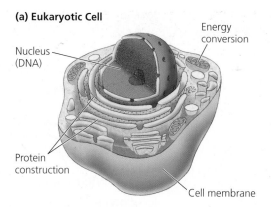

(a) Eukaryotic Cell

Nucleus (DNA)

Energy conversion

Protein construction

Cell membrane

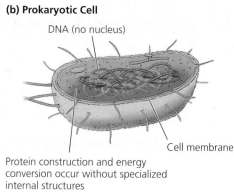

(b) Prokaryotic Cell

DNA (no nucleus)

Cell membrane

Protein construction and energy conversion occur without specialized internal structures

---

Biosphere — Parts of the earth's air, water, and soil where life is found

Ecosystem — A community of different species interacting with one another and with their nonliving environment of matter and energy

Community — Populations of different species living in a particular place, and potentially interacting with each other

Population — A group of individuals of the same species living in a particular place

Organism — An individual living being

Cell — The fundemental structural and functional unit of life

Molecule — Chemical combination of two or more atoms of the same or different elements

Water

Atom — Smallest unit of a chemical element that exhibits its chemical properties

H   O
Hydrogen   Oxygen

---

distinctive system for classifying and naming each species, as discussed in Supplement 7 on p. S46.

We do not know how many species are on the earth. Estimates range from 4 million to 100 million. The best guess is that there are 10–14 million species. So far biologists have identified about 1.8 million species. These and millions of species still to be classified are the entries in the encyclopedia of life found on the earth. Up to half of the world's plant and animal species live in tropical rain forests that are being cleared rapidly (**Core Case Study**). Insects make up most of the world's known species (Science Focus, p. 54). In 2007, scientists began a $100 million, 10-year project to list and describe all 1.8 million known species in a free Internet encyclopedia (**www.eol.org**).

CORE CASE STUDY

Ecologists Study Connections in Nature

Ecology (from the Greek words *oikos,* meaning "house" or "place to live," and *logos,* meaning "study of") is the study of how organisms interact with their living (biotic) environment of other organisms and with their nonliving (abiotic) environment of soil, water, other forms of matter, and energy mostly from the sun (**Concept 3-1**). In effect, it is a study of *connections in nature*.

To enhance their understanding of nature, scientists classify matter into levels of organization from atoms to the biosphere (Figure 3-3). Ecologists focus on organisms, populations, communities, ecosystems, and the biosphere.

A **population** is a group of individuals of the same species that live in the same place at the same time. Examples include a school of glassfish in the Red Sea (Figure 3-4), the field mice living in a cornfield, monarch butterflies clustered in a tree, and people in a country.

CENGAGENOW™ **Active Figure 3-3** Some levels of organization of matter in nature. Ecology focuses on the top five of these levels. *See an animation based on this figure at CengageNOW.*

Figure 3-4 Population (school) of glassfish in a cave in the Red Sea.

Wolfgang Poelzer/Peter Arnold, Inc.

In most natural populations, individuals vary slightly in their genetic makeup, which is why they do not all look or act alike. This variation in a population is called **genetic diversity** (Figure 3-5).

The place where a population or an individual organism normally lives is its **habitat.** It may be as large as an ocean or as small as the intestine of a termite. An organism's habitat can be thought of as its natural "address." Each habitat, such as a tropical rain forest (Figure 3-1, **Core Case Study**), a desert, or a pond, has certain resources, such as water, and environmental conditions, such as temperature and light, that its organisms need in order to survive.

CORE CASE STUDY

A **community,** or **biological community,** consists of all the populations of different species that live in a particular place. For example, a catfish species in a pond usually shares the pond with other fish species, and with plants, insects, ducks, and many other species that make up the community. Many of the organisms in a community interact with one another in feeding and other relationships.

An **ecosystem** is a community of different species interacting with one another and with their nonliving environment of soil, water, other forms of matter, and energy, mostly from the sun. Ecosystems can range in size from a puddle of water to an ocean, or from a patch of woods to a forest. Ecosystems can be natural or artificial (human created). Examples of artificial ecosystems are crop fields, tree farms, and reservoirs.

Ecosystems do not have clear boundaries and are not isolated from one another. Matter and energy move from one ecosystem to another. For example, soil can wash from a grassland or crop field into a nearby river or lake. Water flows from forests into nearby rivers

and crop fields. Birds and various other species migrate from one ecosystem to another. And winds can blow pollen from a forest into a grassland.

The **biosphere** consists of the parts of the earth's air, water, and soil where life is found. In effect, it is the global ecosystem in which all organisms exist and can interact with one another.

CENGAGENOW™ Learn more about how the earth's life is organized in five levels in the study of ecology at CengageNOW™.

Figure 3-5 *Genetic diversity* among individuals in a population of a species of Caribbean snail is reflected in the variations in shell color and banding patterns. Genetic diversity can also include other variations such as slight differences in chemical makeup, sensitivity to various chemicals, and behavior.

SCIENCE FOCUS

Insects are an important part of the earth's natural capital, although they generally have a bad reputation. We classify many insect species as *pests* because they compete with us for food, spread human diseases such as malaria, bite or sting us, and invade our lawns, gardens, and houses. Some people fear insects and think the only good bug is a dead bug. They fail to recognize the vital roles insects play in helping to sustain life on earth.

For example, pollination is a natural service that allows plants to reproduce sexually when pollen grains are transferred from one plant to a receptive part of another plant. Some plants are pollinated by species such as hummingbirds and bats, and by pollen being transmitted by wind or flowing water. But many of the earth's plant species depend on insects to pollinate their flowers (Figure 3-A, left).

Insects that eat other insects—such as the praying mantis (Figure 3-A, right)—help control the populations of at least half the species of insects we call pests. This free pest control service is an important part of the earth's natural capital. Some insects also play a key role in loosening and renewing the soil that supports terrestrial plant life. In 2006, scientists John Losey and Mace Vaughan estimated that the value of the ecological services provided by insects in the United States is at least $57 billion a year.

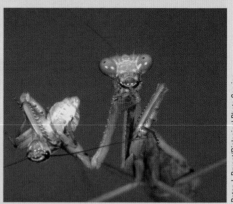

Figure 3-A *Importance of insects:* The monarch butterfly, which feeds on pollen in a flower (left), and other insects pollinate flowering plants that serve as food for many plant eaters. The praying mantis, which is eating a house cricket (right), and many other insect species help to control the populations of most of the insect species we classify as pests.

Insects have been around for at least 400 million years—about 4,000 times longer than the latest version of the human species that we belong to. They are phenomenally successful forms of life. Some reproduce at an astounding rate and can rapidly develop new genetic traits, such as resistance to pesticides. They also have an exceptional ability to evolve into new species when faced with new environmental conditions, and they are very resistant to extinction. This is fortunate because, according to ant specialist and biodiversity expert E. O. Wilson, if all insects disappeared, parts of the life support systems for us and other species would be greatly disrupted.

The environmental lesson: although insects do not need newcomer species such as us, we and most other land organisms need them.

Critical Thinking

Identify three insect species not discussed above that benefit your life.

3-2 What Keeps Us and Other Organisms Alive?

▶ **CONCEPT 3-2** Life is sustained by the flow of energy from the sun through the biosphere, the cycling of nutrients within the biosphere, and gravity.

The Earth's Life–Support System Has Four Major Components

Scientific studies reveal that the earth's life-support system consists of four main spherical systems that interact with one another—the atmosphere (air), the hydrosphere (water), the geosphere (rock, soil, sediment), and the biosphere (living things) (Figure 3-6).

The **atmosphere** is a thin spherical envelope of gases surrounding the earth's surface. Its inner layer, the **troposphere,** extends only about 17 kilometers (11 miles) above sea level at the tropics and about 7 kilometers (4 miles) above the earth's north and south poles. It contains the majority of the air that we breathe, consisting mostly of nitrogen (78% of the total volume) and oxygen (21%). The remaining 1% of the air includes water vapor, carbon dioxide, and methane, all of which are called **greenhouse gases,** because they trap heat and thus warm the lower atmosphere. Almost all of the earth's weather occurs in this layer.

The next layer, stretching 17–50 kilometers (11–31 miles) above the earth's surface, is the **stratosphere.** Its lower portion contains enough ozone (O_3) gas to filter out most of the sun's harmful ultraviolet radiation. This global sunscreen allows life to exist on land and in the surface layers of bodies of water.

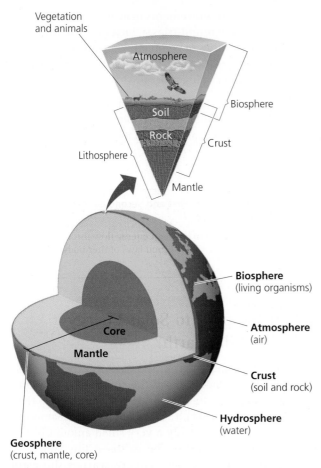

Figure 3-6 **Natural capital:** general structure of the earth showing that it consists of a land sphere, air sphere, water sphere, and life sphere.

vapor in the atmosphere. Most of this water is in the oceans, which cover about 71% of the globe.

The **geosphere** consists of the earth's intensely hot *core*, a thick *mantle* composed mostly of rock, and a thin outer *crust*. Most of the geosphere is located in the earth's interior. Its upper portion contains nonrenewable fossil fuels and minerals that we use, as well as renewable soil chemicals that organisms need in order to live, grow, and reproduce.

The *biosphere* occupies those parts of the atmosphere, hydrosphere, and geosphere where life exists. This thin layer of the earth extends from about 9 kilometers (6 miles) above the earth's surface down to the bottom of the ocean, and it includes the lower part of the atmosphere, most of the hydrosphere, and the uppermost part of the geosphere. If the earth were an apple, the biosphere would be no thicker than the apple's skin. *The goal of ecology is to understand the interactions in this thin layer of air, water, soil, and organisms.*

Life Exists on Land and in Water

Biologists have classified the terrestrial (land) portion of the biosphere into **biomes**—large regions such as forests, deserts, and grasslands, with distinct climates and certain species (especially vegetation) adapted to them. Figure 3-7 shows different major biomes along the 39th parallel spanning the United States (see Figure 5 on p. S27 in Supplement 4 for a map of the major biomes

The **hydrosphere** consists of all of the water on or near the earth's surface. It is found as *liquid water* (on the surface and underground), *ice* (polar ice, icebergs, and ice in frozen soil layers called *permafrost*), and *water*

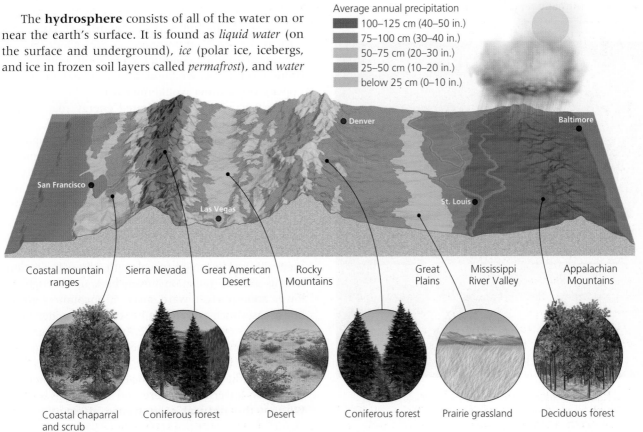

Average annual precipitation

- 100–125 cm (40–50 in.)
- 75–100 cm (30–40 in.)
- 50–75 cm (20–30 in.)
- 25–50 cm (10–20 in.)
- below 25 cm (0–10 in.)

Coastal mountain ranges — Coastal chaparral and scrub

Sierra Nevada — Coniferous forest

Great American Desert — Desert

Rocky Mountains — Coniferous forest

Great Plains — Prairie grassland

Mississippi River Valley

Appalachian Mountains — Deciduous forest

Figure 3-7 Major biomes found along the 39th parallel across the United States. The differences reflect changes in climate, mainly differences in average annual precipitation and temperature.

of North America). We discuss biomes in more detail in Chapter 7.

Scientists divide the watery parts of the biosphere into **aquatic life zones,** each containing numerous ecosystems. There are *freshwater life zones* (such as lakes and streams) and *ocean* or *marine life zones* (such as coral reefs and coastal estuaries). The earth is mostly a water planet with saltwater covering about 71% of its surface and freshwater covering just 2%.

Three Factors Sustain Life on Earth

Life on the earth depends on three interconnected factors (**Concept 3-2**):

- The *one-way flow of high-quality energy* from the sun, through living things in their feeding interactions, into the environment as low-quality energy (mostly heat dispersed into air or water at a low temperature), and eventually back into space as heat. No round-trips are allowed because high-quality energy cannot be recycled. The first and second laws of thermodynamics (**Concepts 2-4A** and **2-4B**, p. 40) govern this energy flow. CONCEPT LINKS

- The *cycling of matter* or *nutrients* (the atoms, ions, and compounds needed for survival by living organisms) through parts of the biosphere. Because the earth is closed to significant inputs of matter from space, its essentially fixed supply of nutrients must be continually recycled to support life (Figure 1-4,

p. 9). Nutrient movements in ecosystems and in the biosphere are round-trips, which can take from seconds to centuries to complete. The law of conservation of matter (**Concept 2-3**, p. 39) governs this nutrient cycling process. CONCEPT LINK

- *Gravity,* which allows the planet to hold onto its atmosphere and helps to enable the movement and cycling of chemicals through the air, water, soil, and organisms.

THINKING ABOUT
Energy Flow and the First and Second Laws of Thermodynamics

Explain the relationship between energy flow through the biosphere and the first and second laws of thermodynamics (pp. 42–43).

What Happens to Solar Energy Reaching the Earth?

Millions of kilometers from the earth, in the immense nuclear fusion reactor that is the sun, nuclei of hydrogen fuse together to form larger helium nuclei (Figure 2-7, bottom, p. 41), releasing tremendous amounts of energy into space. Only a very small amount of this output of energy reaches the earth—a tiny sphere in the vastness of space. This energy reaches the earth in the form of electromagnetic waves, mostly as visible light, ultraviolet (UV) radiation, and heat (infrared radiation) (Figure 2-8, p. 42). Much of this energy is absorbed or reflected back into space by the earth's atmosphere, clouds, and surface (Figure 3-8). Ozone gas (O_3) in the lower stratosphere absorbs about 95% of the sun's harmful incoming UV radiation. Without this ozone layer, life as we know it on the land and in the upper layer of water would not exist.

The UV, visible, and infrared energy that reaches the atmosphere lights the earth during daytime, warms the air, and evaporates and cycles water through the biosphere. Approximately 1% of this incoming energy generates winds. Green plants, algae, and some types of bacteria use less than 0.1% of it to produce the nutrients they need through photosynthesis and in turn to feed animals that eat plants and flesh.

Of the total solar radiation intercepted by the earth, about 1% reaches the earth's surface, and most of it is then reflected as longer-wavelength infrared radiation. As this infrared radiation travels back up through the lower atmosphere toward space, it encounters greenhouse gases such as water vapor, carbon dioxide, methane, nitrous oxide, and ozone. It causes these gaseous molecules to vibrate and release infrared radiation with even longer wavelengths. The vibrating gaseous molecules then have higher kinetic energy, which helps to warm the lower atmosphere and the earth's surface. Without this **natural greenhouse effect,** the earth

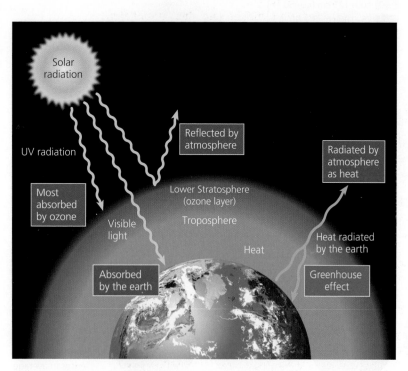

CENGAGENOW™ **Active Figure 3-8** *Solar capital:* flow of energy to and from the earth. *See an animation based on this figure at* CengageNOW.

would be too cold to support the forms of life we find here today. (See *The Habitable Planet,* Video 2, **www.learner.org/resources/series209.html**.)

Human activities add greenhouse gases to the atmosphere. For example, burning carbon-containing fuels releases huge amounts of carbon dioxide (CO_2) into the atmosphere. Growing crops and raising livestock release large amounts of methane (CH_4) and nitrous oxide (N_2O). Clearing CO_2-absorbing tropical rain forests (**Core Case Study**) faster than they can

CORE CASE STUDY

grow back also increases the amount of CO_2 in the atmosphere. There is considerable and growing evidence that these activities are increasing the natural greenhouse effect and warming the earth's atmosphere (Science Focus, p. 33). This in turn is changing the earth's climate as we discuss at length in Chapter 19.

CENGAGENOW™ Learn more about the flow of energy—from sun to earth and within the earth's systems—at CengageNOW.

3-3 What Are the Major Components of an Ecosystem?

▶ **CONCEPT 3-3A** Ecosystems contain living (biotic) and nonliving (abiotic) components.

▶ **CONCEPT 3-3B** Some organisms produce the nutrients they need, others get their nutrients by consuming other organisms, and some recycle nutrients back to producers by decomposing the wastes and remains of organisms.

Ecosystems Have Living and Nonliving Components

Two types of components make up the biosphere and its ecosystems: One type, called **abiotic,** consists of nonliving components such as water, air, nutrients, rocks, heat, and solar energy. The other type, called **biotic,** consists of living and once living biological components—plants, animals, and microbes (**Concept 3-3A**). Biotic factors also include dead organisms, dead parts of organisms, and the waste products of organisms. Figure 3-9 is a greatly simplified diagram of some of the biotic and abiotic components of a terrestrial ecosystem.

Different species and their populations thrive under different physical and chemical conditions. Some need bright sunlight; others flourish in shade. Some need a hot environment; others prefer a cool or cold one. Some do best under wet conditions; others thrive under dry conditions.

Each population in an ecosystem has a **range of tolerance** to variations in its physical and chemical environment, as shown in Figure 3-10 (p. 58). Individuals within a population may also have slightly different tolerance ranges for temperature or other factors because of small differences in genetic makeup, health, and age. For example, a trout population may do best within

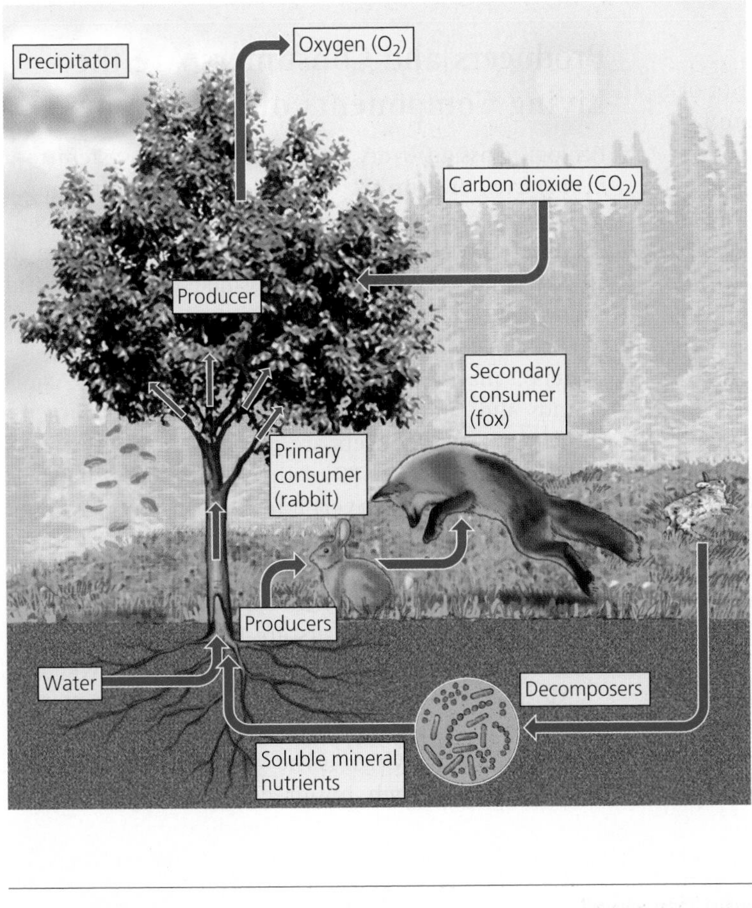

CENGAGENOW™ **Active Figure 3-9** Major living (biotic) and nonliving (abiotic) components of an ecosystem in a field. *See an animation based on this figure* at CengageNOW.

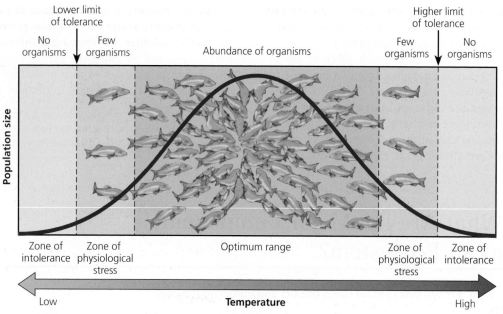

Figure 3-10 Range of tolerance for a population of organisms, such as fish, to an abiotic environmental factor—in this case, temperature. These restrictions keep particular species from taking over an ecosystem by keeping their population size in check. **Question:** Which **scientific principle of sustainability** (see back cover) is related to the range of tolerance concept?

a narrow band of temperatures (*optimum level* or *range*), but a few individuals can survive above and below that band. Of course, if the water becomes much too hot or too cold, none of the trout can survive.

Several Abiotic Factors Can Limit Population Growth

A variety of abiotic factors can affect the number of organisms in a population. Sometimes one or more factors, known as **limiting factors,** are more important in regulating population growth than other factors are. This ecological principle is called the **limiting factor principle:** *Too much or too little of any abiotic factor can limit or prevent growth of a population, even if all other factors are at or near the optimal range of tolerance.* This principle describes one way in which population control—a **scientific principle of sustainability** (see back cover)—is achieved.

On land, precipitation often is the limiting abiotic factor. Lack of water in a desert limits plant growth. Soil nutrients also can act as a limiting factor on land. Suppose a farmer plants corn in phosphorus-poor soil. Even if water, nitrogen, potassium, and other nutrients are at optimal levels, the corn will stop growing when it uses up the available phosphorus. Too much of an abiotic factor can also be limiting. For example, too much water or fertilizer can kill plants. Temperature can also be a limiting factor. Both high and low temperatures can limit the survival and population sizes of various terrestrial species, especially plants.

Important limiting abiotic factors in aquatic life zones include temperature, sunlight, nutrient availability, and the low solubility of oxygen gas in water (*dissolved oxygen content*). Another such factor is *salinity*—the amounts of various inorganic minerals or salts dissolved in a given volume of water.

Producers and Consumers Are the Living Components of Ecosystems

Ecologists assign every organism in an ecosystem to a *feeding level,* or **trophic level,** depending on its source of food or nutrients. The organisms that transfer energy and nutrients from one trophic level to another in an ecosystem can be broadly classified as producers and consumers.

Producers, sometimes called **autotrophs** (self-feeders), make the nutrients they need from compounds and energy obtained from their environment. On land, most producers are green plants, which generally capture about 1% of the solar energy that falls on their leaves and convert it to chemical energy stored in organic molecules such as carbohydrates. In freshwater and marine ecosystems, algae and aquatic plants are the major producers near shorelines. In open water, the dominant producers are *phytoplankton*—mostly microscopic organisms that float or drift in the water.

Most producers capture sunlight to produce energy-rich carbohydrates (such as glucose, $C_6H_{12}O_6$) by **photosynthesis,** which is the way energy enters most ecosystems. Although hundreds of chemical changes take place during photosynthesis, the overall reaction can be summarized as follows:

carbon dioxide + water + solar energy $\longrightarrow$ glucose + oxygen

$6\,CO_2 + 6\,H_2O + \text{solar energy} \longrightarrow C_6H_{12}O_6 + 6\,O_2$

(See p. S44 in Supplement 6 for information on how to balance chemical equations such as this one and p. S44 in Supplement 6 for more details on photosynthesis.)

A few producers, mostly specialized bacteria, can convert simple inorganic compounds from their environment into more complex nutrient compounds without using sunlight, through a process called **chemosynthesis.** In 1977, scientists discovered a community of bacteria living in the extremely hot water around *hydrothermal vents* on the deep ocean floor. These bacteria serve as producers for their ecosystems without the use of sunlight. They draw energy and produce carbohydrates from hydrogen sulfide (H_2S) gas escaping through fissures in the ocean floor. Most of the earth's organisms get their energy indirectly from the sun. But chemosynthetic organisms in these dark and deep-sea habitats survive indirectly on *geothermal energy* from the earth's interior and represent an exception to the first **scientific principle of sustainability.**

All other organisms in an ecosystem are **consumers,** or **heterotrophs** ("other-feeders"), that cannot produce the nutrients they need through photosynthesis or other processes and must obtain their nutrients by feeding on other organisms (producers or other consumers) or their remains. In other words, all consumers (including humans) are directly or indirectly dependent on producers for their food or nutrients.

There are several types of consumers:

- **Primary consumers,** or **herbivores** (plant eaters), are animals such as rabbits, grasshoppers, deer, and zooplankton that eat producers, mostly by feeding on green plants.

- **Secondary consumers,** or **carnivores** (meat eaters), are animals such as spiders, hyenas, birds, frogs, and some zooplankton-eating fish, all of which feed on the flesh of herbivores.

- **Third-** and **higher-level consumers** are carnivores such as tigers, wolves, mice-eating snakes, hawks, and killer whales (orcas) that feed on the flesh of other carnivores. Some of these relationships are shown in Figure 3-9.

- **Omnivores** such as pigs, foxes, cockroaches, and humans, play dual roles by feeding on both plants and animals.

- **Decomposers,** primarily certain types of bacteria and fungi, are consumers that release nutrients from the dead bodies of plants and animals and return them to the soil, water, and air for reuse by producers. They feed by secreting enzymes that speed up the break down of bodies of dead organisms into nutrient compounds such as water, carbon dioxide, minerals, and simpler organic compounds.

- **Detritus feeders,** or **detritivores,** feed on the wastes or dead bodies of other organisms, called **detritus** ("di-TRI-tus," meaning debris). Examples include small organisms such as mites and earthworms, some insects, catfish, and larger scavenger organisms such as vultures.

Hordes of decomposers and detritus feeders can transform a fallen tree trunk into a powder and finally into simple inorganic molecules that plants can absorb as nutrients (Figure 3-11, p. 60). In summary, some organisms produce the nutrients they need, others get their nutrients by consuming other organisms, and still others recycle the nutrients in the wastes and remains of organisms so that producers can use them again (**Concept 3-3B**).

THINKING ABOUT
What You Eat

When you had your most recent meal, were you an herbivore, a carnivore, or an omnivore?

Producers, consumers, and decomposers use the chemical energy stored in glucose and other organic compounds to fuel their life processes. In most cells this energy is released by **aerobic respiration,** which uses oxygen to convert glucose (or other organic nutrient molecules) back into carbon dioxide and water. The net effect of the hundreds of steps in this complex process is represented by the following reaction:

glucose + oxygen $\longrightarrow$ carbon dioxide + water + energy

$C_6H_{12}O_6 + 6\,O_2 \longrightarrow 6\,CO_2 + 6\,H_2O + \text{energy}$

Although the detailed steps differ, the net chemical change for aerobic respiration is the opposite of that for photosynthesis.

Some decomposers get the energy they need by breaking down glucose (or other organic compounds) in the *absence* of oxygen. This form of cellular respiration is called **anaerobic respiration,** or **fermentation.** Instead of carbon dioxide and water, the end products of this process are compounds such as methane gas (CH_4, the main component of natural gas), ethyl alcohol (C_2H_6O), acetic acid ($C_2H_4O_2$, the key component of vinegar), and hydrogen sulfide (H_2S, when sulfur compounds are broken down). Note that all organisms get their energy from aerobic or anaerobic respiration but only plants carry out photosynthesis.

CENGAGE**NOW** Explore the components of ecosystems, how they interact, the roles of bugs and plants, and what a fox will eat at CengageNOW.

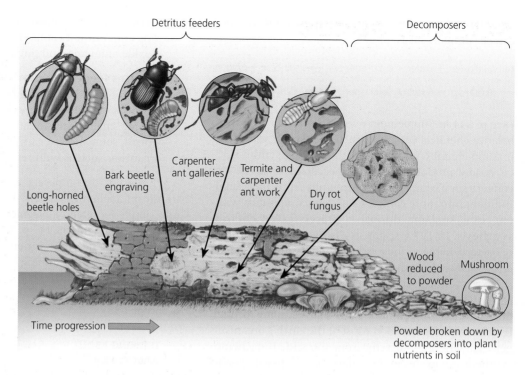

Figure 3-11 Various detritivores and decomposers (mostly fungi and bacteria) can "feed on" or digest parts of a log and eventually convert its complex organic chemicals into simpler inorganic nutrients that can be taken up by producers.

Detritus feeders

Decomposers

Long-horned beetle holes

Bark beetle engraving

Carpenter ant galleries

Termite and carpenter ant work

Dry rot fungus

Wood reduced to powder

Mushroom

Time progression

Powder broken down by decomposers into plant nutrients in soil

Energy Flow and Nutrient Cycling Sustain Ecosystems and the Biosphere

Ecosystems and the biosphere are sustained through a combination of *one-way energy flow* from the sun through these systems and *nutrient cycling* of key materials within them—two important natural services that are components of the earth's natural capital. These two **scientific principles of sustainability** arise from the structure and function of natural ecosystems (Figure 3-12), the law of conservation of matter (**Concept 2-3**, p. 39), and the two laws of thermodynamics (**Concepts 2-4A** and **2-4B**, p. 40).

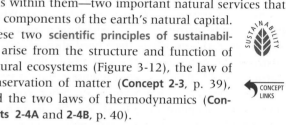

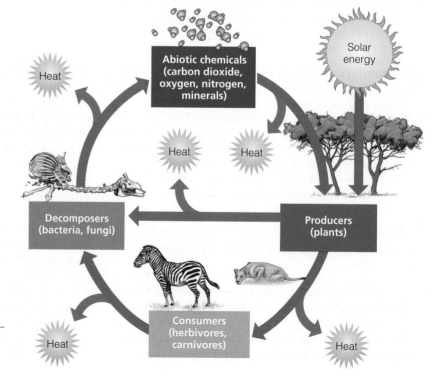

CENGAGENOW™ **Active Figure 3-12 Natural capital:** the main structural components of an ecosystem (energy, chemicals, and organisms). Nutrient cycling and the flow of energy—first from the sun, then through organisms, and finally into the environment as low-quality heat—link these components. *See an animation based on this figure at* CengageNOW.

Heat

Abiotic chemicals (carbon dioxide, oxygen, nitrogen, minerals)

Solar energy

Heat

Heat

Decomposers (bacteria, fungi)

Producers (plants)

Consumers (herbivores, carnivores)

Heat

Heat

SCIENCE FOCUS

Many of the World's Most Important Species Are Invisible to Us

They are everywhere. Billions of them can be found inside your body, on your body, in a handful of soil, and in a cup of ocean water.

These mostly invisible rulers of the earth are *microbes,* or *microorganisms,* catchall terms for many thousands of species of bacteria, protozoa, fungi, and floating phytoplankton—most too small to be seen with the naked eye.

Microbes do not get the respect they deserve. Most of us view them primarily as threats to our health in the form of infectious bacteria or "germs," fungi that cause athlete's foot and other skin diseases, and protozoa that cause diseases such as malaria. But these harmful microbes are in the minority.

We are alive because of multitudes of microbes toiling away mostly out of sight.

Bacteria in our intestinal tracts help to break down the food we eat and microbes in our noses help to prevent harmful bacteria from reaching our lungs.

Bacteria and other microbes help to purify the water we drink by breaking down wastes. Bacteria also help to produce foods such as bread, cheese, yogurt, soy sauce, beer, and wine. Bacteria and fungi in the soil decompose organic wastes into nutrients that can be taken up by plants that we and most other animals eat. Without these tiny creatures, we would go hungry and be up to our necks in waste matter.

Microbes, particularly phytoplankton in the ocean, provide much of the planet's oxygen, and help slow global warming by removing some of the carbon dioxide produced when we burn coal, natural gas, and

gasoline. (See *The Habitable Planet,* Video 3, www.learner.org/resources/series209 .html.) Scientists are working on using microbes to develop new medicines and fuels. Genetic engineers are inserting genetic material into existing microbes to convert them to microbes that can help clean up polluted water and soils.

Some microbes help control diseases that affect plants and populations of insect species that attack our food crops. Relying more on these microbes for pest control could reduce the use of potentially harmful chemical pesticides. In other words, microbes are a vital part of the earth's natural capital.

Critical Thinking

What are three advantages that microbes have over us for thriving in the world?

Decomposers and detritus feeders, many of which are microscopic organisms, (Science Focus, above) are the key to nutrient cycling because they break down organic matter into simpler nutrients that can be reused by producers. Without decomposers and detritus feeders, there would be little, if any, nutrient cycling and the planet would be overwhelmed with plant litter, dead animal bodies, animal wastes, and garbage. In addition, most life as we know it could not exist because the nutrients stored in such wastes and dead bodies would be locked up and unavailable for use by other organisms.

> **THINKING ABOUT**
> **Chemical Cycling and the Law of Conservation of Matter**
> CONCEPT LINK
>
> Explain the relationship between chemical cycling in ecosystems and in the biosphere and the law of conservation of matter (**Concept 2-3**, p. 39).

3-4 What Happens to Energy in an Ecosystem?

▶ **CONCEPT 3-4A** Energy flows through ecosystems in food chains and webs.

▶ **CONCEPT 3-4B** As energy flows through ecosystems in food chains and webs, the amount of chemical energy available to organisms at each succeeding feeding level decreases.

Energy Flows through Ecosystems in Food Chains and Food Webs

The chemical energy stored as nutrients in the bodies and wastes of organisms flows through ecosystems from one trophic (feeding) level to another. For example, a plant uses solar energy to store chemical energy in a leaf. A caterpillar eats the leaf, a robin eats the caterpillar, and a hawk eats the robin. Decomposers and detritus feeders consume the leaf, caterpillar, robin, and hawk after they die and return their nutrients to the soil for reuse by producers.

A sequence of organisms, each of which serves as a source of food or energy for the next, is called a **food chain.** It determines how chemical energy and nutrients move from one organism to another through the trophic levels in an ecosystem—primarily through photosynthesis, feeding, and decomposition—as shown

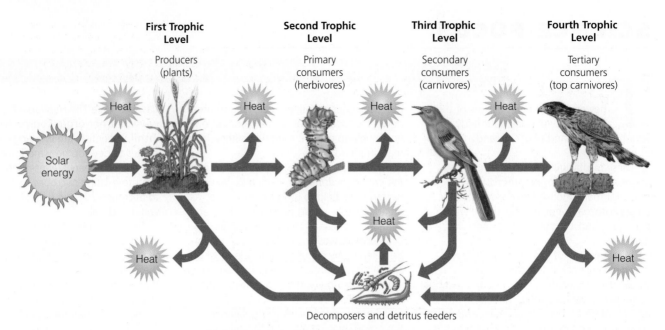

First Trophic Level
Producers (plants)

Second Trophic Level
Primary consumers (herbivores)

Third Trophic Level
Secondary consumers (carnivores)

Fourth Trophic Level
Tertiary consumers (top carnivores)

Solar energy

Heat

Decomposers and detritus feeders

CENGAGENOW™ **Active Figure 3-13** A *food chain*. The arrows show how chemical energy in nutrients flows through various *trophic levels* in energy transfers; most of the energy is degraded to heat, in accordance with the second law of thermodynamics. *See an animation based on this figure at* CengageNOW. **Question:** Think about what you ate for breakfast. At what level or levels on a food chain were you eating?

in Figure 3-13. Every use and transfer of energy by organisms involves a loss of some useful energy to the environment as heat. Thus, eventually an ecosystem and the biosphere would run out of energy if they were not powered by a continuous inflow of energy from an outside source, ultimately the sun.

In natural ecosystems, most consumers feed on more than one type of organism, and most organisms are eaten or decomposed by more than one type of consumer. Because of this, organisms in most ecosystems form a complex network of interconnected food chains called a **food web** (Figure 3-14). Trophic levels can be assigned in food webs just as in food chains. Food chains and webs show how producers, consumers, and decomposers are connected to one another as energy flows through trophic levels in an ecosystem.

> **THINKING ABOUT**
> **Energy Flow and Tropical Rain forests**
> What happens to the flow of energy through tropical rain forest ecosystems when such forests are degraded (**Core Case Study**)?
>
> CORE CASE STUDY

Usable Energy Decreases with Each Link in a Food Chain or Web

Each trophic level in a food chain or web contains a certain amount of **biomass,** the dry weight of all organic matter contained in its organisms. In a food chain or web, chemical energy stored in biomass is transferred from one trophic level to another.

Energy transfer through food chains and food webs is not very efficient because, with each transfer, some usable chemical energy is degraded and lost to the environment as low-quality heat, as a result of the second law of thermodynamics. In other words, as energy flows through ecosystems in food chains and webs, there is a decrease in the amount of chemical energy available to organisms at each succeeding feeding level (**Concept 3-4B**).

The percentage of usable chemical energy transferred as biomass from one trophic level to the next is called **ecological efficiency.** It ranges from 2% to 40% (that is, a loss of 60–98%) depending on what types of species and ecosystems are involved, but 10% is typical.

Assuming 10% ecological efficiency (90% loss of usable energy) at each trophic transfer, if green plants in an area manage to capture 10,000 units of energy from the sun, then only about 1,000 units of chemical energy will be available to support herbivores, and only about 100 units will be available to support carnivores.

The more trophic levels there are in a food chain or web, the greater is the cumulative loss of usable chemical energy as it flows through the trophic levels. The **pyramid of energy flow** in Figure 3-15 illustrates this energy loss for a simple food chain, assuming a 90% energy loss with each transfer.

> **THINKING ABOUT**
> **Energy Flow and the Second Law of Thermodynamics**
> Explain the relationship between the second law of thermodynamics (**Concept 2-4B**, p. 40) and the flow of energy through a food chain or web.
>
> CONCEPT LINK

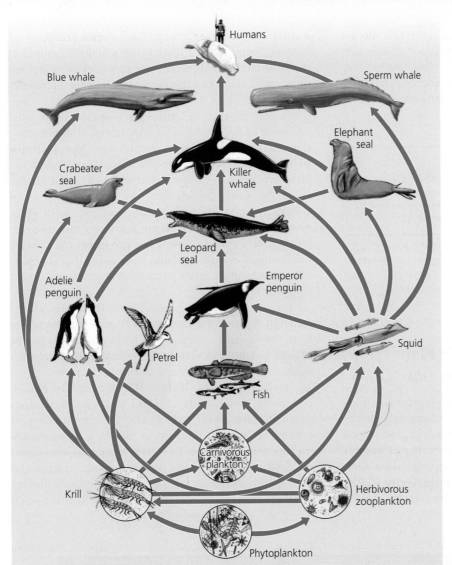

Figure 3-14 *Greatly simplified food web* in the Antarctic. Many more participants in the web, including an array of decomposer and detritus feeder organisms, are not depicted here. **Question:** Can you imagine a food web of which you are a part? Try drawing a simple diagram of it.

Energy flow pyramids explain why the earth can support more people if they eat at lower trophic levels by consuming grains, vegetables, and fruits directly, rather than passing such crops through another trophic level and eating grain eaters or herbivores such as cattle. About two-thirds of the world's people survive primarily by eating wheat, rice, and corn at the first trophic level, mostly because they cannot afford meat.

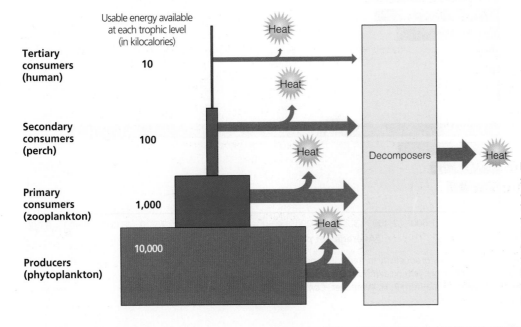

Figure 3-15 Generalized *pyramid of energy flow* showing the decrease in usable chemical energy available at each succeeding trophic level in a food chain or web. In nature, ecological efficiency varies from 2% to 40%, with 10% efficiency being common. This model assumes a 10% ecological efficiency (90% loss of usable energy to the environment, in the form of low-quality heat) with each transfer from one trophic level to another. **Question:** Why is a vegetarian diet more energy efficient than a meat-based diet?

The large loss in chemical energy between successive trophic levels also explains why food chains and webs rarely have more than four or five trophic levels. In most cases, too little chemical energy is left after four or five transfers to support organisms feeding at these high trophic levels.

THINKING ABOUT
Food Webs, Tigers, and Insects

Use Figure 3-15 to help explain **(a)** why there are not many tigers in the world and why they are vulnerable to premature extinction because of human activities, and **(b)** why there are so many insects (Science Focus, p. 54) in the world.

CENGAGENOW™ Examine how energy flows among organisms at different trophic levels and through food webs in tropical rain forests, prairies, and other ecosystems at CengageNOW.

Some Ecosystems Produce Plant Matter Faster Than Others Do

The amount, or mass, of living organic material (biomass) that a particular ecosystem can support is determined by the amount of energy captured and stored as chemical energy by the producers of that ecosystem and by how rapidly they can produce and store such chemical energy. **Gross primary productivity (GPP)** is the *rate* at which an ecosystem's producers (usually plants) convert solar energy into chemical energy as biomass found in their tissues. It is usually measured in terms of energy production per unit area over a given time span, such as kilocalories per square meter per year ($kcal/m^2/yr$).

To stay alive, grow, and reproduce, producers must use some of the chemical energy stored in the biomass they make for their own respiration. **Net primary productivity (NPP)** is the *rate* at which producers use photosynthesis to produce and store chemical energy *minus* the *rate* at which they use some of this stored chemical energy through aerobic respiration. In other words, NPP = GPP − R, where R is energy used in respiration. NPP measures how fast producers can provide the chemical energy stored in their tissue that is potentially available to other organisms (consumers) in an ecosystem.

Ecosystems and life zones differ in their NPP as illustrated in Figure 3-16. On land, NPP generally decreases from the equator toward the poles because the amount of solar radiation available to terrestrial plant producers is highest at the equator and lowest at the poles.

In the ocean, the highest NPP is found in estuaries where high inputs of plant nutrients flow from nutrient-laden rivers, which also stir up nutrients in bottom sediments. Because of the lack of nutrients, the open ocean has a low NPP, except at occasional areas where an *upwelling* (water moving up from the depths toward the surface) brings nutrients in bottom sediments to the surface. Despite its low NPP, the open

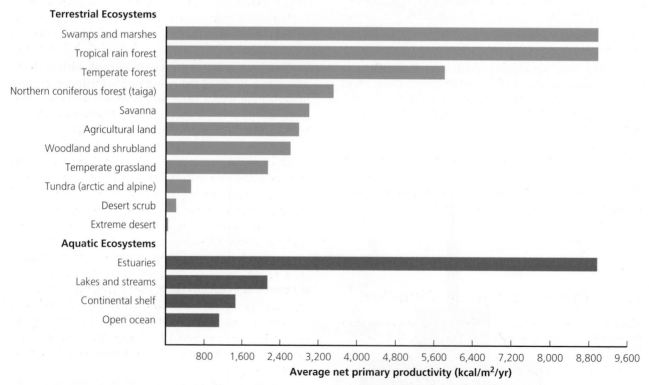

Figure 3-16 Estimated annual average *net primary productivity* in major life zones and ecosystems, expressed as kilocalories of energy produced per square meter per year ($kcal/m^2/yr$). **Question:** What are nature's three most productive and three least productive systems? (Data from R. H. Whittaker, *Communities and Ecosystems*, 2nd ed., New York: Macmillan, 1975)

ocean produces more of the earth's biomass per year than any other ecosystem or life zone, simply because there is so much open ocean.

As we have seen, producers are the source of all nutrients or chemical energy in an ecosystem for themselves and for the animals and decomposers that feed on them. Only the biomass represented by NPP is available as nutrients for consumers, and they use only a portion of this amount. Thus, *the planet's NPP ultimately limits the number of consumers (including humans) that can survive on the earth.* This is an important lesson from nature.

Peter Vitousek, Stuart Rojstaczer, and other ecologists estimate that humans now use, waste, or destroy about 20–32% of the earth's total potential NPP. This is a remarkably high value, considering that the human population makes up less than 1% of the total biomass of all of the earth's consumers that depend on producers for their nutrients.

3-5   What Happens to Matter in an Ecosystem?

▶ **CONCEPT 3-5** Matter, in the form of nutrients, cycles within and among ecosystems and the biosphere, and human activities are altering these chemical cycles.

Nutrients Cycle in the Biosphere

The elements and compounds that make up nutrients move continually through air, water, soil, rock, and living organisms in ecosystems and in the biosphere in cycles called **biogeochemical cycles** (literally, life-earth-chemical cycles) or **nutrient cycles**— prime examples of one of the four **scientific principles of sustainability** (see back cover). These cycles, driven directly or indirectly by incoming solar energy and gravity, include the hydrologic (water), carbon, nitrogen, phosphorus, and sulfur cycles. These cycles are an important component of the earth's natural capital (Figure 1-3, p. 8), and human activities are altering them (**Concept 3-5**).

As nutrients move through the biogeochemical cycles, they may accumulate in one portion of the cycle and remain there for different lengths of time. These temporary storage sites such as the atmosphere, the oceans and other waters, and underground deposits are called *reservoirs.*

Nutrient cycles connect past, present, and future forms of life. Some of the carbon atoms in your skin may once have been part of an oak leaf, a dinosaur's skin, or a layer of limestone rock. Your grandmother, Attila the Hun, or a hunter–gatherer who lived 25,000 years ago may have inhaled some of the oxygen molecules you just inhaled.

Water Cycles through the Biosphere

The **hydrologic cycle,** or **water cycle,** collects, purifies, and distributes the earth's fixed supply of water, as shown in Figure 3-17 (p. 66). The water cycle is a global cycle because there is a large reservoir of water in the atmosphere as well as in the hydrosphere, especially the oceans. Water is an amazing substance (Science Focus, p. 67), which makes the water cycle critical to life on earth.

The water cycle is powered by energy from the sun and involves three major processes—evaporation, precipitation, and transpiration. Incoming solar energy causes *evaporation* of water from the oceans, lakes, rivers, and soil. Evaporation changes liquid water into water vapor in the atmosphere, and gravity draws the water back to the earth's surface as *precipitation* (rain, snow, sleet, and dew). About 84% of water vapor in the atmosphere comes from the oceans, which cover almost three-fourths of the earth's surface; the rest comes from land. Over land, about 90% of the water that reaches the atmosphere evaporates from the surfaces of plants through a process called **transpiration.**

Water returning to the earth's surface as precipitation takes various paths. Most precipitation falling on terrestrial ecosystems becomes *surface runoff.* This water flows into streams and lakes, which eventually carry water back to the oceans, from which it can evaporate to repeat the cycle. Some surface water also seeps into the upper layer of soils and some evaporates from soil, lakes, and streams back into the atmosphere.

Some precipitation is converted to ice that is stored in *glaciers,* usually for long periods of time. Some precipitation sinks through soil and permeable rock formations to underground layers of rock, sand, and gravel called *aquifers,* where it is stored as *groundwater.*

A small amount of the earth's water ends up in the living components of ecosystems. Roots of plants absorb some of this water, most of which evaporates from plant leaves back into the atmosphere. Some combines with carbon dioxide during photosynthesis to produce high-energy organic compounds such as carbohydrates. Eventually these compounds are broken down in plant cells, which release water back into the environment. Consumers get their water from their food or by drinking it.

Surface runoff replenishes streams and lakes, but also causes soil erosion, which moves soil and rock fragments from one place to another. Water is the primary sculptor of the earth's landscape. Because water

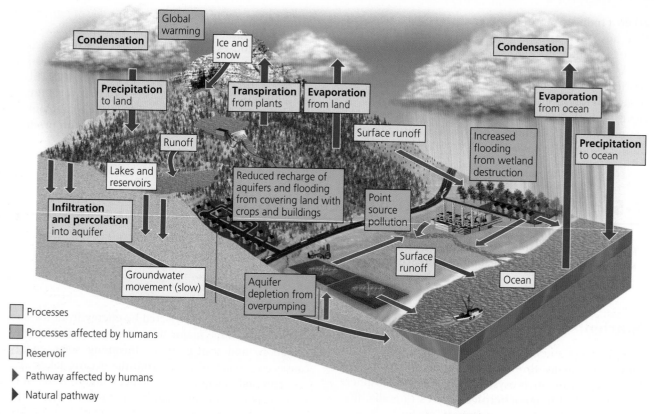

Processes

Processes affected by humans

Reservoir

▶ Pathway affected by humans

▶ Natural pathway

CENGAGENOW™ **Active Figure 3-17 Natural capital:** simplified model of the *hydrologic cycle* with major harmful impacts of human activities shown in red. *See an animation based on this figure at* CengageNOW. **Question:** What are three ways in which your lifestyle directly or indirectly affects the hydrologic cycle?

dissolves many nutrient compounds, it is a major medium for transporting nutrients within and between ecosystems.

Throughout the hydrologic cycle, many natural processes purify water. Evaporation and subsequent precipitation act as a natural distillation process that removes impurities dissolved in water. Water flowing above ground through streams and lakes and below ground in aquifers is naturally filtered and partially purified by chemical and biological processes—mostly by the actions of decomposer bacteria—as long as these natural processes are not overloaded. Thus, *the hydrologic cycle can be viewed as a cycle of natural renewal of water quality.*

Only about 0.024% of the earth's vast water supply is available to us as liquid freshwater in accessible groundwater deposits and in lakes, rivers, and streams. The rest is too salty for us to use, is stored as ice, or is too deep underground to extract at affordable prices using current technology.

We alter the water cycle in three major ways (see red arrows and boxes in Figure 3-17). *First,* we withdraw large quantities of freshwater from streams, lakes, and underground sources, sometimes at rates faster than nature can replace it.

Second, we clear vegetation from land for agriculture, mining, road building, and other activities, and cover much of the land with buildings, concrete, and asphalt. This increases runoff, reduces infiltration that would normally recharge groundwater supplies, in-

creases the risk of flooding, and accelerates soil erosion and landslides.

Clearing vegetation can also alter weather patterns by reducing transpiration. This is especially important in dense tropical rain forests (**Core Case Study** and Figure 3-1). Because so many plants in a tropical rain forest transpire water into the atmosphere, vegetation is the primary source of local rainfall. In other words, as part of the water cycle, these plants create their own rain.

Cutting down the forest raises ground temperatures (because it reduces shade) and can reduce local rainfall so much that the forest cannot grow back. When such a tipping point is reached, these biologically diverse forests are converted into much less diverse tropical grasslands, as a 2005 study showed in parts of Brazil's huge Amazon basin. Models project that if current burning and deforestation rates continue, 20–30% of the Amazon rain forests will turn into tropical grassland in the next 50 years.

The *third* way in which we alter the water cycle is by increasing flooding. This happens when we drain wetlands for farming and other purposes. Left undisturbed, wetlands provide the natural service of flood control, acting like sponges to absorb and hold overflows of water from drenching rains or rapidly melting snow. We also cover much of the land with roads, parking lots, and buildings, eliminating the land's ability to absorb water and dramatically increasing runoff and flooding.

SCIENCE FOCUS

Water's Unique Properties

Water is a remarkable substance with a unique combination of properties:

- *Forces of attraction, called hydrogen bonds* (see Figure 7 on p. S42 in Supplement 6), *hold water molecules together*—the major factor determining water's distinctive properties.

- *Water exists as a liquid over a wide temperature range because of the hydrogen bonds.* Without water's high boiling point the oceans would have evaporated long ago.

- *Liquid water changes temperature slowly because it can store a large amount of heat without a large change in temperature.* This high heat storage capacity helps protect living organisms from temperature changes, moderates the earth's climate, and makes water an excellent coolant for car engines and power plants.

- *It takes a large amount of energy to evaporate water because of the hydrogen bonds.* Water absorbs large amounts of heat as it changes into water vapor and releases this heat as the vapor condenses back to liquid water. This helps to distribute heat throughout the world and to determine regional and local climates. It also makes evaporation a cooling process—explaining why you feel cooler when perspiration evaporates from your skin.

- *Liquid water can dissolve a variety of compounds* (see Figure 3, p. S40, in Supplement 6). It carries dissolved nutrients into the tissues of living organisms, flushes waste products out of those tissues, serves as an all-purpose cleanser, and helps remove and dilute the water-soluble wastes of civilization. This property also means that water-soluble wastes can easily pollute water.

- *Water filters out some of the sun's ultraviolet radiation* (Figure 2-8, p. 42) *that would harm some aquatic organisms.* However, up to a certain depth it is transparent to visible light needed for photosynthesis.

- *Hydrogen bonds allow water to adhere to a solid surface.* This enables narrow columns of water to rise through a plant from its roots to its leaves (a process called capillary action).

- *Unlike most liquids, water expands when it freezes.* This means that ice floats on water because it has a lower density (mass per unit of volume) than liquid water. Otherwise, lakes and streams in cold climates would freeze solid, losing most of their aquatic life. Because water expands upon freezing, it can break pipes, crack a car's engine block (if it doesn't contain antifreeze), break up street pavements, and fracture rocks.

Critical Thinking

Water is a bent molecule (see Figure 4 on p. S40 in Supplement 6) and this allows it to form hydrogen bonds (Figure 7, p. S42, in Supplement 6) between its molecules. What are three ways in which your life would be different if water were a linear or straight molecule?

Carbon Cycles through the Biosphere and Depends on Photosynthesis and Respiration

Carbon is the basic building block of the carbohydrates, fats, proteins, DNA, and other organic compounds necessary for life. It circulates through the biosphere, the atmosphere, and parts of the hydrosphere, in the **carbon cycle** shown in Figure 3-18 (p. 68). It depends on photosynthesis and aerobic respiration by the earth's living organisms.

The carbon cycle is based on carbon dioxide (CO_2) gas, which makes up 0.038% of the volume of the atmosphere and is also dissolved in water. Carbon dioxide is a key component of nature's thermostat. If the carbon cycle removes too much CO_2 from the atmosphere, the atmosphere will cool, and if it generates too much CO_2, the atmosphere will get warmer. Thus, even slight changes in this cycle caused by natural or human factors can affect climate and ultimately help determine the types of life that can exist in various places.

Terrestrial producers remove CO_2 from the atmosphere and aquatic producers remove it from the water. (See *The Habitable Planet*, Video 3, **www.learner.org/resources/series209.html**.) These producers then use photosynthesis to convert CO_2 into complex carbohydrates such as glucose ($C_6H_{12}O_6$).

The cells in oxygen-consuming producers, consumers, and decomposers then carry out aerobic respiration. This process breaks down glucose and other complex organic compounds and converts the carbon back to CO_2 in the atmosphere or water for reuse by producers. This linkage between *photosynthesis* in producers and *aerobic respiration* in producers, consumers, and decomposers circulates carbon in the biosphere. Oxygen and hydrogen—the other elements in carbohydrates—cycle almost in step with carbon.

Some carbon atoms take a long time to recycle. Decomposers release the carbon stored in the bodies of dead organisms on land back into the air as CO_2. However, in water, decomposers can release carbon that is stored as insoluble carbonates in bottom sediment. Indeed, marine sediments are the earth's largest store of carbon. Over millions of years, buried deposits of dead plant matter and bacteria are compressed between layers of sediment, where high pressure and heat convert them to carbon-containing *fossil fuels* such as coal, oil, and natural gas (Figure 3-18). This carbon is not released to the atmosphere as CO_2 for recycling until these fuels are extracted and burned, or until long-term geological processes expose these deposits to air. In only a few hundred years, we have extracted and burned large quantities of fossil fuels that took millions of years to form. This is why, on a human time scale, fossil fuels are nonrenewable resources.

Active Figure 3-18
Natural capital: simplified model
of the global *carbon cycle*, with
major harmful impacts of human
activities shown by red arrows.
*See an animation based on
this figure at CengageNOW.*
Question: What are three
ways in which you directly
or indirectly affect the
carbon cycle?

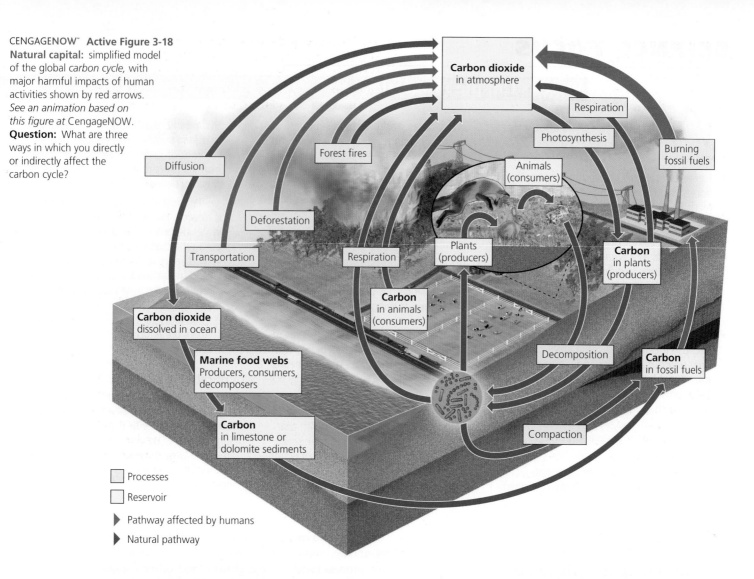

□ Processes

□ Reservoir

▶ Pathway affected by humans

▶ Natural pathway

Since 1800, and especially since 1950, we have been intervening in the earth's carbon cycle by adding carbon dioxide to the atmosphere in two ways (shown by red arrows in Figure 3-18). *First,* in some areas, especially in tropical forests, we clear trees and other plants, which absorb CO_2 through photosynthesis, faster than they can grow back (**Core Case Study**). *Second,* we add large amounts of CO_2 to the atmosphere by burning carbon-containing fossil fuels and wood.

> **THINKING ABOUT**
> **The Carbon Cycle, Tropical Deforestation, and Global Warming**
>
> Use Figure 3-18 and Figure 3-8 to explain why clearing tropical rain forests faster than they can grow back (**Core Case Study**) can warm the earth's atmosphere. What are two ways in which this could affect the survival of remaining tropical forests? What are two ways in which it could affect your lifestyle?

Computer models of the earth's climate systems indicate that increased concentrations of atmospheric CO_2 and other gases are very likely (90–99% probability) to enhance the planet's natural greenhouse effect, which

will cause global warming and change the earth's climate (see Science Focus, p. 33).

Nitrogen Cycles through the Biosphere: Bacteria in Action

The major reservoir for nitrogen is the atmosphere. Chemically unreactive nitrogen gas (N_2) makes up 78% of the volume of the atmosphere. Nitrogen is a crucial component of proteins, many vitamins, and nucleic acids such as DNA. However, N_2 cannot be absorbed and used directly as a nutrient by multicellular plants or animals.

Fortunately, two natural processes convert or *fix* N_2 into compounds useful as nutrients for plants and animals. One is electrical discharges, or lightning, taking place in the atmosphere. The other takes place in aquatic systems, soil, and the roots of some plants, where specialized bacteria, called *nitrogen-fixing bacteria,* complete this conversion as part of the **nitrogen cycle,** which is depicted in Figure 3-19.

The nitrogen cycle consists of several major steps. In *nitrogen fixation,* specialized bacteria in soil and blue-green algae (cyanobacteria) in aquatic environments combine gaseous N_2 with hydrogen to make ammonia (NH_3). The bacteria use some of the ammonia they produce as a nutrient and excrete the rest to the soil or water. Some of the ammonia is converted to ammonium ions (NH_4^+) that can be used as a nutrient by plants.

Ammonia not taken up by plants may undergo *nitrification*. In this two-step process, specialized soil bacteria convert most of the NH_3 and NH_4^+ in soil to *nitrate ions* (NO_3^-), which are easily taken up by the roots of plants. The plants then use these forms of nitrogen to produce various amino acids, proteins, nucleic acids, and vitamins (see Supplement 6, p. S39). Animals that eat plants eventually consume these nitrogen-containing compounds, as do detritus feeders, or decomposers.

Plants and animals return nitrogen-rich organic compounds to the environment as wastes, cast-off particles, and through their bodies when they die and are decomposed or eaten by detritus feeders. In *ammonification*, vast armies of specialized decomposer bacteria convert this detritus into simpler nitrogen-containing inorganic compounds such as ammonia (NH_3) and water-soluble salts containing ammonium ions (NH_4^+).

In *denitrification*, specialized bacteria in waterlogged soil and in the bottom sediments of lakes, oceans, swamps, and bogs convert NH_3 and NH_4^+ back into nitrite and nitrate ions, and then into nitrogen gas (N_2) and nitrous oxide gas (N_2O). These gases are released to the atmosphere to begin the nitrogen cycle again.

We intervene in the nitrogen cycle in several ways (as shown by red arrows in Figure 3-19). *First*, we add large amounts of nitric oxide (NO) into the atmosphere when N_2 and O_2 combine as we burn any fuel at high temperatures, such as in car, truck, and jet engines. In the atmosphere, this gas can be converted to nitrogen dioxide gas (NO_2) and nitric acid vapor (HNO_3), which can return to the earth's surface as damaging *acid deposition*, commonly called *acid rain*.

Second, we add nitrous oxide (N_2O) to the atmosphere through the action of anaerobic bacteria on livestock wastes and commercial inorganic fertilizers applied to the soil. This greenhouse gas can warm the atmosphere and deplete stratospheric ozone, which keeps most of the sun's harmful ultraviolet radiation from reaching the earth's surface.

Third, we release large quantities of nitrogen stored in soils and plants as gaseous compounds into the atmosphere through destruction of forests, grasslands, and wetlands.

Fourth, we upset the nitrogen cycle in aquatic ecosystems by adding excess nitrates to bodies of water through agricultural runoff and discharges from municipal sewage systems.

Fifth, we remove nitrogen from topsoil when we harvest nitrogen-rich crops, irrigate crops (washing nitrates out of the soil), and burn or clear grasslands and forests before planting crops.

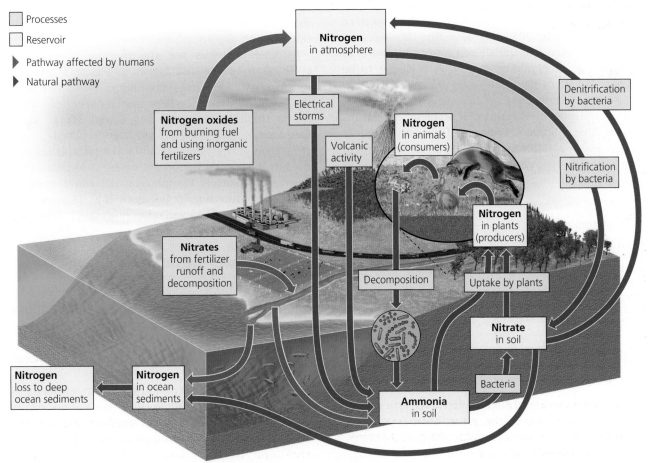

Processes

Reservoir

▶ Pathway affected by humans

▶ Natural pathway

Nitrogen in atmosphere

Nitrogen oxides from burning fuel and using inorganic fertilizers

Electrical storms

Volcanic activity

Nitrogen in animals (consumers)

Denitrification by bacteria

Nitrification by bacteria

Nitrogen in plants (producers)

Nitrates from fertilizer runoff and decomposition

Decomposition

Uptake by plants

Nitrate in soil

Nitrogen loss to deep ocean sediments

Nitrogen in ocean sediments

Bacteria

Ammonia in soil

CENGAGENOW™
Figure 3-19
Natural capital: simplified model of the *nitrogen cycle* with major harmful human impacts shown by red arrows. *See an animation based on this figure at* CengageNOW.
Question: What are three ways in which you directly or indirectly affect the nitrogen cycle?

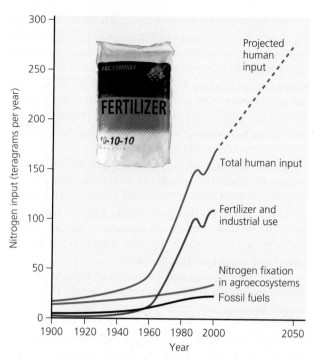

Figure 3-20 Global trends in the annual inputs of nitrogen into the environment from human activities, with projections to 2050. (Data from 2005 Millennium Ecosystem Assessment)

According to the 2005 Millennium Ecosystem Assessment, since 1950, human activities have more than doubled the annual release of nitrogen from the land into the rest of the environment. Most of this is from the greatly increased use of inorganic fertilizer to grow crops, and the amount released is projected to double again by 2050 (Figure 3-20). This excessive input of nitrogen into the air and water contributes to pollution, acid deposition, and other problems to be discussed in later chapters.

Nitrogen overload is a serious and growing local, regional, and global environmental problem that has attracted little attention. Princeton University physicist Robert Socolow calls for countries around the world to work out some type of nitrogen management agreement to help prevent this problem from reaching crisis levels.

THINKING ABOUT

The Nitrogen Cycle and Tropical Deforestation   CORE CASE STUDY

What effects might the clearing and degrading of tropical rain forests (**Core Case Study**) have on the nitrogen cycle in such ecosystems and on any nearby water systems (see Figure 2-1, p. 28, and Figure 2-4, p. 37)?

Phosphorus Cycles through the Biosphere

Phosphorus circulates through water, the earth's crust, and living organisms in the **phosphorus cycle,** depicted in Figure 3-21. In contrast to the cycles of water, carbon, and nitrogen, the phosphorus cycle does not include the atmosphere. The major reservoir for phosphorous is phosphate salts containing phosphate ions (PO_4^{3-}) in terrestrial rock formations and ocean bottom sediments. The phosphorus cycle is slow compared to the water, carbon, and nitrogen cycles.

As water runs over exposed phosphorus-containing rocks, it slowly erodes away inorganic compounds that contain phosphate ions (PO_4^{3-}). The dissolved phosphate can be absorbed by the roots of plants and by other producers. Phosphorous is transferred by food webs from such producers to consumers, eventually including detritus feeders and decomposers. In both producers and consumers, phosphorous is a component of biologically important molecules such as nucleic acids (Figure 10, p. S43, in Supplement 6) and energy transfer molecules such as ADP and ATP (Figure 14, p. S44, in Supplement 6). It is also a major component of vertebrate bones and teeth.

Phosphate can be lost from the cycle for long periods when it washes from the land into streams and rivers and is carried to the ocean. There it can be deposited as marine sediment and remain trapped for millions of years. Someday, geological processes may uplift and expose these seafloor deposits, from which phosphate can be eroded to start the cycle again.

Because most soils contain little phosphate, it is often the *limiting factor* for plant growth on land unless phosphorus (as phosphate salts mined from the earth) is applied to the soil as an inorganic fertilizer. Phosphorus also limits the growth of producer populations in many freshwater streams and lakes because phosphate salts are only slightly soluble in water.

Human activities are affecting the phosphorous cycle (as shown by red arrows in Figure 3-21). This includes removing large amounts of phosphate from the earth to make fertilizer and reducing phosphorus in tropical soils by clearing forests (**Core Case Study**). Soil that is eroded from fertilized crop fields carries large quantities of phosphates into streams, lakes, and the ocean, where it stimulates the growth of producers. Phosphorous-rich runoff from the land can produce huge populations of algae, which can upset chemical cycling and other processes in lakes.

Sulfur Cycles through the Biosphere

Sulfur circulates through the biosphere in the **sulfur cycle,** shown in Figure 3-22 (p. 72). Much of the earth's sulfur is stored underground in rocks and minerals, including sulfate (SO_4^{2-}) salts buried deep under ocean sediments.

Sulfur also enters the atmosphere from several natural sources. Hydrogen sulfide (H_2S)—a colorless, highly poisonous gas with a rotten-egg smell—is released from active volcanoes and from organic matter broken down

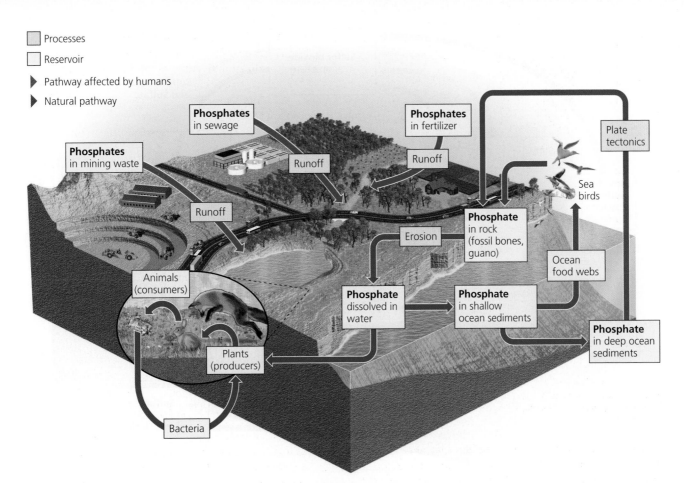

Processes

Reservoir

▶ Pathway affected by humans

▶ Natural pathway

Phosphates in sewage

Phosphates in fertilizer

Phosphates in mining waste

Plate tectonics

Runoff

Runoff

Sea birds

Runoff

Phosphate in rock (fossil bones, guano)

Erosion

Ocean food webs

Animals (consumers)

Phosphate dissolved in water

Phosphate in shallow ocean sediments

Phosphate in deep ocean sediments

Plants (producers)

Bacteria

Figure 3-21 Natural capital: simplified model of the *phosphorus cycle,* with major harmful human impacts shown by red arrows. **Question:** What are three ways in which you directly or indirectly affect the phosphorus cycle?

by anaerobic decomposers in flooded swamps, bogs, and tidal flats. Sulfur dioxide (SO_2), a colorless and suffocating gas, also comes from volcanoes.

Particles of sulfate (SO_4^{2-}) salts, such as ammonium sulfate, enter the atmosphere from sea spray, dust storms, and forest fires. Plant roots absorb sulfate ions and incorporate the sulfur as an essential component of many proteins.

Certain marine algae produce large amounts of volatile dimethyl sulfide, or DMS (CH_3SCH_3). Tiny droplets of DMS serve as nuclei for the condensation of water into droplets found in clouds. In this way, changes in DMS emissions can affect cloud cover and climate.

In the atmosphere, DMS is converted to sulfur dioxide, some of which in turn is converted to sulfur trioxide gas (SO_3) and to tiny droplets of sulfuric acid (H_2SO_4). DMS also reacts with other atmospheric chemicals such as ammonia to produce tiny particles of sulfate salts. These droplets and particles fall to the earth as components of *acid deposition,* which along with other air pollutants can harm trees and aquatic life.

In the oxygen-deficient environments of flooded soils, freshwater wetlands, and tidal flats, specialized bacteria convert sulfate ions to sulfide ions (S^{2-}). The sulfide ions can then react with metal ions to form insol-

uble metallic sulfides, which are deposited as rock, and the cycle continues.

Human activities have affected the sulfur cycle primarily by releasing large amounts of sulfur dioxide (SO_2) into the atmosphere (as shown by red arrows in Figure 3-22). We add sulfur dioxide to the atmosphere in three ways. *First,* we burn sulfur-containing coal and oil to produce electric power. *Second,* we refine sulfur-containing petroleum to make gasoline, heating oil, and other useful products. *Third,* we convert sulfur-containing metallic mineral ores into free metals such as copper, lead, and zinc. Once in the atmosphere, SO_2 is converted to droplets of sulfuric acid (H_2SO_4) and particles of sulfate (SO_4^{2-}) salts, which return to the earth as acid deposition.

─── **RESEARCH FRONTIER** ───

The effects of human activities on the major nutrient cycles and how we can reduce these effects. See **academic .cengage.com/biology/miller**.

CENGAGENOW™ Learn more about the water, carbon, nitrogen, phosphorus, and sulfur cycles using interactive animations at CengageNOW.

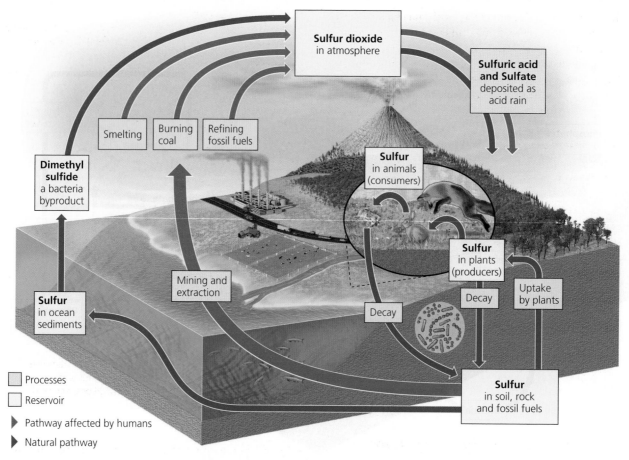

Processes

Reservoir

▶ Pathway affected by humans

▶ Natural pathway

CENGAGENOW™ Active Figure 3-22 Natural capital: simplified model of the *sulfur cycle,* with major harmful impacts of human activities shown by red arrows. *See an animation based on this figure at* CengageNOW. **Question:** What are three ways in which your lifestyle directly or indirectly affects the sulfur cycle?

3-6 How Do Scientists Study Ecosystems?

▶ **CONCEPT 3-6** Scientists use field research, laboratory research, and mathematical and other models to learn about ecosystems.

Some Scientists Study Nature Directly

Scientists use field research, laboratory research, and mathematical and other models to learn about ecosystems (**Concept 3-6**). *Field research,* sometimes called "muddy-boots biology," involves observing and measuring the structure of natural ecosystems and what happens in them. Most of what we know about structure and functioning of ecosystems has come from such research. **GREEN CAREER:** Ecologist. See **academic .cengage.com/biology/miller** for details on various green careers.

Ecologists trek through forests, deserts, and grasslands and wade or boat through wetlands, lakes, streams, and oceans collecting and observing species.

Sometimes they carry out controlled experiments by isolating and changing a variable in part of an area and comparing the results with nearby unchanged areas (Chapter 2 Core Case Study, p. 28, and see *The Habitable Planet,* Videos 4 and 9, **www.learner.org/resources/ series209.html**). Tropical ecologists have erected tall construction cranes over the canopies of tropical forests from which they observe the rich diversity of species living or feeding in these treetop habitats.

Increasingly, new technologies are being used to collect ecological data. Scientists use aircraft and satellites equipped with sophisticated cameras and other *remote sensing* devices to scan and collect data on the earth's surface. Then they use *geographic information system* (GIS) software to capture, store, analyze, and display such geographically or spatially based information.

In a GIS, geographic and ecological data can be stored electronically as numbers or as images in computer databases. For example, a GIS can convert digital satellite images generated through remote sensing into global, regional, and local maps showing variations in vegetation (Figure 1, pp. S20–S21, and Figure 2, pp. S22–S23, in Supplement 4), gross primary productivity (Figure 6, p. S27, in Supplement 4), temperature patterns, air pollution emissions, and other variables.

Scientists also use GIS programs and digital satellite images to produce two- or three-dimensional maps combining information about a variable such as land use with other data. Separate layers within such maps, each showing how a certain factor varies over an area, can be combined to show a composite effect (Figure 3-23). Such composites of information can lead to a better understanding of environmental problems and to better decision making about how to deal with such problems.

In 2005, scientists launched the Global Earth Observation System of Systems (GEOSS)—a 10-year program to integrate data from sensors, gauges, buoys, and satellites that monitor the earth's surface, atmosphere, and oceans. **GREEN CAREERS:** GIS analyst; remote sensing analyst

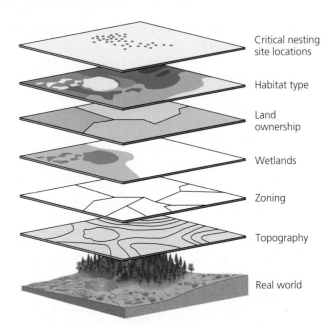

Critical nesting site locations

Habitat type

Land ownership

Wetlands

Zoning

Topography

Real world

Figure 3-23 *Geographic information systems* (*GIS*) provide the computer technology for storing, organizing, and analyzing complex data collected over broad geographic areas. They enable scientists to produce maps of various geographic data sets and then to overlay and compare the layers of data (such as soils, topography, distribution of endangered populations, and land protection status).

Some Scientists Study Ecosystems in the Laboratory

During the past 50 years, ecologists have increasingly supplemented field research by using *laboratory research* to set up, observe, and make measurements of model ecosystems and populations under laboratory conditions. Such simplified systems have been created in containers such as culture tubes, bottles, aquaria tanks, and greenhouses, and in indoor and outdoor chambers where temperature, light, CO_2, humidity, and other variables can be controlled.

Such systems make it easier for scientists to carry out controlled experiments. In addition, laboratory experiments often are quicker and less costly than similar experiments in the field.

THINKING ABOUT

Greenhouse Experiments and Tropical Rain Forests

 CORE CASE STUDY

How would you design an experiment, including an experimental group and a control group, that uses a greenhouse to determine the effect of clearing a patch of tropical rain forest vegetation (**Core Case Study**) on the temperature above the cleared patch?

But there is a catch. Scientists must consider how well their scientific observations and measurements in a simplified, controlled system under laboratory conditions reflect what takes place under the more complex and dynamic conditions found in nature. Thus, the re-

sults of laboratory research must be coupled with and supported by field research. (See *The Habitable Planet*, Videos, 2, 3, and 12, **www.learner.org/resources/series209.html**.)

Some Scientists Use Models to Simulate Ecosystems

Since the late 1960s, ecologists have developed mathematical and other models that simulate ecosystems. Computer simulations can help scientists understand large and very complex systems that cannot be adequately studied and modeled in field and laboratory research. Examples include rivers, lakes, oceans, forests, grasslands, cities, and the earth's climate system. Scientists are learning a lot about how the earth works by feeding data into increasingly sophisticated models of the earth's systems and running them on supercomputers.

Researchers can change values of the variables in their computer models to project possible changes in environmental conditions, to help them anticipate environmental surprises, and to analyze the effectiveness of various alternative solutions to environmental problems. **GREEN CAREER:** Ecosystem modeler

Of course, simulations and projections made with ecosystem models are no better than the data and assumptions used to develop the models. Ecologists must do careful field and laboratory research to get *baseline data*, or beginning measurements, of variables being studied. They also must determine the relationships among key variables that they will use to develop and test ecosystem models.

We Need to Learn More about the Health of the World's Ecosystems

We need baseline data on the condition of the world's ecosystems to see how they are changing and to develop effective strategies for preventing or slowing their degradation.

By analogy, your doctor needs baseline data on your blood pressure, weight, and functioning of your organs and other systems, as revealed through basic tests. If your health declines in some way, the doctor can run new tests and compare the results with the baseline data to identify changes and come up with a treatment.

According to a 2002 ecological study published by the Heinz Foundation and the 2005 Millennium Ecosystem Assessment, scientists have less than half of the basic ecological data they need to evaluate the status of ecosystems in the United States. Even fewer data are available for most other parts of the world. Ecologists call for a massive program to develop baseline data for the world's ecosystems.

REVISITING Tropical Rain Forests and Sustainability

This chapter applied two of the **scientific principles of sustainability** (see back cover and **Concept 1-6**, p. 23) by which the biosphere and the ecosystems it contains have been sustained over the long term. *First,* the biosphere and almost all of its ecosystems use *solar energy* as their energy source, and this energy flows through the biosphere. *Second,* they *recycle the chemical nutrients* that their organisms need for survival, growth, and reproduction.

These two principles arise from the structure and function of natural ecosystems (Figure 3-12), the law of conservation of matter (**Concept 2-3**, p. 39), and the two laws of thermodynamics (**Concepts 2-4A** and **2-4B**, p. 40). Nature's required adherence to these principles is enhanced by *biodiversity,* another sustainability principle, which also helps to *regulate population levels* of interacting species in the world's ecosystems—yet another of the sustainability principles.

This chapter started with a discussion of the importance of incredibly diverse tropical rain forests (**Core Case Study**), which showcase the functioning of the four **scientific principles of sustainability**. Producers within rain forests rely on solar energy to produce a vast amount of biomass through photosynthesis. Species living in the forests take part in, and depend on cycling of nutrients in the biosphere and the flow of energy through the biosphere. Tropical forests contain a huge and vital part of the earth's biodiversity, and interactions among species living in these forests help to control the populations of the species living there.

All things come from earth,
and to earth they all return.

MENANDER (342–290 B.C.)

REVIEW

1. Review the Key Questions and Concepts for this chapter on p. 51. What are three harmful effects resulting from the clearing and degradation of tropical rain forests?

2. What is a **cell**? What is the **cell theory?** Distinguish between a **eukaryotic cell** and a **prokaryotic cell.** What is a **species?** Explain the importance of insects.

Define **ecology.** What is **genetic diversity?** Distinguish among a **species, population, community (biological community), habitat, ecosystem,** and the **biosphere.**

3. Distinguish among the **atmosphere, troposphere, stratosphere, greenhouse gases, hydrosphere,** and

geosphere. Distinguish between **biomes** and **aquatic life zones** and give an example of each. What three interconnected factors sustain life on earth?

4. Describe what happens to solar energy as it flows to and from the earth. What is the **natural greenhouse effect** and why is it important for life on earth?

5. Distinguish between the **abiotic** and **biotic components** in ecosystems and give two examples of each. What is the **range of tolerance** for an abiotic factor? Define and give an example of a **limiting factor**. What is the **limiting factor principle?**

6. What is a **trophic level?** Distinguish among **producers (autotrophs), consumers (heterotrophs),** and **decomposers** and give an example of each in an ecosystem. Distinguish among **primary consumers (herbivores), secondary consumers (carnivores), high-level (third-level) consumers, omnivores, decomposers,** and **detritus feeders (detritivores),** and give an example of each.

7. Distinguish among **photosynthesis, chemosynthesis, aerobic respiration,** and **anaerobic respiration (fermentation).** What two processes sustain ecosystems and the biosphere and how are they linked? Explain the importance of microbes.

8. Explain what happens to energy as it flows through the food chains and food webs of an ecosystem. Distinguish between a **food chain** and a **food web.** What is **biomass?** What is **ecological efficiency?** What is the **pyramid of energy flow?** Discuss the difference between **gross primary productivity (GPP)** and **net primary productivity (NPP)** and explain their importance.

9. What happens to matter in an ecosystem? What is a **biogeochemical cycle (nutrient cycle)?** Describe the unique properties of water. What is **transpiration?** Describe the **hydrologic (water), carbon, nitrogen, phosphorus,** and **sulfur cycles** and describe how human activities are affecting each cycle.

10. Describe three ways in which scientists study ecosystems. Explain why we need much more basic data about the structure and condition of the world's ecosystems. How are the four **scientific principles of sustainability** showcased in tropical rain forests (**Core Case Study**)?

Note: Key Terms are in **bold** type.

CRITICAL THINKING

1. List three ways in which you could apply **Concept 3-4B** and **Concept 3-5** to making your lifestyle more environmentally sustainable.

2. How would you explain the importance of tropical rain forests (**Core Case Study**) to people who think that such forests have no connection to their lives?

3. Explain why **(a)** the flow of energy through the biosphere (**Concept 3-2**) depends on the cycling of nutrients, and **(b)** the cycling of nutrients depends on gravity.

4. Explain why microbes are so important. List two beneficial and two harmful effects of microbes on your health and lifestyle.

5. Make a list of the food you ate for lunch or dinner today. Trace each type of food back to a particular producer species.

6. Use the second law of thermodynamics (**Concept 2-5B**, p. 44) to explain why many poor people in developing countries live on a mostly vegetarian diet.

7. Why do farmers not need to apply carbon to grow their crops but often need to add fertilizer containing nitrogen and phosphorus?

8. What changes might take place in the hydrologic cycle if the earth's climate becomes **(a)** hotter or **(b)** cooler? In each case, what are two ways in which these changes might affect your lifestyle?

9. What would happen to an ecosystem if **(a)** all its decomposers and detritus feeders were eliminated, **(b)** all its producers were eliminated, or **(c)** all of its insects were eliminated? Could a balanced ecosystem exist with only producers and decomposers and no consumers such as humans and other animals? Explain.

10. List two questions that you would like to have answered as a result of reading this chapter.

Note: See Supplement 13 (p. S78) for a list of Projects related to this chapter.

Based on the following carbon dioxide emissions data and 2007 population data, answer the questions below.

Country	Total Carbon Footprint— Carbon Dioxide Emissions (in metric gigatons per year*)	Population (in billions, 2007)	Per Capita Carbon Footprint— Per Capita Carbon Dioxide Emissions Per Year
China	5.0 (5.5)	1.3	
India	1.3 (1.4)	1.1	
Japan	1.3 (1.4)	0.13	
Russia	1.5 (1.6)	1.14	
United States	6.0 (6.6)	0.30	
WORLD	29 (32)	6.6	

Souce: Data from World Resources Institute and International Energy Agency
*The prefix *giga* stands for "1 billion."

1. Calculate the per capita carbon footprint for each country and the world and complete the table.

2. It has been suggested that a sustainable average world-wide carbon footprint per person should be no more than 2.0 metric tons per person per year (2.2 tons per person per year). How many times larger is the U.S. carbon footprint per person than are **(a)** the sustainable level, and **(b)** the world average?

3. By what percentage will China, Japan, Russia, the United States, and the world each have to reduce their carbon footprints per person to achieve the estimated maximum sustainable carbon footprint per person of 2.0 metric tons (2.2 tons) per person per year?

Biodiversity and Evolution

Why Should We Care about the American Alligator?

The American alligator (Figure 4-1), North America's largest reptile, has no natural predators except for humans, and it plays a number of important roles in the ecosystems where it is found. This species outlived the dinosaurs and has been able to survive numerous dramatic changes in the earth's environmental conditions.

But starting in the 1930s, these alligators faced a new challenge. Hunters began killing them in large numbers for their exotic meat and their supple belly skin, used to make shoes, belts, and pocketbooks. Other people hunted alligators for sport or out of hatred. By the 1960s, hunters and poachers had wiped out 90% of the alligators in the U.S. state of Louisiana, and the alligator population in the Florida Everglades was also near extinction.

Those who did not care much for the American alligator were probably not aware of its important ecological role—its *niche*—in subtropical wetland communities. These alligators dig deep depressions, or gator holes, which hold freshwater during dry spells, serve as refuges for aquatic life, and supply freshwater and food for fish, insects, snakes, turtles, birds, and other animals. Large alligator nesting mounds provide nesting and feeding sites for species of herons and egrets, and red-bellied turtles use old gator nests for incubating their eggs. These alligators eat large numbers of gar, a predatory fish. This helps maintain populations of game fish such as bass and bream.

As alligators move from gator holes to nesting mounds, they help keep areas of open water free of invading vegetation. Without these free ecosystem services, freshwater ponds and coastal wetlands where these alligators live would be filled in with shrubs and trees, and dozens of species would disappear from these ecosystems. Some ecologists classify the American alligator as a *keystone species* because of its important ecological role in helping to maintain the structure, function, and sustainability of the ecosystems where it is found. And, in 2008, scientists began analyzing the blood of the American alligator to identify compounds that could kill a variety of harmful bacteria, including those that have become resistant to commonly used antibiotics.

In 1967, the U.S. government placed the American alligator on the endangered species list. Protected from hunters, the population made a strong comeback in many areas by 1975—too strong, according to those who find alligators in their backyards and swimming pools, and to duck hunters whose retriever dogs are sometimes eaten by alligators. In 1977, the U.S. Fish and Wildlife Service reclassified the American alligator as a *threatened* species in the U.S. states of Florida, Louisiana, and Texas, where 90% of the animals live. Today there are 1–2 million American alligators in Florida, and the state now allows property owners to kill alligators that stray onto their land.

To biologists, the comeback of the American alligator is an important success story in wildlife conservation. This tale illustrates how each species in a community or ecosystem fills a unique role, and it highlights how interactions between species can affect ecosystem structure and function. In this chapter, we will examine biodiversity, with an emphasis on species diversity, and the theory of how the earth's diverse species arose.

Figure 4-1 The American alligator plays an important ecological role in its marsh and swamp habitats in the southeastern United States. Since being classified as an endangered species in 1967, it has recovered enough to have its status changed from endangered to threatened—an outstanding success story in wildlife conservation.

A. & J. Visage/Peter Arnold, Inc.

Key Questions and Concepts

4-1 What is biodiversity and why is it important?

CONCEPT 4-1 The biodiversity found in genes, species, ecosystems, and ecosystem processes is vital to sustaining life on earth.

4-2 Where do species come from?

CONCEPT 4-2A The scientific theory of evolution explains how life on earth changes over time through changes in the genes of populations.

CONCEPT 4-2B Populations evolve when genes mutate and give some individuals genetic traits that enhance their abilities to survive and to produce offspring with these traits (natural selection).

4-3 How do geological processes and climate change affect evolution?

CONCEPT 4-3 Tectonic plate movements, volcanic eruptions, earthquakes, and climate change have shifted wildlife habitats, wiped out large numbers of species, and created opportunities for the evolution of new species.

4-4 How do speciation, extinction, and human activities affect biodiversity?

CONCEPT 4-4A As environmental conditions change, the balance between formation of new species and extinction of existing species determines the earth's biodiversity.

CONCEPT 4-4B Human activities can decrease biodiversity by causing the premature extinction of species and by destroying or degrading habitats needed for the development of new species.

4-5 What is species diversity and why is it important?

CONCEPT 4-5 Species diversity is a major component of biodiversity and tends to increase the sustainability of ecosystems.

4-6 What roles do species play in ecosystems?

CONCEPT 4-6A Each species plays a specific ecological role called its niche.

CONCEPT 4-6B Any given species may play one or more of five important roles—native, nonnative, indicator, keystone, or foundation roles—in a particular ecosystem.

Note: Supplements 2 (p. S4), 4 (p. S20), 6 (p. S39), 7 (p. S46), 8 (p. S47), and 13 (p. S78) can be used with this chapter.

There is grandeur to this view of life…
that, whilst this planet has gone cycling on…
endless forms most beautiful and most wonderful
have been, and are being, evolved.

CHARLES DARWIN

4-1   What Is Biodiversity and Why Is It Important?

▶ **CONCEPT 4-1** The biodiversity found in genes, species, ecosystems, and ecosystem processes is vital to sustaining life on earth.

Biodiversity Is a Crucial Part of the Earth's Natural Capital

Biological diversity, or **biodiversity,** is the variety of the earth's species, the genes they contain, the ecosystems in which they live, and the ecosystem processes such as energy flow and nutrient cycling that sustain all life (Figure 4-2). Biodiversity is a vital renewable resource (**Concept 4-1**).

So far, scientists have identified about 1.8 million of the earth's 4 million to 100 million species, and every year, thousands of new species are identified. The identified species include almost a million species of insects, 270,000 plant species, and 45,000 vertebrate animal species. Later in this chapter, we look at the scientific theory of how such a great variety of life forms came to be, and we consider the importance of species diversity.

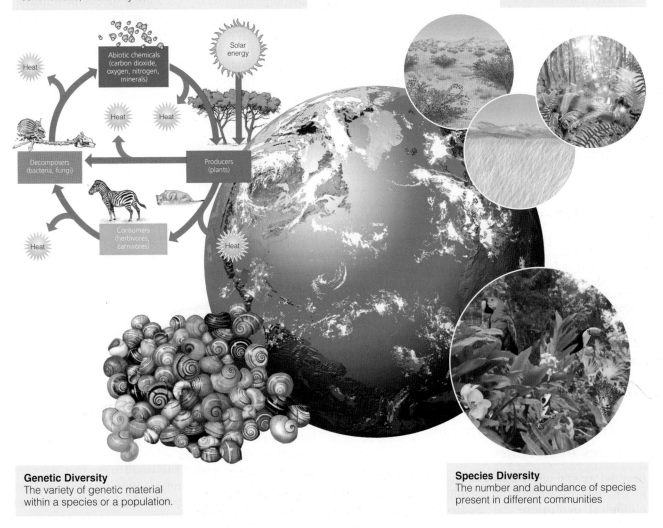

CENGAGENOW™ **Active Figure 4-2 Natural capital:** the major components of the earth's *biodiversity*—one of the earth's most important renewable resources. *See an animation based on this figure at* CengageNOW™. **Question:** What are three examples of how people, in their daily living, intentionally or unintentionally degrade each of these types of biodiversity?

Species diversity is the most obvious, but not the only, component of biodiversity. Another important component is *genetic diversity* (Figure 3-5, p. 53). The earth's variety of species contains an even greater variety of genes. Genetic diversity enables life on the earth to adapt to and survive dramatic environmental changes. In other words, genetic diversity is vital to the sustainability of life on earth.

Ecosystem diversity—the earth's variety of deserts, grasslands, forests, mountains, oceans, lakes, rivers, and wetlands is another major component of biodiversity. Each of these ecosystems is a storehouse of genetic and species diversity.

Yet another important component of biodiversity is *functional diversity*—the variety of processes such as

matter cycling and energy flow taking place within ecosystems (Figure 3-12, p. 60) as species interact with one another in food chains and webs. Part of the importance of the American alligator (Figure 4-1) is its role in supporting these processes within its ecosystems, which help to maintain other species of animals and plants that live there.

THINKING ABOUT

Alligators and Biodiversity

What are three ways in which the American alligator (**Core Case Study**) supports one or more of the four components of biodiversity within its environment?

CORE CASE STUDY

The earth's biodiversity is a vital part of the natural capital that keeps us alive. It supplies us with food, wood, fibers, energy, and medicines—all of which represent hundreds of billions of dollars in the world economy each year. Biodiversity also plays a role in preserving the quality of the air and water and maintaining the fertility of soils. It helps us to dispose of wastes and to control populations of pests. In carrying out these free ecological services, which are also part of the earth's natural capital (**Concept 1-1A**, p. 6), ◀ CONCEPT LINK biodiversity helps to sustain life on the earth.

Because biodiversity is such an important concept and so vital to sustainability, we are going to take a grand tour of biodiversity in this and the next seven chapters. This chapter focuses on the earth's variety of species, how these species evolved, and the major roles that species play in ecosystems. Chapter 5 examines how different interactions among species help to control population sizes and promote biodiversity. Chapter 6 uses principles of population dynamics developed in Chapter 5 to look at human population growth and its effects on biodiversity. Chapters 7 and 8, respectively, look at the major types of terrestrial and aquatic ecosystems that make up a key component of biodiversity. Then, the next three chapters examine major threats to species diversity (Chapter 9), terrestrial biodiversity (Chapter 10), and aquatic biodiversity (Chapter 11), and solutions for dealing with these threats.

4-2   Where Do Species Come From?

▶ **CONCEPT 4-2A** The scientific theory of evolution explains how life on earth changes over time through changes in the genes of populations.

▶ **CONCEPT 4-2B** Populations evolve when genes mutate and give some individuals genetic traits that enhance their abilities to survive and to produce offspring with these traits (natural selection).

Biological Evolution by Natural Selection Explains How Life Changes over Time

How did we end up with an amazing array of 4 million to 100 million species? The scientific answer involves **biological evolution:** the process whereby earth's life changes over time through changes in the genes of populations (**Concept 4-2A**).

The idea that organisms change over time and are descended from a single common ancestor has been around in one form or another since the early Greek philosophers. But no one had come up with a credible explanation of how this could happen until 1858 when naturalists Charles Darwin (1809–1882) and Alfred Russel Wallace (1823–1913) independently proposed the concept of *natural selection* as a mechanism for biological evolution. Although Wallace also proposed the idea of natural selection, it was Darwin, who meticulously gathered evidence for this idea and published it in 1859 in his book, *On the Origin of Species by Means of Natural Selection.*

Darwin and Wallace observed that organisms must constantly struggle to obtain enough food and other resources to survive and reproduce. They also observed that individuals in a population with a specific advantage over other individuals are more likely to survive, reproduce, and have offspring with similar survival skills. The advantage was due to a characteristic, or *trait,* possessed by these individuals but not by others.

Darwin and Wallace concluded that these survival traits would become more prevalent in future populations of the species through a process called **natural selection,** which occurs when some individuals of a population have genetically based traits that enhance their ability to survive and produce offspring with the same traits. A change in the genetic characteristics of a population from one generation to another is known as *biological evolution,* or simply *evolution.*

A huge body of field and laboratory evidence has supported this idea. As a result, biological evolution through natural selection has become an important scientific theory. According to this theory, life has evolved into six major groups of species, called *kingdoms,* as a result of natural selection. This view sees the development of life as an ever-branching tree of species diversity, sometimes called the *tree of life* (Figure 4-3).

This scientific theory generally explains how life has changed over the past 3.7 billion years and why life is so diverse today. However, there are still many unanswered questions and scientific debates about the details of evolution by natural selection. Such continual questioning and discussion is an important way in which science advances our knowledge of how the earth works.

CENGAGENOW™ Get a detailed look at early biological evolution by natural selection—the roots of the tree of life—at CengageNOW™.

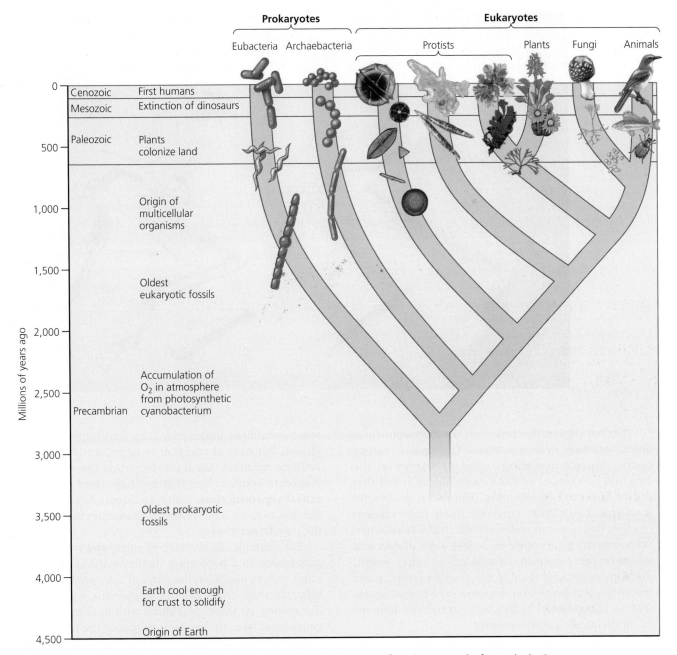

Figure 4-3 Overview of the evolution of life on the earth into six major kingdoms of species as a result of natural selection. For more details, see p. S46 in Supplement 7.

The Fossil Record Tells Much of the Story of Evolution

Most of what we know of the earth's life history comes from **fossils:** mineralized or petrified replicas of skeletons, bones, teeth, shells, leaves, and seeds, or impressions of such items found in rocks. Also, scientists drill cores from glacial ice at the earth's poles and on mountaintops and examine the kinds of life found at different layers. Fossils provide physical evidence of ancient organisms and reveal what their internal structures looked like (Figure 4-4, p. 82).

The world's cumulative body of fossils found is called the *fossil record*. This record is uneven and incomplete. Some forms of life left no fossils, and some fossils have decomposed. The fossils found so far probably represent only 1% of all species that have ever lived. Trying to reconstruct the development of life with so little evidence—a challenging scientific detective game—is the work of paleontologists. **GREEN CAREER:** Paleontologist

The Genetic Makeup of a Population Can Change

The process of biological evolution by natural selection involves changes in a population's genetic makeup through successive generations. Note that *populations— not individuals—evolve by becoming genetically different.*

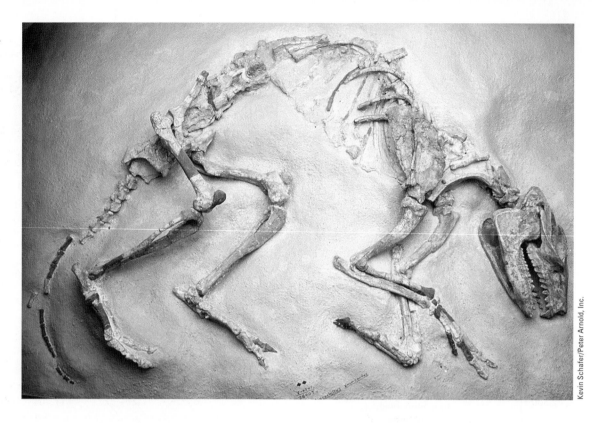

Figure 4-4 Fossilized skeleton of an herbivore that lived during the Cenozoic era from 26–66 million years ago.

Kevin Schafer/Peter Arnold, Inc.

The first step in this process is the development of *genetic variability* in a population. This genetic variety occurs through **mutations:** *random* changes in the structure or number of DNA molecules in a cell that can be inherited by offspring (Figure 11, p. S43, in Supplement 6). Most mutations result from random changes that occur in coded genetic instructions when DNA molecules are copied each time a cell divides and whenever an organism reproduces. In other words, this copying process is subject to random errors. Some mutations also occur from exposure to external agents such as radioactivity, X rays, and natural and human-made chemicals (called *mutagens*).

Mutations can occur in any cell, but only those taking place in reproductive cells are passed on to offspring. Sometimes a mutation can result in a new genetic trait that gives an individual and its offspring better chances for survival and reproduction under existing environmental conditions or when such conditions change.

Individuals in Populations with Beneficial Genetic Traits Can Leave More Offspring

The next step in biological evolution is *natural selection,* which occurs when some individuals of a population have genetically based traits (resulting from mutations) that enhance their ability to survive and produce offspring with these traits (**Concept 4-2B**).

An **adaptation,** or **adaptive trait,** is any heritable trait that enables an individual organism to survive through natural selection and to reproduce more than other individuals under prevailing environmental conditions. For natural selection to occur, a trait must be *heritable,* meaning that it can be passed from one generation to another. The trait must also lead to **differential reproduction,** which enables individuals with the trait to leave more offspring than other members of the population leave.

For example, in the face of snow and cold, a few gray wolves in a population that have thicker fur than other wolves might live longer and thus produce more offspring than those without thicker fur who do not live as long. As those individuals with thicker fur mate, genes for thicker fur spread throughout the population and individuals with those genes increase in number and pass this helpful trait on to their offspring. Thus, the concept of natural selection explains how populations adapt to changes in environmental conditions.

Genetic resistance is the ability of one or more organisms in a population to tolerate a chemical designed to kill it. For example, an organism might have a gene that allows it to break the chemical down into harmless substances. Another important example of natural selection at work is the evolution of antibiotic resistance in disease-causing bacteria. Scientists have developed antibacterial drugs (antibiotics) to fight these bacteria, and the drugs have become a driving force of natural selection. The few bacteria that are genetically resistant to the drugs (because of some trait they possess) survive and produce more offspring than the bacteria that were killed by the drugs could have produced. Thus, the antibiotic eventually loses its effectiveness, as resistant bacteria rapidly reproduce and those that are susceptible to the drug die off (Figure 4-5).

(a)
A group of bacteria, including genetically resistant ones, are exposed to an antibiotic

(b)
Most of the normal bacteria die

(c)
The genetically resistant bacteria start multiplying

(d)
Eventually the resistant strain replaces the strain affected by the antibiotic

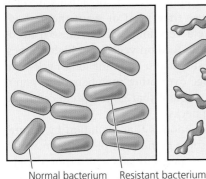
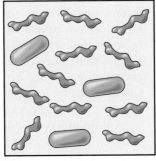
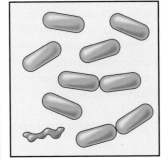
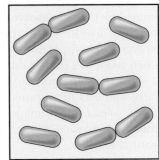

Normal bacterium   Resistant bacterium

Figure 4-5 *Evolution by natural selection.* **(a)** A population of bacteria is exposed to an antibiotic, which **(b)** kills all but those possessing a trait that makes them resistant to the drug. **(c)** The resistant bacteria multiply and eventually **(d)** replace the nonresistant bacteria.

Note that natural selection acts on individuals, but evolution occurs in populations. In other words, populations can evolve when genes change or mutate and give some individuals genetic traits that enhance their ability to survive and to produce offspring with these traits (natural selection) (**Concept 4-2B**).

CENGAGENOW™ How many moths can you eat? Find out and learn more about adaptation at CengageNOW.

Another way to summarize the process of biological evolution by natural selection is: *Genes mutate, individuals are selected, and populations evolve that are better adapted to survive and reproduce under existing environmental conditions.*

When environmental conditions change, a population of a species faces three possible futures: *adapt* to the new conditions through natural selection, *migrate* (if possible) to an area with more favorable conditions, or *become extinct.*

A remarkable example of evolution by natural selection is human beings. We have evolved certain traits that have allowed us to take over much of the world (see Case Study below).

■ CASE STUDY

How Did Humans Become Such a Powerful Species?

Like many other species, humans have survived and thrived because we have certain traits that allow us to adapt to and modify parts of the environment to increase our survival chances.

Evolutionary biologists attribute our success to three adaptations: *strong opposable thumbs* that allow us to grip and use tools better than the few other animals that have thumbs can do, an ability to *walk upright,* and a *complex brain.* These adaptations have helped us develop

weapons, protective devices, and technologies that extend our limited senses and make up for some of our deficiencies. Thus, in just a twitch of the 3.56-billion-year history of life on earth, we have developed powerful technologies and taken over much of the earth's life-support systems and net primary productivity.

But adaptations that make a species successful during one period of time may not be enough to ensure the species' survival when environmental conditions change. This is no less true for humans, and some environmental conditions are now changing rapidly, largely due to our own actions.

The *good news* is that one of our adaptations—our powerful brain—may enable us to live more sustainably by understanding and copying the ways in which nature has sustained itself for billions of years, despite major changes in environmental conditions (**Concept 1-6**, p. 23).

CONCEPT LINK

┌─ **THINKING ABOUT**
Human Adaptations

An important adaptation of humans is strong opposable thumbs, which allow us to grip and manipulate things with our hands. Make a list of the things you could not do without the use of your thumbs.

Adaptation through Natural Selection Has Limits

In the not-too-distant future, will adaptations to new environmental conditions through natural selection allow our skin to become more resistant to the harmful effects of ultraviolet radiation, our lungs to cope with air pollutants, and our livers to better detoxify pollutants?

According to scientists in this field, the answer is *no* because of two limits to adaptations in nature through natural selection. *First,* a change in environmental conditions can lead to such an adaptation only for genetic traits already present in a population's gene pool or for traits resulting from mutations.

Second, even if a beneficial heritable trait is present in a population, the population's ability to adapt may be limited by its reproductive capacity. Populations of genetically diverse species that reproduce quickly—such as weeds, mosquitoes, rats, bacteria, or cockroaches—often adapt to a change in environmental conditions in a short time. In contrast, species that cannot produce large numbers of offspring rapidly—such as elephants, tigers, sharks, and humans—take a long time (typically thousands or even millions of years) to adapt through natural selection.

Three Common Myths about Evolution through Natural Selection

According to evolution experts, there are three common misconceptions about biological evolution through natural selection. One is that "survival of the fittest" means "survival of the strongest." To biologists, *fitness* is a measure of reproductive success, not strength. Thus, the fittest individuals are those that leave the most descendants.

Another misconception is that organisms develop certain traits because they need or want them. A giraffe does not have a very long neck because it needs or wants it in order to feed on vegetation high in trees. Rather, some ancestor had a gene for long necks that gave it an advantage over other members of its population in getting food, and that giraffe produced more offspring with long necks.

A third misconception is that evolution by natural selection involves some grand plan of nature in which species become more perfectly adapted. From a scientific standpoint, no plan or goal of genetic perfection has been identified in the evolutionary process. Rather, it appears to be a random, branching process that results in a great variety of species (Figure 4-3).

4-3 How Do Geological Processes and Climate Change Affect Evolution?

▶ **CONCEPT 4-3** Tectonic plate movements, volcanic eruptions, earthquakes, and climate change have shifted wildlife habitats, wiped out large numbers of species, and created opportunities for the evolution of new species.

Geological Processes Affect Natural Selection

The earth's surface has changed dramatically over its long history. Scientists have discovered that huge flows of molten rock within the earth's interior break its surface into a series of gigantic solid plates, called *tectonic plates*. For hundreds of millions of years, these plates have drifted slowly atop the planet's mantle (Figure 4-6).

This process has had two important effects on the evolution and location of life on the earth. *First,* the locations of continents and oceanic basins greatly influence the earth's climate and thus help determine where plants and animals can live.

Second, the movement of continents has allowed species to move, adapt to new environments, and form new species through natural selection. When continents join together, populations can disperse to new areas and adapt to new environmental conditions. And when continents separate, populations either evolve under the new conditions or become extinct.

Earthquakes can also affect biological evolution by causing fissures in the earth's crust that can separate and isolate populations of species. Over long periods of time, this can lead to the formation of new species as each isolated population changes genetically in response to new environmental conditions. And *volcanic eruptions* affect biological evolution by destroying habitats and reducing or wiping out populations of species (**Concept 4-3**).

Climate Change and Catastrophes Affect Natural Selection

Throughout its long history, the earth's climate has changed drastically. Sometimes it has cooled and covered much of the earth with ice. At other times it has warmed, melted ice, and drastically raised sea levels. Such alternating periods of cooling and heating have led to advances and retreats of ice sheets at high latitudes over much of the northern hemisphere, most recently, about 18,000 years ago (Figure 4-7).

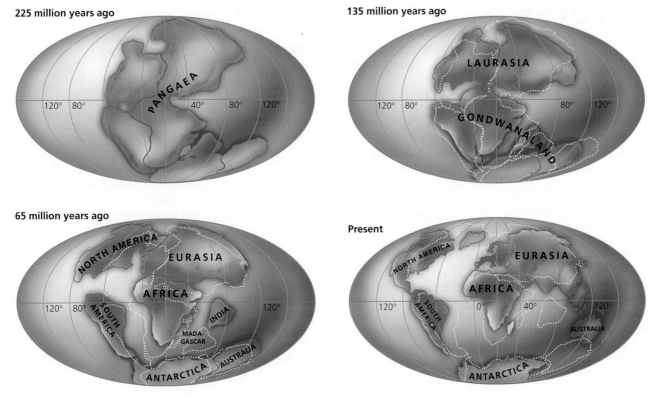

Figure 4-6 Over millions of years, the earth's continents have moved very slowly on several gigantic tectonic plates. This process plays a role in the extinction of species, as land areas split apart, and also in the rise of new species when isolated land areas combine. Rock and fossil evidence indicates that 200–250 million years ago, all of the earth's present-day continents were locked together in a supercontinent called Pangaea (top left). About 180 million years ago, Pangaea began splitting apart as the earth's tectonic plates separated, eventually resulting in today's locations of the continents (bottom right). **Question:** How might an area of land splitting apart cause the extinction of a species?

These long-term climate changes have a major effect on biological evolution by determining where different types of plants and animals can survive and thrive and by changing the locations of different types of ecosystems such as deserts, grasslands, and forests (**Concept 4-3**). Some species became extinct because the climate changed too rapidly for them to survive, and new species evolved to fill their ecological roles.

Another force affecting natural selection has been catastrophic events such as collisions between the earth and large asteroids. There have probably been many of these collisions during the earth's 4.5 billion years. Such impacts have caused widespread destruction of ecosystems and wiped out large numbers of species. But they have also caused shifts in the locations of ecosystems and created opportunities for the evolution of new species. On a long-term basis, the four **scientific principles of sustainability** (see back cover), especially biodiversity (Figure 4-2) have enabled life on earth to adapt to drastic changes in environmental conditions (Science Focus, p. 86). In other words, we live on a habitable planet. (See *The Habitable Planet*, Video 1, **www.learner.org/resources/series209.html**.)

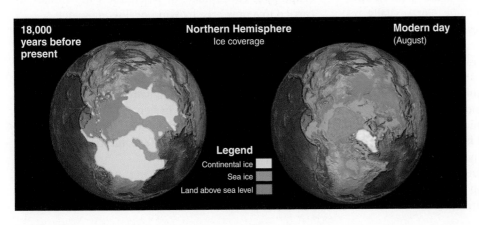

Figure 4-7 Changes in ice coverage in the northern hemisphere during the past 18,000 years. **Question:** What are two characteristics of an animal and two characteristics of a plant that natural selection would have favored as these ice sheets (left) advanced? (Data from the National Oceanic and Atmospheric Administration)

Earth Is Just Right for Life to Thrive

Life on the earth, as we know it, can thrive only within a certain temperature range, which depends on the liquid water that dominates the earth's surface. Most life on the earth requires average temperatures between the freezing and boiling points of water.

The earth's orbit is the right distance from the sun to provide these conditions. If the earth were much closer to the sun, it would be too hot—like Venus—for water vapor to condense and form rain. If it were much farther away, the earth's surface would be so cold—like Mars—that its water would exist only as ice. The earth also spins; if it did not, the side facing the sun would be too hot and the other side too cold for water-based life to exist.

The size of the earth is also just right for life. It has enough gravitational mass to keep its iron and nickel core molten and to keep the atmosphere—made up of light gaseous molecules required for life (such as N_2, O_2, CO_2, and H_2O)—from flying off into space.

Although life on earth has been enormously resilient and adaptive, it has benefitted from a favorable temperature range. During the 3.7 billion years since life arose, the average surface temperature of the earth has remained within the narrow range of 10–20 °C (50–68 °F), even with a 30–40% increase in the sun's energy output. One reason for this is the evolution of organisms that modify levels of the temperature-regulating gas carbon dioxide in the atmosphere as a part of the carbon cycle (Figure 3-18, p. 68)

For almost 600 million years, oxygen has made up about 21% of the volume of earth's atmosphere. If this oxygen content dropped to about 15%, it would be lethal for most forms of life. If it increased to about 25%, oxygen in the atmosphere would probably ignite into a giant fireball. The current oxygen content of the atmosphere is largely the result of producer and consumer organisms interacting in the carbon cycle. Also, because of the development of photosynthesizing bacteria that have been adding oxygen to the atmosphere for more than 2 billion years, an ozone sunscreen in the stratosphere protects us and many other forms of life from an overdose of ultraviolet radiation.

In short, this remarkable planet we live on is uniquely suited for life as we know it.

Critical Thinking

Design an experiment to test the hypothesis that various forms of life can maintain the oxygen content in the atmosphere at around 21% of its volume.

4-4 How Do Speciation, Extinction, and Human Activities Affect Biodiversity?

▶ **CONCEPT 4-4A** As environmental conditions change, the balance between formation of new species and extinction of existing species determines the earth's biodiversity.

▶ **CONCEPT 4-4B** Human activities can decrease biodiversity by causing the premature extinction of species and by destroying or degrading habitats needed for the development of new species.

How Do New Species Evolve?

Under certain circumstances, natural selection can lead to an entirely new species. In this process, called **speciation,** two species arise from one. For sexually reproducing species, a new species is formed when some members of a population have evolved to the point where they no longer can breed with other members to produce fertile offspring.

The most common mechanism of speciation (especially among sexually reproducing animals) takes place in two phases: geographic isolation and reproductive isolation. **Geographic isolation** occurs when different groups of the same population of a species become physically isolated from one another for long periods. For example, part of a population may migrate in search of food and then begin living in another area with different environmental conditions. Separation of populations can occur because of a physical barrier (such as a mountain range, stream, or road), a volcanic eruption or earthquake, or when a few individuals are carried to a new area by wind or flowing water.

In **reproductive isolation,** mutation and change by natural selection operate independently in the gene pools of geographically isolated populations. If this process continues long enough, members of the geographically and reproductively isolated populations may become so different in genetic makeup that they cannot produce live, fertile offspring if they are rejoined. Then one species has become two, and speciation has occurred (Figure 4-8).

For some rapidly reproducing organisms, this type of speciation may occur within hundreds of years. For most species, it takes from tens of thousands to millions

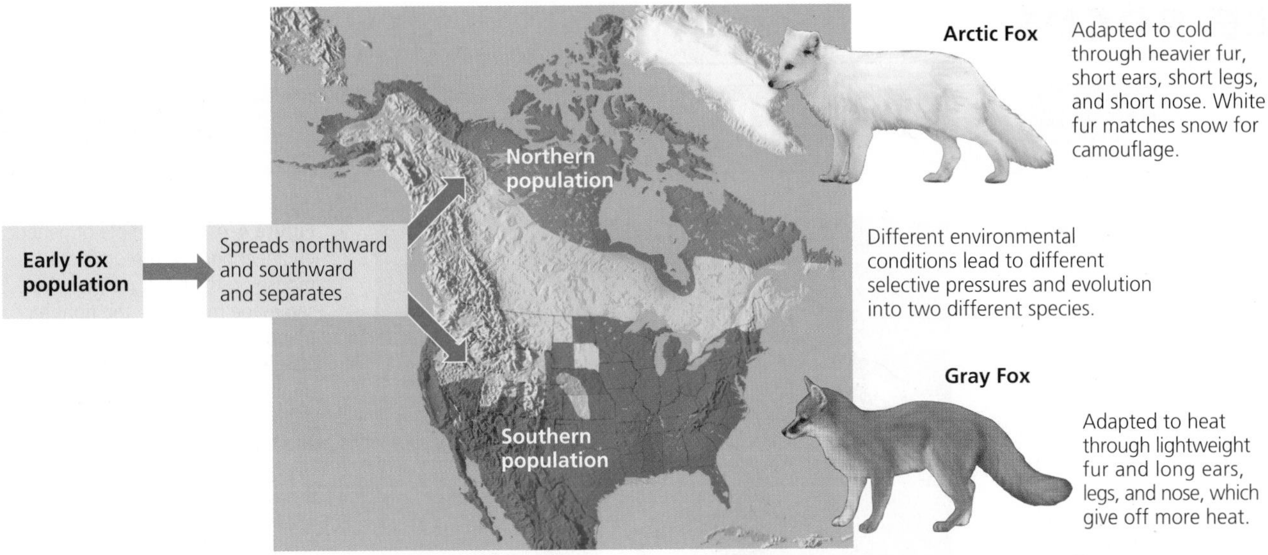

Figure 4-8
Geographic isolation can lead to reproductive isolation, divergence of gene pools, and speciation.

Arctic Fox Adapted to cold through heavier fur, short ears, short legs, and short nose. White fur matches snow for camouflage.

Different environmental conditions lead to different selective pressures and evolution into two different species.

Gray Fox Adapted to heat through lightweight fur and long ears, legs, and nose, which give off more heat.

Early fox population

Spreads northward and southward and separates

Northern population

Southern population

---

of years—making it difficult to observe and document the appearance of a new species.

CENGAGE**NOW** Learn more about different types of speciation and ways in which they occur at CengageNOW.

Humans are playing an increasing role in the process of speciation. We have learned to shuffle genes from one species to another though artificial selection and more recently through genetic engineering (Science Focus, p. 88).

> **THINKING ABOUT**
> **Speciation and American Alligators**
>
> Imagine how a population of American alligators (**Core Case Study**) might have evolved into two species had they become separated, with one group evolving in a more northern climate. Describe some of the traits of the hypothetical northern species.

CORE CASE STUDY

Figure 4-9 Male golden toad in Costa Rica's high-altitude Monteverde Cloud Forest Reserve. This species has recently become extinct, primarily because changes in climate dried up its habitat.

Extinction Is Forever

Another process affecting the number and types of species on the earth is **extinction,** in which an entire species ceases to exist. Species that are found in only one area are called **endemic species** and are especially vulnerable to extinction. They exist on islands and in other unique small areas, especially in tropical rain forests where most species are highly specialized.

One example is the brilliantly colored golden toad (Figure 4-9) once found only in a small area of lush cloud rain forests in Costa Rica's mountainous region. Despite living in the country's well-protected Monteverde Cloud Forest Reserve, by 1989, the golden toad had apparently become extinct. Much of the moisture that supported its rain forest habitat came in the form of moisture-laden clouds blowing in from the Caribbean Sea. But warmer air from global climate change caused

these clouds to rise, depriving the forests of moisture, and the habitat for the golden toad and many other species dried up. The golden toad appears to be one of the first victims of climate change caused largely by global warming. A 2007 study found that global warming has also contributed to the extinction of five other toad and frog species in the jungles of Costa Rica.

Extinction Can Affect One Species or Many Species at a Time

All species eventually become extinct, but drastic changes in environmental conditions can eliminate large groups of species. Throughout most of history, species have disappeared at a low rate, called **background**

SCIENCE FOCUS

We Have Developed Two Ways to Change the Genetic Traits of Populations

We have used **artificial selection** to change the genetic characteristics of populations with similar genes. In this process, we select one or more desirable genetic traits in the population of a plant or animal, such as a type of wheat, fruit, or dog. Then we use *selective breeding* to generate populations of the species containing large numbers of individuals with the desired traits. Note that artificial selection involves crossbreeding between genetic varieties of the same species and thus is not a form of speciation. Most, of the grains, fruits, and vegetables we eat are produced by artificial selection.

Artificial selection has given us food crops with higher yields, cows that give more milk, trees that grow faster, and many different types of dogs and cats. But traditional crossbreeding is a slow process. Also, it can combine traits only from species that are close to one another genetically.

Now scientists are using genetic engineering to speed up our ability to manipulate genes. **Genetic engineering,** or **gene splicing,** is the alteration of an organism's genetic material, through adding, deleting, or changing segments of its DNA (Figure 11, p. S43, in Supplement 6), to produce desirable traits or eliminate undesirable ones. It enables scientists to transfer genes between different species that would not interbreed in nature. For example, genes from a fish species can be put into a tomato plant to give it certain properties.

Scientists have used gene splicing to develop modified crop plants, new drugs, pest-resistant plants, and animals that grow rapidly (Figure 4-A). They have also created genetically engineered bacteria to extract minerals such as copper from their underground ores and to clean up spills of oil and other toxic pollutants.

Application of our increasing genetic knowledge is filled with great promise, but it raises some serious ethical and privacy issues. For example, some people have genes that make them more likely to develop certain genetic diseases or disorders. We now have the power to detect these genetic deficiencies, even before birth. Questions of ethics and morality arise over how this knowledge and technology will be applied, who will benefit, and who might suffer from it.

Further, what will be the environmental impacts of such applications? If genetic engineering could help all humans live in good health much longer than we do now, it might increase pollution, environmental degradation, and the strain on natural resources. More and more affluent people living longer and longer could create an enormous and ever-growing ecological footprint.

Some people dream of a day when our genetic engineering prowess could eliminate death and aging altogether. As one's cells, organs, or other parts wear out or become damaged, they could be replaced with new ones grown in genetic engineering facilities.

Assuming this is scientifically possible, is it morally acceptable to take this path? Who will decide? Who will regulate this new industry? Sometime in the not-too-distant future, will we be able to change the nature of what it means to be human? If so, how will it change? These are some of the most important and controversial ethical questions of the 21st century.

Another concern is that most new technologies have had unintended harmful consequences. For example, pesticides have helped protect crops from insect pests and disease. But their overuse has accelerated the evolution of pesticide-resistant species and has wiped out many natural predator insects that had helped to keep pest populations under control.

For these and other reasons, a backlash developed in the 1990s against the increasing use of genetically modified food plants and animals. Some protesters argue against using this new technology, mostly for ethical reasons. Others advocate slowing down the technological rush and taking a closer look at the short- and long-term advantages and disadvantages of genetic technologies.

Critical Thinking

What might be some beneficial and harmful effects on the evolutionary process if genetic engineering is widely applied to plants and animals?

Figure 4-A An example of genetic engineering. The 6-month-old mouse on the left is normal; the same-age mouse on the right has a human growth hormone gene inserted in its cells. Mice with the human growth hormone gene grow two to three times faster and twice as large as mice without the gene. **Question:** How do you think the creation of such species might change the process of evolution by natural selection?

R. L. Brinster and R. E. Hammer/School of Veterinary Medicine, University of Pennsylvania

---

extinction. Based on the fossil record and analysis of ice cores, biologists estimate that the average annual background extinction rate is one to five species for each million species on the earth.

In contrast, **mass extinction** is a significant rise in extinction rates above the background level. In such a catastrophic, widespread (often global) event, large groups of species (perhaps 25–70%) are wiped out in a geological period lasting up to 5 million years. Fossil and geological evidence indicate that the earth's species have experienced five mass extinctions (20–60 million years apart) during the past 500 million years. For example, about 250 million years ago, as much as 95% of all existing species became extinct.

Some biologists argue that a mass extinction should be distinguished by a low speciation rate as well as by a high rate of extinction. Under this more strict definition, there have been only three mass extinctions. As this subject is debated, the definitions will be refined, and one argument or the other will be adopted as the working hypothesis. Either way, there is substantial evidence that large numbers of species have become extinct several times in the past.

A mass extinction provides an opportunity for the evolution of new species that can fill unoccupied ecological roles or newly created ones. As environmental conditions change, the balance between formation of new species (speciation) and extinction of existing species determines the earth's biodiversity (**Concept 4-4A**). The existence of millions of species today means that speciation, on average, has kept ahead of extinction.

Extinction is a natural process. But much evidence indicates that humans have become a major force in the premature extinction of a growing number of species, as discussed further in Chapter 9.

4-5 What Is Species Diversity and Why Is It Important?

▶ **CONCEPT 4-5** Species diversity is a major component of biodiversity and tends to increase the sustainability of ecosystems.

Species Diversity Includes the Variety and Abundance of Species in a Particular Place

An important characteristic of a community and the ecosystem to which it belongs is its **species diversity:** the number of different species it contains (**species richness**) combined with the relative abundance of individuals within each of those species (**species evenness**).

For example, a biologically diverse community such as a tropical rain forest or a coral reef (Figure 4-10, left) with a large number of different species (high species richness) generally has only a few members of each

Reinhard Dirscherl/Bruce Coleman USA

Alan Majchrowicz/Peter Arnold

Figure 4-10 *Variations in species richness and species evenness.* A coral reef (left), with a large number of different species (high species richness), generally has only a few members of each species (low species evenness). In contrast, a grove of aspen trees in Alberta, Canada, in the fall (right) has a small number of different species (low species richness), but large numbers of individuals of each species (high species evenness).

Species Richness on Islands

In the 1960s, ecologists Robert MacArthur and Edward O. Wilson began studying communities on islands to discover why large islands tend to have more species of a certain category such as insects, birds, or ferns than do small islands.

To explain these differences in species richness among islands of varying sizes, MacArthur and Wilson carried out research and used their findings to propose what is called the **species equilibrium model,** or the **theory of island biogeography.** According to this widely accepted scientific theory, the number of different species (species richness) found on an island is determined by the interactions of two factors: the rate at which new species immigrate to the island and the rate at which species become *extinct,* or cease to exist, on the island.

The model projects that, at some point, the rates of species immigration and species extinction should balance so that neither rate is increasing or decreasing sharply. This balance point is the equilibrium point that determines the island's average number of different species (species richness) over time.

(The website **CengageNOW** has a great interactive animation of this model. Go to the end of any chapter for instructions on how to use it.)

According to the model, two features of an island affect the immigration and extinction rates of its species and thus its species diversity. One is the island's *size.* Small islands tend to have fewer species than large islands do because they make smaller targets for potential colonizers flying or floating toward them. Thus, they have lower immigration rates than larger islands do. In addition, a small island should have a higher extinction rate because it usually has fewer resources and less diverse habitats for its species.

A second factor is an island's *distance from the nearest mainland.* Suppose we have two islands about equal in size, extinction rates, and other factors. According to the model, the island closer to a mainland source of immigrant species should have the higher immigration rate and thus a higher species richness. The farther a potential colonizing species has to travel, the less likely it is to reach the island.

These factors interact to influence the relative species richness of different islands. Thus, larger islands closer to a mainland tend to have the most species, while smaller islands farther away from a mainland tend to have the fewest. Since MacArthur and Wilson presented their hypothesis and did their experiments, others have conducted more scientific studies that have born out their hypothesis, making it a widely accepted scientific theory.

Scientists have used this theory to study and make predictions about wildlife in *habitat islands*—areas of natural habitat, such as national parks and mountain ecosystems, surrounded by developed and fragmented land. These studies and predictions have helped scientists to preserve these ecosystems and protect their resident wildlife.

Critical Thinking

Suppose we have two national parks surrounded by development. One is a large park and the other is much smaller. Which park is likely to have the highest species richness? Why?

---

species (low species evenness). Biologist Terry Erwin found an estimated 1,700 different beetle species in a single tree in a tropical forest in Panama but only a few individuals of each species. On the other hand, an aspen forest community in Canada (Figure 4-10, right) may have only a few plant species (low species richness) but large numbers of each species (high species evenness).

The species diversity of communities varies with their *geographical location.* For most terrestrial plants and animals, species diversity (primarily species richness) is highest in the tropics and declines as we move from the equator toward the poles (see Figure 2, pp. S22–S23, in Supplement 4). The most species-rich environments are tropical rain forests, coral reefs (Figure 4-10, left), the ocean bottom zone, and large tropical lakes.

Scientists have sought to learn more about species richness by studying species on islands (Science Focus, above). Islands make good study areas because they are relatively isolated, and it is easier to observe species arriving and disappearing from islands than it would be to make such a study in other less isolated ecosystems.

CENGAGENOW™ Learn about how latitude affects species diversity and about the differences between big and small islands at CengageNOW.

Species-Rich Ecosystems Tend to Be Productive and Sustainable

How does species richness affect an ecosystem? In trying to answer this question, ecologists have been conducting research to answer two related questions: Is plant productivity higher in species-rich ecosystems? And does species richness enhance the *stability,* or sustainability of an ecosystem? Research suggests that the answers to both questions may be "yes" but more research is needed before these scientific hypotheses can be accepted as scientific theories.

According to the first hypothesis, the more diverse an ecosystem is, the more productive it will be. That is, with a greater variety of producer species, an ecosystem will produce more plant biomass, which in turn will support a greater variety of consumer species.

A related hypothesis is that greater species richness and productivity will make an ecosystem more stable or sustainable. In other words, the greater the species richness and the accompanying web of feeding and biotic interactions in an ecosystem, the greater its sustainability, or ability to withstand environmental disturbances such as drought or insect infestations. According to this hypothesis, a complex ecosystem with many different species (high species richness) and the resulting variety of feeding paths has more ways to respond to most environmental stresses because it does not have "all its eggs in one basket."

Many studies support the idea that some level of species richness and productivity can provide insurance against catastrophe. In one prominent 11-year study, David Tilman and his colleagues at the University of Minnesota found that communities with high plant species richness produced a certain amount of biomass more consistently than did communities with fewer species. The species-rich communities were also less affected by drought and more resistant to invasions by new insect species. Because of their higher level of biomass, the species-rich communities also consumed more carbon dioxide and took up more nitrogen, thus taking more robust roles in the carbon and nitrogen cycles (**Concept 3-5**, p. 65). Later laboratory studies involved setting up artificial ecosystems in growth chambers where key variables such as temperature, light, and atmospheric gas concentrations could be controlled and varied. These studies have supported Tilman's findings.

Ecologists hypothesize that in a species-rich ecosystem, each species can exploit a different portion of the resources available. For example, some plants will bloom early and others will bloom late. Some have shallow roots to absorb water and nutrients in shallow soils, and others use deeper roots to tap into deeper soils.

There is some debate among scientists about how much species richness is needed to help sustain various ecosystems. Some research suggests that the average annual net primary productivity of an ecosystem reaches a peak with 10–40 producer species. Many ecosystems contain more than 40 producer species, but do not necessarily produce more biomass or reach a higher level of stability. Scientists are still trying to determine how many producer species are needed to enhance the sustainability of particular ecosystems and which producer species are the most important in providing such stability.

Bottom line: species richness appears to increase the productivity and stability or sustainability of an ecosystem (**Concept 4-5**). While there may be some exceptions to this, most ecologists now accept it as a useful hypothesis.

RESEARCH FRONTIER

Learning more about how biodiversity is related to ecosystem stability and sustainability. See **academic.cengage.com/ biology/miller**.

4-6   What Roles Do Species Play in Ecosystems?

▶ **CONCEPT 4-6A** Each species plays a specific ecological role called its niche.

▶ **CONCEPT 4-6B** Any given species may play one or more of five important roles—native, nonnative, indicator, keystone, or foundation roles—in a particular ecosystem.

Each Species Plays a Unique Role in Its Ecosystem

An important principle of ecology is that *each species has a distinct role to play in the ecosystems where it is found* (**Concept 4-6A**). Scientists describe the role that a species plays in its ecosystem as its **ecological niche,** or simply **niche** (pronounced "nitch"). It is a species' way of life in a community and includes everything that affects its survival and reproduction, such as how much water and sunlight it needs, how much space it requires, and the temperatures it can tolerate. A species' niche should not be confused with its *habitat,* which is the place where it lives. Its niche is its pattern of living.

Scientists use the niches of species to classify them broadly as *generalists* or *specialists.* **Generalist species** have broad niches (Figure 4-11, right curve). They can live in many different places, eat a variety of foods, and

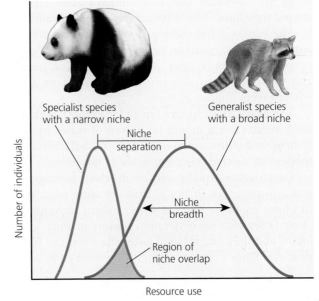

Figure 4-11
Specialist species such as the giant panda have a narrow niche (left) and generalist species such as a raccoon have a broad niche (right).

often tolerate a wide range of environmental conditions. Flies, cockroaches (see Case Study below), mice, rats, white-tailed deer, raccoons, and humans are generalist species.

In contrast, **specialist species** occupy narrow niches (Figure 4-11, left curve). They may be able to live in only one type of habitat, use one or a few types of food, or tolerate a narrow range of climatic and other environmental conditions. This makes specialists more prone to extinction when environmental conditions change.

For example, *tiger salamanders* breed only in fishless ponds where their larvae will not be eaten. China's *giant panda* (Figure 4-11, left) is highly endangered because of a combination of habitat loss, low birth rate, and its specialized diet consisting mostly of bamboo. Some shorebirds occupy specialized niches, feeding on crustaceans, insects, and other organisms on sandy beaches and their adjoining coastal wetlands (Figure 4-13).

Is it better to be a generalist or a specialist? It depends. When environmental conditions are fairly constant, as in a tropical rain forest, specialists have an advantage because they have fewer competitors. But under rapidly changing environmental conditions, the generalist usually is better off than the specialist.

THINKING ABOUT

The American Alligator's Niche

 CORE CASE STUDY

Does the American alligator (**Core Case Study**) have a specialist or a generalist niche? Explain.

■ CASE STUDY

Cockroaches: Nature's Ultimate Survivors

Cockroaches (Figure 4-12), the bugs many people love to hate, have been around for 350 million years, outliving the dinosaurs. One of evolution's great success stories, they have thrived because they are *generalists.*

The earth's 3,500 cockroach species can eat almost anything, including algae, dead insects, fingernail clippings, salts in tennis shoes, electrical cords, glue, paper, and soap. They can also live and breed almost anywhere except in polar regions.

Some cockroach species can go for a month without food, survive for a month on a drop of water from a dishrag, and withstand massive doses of radiation. One species can survive being frozen for 48 hours.

Cockroaches usually can evade their predators—and a human foot in hot pursuit—because most species have antennae that can detect minute movements of air. They also have vibration sensors in their knee joints, and they can respond faster than you can blink your eye. Some even have wings. They have compound eyes that allow them to see in almost all directions at once. Each eye has about 2,000 lenses, compared to one in each of your eyes.

And, perhaps most significantly, they have high reproductive rates. In only a year, a single Asian cockroach and its offspring can add about 10 million new cockroaches to the world. Their high reproductive rate also helps them to quickly develop genetic resistance to almost any poison we throw at them.

Most cockroaches sample food before it enters their mouths and learn to shun foul-tasting poisons. They also clean up after themselves by eating their own dead and, if food is scarce enough, their living.

About 25 species of cockroach live in homes and can carry viruses and bacteria that cause diseases. On the other hand, cockroaches play a role in nature's food webs. They make a tasty meal for birds and lizards.

Niches Can Be Occupied by Native and Nonnative Species

Niches can be classified further in terms of specific roles that certain species play within ecosystems. Ecologists describe *native, nonnative, indicator, keystone,* and *foundation species.* Any given species may play one or more of these five roles in a particular community (**Concept 4-6B**).

Native species are those species that normally live and thrive in a particular ecosystem. Other species that migrate into or are deliberately or accidentally introduced into an ecosystem are called **nonnative species,** also referred to as *invasive, alien,* or *exotic species.*

Figure 4-12 As generalists, cockroaches are among the earth's most adaptable and prolific species. This is a photo of an American cockroach.

Clemson University–USDA Cooperative Extension Slide Series

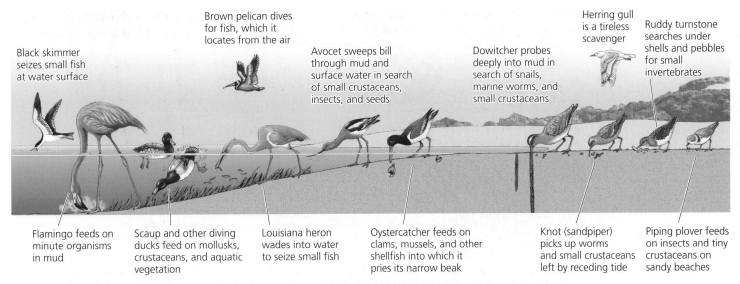

Figure 4-13 Specialized feeding niches of various bird species in a coastal wetland. This specialization reduces competition and allows sharing of limited resources.

Black skimmer seizes small fish at water surface

Brown pelican dives for fish, which it locates from the air

Avocet sweeps bill through mud and surface water in search of small crustaceans, insects, and seeds

Dowitcher probes deeply into mud in search of snails, marine worms, and small crustaceans

Herring gull is a tireless scavenger

Ruddy turnstone searches under shells and pebbles for small invertebrates

Flamingo feeds on minute organisms in mud

Scaup and other diving ducks feed on mollusks, crustaceans, and aquatic vegetation

Louisiana heron wades into water to seize small fish

Oystercatcher feeds on clams, mussels, and other shellfish into which it pries its narrow beak

Knot (sandpiper) picks up worms and small crustaceans left by receding tide

Piping plover feeds on insects and tiny crustaceans on sandy beaches

Some people tend to think of nonnative species as villains. In fact, most introduced and domesticated species of crops and animals, such as chickens, cattle, and fish from around the world, are beneficial to us. However, some nonnative species can threaten a community's native species and cause unintended and unexpected consequences. In 1957, for example, Brazil imported wild African bees to help increase honey production. Instead, the bees displaced domestic honeybees and reduced the honey supply.

Since then, these nonnative bee species—popularly known as "killer bees"—have moved northward into Central America and parts of the southwestern and southeastern United States. The wild African bees are not the fearsome killers portrayed in some horror movies, but they are aggressive and unpredictable. They have killed thousands of domesticated animals and an estimated 1,000 people in the western hemisphere, many of whom were allergic to bee stings.

Nonnative species can spread rapidly if they find a new more favorable niche. In their new niches, these species often do not face the predators and diseases they faced before, or they may be able to out-compete some native species in their new niches. We will examine this environmental threat in greater detail in Chapter 9.

Indicator Species Serve as Biological Smoke Alarms

Species that provide early warnings of damage to a community or an ecosystem are called **indicator species.** For example, the presence or absence of trout species in water at temperatures within their range of tolerance (Figure 3-10, p. 58) is an indicator of water quality because trout need clean water with high levels of dissolved oxygen.

Birds are excellent biological indicators because they are found almost everywhere and are affected quickly by environmental changes such as loss or fragmentation of their habitats and introduction of chemical pesticides. The populations of many bird species are declining. Butterflies are also good indicator species because their association with various plant species makes them vulnerable to habitat loss and fragmentation. Some amphibians are also classified as indicator species (Case Study below).

Using a living organism to monitor environmental quality is not new. Coal mining is a dangerous occupation, partly because of the underground presence of poisonous and explosive gases, many of which have no detectable odor. In the 1800s and early 1900s, coal miners took caged canaries into mines to act as early-warning sentinels. These birds sing loudly and often. If they quit singing for a long period and appeared to be distressed, miners took this as an indicator of the presence of poisonous or explosive gases and got out of the mine.

■ CASE STUDY
Why Are Amphibians Vanishing?

Amphibians (frogs, toads, and salamanders) live part of their lives in water and part on land. Populations of some amphibians, also believed to be indicator species, are declining throughout the world.

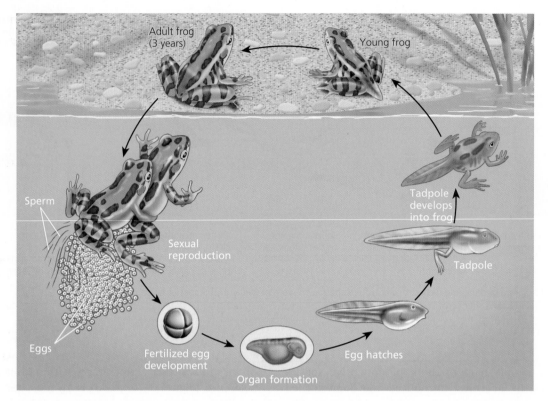

Figure 4-14 *Life cycle of a frog.* Populations of various frog species can decline because of the effects of harmful factors at different points in their life cycle. Such factors include habitat loss, drought, pollution, increased ultraviolet radiation, parasitism, disease, overhunting by humans, and nonnative predators and competitors.

Amphibians were the first vertebrates to set foot on the earth. Historically, they have been better than many other species have been at adapting to environmental changes through evolution. But many amphibian species apparently are having difficulty adapting to some of the rapid environmental changes that have taken place in the air and water and on the land during the past few decades—changes resulting mostly from human activities. Evolution takes time and some amphibians have traits that can make them vulnerable to certain changes in environmental conditions. Frogs, for example, are especially vulnerable to environmental disruption at various points in their life cycle (Figure 4-14).

As tadpoles, frogs live in water and eat plants; as adults, they live mostly on land and eat insects, which can expose them to pesticides. The eggs of frogs have no protective shells to block UV radiation or pollution. As adults, they take in water and air through their thin, permeable skins, which can readily absorb pollutants from water, air, or soil. During their life cycle, frogs and many other amphibian species also seek sunlight, which warms them and helps them to grow and develop, but which also increases their exposure to UV radiation.

Since 1980, populations of hundreds of the world's almost 6,000 amphibian species have been vanishing or declining in almost every part of the world, even in protected wildlife reserves and parks. According to the 2004 Global Amphibian Assessment, about 33% of all known amphibian species (and more than 80% of those in the Caribbean) are threatened with extinction, and populations of 43% of the species are declining.

No single cause has been identified to explain these amphibian declines. However, scientists have identified a number of factors that can affect frogs and other amphibians at various points in their life cycles:

- *Habitat loss and fragmentation,* especially from draining and filling of inland wetlands, deforestation, and urban development

- *Prolonged drought,* which can dry up breeding pools so that few tadpoles survive

- *Pollution,* especially exposure to pesticides, which can make frogs more vulnerable to bacterial, viral, and fungal diseases

- *Increases in UV radiation* caused by reductions in stratospheric ozone during the past few decades, caused by chemicals we have put into the air that have ended up in the stratosphere

- *Parasites* such as flatbed worms, which feed on the amphibian eggs laid in water, apparently have caused an increase in births of amphibians with missing or extra limbs

- *Viral and fungal diseases,* especially the chytrid fungus, which attacks the skin of frogs, apparently reducing their ability to take in water, which leads to death from dehydration. Such diseases spread when adults of many amphibian species congregate in large numbers to breed.

- *Climate change.* A 2005 study found an apparent correlation between global warming and the extinction of about two-thirds of the 110 known species of harlequin frog in tropical forests in Central and South America by creating favorable conditions for the spread of the deadly chytrid fungus to the frogs. But a 2008 study cast doubt on this hypothesis—another example of how science works. Climate change from global warming has also been identified as the primary cause of the extinction of the golden toad in Costa Rica (Figure 4-9).

- *Overhunting,* especially in Asia and France, where frog legs are a delicacy

- *Natural immigration of, or deliberate introduction of, nonnative predators and competitors* (such as certain fish species)

A combination of such factors probably is responsible for the decline or disappearance of most amphibian species.

┌─ **RESEARCH FRONTIER** ─────────────────┐
│ Learning more about why amphibians are disappearing │
│ and applying this knowledge to other threatened species. See │
│ **academic.cengage.com/biology/miller**. │
└──┘

Why should we care if some amphibian species become extinct? Scientists give three reasons. *First,* amphibians are sensitive biological indicators of changes in environmental conditions such as habitat loss and degradation, air and water pollution, exposure to ultraviolet light, and climate change. Their possible extinction suggests that environmental health is deteriorating in parts of the world.

Second, adult amphibians play important ecological roles in biological communities. For example, amphibians eat more insects (including mosquitoes) than do birds. In some habitats, extinction of certain amphibian species could lead to extinction of other species, such as reptiles, birds, aquatic insects, fish, mammals, and other amphibians that feed on them or their larvae.

Third, amphibians are a genetic storehouse of pharmaceutical products waiting to be discovered. Compounds in secretions from amphibian skin have been isolated and used as painkillers and antibiotics and as treatment for burns and heart disease.

The rapidly increasing global extinction of a variety of amphibian species is a warning about the harmful effects of an array of environmental threats to biodiversity. Like canaries in a coal mine, these indicator species are sending us urgent distress signals.

Keystone and Foundation Species Help Determine the Structure and Functions of Their Ecosystems

A keystone is the wedge-shaped stone placed at the top of a stone archway. Remove this stone and the arch collapses. In some communities and ecosystems, ecologists hypothesize that certain species play a similar role. **Keystone species** have a large effect on the types and abundances of other species in an ecosystem.

The effects that keystone species have in their ecosystems is often much larger than their numbers would suggest, and because of their relatively limited numbers, some keystone species are more vulnerable to extinction than others are. As was shown by the near extinction of the American alligator (**Core Case Study**) in the southeastern United States, eliminating a keystone species may dramatically alter the structure and function of a community.

Keystone species can play several critical roles in helping to sustain an ecosystem. One such role is *pollination* of flowering plant species by bees, butterflies (Figure 3-A, left, p. 54), hummingbirds, bats, and other species. In addition, *top predator* keystone species feed on and help regulate the populations of other species. Examples are the alligator, wolf, leopard, lion, and some shark species (see Case Study, p. 96).

Ecologist Robert Paine conducted a controlled experiment along the rocky Pacific coast of the U.S. state of Washington that demonstrated the keystone role of the top-predator sea star *Piaster orchaceus* in an intertidal zone community. Paine removed the mussel-eating *Piaster* sea stars from one rocky shoreline community but not from an adjacent community, which served as a control group. Mussels took over and crowded out most other species in the community without the *Piaster* sea stars.

The loss of a keystone species can lead to population crashes and extinctions of other species in a community that depends on it for certain services, as we saw in the **Core Case Study** that opens this chapter. This explains why it so important for scientists to identify and protect keystone species.

Another important type of species in some ecosystems is a **foundation species,** which plays a major role in shaping communities by creating and enhancing their habitats in ways that benefit other species. For example, elephants push over, break, or uproot trees, creating forest openings in the grasslands and woodlands of Africa. This promotes the growth of grasses and other forage plants that benefit smaller grazing species such as antelope. It also accelerates nutrient cycling rates.

Beavers are another good example of a foundation species. Acting as "ecological engineers," they build dams in streams to create ponds and other wetlands used by other species. Some bat and bird foundation species help to regenerate deforested areas and spread fruit plants by depositing plant seeds in their droppings.

Keystone and foundation species play similar roles. In general, the major difference between the two types of species is that foundation species help to create habitats and ecosystems. They often do this almost literally by providing the foundation for the ecosystem (as beavers do, for example). On the other hand, keystone species can do this and more. They sometimes play this foundation role (as do American alligators, for example), but they also play an active role in maintaining the ecosystem and keeping it functioning in a way that serves the other species living there. (Recall that the American alligator helps to keep the waters in its habitat clear of invading vegetation for use by other species that need open water.)

RESEARCH FRONTIER

Identifying and protecting keystone and foundation species. See **academic.cengage.com/biology/miller**.

THINKING ABOUT

The American Alligator

What species might disappear or suffer sharp population declines if the American alligator (**Core Case Study**) became extinct in subtropical wetland ecosystems?

 CORE CASE STUDY

■ **CASE STUDY**

Why Should We Protect Sharks?

The world's 370 shark species vary widely in size. The smallest is the dwarf dog shark, about the size of a large goldfish. The largest, the whale shark, can grow to 15 meters (50 feet) long and weigh as much as two full-grown African elephants.

Shark species that feed at or near the tops of food webs (Figure 3-14, p. 63) remove injured and sick animals from the ocean, and thus play an important ecological role. Without the services provided by these *keystone species,* the oceans would be teeming with dead and dying fish.

In addition to their important ecological roles, sharks could save human lives. If we can learn why they almost never get cancer, we could possibly use this information to fight cancer in our own species. Scientists are also studying their highly effective immune system, which allows wounds to heal without becoming infected.

Many people—influenced by movies, popular novels, and widespread media coverage of a fairly small number of shark attacks per year—think of sharks as

people-eating monsters. In reality, the three largest species—the whale shark, basking shark, and megamouth shark—are gentle giants. They swim through the water with their mouths open, filtering out and swallowing huge quantities of plankton.

Media coverage of shark attacks greatly distorts the danger from sharks. Every year, members of a few species—mostly great white, bull, tiger, gray reef, lemon, hammerhead, shortfin mako, and blue sharks—injure 60–100 people worldwide. Since 1990, sharks have killed an average of seven people per year. Most attacks involve great white sharks, which feed on sea lions and other marine mammals and sometimes mistake divers and surfers for their usual prey. Compare the risks: poverty prematurely kills about 11 million people a year, tobacco 5 million a year, and air pollution 3 million a year.

For every shark that injures a person, we kill at least 1 million sharks. Sharks are caught mostly for their valuable fins and then thrown back alive into the water, fins removed, to bleed to death or drown because they can no longer swim. The fins are widely used in Asia as a soup ingredient and as a pharmaceutical cure-all. A top (dorsal) fin from a large whale shark can fetch up to $10,000. In high-end restaurants in China, a bowl of shark fin soup can cost $100 or more. Ironically, shark fins have been found to contain dangerously high levels of toxic mercury.

Sharks are also killed for their livers, meat, hides, and jaws, and because we fear them. Some sharks die when they are trapped in nets or lines deployed to catch swordfish, tuna, shrimp, and other species. Sharks are especially vulnerable to overfishing because they grow slowly, mature late, and have only a few offspring per generation. Today, they are among the most vulnerable and least protected animals on earth.

In 2008, The IUCN–World Conservation Union reported that the populations of many large shark species have declined by half since the 1970s. Because of the increased demand for shark fins and meat, eleven of the world's open ocean shark species are considered critically endangered or endangered, and 81 species are threatened with extinction. In response to a public outcry over depletion of some species, the United States and several other countries have banned the hunting of sharks for their fins. But such bans apply only in territorial waters and are difficult to enforce.

Scientists call for banning shark finning in international waters and establishing a network of fully protected marine reserves to help protect coastal shark and other aquatic species from overfishing. Between 1970 and 2005, overfishing of hammerhead, bull, dusky, and other large predatory sharks in the northwest Atlantic for their fins and meat cut their numbers by 99%. In 2007, scientists Charles "Pete" Peterson and Julia Baum reported that this decline may be indirectly decimating the bay scallop fishery along the eastern coast of the United States. With fewer sharks around, populations of rays and skates, which sharks normally feed on, have exploded and are feasting on bay scallops in seagrass beds along the Atlantic coast.

Sharks have been around for more than 400 million years. Sustaining this portion of the earth's biodiversity begins with the knowledge that sharks may not need us, but that we and other species need them.

REVISITING   The American Alligator and Sustainability

The **Core Case Study** of the American alligator at the beginning of this chapter illustrates the power humans have over the environment—the power both to do harm and to make amends. As most American alligators were eliminated from their natural areas in the 1950s, scientists began pointing out the ecological benefits these animals had been providing to their ecosystems (such as building water holes, nesting mounds, and feeding sites for other species). Scientific understanding of these ecological connections led to protection of this species and to its recovery.

In this chapter, we studied the importance of biodiversity, especially the numbers and varieties of species found in different parts of the world (species richness), along with the other forms of biodiversity—functional, ecosystem, and genetic diversity. We also studied the process whereby all species came to be, according to scientific theory of biological evolution through natural selection. Taken together, these two great assets, biodiversity and evolution, represent irreplaceable natural capital. Each depends upon the other and upon whether humans can respect and preserve this natural capital. Finally, we examined the variety of roles played by species in ecosystems.

Ecosystems and the variety of species they contain are functioning examples of the four **scientific principles of sustainability** (see back cover) in action. They depend on solar energy and provide functional biodiversity in the form of energy flow and the chemical cycling of nutrients. In addition, ecosystems sustain biodiversity in all its forms, and population sizes are controlled by interactions among diverse species. In the next chapter, we delve further into this natural regulation of populations and the biodiversity of ecosystems.

All we have yet discovered is but a trifle in comparison with what lies hid in the great treasury of nature.

ANTOINE VAN LEEUWENHOEK

REVIEW

1. Review the Key Questions and Concepts for this chapter on p. 78. Explain why we should protect the American alligator (**Core Case Study**) from being driven to extinction as a result of our activities.

2. What are the four major components of **biodiversity (biological diversity)?** What is the importance of biodiversity?

3. What is **biological evolution?** What is **natural selection?** What is a **fossil** and why are fossils important in understanding biological evolution? What is a **mutation** and what role do mutations play in evolution by natural selection? What is an **adaptation (adaptive trait)?** What is **differential reproduction?** How did we become such a powerful species?

4. What are two limits to evolution by natural selection? What are three myths about evolution through natural selection?

5. Describe how geologic processes and climate change can affect natural selection. Describe conditions on the earth that favor the development of life as we know it.

6. What is **speciation?** Distinguish between **geographic isolation** and **reproductive isolation** and explain how they can lead to the formation of a new species. Distinguish between **artificial selection** and **genetic engineering (gene splicing)** and give an example of each. What are some possible social, ethical, and environmental problems with the widespread use of genetic

engineering? What is **extinction?** What is an **endemic species** and why is it vulnerable to extinction? Distinguish between **background extinction** and **mass extinction.**

7. What is **species diversity?** Distinguish between **species richness** and **species evenness** and give an example of each. Describe the **theory of island biogeography (species equilibrium model).** Explain why species-rich ecosystems tend to be productive and sustainable.

8. What is an **ecological niche?** Distinguish between **specialist species** and **generalist species** and give an example of each.

9. Distinguish among **native, nonnative, indicator, keystone,** and **foundation** species and give an example of each type. Explain why birds are excellent indicator species. Why are amphibians vanishing and why should we protect them? Why should we protect shark species from being driven to extinction as a result of our activities? Describe the role of the beaver as a foundation species.

10. Explain how the role of the American alligator in its ecosystem (**Core Case Study**) illustrates the biodiversity **principles of sustainability?**

Note: Key Terms are in **bold** type.

CRITICAL THINKING

1. List three ways in which you could apply **Concept 4-4B** in order to live a more environmentally sustainable lifestyle.

2. Explain what could happen to the ecosystem where American alligators (**Core Case Study**) live if the alligators went extinct. Name a plant species and an animal species that would be seriously affected, and describe how each might respond to these changes in their environmental conditions.

3. What role does each of the following processes play in helping implement the four **scientific principles of sustainability** (see back cover): **(a)** natural selection, **(b)** speciation, and **(c)** extinction?

4. Describe the major differences between the ecological niches of humans and cockroaches. Are these two species in competition? If so, how do they manage to coexist?

5. How would you experimentally determine whether an organism is a keystone species?

6. Is the human species a keystone species? Explain. If humans were to become extinct, what are three species that might also become extinct and three species whose populations would probably grow?

7. How would you respond to someone who tells you:
   a. that he or she does not believe in biological evolution because it is "just a theory"?
   b. that we should not worry about air pollution because natural selection will enable humans to develop lungs that can detoxify pollutants?

8. How would you respond to someone who says that because extinction is a natural process, we should not worry about the loss of biodiversity when species become prematurely extinct as a result of our activities?

9. Congratulations! You are in charge of the future evolution of life on the earth. What are the three most important things you would do?

10. List two questions that you would like to have answered as a result of reading this chapter.

Note: See Supplement 13 (p. S78) for a list of Projects related to this chapter.

DATA ANALYSIS

Injuries and deaths from shark attacks are highly publicized by the media. However, the risk of injury or death from a shark attack for people going into coastal waters as swimmers, surfers, or divers is extremely small (see Case Study, p. 96). For example, according to the National Safety Council, the Centers for Disease Control and Prevention, and the International Shark Attack File, the estimated lifetime risk of dying from a shark attack in the United States is about 1 in 3,750,000 compared to risks of 1 in 1,130 from drowning, 1 in 218 from a fall, 1 in 84 from a car accident, 1 in 63 from the flu, and 1 in 38 from a hospital infection.

Between 1998 and 2007, the United States had the world's highest percentage of deaths and injuries from unprovoked shark attacks, and the U.S. state of Florida had the country's highest percentage of deaths and injuries from unprovoked shark attacks, as shown by the following data about shark attacks in the world, in the United States, and in Florida.

Year	World			United States			Florida		
	Total Attacks	Fatal Attacks	Non-fatal Attacks	Total Attacks	Fatal Attacks	Non-fatal Attacks	Total Attacks	Fatal Attacks	Non-fatal Attacks
1998	51	6	45	26	1	25	21	1	20
1999	56	4	52	38	0	38	26	0	26
2000	79	11	68	53	1	52	37	1	36
2001	68	4	64	50	3	47	34	1	33
2002	62	3	59	47	0	47	29	0	29
2003	57	4	53	40	1	39	30	0	30
2004	65	7	58	30	2	28	12	0	12
2005	61	4	57	40	1	39	20	1	19
2006	63	4	59	40	0	40	23	0	23
2007	71	1	70	50	0	50	32	0	32

Source: Data from *International Shark Attack File*, Florida Museum of Natural History, University of Florida.

1. What is the average number for each of the nine columns of data—unprovoked shark attacks, deaths, and non-fatal injuries between 1998 and 2007 for the world, the United States, and Florida?

2. What percentage of the world's average annual unprovoked shark attacks between 1998 and 2007 occurred in (a) the United States and (b) Florida?

2. What percentage of the average annual unprovoked shark attacks in the United States between 1998 and 2007 occurred in Florida?

LEARNING ONLINE

Log on to the Student Companion Site for this book at **academic.cengage.com/biology/miller**, and choose Chapter 4 for many study aids and ideas for further reading and research. These include flash cards, practice quizzing, Weblinks, information on Green Careers, and InfoTrac® College Edition articles.

5 Biodiversity, Species Interactions, and Population Control

Southern Sea Otters: Are They Back from the Brink of Extinction?

Southern sea otters (Figure 5-1, top left) live in giant kelp forests (Figure 5-1, right) in shallow waters along part of the Pacific coast of North America. Most remaining members of this endangered species are found between the U.S. state of California's coastal cities of Santa Cruz and Santa Barbara.

Southern sea otters are fast and agile swimmers that dive to the ocean bottom looking for shellfish and other prey. These tool-using marine mammals use stones to pry shellfish off rocks under water. When they return to the surface they break open the shells while swimming on their backs, using their bellies as a table (Figure 5-1, top left). Each day a sea otter consumes about a fourth of its weight in clams, mussels, crabs, sea urchins, abalone, and about 40 other species of bottom-dwelling organisms.

Historically, between 16,000 and 17,000 southern sea otters are believed to have populated the waters along their habitat area of the California coast before fur traders began killing them for their thick, luxurious fur. For that reason, and because the otters competed with humans for valuable abalone and other shellfish, the species was hunted almost to extinction in this region by the early 1900s.

However, between 1938 and 2007 the population of southern sea otters off California's coast increased from about 50 to almost 3,026. This partial recovery was helped when, in 1977, the U.S. Fish and Wildlife Service declared the species endangered in most of its range. But this species has a long way to go before its population increases enough to allow removing it from the endangered species list.

Why should we care about this species? One reason is that people love to look at these charismatic, cute, and cuddly animals as they play in the water. As a result, they help to generate millions of dollars a year in tourism income in coastal areas where they are found. Another reason is *ethical*. Some people believe it is wrong to cause the premature extinction of any species.

A third reason to care about otters—and a key reason in our study of environmental science—is that biologists classify them as a *keystone species* (p. 95), which play an important ecological role through its interactions with other species. The otters help to keep sea urchins and other kelp-eating species from depleting highly productive and rapidly growing kelp forests, which provide habitats for a number of species in offshore coastal waters, as discussed in more detail later in this chapter. Without southern sea otters, sea urchins would probably destroy the kelp forests and much of the rich biodiversity associated with them.

Biodiversity, an important part of the earth's natural capital, is the focus of one of the four **scientific principles of sustainability** (see back cover).

One of its components is species diversity (Figure 4-2, p. 79), which is affected by how species interact with one another and, in the process, help control each others' population sizes.

Figure 5-1 An endangered southern sea otter in Monterey Bay, California (USA), uses a stone to crack the shell of a clam (top left). It lives in a giant kelp bed near San Clemente Island, California (right). Scientific studies indicate that the otters act as a keystone species in a kelp forest system by helping to control the populations of sea urchins and other kelp-eating species.

Tom and Pat Leeson, Ardea London Ltd

Bruce Coleman USA

5-1 How do species interact?

CONCEPT 5-1 Five types of species interactions—competition, predation, parasitism, mutualism, and commensalism—affect the resource use and population sizes of the species in an ecosystem.

5-2 How can natural selection reduce competition between species?

CONCEPT 5-2 Some species develop adaptations that allow them to reduce or avoid competition with other species for resources.

5-3 What limits the growth of populations?

CONCEPT 5-3 No population can continue to grow indefinitely because of limitations on resources and because of competition among species for those resources.

5-4 How do communities and ecosystems respond to changing environmental conditions?

CONCEPT 5-4 The structure and species composition of communities and ecosystems change in response to changing environmental conditions through a process called ecological succession.

Note: Supplements 2 (p. S4), 4 (p. S20), 5 (p. S31), 6 (p. S39), and 13 (p. S78) can be used with this chapter.

In looking at nature, never forget that every single organic being around us may be said to be striving to increase its numbers.

CHARLES DARWIN, 1859

5-1 How Do Species Interact?

> **CONCEPT 5-1** Five types of species interactions—competition, predation, parasitism, mutualism, and commensalism—affect the resource use and population sizes of the species in an ecosystem.

Species Interact in Five Major Ways

Ecologists identify five basic types of interactions between species that share limited resources such as food, shelter, and space:

- **Interspecific competition** occurs when members of two or more species interact to gain access to the same limited resources such as food, light, or space.

- **Predation** occurs when a member of one species (the *predator*) feeds directly on all or part of a member of another species (the *prey*).

- **Parasitism** occurs when one organism (the *parasite*) feeds on the body of, or the energy used by, another organism (the *host*), usually by living on or in the host.

- **Mutualism** is an interaction that benefits both species by providing each with food, shelter, or some other resource.

- **Commensalism** is an interaction that benefits one species but has little, if any, effect on the other.

These interactions have significant effects on the resource use and population sizes of the species in an ecosystem (**Concept 5-1**). Interactions that help to limit population size illustrate one of the four **scientific principles of sustainability** (see back cover). These interactions also influence the abilities of the interacting species to survive and reproduce; thus the interactions serve as agents of natural selection (**Concept 4-2B**, p. 80).

Most Species Compete with One Another for Certain Resources

The most common interaction between species is *competition* for limited resources. While fighting for resources does occur, most competition involves the ability of one species to become more efficient than another species in acquiring food or other resources.

Recall that each species plays a unique role in its ecosystem called its *ecological niche* (p. 91). Some species are generalists with broad niches and some are specialists with narrow niches. When two species compete with one another for the same resources such as food, light, or space, their niches overlap (Figure 4-11, p. 91).

Links: **CORE CASE STUDY** refers to the Core Case Study. ❀ **SUSTAINABILITY** refers to the book's sustainability theme. ↰ **CONCEPT LINK** indicates links to key concepts in earlier chapters.

The greater this overlap the more intense their competition for key resources.

Although different species may share some aspects of their niches, no two species can occupy exactly the same ecological niche for very long—a concept known as the *competitive exclusion principle*. When there is intense competition between two species for the same resources, both species suffer harm by having reduced access to important resources. If one species can take over the largest share of one or more key resources, the other competing species must migrate to another area (if possible), shift its feeding habits or behavior through natural selection to reduce or alter its niche, suffer a sharp population decline, or become extinct in that area.

Humans compete with many other species for space, food, and other resources. As our ecological footprints grow and spread (Figure 1-10, p. 15) and we convert more of the earth's land, aquatic resources, and net primary productivity (Figure 3-16, p. 64) to our uses, we are taking over the habitats of many other species and depriving them of resources they need to survive.

THINKING ABOUT

Humans and the Southern Sea Otter

CORE CASE STUDY

What human activities have interfered with the ecological niche of the southern sea otter (**Core Case Study**)?

Most Consumer Species Feed on Live Organisms of Other Species

All organisms must have a source of food to survive. Recall that members of producer species, such as plants and floating phytoplankton, make their own food, mostly through photosynthesis (p. 58). Other species are consumers that interact with some species by feeding on them. Some consumers feed on live individuals of other species. They include herbivores that feed on plants, carnivores that feed on the flesh of other animals, and omnivores that feed on plants and animals. Other consumers, such as detritus feeders and decomposers, feed on the wastes or dead bodies of organisms.

In **predation,** a member of one species (the **predator**) feeds directly on all or part of a living organism of another plant or animal species (the **prey**) as part of a food web (**Concept 3-4A**, p. 61). Together, the two different species, such as lions (the predator or hunter) and zebras (the prey or hunted), form a **predator–prey relationship.** Such relationships are shown in Figures 3-13 (p. 62) and 3-14 (p. 63).

CONCEPT LINK

Herbivores, carnivores, and omnivores are predators. However, detritus feeders and decomposers, while they do feed on other organisms after they have died, are not considered predators because they do not feed on live organisms.

Sometimes predator–prey relationships can surprise us. During the summer months, the grizzly bears of the Greater Yellowstone ecosystem in the western United States eat huge amounts of army cutworm moths, which huddle in masses high on remote mountain slopes. One grizzly bear can dig out and lap up as many as 40,000 of these moths in a day. Consisting of 50–70% fat, the moths offer a nutrient that the bear can store in its fatty tissues and draw on during its winter hibernation.

In giant kelp forest ecosystems, sea urchins prey on giant kelp, a form of seaweed (**Core Case Study**, Figure 5-1, right). However, as keystone species, southern sea otters (Figure 5-1, top left) prey on the sea urchins and help to keep them from destroying the kelp forest ecosystems (Science Focus, p. 104).

CORE CASE STUDY

Predators have a variety of methods that help them capture prey. *Herbivores* can simply walk, swim, or fly up to the plants they feed on. For example, sea urchins (Science Focus, Figure 5-A, p. 104) can move along the ocean bottom to feed on the base of giant kelp plants. *Carnivores* feeding on mobile prey have two main options: *pursuit* and *ambush.* Some, such as the cheetah, catch prey by running fast; others, such as the American bald eagle, can fly and have keen eyesight; still others, such as wolves and African lions, cooperate in capturing their prey by hunting in packs.

Other predators use *camouflage* to hide in plain sight and ambush their prey. For example, praying mantises (Figure 3-A, right, p. 54) sit in flowers of a similar color and ambush visiting insects. White ermines (a type of weasel) and snowy owls hunt in snow-covered areas. People camouflage themselves to hunt wild game and use camouflaged traps to ambush wild game.

Some predators use *chemical warfare* to attack their prey. For example, spiders and poisonous snakes use venom to paralyze their prey and to deter their predators.

Prey species have evolved many ways to avoid predators, including abilities to run, swim, or fly fast, and highly developed senses of sight or smell that alert them to the presence of predators. Other avoidance adaptations include protective shells (as on armadillos and turtles), thick bark (giant sequoia), spines (porcupines), and thorns (cacti and rosebushes). Many lizards have brightly colored tails that break off when they are attacked, often giving them enough time to escape.

Other prey species use the camouflage of certain shapes or colors or the ability to change color (chameleons and cuttlefish). Some insect species have shapes that make them look like twigs (Figure 5-2a), bark, thorns, or even bird droppings on leaves. A leaf insect can be almost invisible against its background (Figure 5-2b), as can an arctic hare in its white winter fur.

Chemical warfare is another common strategy. Some prey species discourage predators with chemicals that are *poisonous* (oleander plants), *irritating* (stinging nettles and bombardier beetles, Figure 5-2c), *foul smelling* (skunks, skunk cabbages, and stinkbugs), or *bad tast-*

ing (buttercups and monarch butterflies, Figure 5-2d). When attacked, some species of squid and octopus emit clouds of black ink, allowing them to escape by confusing their predators.

Many bad-tasting, bad-smelling, toxic, or stinging prey species have evolved *warning coloration,* brightly colored advertising that enables experienced predators to recognize and avoid them. They flash a warning: "Eating me is risky." Examples are brilliantly colored poisonous frogs (Figure 5-2e); and foul-tasting monarch butterflies (Figure 5-2d). For example, when a bird such as a blue jay eats a monarch butterfly it usually vomits and learns to avoid them.

Biologist Edward O. Wilson gives us two rules, based on coloration, for evaluating possible danger from an unknown animal species we encounter in nature. *First,* if it is small and strikingly beautiful, it is probably poisonous. *Second,* if it is strikingly beautiful and easy to catch, it is probably deadly.

Some butterfly species, such as the nonpoisonous viceroy (Figure 5-2f), gain protection by looking and acting like the monarch, a protective device known as *mimicry.* Other prey species use *behavioral strategies* to avoid predation. Some attempt to scare off predators by puffing up (blowfish), spreading their wings (peacocks), or mimicking a predator (Figure 5-2h). Some moths have wings that look like the eyes of much larger animals (Figure 5-2g). Other prey species gain some protection by living in large groups such as schools of fish and herds of antelope.

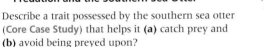

THINKING ABOUT

Predation and the Southern Sea Otter

CORE CASE STUDY

Describe a trait possessed by the southern sea otter (**Core Case Study**) that helps it **(a)** catch prey and **(b)** avoid being preyed upon?

At the individual level, members of the predator species benefit and members of the prey species are harmed. At the population level, predation plays a role in evolution by natural selection (**Concept 4-2B**, p. 80). Animal predators, for example, tend to kill the sick, weak, aged, and least fit members of a population because they are the easiest to catch. This leaves behind individuals with better defenses against predation. Such individuals tend to survive longer and leave more offspring with adaptations that can help them avoid predation. Thus, predation can help increase biodiversity by promoting natural selection in which species evolve with the ability to share limited resources by reducing their niche overlap.

Some people tend to view certain animal predators with contempt. When a hawk tries to capture and feed on a rabbit, some root for the rabbit. Yet the hawk, like all predators, is merely trying to get enough food for itself and its young. In doing so, it plays an important ecological role in controlling rabbit populations.

(a) Span worm

(b) Wandering leaf insect

(c) Bombardier beetle

(d) Foul-tasting monarch butterfly

(e) Poison dart frog

(f) Viceroy butterfly mimics monarch butterfly

(g) Hind wings of Io moth resemble eyes of a much larger animal.

(h) When touched, snake caterpillar changes shape to look like head of snake.

Figure 5-2 Some ways in which prey species avoid their predators: **(a, b)** *camouflage,* **(c–e)** *chemical warfare,* **(d, e)** *warning coloration,* **(f)** *mimicry,* **(g)** *deceptive looks,* and **(h)** *deceptive behavior.*

Predator and Prey Species Can Drive Each Other's Evolution

To survive, predators must eat and prey must avoid becoming a meal. As a result, predator and prey populations exert intense natural selection pressures on one another. Over time, as prey develop traits that make them more difficult to catch, predators face selection

SCIENCE FOCUS

Why Should We Care about Kelp Forests?

A kelp forest is a forest of seaweed called giant kelp whose large blades grow straight to the surface (Figure 5-1, right) (**Core Case Study**). The dependence of these plants on photosynthesis restricts their growth to cold, nutrient-rich, and fairly shallow coastal waters, which are found in various areas of the world, such as off the coast of northern California (USA).

Giant kelp is one of the world's fastest growing plants. Under good conditions, its blades can grow 0.6 meter (2 feet) a day. Each blade is held up by a gas-filled bladder at its base. The blades are very flexible and can survive all but the most violent storms and waves.

Kelp forests are one of the most biologically diverse ecosystems found in marine waters, supporting large numbers of marine plants and animals. These forests help reduce shore erosion by blunting the force of incoming waves and helping to trap outgoing sand. People harvest kelp as a renewable resource, extracting a substance called algin from its blades. We use this substance in toothpaste, cosmetics, ice cream, and hundreds of other products.

Sea urchins and pollution are major threats to kelp forests. Large populations

Figure 5-A Purple sea urchin in coastal waters of the U.S. state of California.

of sea urchins (Figure 5-A) can rapidly devastate a kelp forest because they eat the base of young kelp. Male southern sea otters, a keystone species, help to control populations of sea urchins. An adult male southern sea otter (Figure 5-1, top left) can eat up to 50 sea urchins a day—equivalent to a 68-kilogram (150-pound) person eating 160 quarter-pound hamburgers a day. Without southern sea otters, giant kelp for-

est ecosystems would collapse and reduce aquatic biodiversity.

A second threat to kelp forests is polluted water running off of the land into the coastal waters where kelp forests grow. The pollutants in this runoff include pesticides and herbicides, which can kill kelp plants and other kelp forest species and upset the food webs in these forests. Another runoff pollutant is fertilizer whose plant nutrients (mostly nitrates) can cause excessive growth of algae and other plants. These growths block some of the sunlight needed to support the growth of giant kelp, and thus upset these aquatic ecosystems.

A third looming threat is global warming. Giant kelp forests require fairly cool water, where the temperature stays between 28–36 °C (50–65 °F). If coastal waters warm up as projected during this century, many—perhaps most—of the world's giant kelp forests will disappear and along with them the southern sea otter and many other species.

Critical Thinking

What are three ways to protect giant kelp forests and southern sea otters?

---

pressures that favor traits that increase their ability to catch prey. Then prey must get better at eluding the more effective predators.

When populations of two different species interact in this way over a such long period of time, changes in the gene pool of one species can lead to changes in the gene pool of the other species. Such changes can help both sides to become more competitive or can help to avoid or reduce competition. Biologists call this process **coevolution.**

Consider the species interaction between bats (the predator) and certain species of moths (the prey). Bats like to eat moths, and they hunt at night (Figure 5-3) and use echolocation to navigate and to locate their prey, emitting pulses of extremely high-frequency and high-intensity sound. They capture and analyze the returning echoes and create a sonic "image" of their prey. (We have copied this natural technology by using sonar to detect submarines, whales, and schools of fish.)

As a countermeasure to this effective prey-detection system, certain moth species have evolved ears that are especially sensitive to the sound frequencies that bats use to find them. When the moths hear the bat fre-

quencies, they try to escape by dropping to the ground or flying evasively.

Some bat species evolved ways to counter this defense by changing the frequency of their sound pulses. In turn, some moths have evolved their own high-frequency clicks to jam the bats' echolocation systems. Some bat species then adapted by turning off their echolocation systems and using the moths' clicks to locate their prey.

Coevolution is like an arms race between interacting populations of different species. Sometimes the predators surge ahead; at other times the prey get the upper hand. Coevolution is one of nature's ways of maintaining long-term sustainability through population control (see back cover and **Concept 1-6**, p. 23), and it can promote biodiversity by increasing species diversity.

However, we should not think of coevolution as a process with which species can design strategies to increase their survival chances. Instead, it is a prime example of populations responding to changes in environmental conditions as part of the process of evolution by natural selection. And, unlike a human arms race, each step in this process takes hundreds to thousands of years.

Some Species Feed off Other Species by Living on or in Them

Parasitism occurs when one species (the *parasite*) feeds on the body of, or the energy used by, another organism (the *host*), usually by living on or in the host. In this relationship, the parasite benefits and the host is harmed but not immediately killed.

Unlike the typical predator, a parasite usually is much smaller than its host and rarely kills its host. Also, most parasites remain closely associated with their hosts, draw nourishment from them, and may gradually weaken them over time.

Some parasites, such as tapeworms and some disease-causing microorganisms (pathogens), live *inside* their hosts. Other parasites attach themselves to the *outsides* of their hosts. Examples of the latter include mosquitoes, mistletoe plants (Figure 5-4, left), and sea lampreys, which use their sucker-like mouths to attach themselves to fish and feed on their blood (Figure 5-4, right). Some parasites move from one host to another, as fleas and ticks do; others, such as tapeworms, spend their adult lives with a single host.

Some parasites have little contact with their hosts. For example, North American cowbirds take over the nests of other birds by laying their eggs in them and then letting the host birds raise their young.

From the host's point of view, parasites are harmful. But at the population level, parasites can promote biodiversity by increasing species richness, and they help to keep their hosts' populations in check.

Like predator–prey interactions, parasite–host interactions can lead to coevolutionary change. For example, malaria is caused by a parasite spread by the bites of a certain mosquito species. The parasite in-

ullstein-Nill/Peter Arnold, Inc.

Figure 5-3 *Coevolution.* A Langohrfledermaus bat hunting a moth. Long-term interactions between bats and their prey such as moths and butterflies can lead to coevolution, as the bats evolve traits that increase their chances of getting a meal and the moths evolve traits that help them avoid being eaten.

vades red blood cells, which are destroyed every few days when they are swept into the spleen. However, through coevolution, the malaria parasite developed an adaptation that keeps it from being swept into the spleen. The parasite produces a sticky protein nodule that attaches the cell it has infected to the wall of a blood vessel.

However, the body's immune system detects the foreign protein on the blood vessel wall and sends antibodies to attack it. Through coevolution, the malaria parasite has in turn developed a defense against this attack. It produces thousands of different versions of the sticky protein that keep it attached to the blood vessel wall. By the time the immune system recognizes and attacks one type of the protein, the parasite has switched to another type.

PhotoAlto/SuperStock

U.S. Fish and Wildlife Service Photo

Figure 5-4 *Parasitism:* **(a)** Healthy tree on the left and an unhealthy one on the right, which is infested with parasitic mistletoe. **(b)** Blood-sucking parasitic sea lampreys attached to an adult lake trout from the Great Lakes (USA).

Figure 5-5 Examples of *mutualism.* **(a)** Oxpeckers (or tickbirds) feed on parasitic ticks that infest large, thick-skinned animals such as the endangered black rhinoceros. **(b)** A clownfish gains protection and food by living among deadly stinging sea anemones and helps protect the anemones from some of their predators. (Oxpeckers and black rhinoceros: Joe McDonald/Tom Stack & Associates; clownfish an sea anemone: Fred Bavendam/Peter Arnold, Inc.)

(a) Oxpeckers and black rhinoceros

(b) Clownfish and sea anemone

In Some Interactions, Both Species Benefit

In **mutualism,** two species behave in ways that benefit both by providing each with food, shelter, or some other resource. For example, honeybees, caterpillars, butterflies, and other insects feed on a male flower's nectar, picking up pollen in the process, and then pollinating female flowers when they feed on them.

Figure 5-5 shows two examples of mutualistic relationships that combine *nutrition* and *protection.* One involves birds that ride on the backs of large animals like African buffalo, elephants, and rhinoceroses (Figure 5-5a). The birds remove and eat parasites and pests (such as ticks and flies) from the animal's body and often make noises warning the larger animals when predators approach.

A second example involves the clownfish species (Figure 5-5b), which live within sea anemones, whose tentacles sting and paralyze most fish that touch them. The clownfish, which are not harmed by the tentacles, gain protection from predators and feed on the detritus left from the anemones' meals. The sea anemones benefit because the clownfish protect them from some of their predators.

In *gut inhabitant mutualism,* vast armies of bacteria in the digestive systems of animals help to break down (digest) their hosts' food. In turn, the bacteria receive a sheltered habitat and food from their host. Hundreds of millions of bacteria in your gut secrete enzymes that help digest the food you eat. Cows and termites are able to digest the cellulose in plant tissues they eat because of the large number of microorganisms, mostly bacteria, that live in their guts.

It is tempting to think of mutualism as an example of cooperation between species. In reality, each species benefits by unintentionally exploiting the other as a result of traits they obtained through natural selection.

In Some Interactions, One Species Benefits and the Other Is Not Harmed

Commensalism is an interaction that benefits one species but has little, if any, effect on the other. For example, in tropical forests certain kinds of silverfish insects move along with columns of army ants to share the food obtained by the ants in their raids. The army ants receive no apparent harm or benefit from the silverfish.

Another example involves plants called *epiphytes* (such as certain types of orchids and bromeliads), which attach themselves to the trunks or branches of large trees in tropical and subtropical forests (Figure 5-6). These *air plants* benefit by having a solid base on which to grow. They also live in an elevated spot that gives them better access to sunlight, water from the humid air and rain, and nutrients falling from the tree's upper leaves and limbs. Their presence apparently does not harm the tree.

Figure 5-6 In an example of *commensalism,* this bromeliad—an epiphyte, or air plant, in Brazil's Atlantic tropical rain forest—roots on the trunk of a tree, rather than in soil, without penetrating or harming the tree. In this interaction, the epiphyte gains access to water, other nutrient debris, and sunlight; the tree apparently remains unharmed.

Luiz C. Marigo/Peter Arnold, Inc.

CENGAGENOW™ Review the way in which species can interact and see the results of an experiment on species interaction at CengageNOW™.

5-2 How Can Natural Selection Reduce Competition between Species?

▶ **CONCEPT 5-2** Some species develop adaptations that allow them to reduce or avoid competition with other species for resources.

Some Species Evolve Ways to Share Resources

Over a time scale long enough for natural selection to occur, populations of some species competing for the same resources develop adaptations through natural selection that allow them to reduce or avoid such competition (**Concept 5-2**). In other words, some species evolve to reduce niche overlap. One way this happens is through **resource partitioning.** It occurs when species competing for similar scarce resources evolve specialized traits that allow them to use shared resources at different times, in different ways, or in different places. For example, through natural selection, the fairly broad and overlapping niches of two competing species (Figure 5-7, top) can reduce their niche overlap by becoming more specialized (Figure 5-7, bottom).

Figure 5-8 shows resource partitioning by some insect-eating bird species. In this case, their adaptations allow them to reduce competition by feeding in different portions of the same tree species and by feeding on different insect species. Figure 4-13 (p. 93) shows how bird species in a coastal wetland have

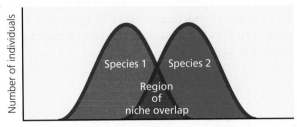

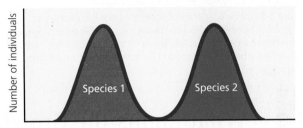

Figure 5-7 Competing species can evolve to reduce niche overlap. The top diagram shows the overlapping niches of two competing species. The bottom diagram shows that through natural selection, the niches of the two species become separated and more specialized (narrower) as the species develop adaptations that allow them to avoid or reduce competition for the same resources.

Figure 5-8 *Sharing the wealth: resource partitioning* of five species of insect-eating warblers in the spruce forests of the U.S. state of Maine. Each species minimizes competition for food with the others by spending at least half its feeding time in a distinct portion (shaded areas) of the spruce trees, and by consuming different insect species. (After R. H. MacArthur, "Population Ecology of Some Warblers in Northeastern Coniferous Forests," *Ecology* 36 (1958): 533–536.)

evolved specialized feeding niches, thereby reducing their competition for resources.

Another example of resource partitioning through natural selection involves birds called honeycreepers that live in the U. S. state of Hawaii. Long ago these birds started from a single ancestor species. But because of evolution by natural selection, there are now numerous honeycreeper species. Each has a different type of beak specialized to feed on certain food sources, such as specific types of insects, nectar from particular types of flowers, and certain types of seeds and fruit (Figure 5-9). This is an example of a process called *evolutionary divergence*.

Figure 5-9 *Specialist species of honeycreepers.* Evolutionary divergence of honeycreepers into species with specialized ecological niches has reduced competition between these species. Each species has evolved a beak specialized to take advantage of certain types of food resources.

5-3   What Limits the Growth of Populations?

▶ **CONCEPT 5-3** No population can continue to grow indefinitely because of limitations on resources and because of competition among species for those resources.

Populations Have Certain Characteristics

Populations differ in factors such as their *distribution, numbers, age structure* (proportions of individuals in different age groups), and *density* (number of individuals in a certain space). **Population dynamics** is a study of how these characteristics of populations change in response to changes in environmental conditions. Examples of such conditions are temperature, presence of disease organisms or harmful chemicals, resource availability, and arrival or disappearance of competing species.

Studying the population dynamics of southern sea otter populations (**Core Case Study**) and their interactions with other species has helped us understand the ecological importance of this keystone species. Let's look at some of the characteristics of populations in more detail.

Most Populations Live Together in Clumps or Patches

Let's begin our study of population dynamics with how individuals in populations are distributed or dispersed within a particular area or volume. Three general patterns of *population distribution* or *dispersion* in a habitat are *clumping, uniform dispersion,* and *random dispersion* (Figure 5-10).

In most populations, individuals of a species live together in clumps or patches (Figure 5-10a). Examples are patches of desert vegetation around springs, cottonwood trees clustered along streams, wolf packs, flocks of birds, and schools of fish. The locations and sizes of these clumps and patches vary with the availability of resources. Southern sea otters (Figure 5-1, top left), for example, are usually found in groups known as rafts or pods ranging in size from a few to several hundred animals.

Why clumping? Several reasons: *First,* the resources a species needs vary greatly in availability from place to place, so the species tends to cluster where the resources are available. *Second,* individuals moving in groups have a better chance of encountering patches or clumps of resources, such as water and vegetation, than they would searching for the resources on their own. *Third,* living in groups protects some animals from predators. *Fourth,* hunting in packs gives some predators a better chance of finding and catching prey. *Fifth,* some species form temporary groups for mating and caring for young.

(a) Clumped (elephants)

(b) Uniform (creosote bush)

(c) Random (dandelions)

Figure 5-10 Generalized *dispersion patterns* for individuals in a population throughout their habitat. The most common pattern is *clumps* of members of a population scattered throughout their habitat, mostly because resources are usually found in patches. **Question:** Why do you think the creosote bushes are uniformly spaced while the dandelions are not?

> **THINKING ABOUT**
> **Population Distribution**
>
> Why do you think living in packs would help some predators to find and kill their prey?

Some species maintain a fairly constant distance between individuals. Because of its sparse distribution pattern, creosote bushes in a desert (Figure 5-10b) have better access to scarce water resources. Organisms with a random distribution (Figure 5-10c) are fairly rare. The living world is mostly clumpy and patchy.

Populations Can Grow, Shrink, or Remain Stable

Over time, the number of individuals in a population may increase, decrease, remain about the same, or go up and down in cycles in response to changes in environmental conditions. Four variables—*births, deaths, immigration,* and *emigration*—govern changes in population size. A population increases by birth and immigration (arrival of individuals from outside the population) and decreases by death and emigration (departure of individuals from the population):

$$\text{Population change} = (\text{Births} + \text{Immigration}) - (\text{Deaths} + \text{Emigration})$$

In natural systems, species that are able to move can leave or *emigrate* from an area where their habitat has been degraded or destroyed and *immigrate* to another area where resources are more plentiful.

A population's **age structure**—the proportions of individuals at various ages—can have a strong effect on how rapidly it increases or decreases in size. Age structures are usually described in terms of organisms not mature enough to reproduce (the *pre-reproductive age*), those capable of reproduction (the *reproductive stage*), and those too old to reproduce (the *post-reproductive stage*).

The size of a population will likely increase if it is made up mostly of individuals in their reproductive stage or soon to enter this stage. In contrast, a population dominated by individuals past their reproductive stage will tend to decrease over time. Excluding emigration and immigration, the size of a population with a fairly even distribution among these three age groups tends to remain stable because reproduction by younger individuals will be roughly balanced by the deaths of older individuals.

No Population Can Grow Indefinitely: J–Curves and S–Curves

Species vary in their **biotic potential** or capacity for population growth under ideal conditions. Generally, populations of species with large individuals, such as elephants and blue whales, have a low biotic potential while those of small individuals, such as bacteria and insects, have a high biotic potential.

The **intrinsic rate of increase (r)** is the rate at which the population of a species would grow if it had unlimited resources. Individuals in populations with a high intrinsic rate of growth typically *reproduce early in life, have short generation times* (the time between successive generations), *can reproduce many times,* and *have many offspring each time they reproduce.*

Some species have an astounding biotic potential. With no controls on population growth, a species of bacteria that can reproduce every 20 minutes would generate enough offspring to form a layer 0.3 meter (1 foot) deep over the entire earth's surface in only 36 hours!

Fortunately, this is not a realistic scenario. Research reveals that no population can grow indefinitely because of limitations on resources and competition with populations of other species for those resources (**Concept 5-3**). In the real world, a rapidly growing population reaches some size limit imposed by one or more *limiting factors,* such as light, water, space, or nutrients, or by exposure to too many competitors, predators, or infectious diseases. *There are always limits to population growth in nature.* This is one of nature's four **scientific principles of sustainability** (see back cover and **Concept 1-6**, p. 23). Sea otters, for example, face extinction because of low biotic potential and other factors (Science Focus, p. 110).

SCIENCE FOCUS

Why Are Protected Sea Otters Making a Slow Comeback?

The southern sea otter (**Core Case Study**) does not have a high biotic potential for several reasons. Female southern sea otters reach sexual maturity between 2 and 5 years of age, can reproduce until age 15, and typically each produce only one pup a year.

The population size of southern sea otters has fluctuated in response to changes in environmental conditions. One such change has been a rise in populations of orcas (killer whales) that feed on them. Scientists hypothesize that orcas feed more on southern sea otters when populations of their normal prey, sea lions and seals, decline.

Another factor may be deaths from parasites known to breed in cats. Scientists hypothesize that some sea otters may be dying because California cat owners flush used cat litter containing these parasites down their toilets or dump it in storm drains that empty into coastal waters.

Thorny-headed worms from seabirds are also known to be killing sea otters, as are toxic algae blooms triggered by urea, a key ingredient in fertilizer that washes into coastal waters. Toxins such as PCBs and other toxic chemicals released by human

Figure 5-B Population size of southern sea otters off the coast of the U.S. state of California, 1983–2007. According to the U.S. Fish and Wildlife Service, the sea otter population would have to reach about 8,400 animals before it can be removed from the endangered species list. (Data from U.S. Geological Survey)

activities accumulate in the tissues of the shellfish on which otters feed and prove fatal to otters. The facts that sea otters feed at high trophic levels and live close to the shore makes them vulnerable to these and other pollutants in coastal waters. In other words, sea otters are *indicator species* that warn us of the condition of coastal waters in their habitat.

Some southern sea otters also die when they encounter oil spilled from ships. The entire California southern sea otter population could be wiped out by a large oil spill from

a single tanker off the state's central coast. These factors plus a fairly low reproductive rate have hindered the ability of the endangered southern sea otter to rebuild its population (Figure 5-B).

Critical Thinking

How would you design a controlled experiment to test the hypothesis that parasites contained in cat litter that is flushed down toilets may be killing sea otters?

Environmental resistance is the combination of all factors that act to limit the growth of a population. Together, biotic potential and environmental resistance determine the **carrying capacity (K)**: the maximum population of a given species that a particular habitat can sustain indefinitely without being degraded. The growth rate of a population decreases as its size nears the carrying capacity of its environment because resources such as food, water, and space begin to dwindle.

A population with few, if any, limitations on its resource supplies can grow exponentially at a fixed rate such as 1% or 2% per year. *Exponential* or *geometric growth* (Figure 1-1, p. 5) starts slowly but then accelerates as the population increases, because the base size of the population is increasing. Plotting the number of individuals against time yields a J-shaped growth curve (Figure 5-11, left half of curve).

Logistic growth involves rapid exponential population growth followed by a steady decrease in population growth until the population size levels off (Figure 5-11, right half of curve). This slowdown occurs as the population encounters environmental resis-

tance from declining resources and other environmental factors and approaches the carrying capacity of its environment. After leveling off, a population with this type of growth typically fluctuates slightly above and below the carrying capacity. The size of such a population may also increase or decrease as the carrying capacity changes because of short- or long-term changes in environmental conditions.

A plot of the number of individuals against time yields a sigmoid, or S-shaped, logistic growth curve (the whole curve in Figure 5-11). Figure 5-12 depicts such a curve for sheep on the island of Tasmania, south of Australia, in the 19th century.

CENGAGENOW™ Learn how to estimate a population of butterflies and see a mouse population growing exponentially at CengageNOW.

Sometimes a species whose population has been kept in check mostly by natural predators is deliberately or accidentally transferred to a different ecosystem where it has few if any predators. For example, the *brown tree snake* is native to the Solomon Islands,

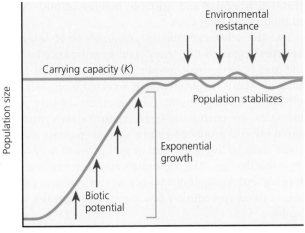

Active Figure 5-11 No population can continue to increase in size indefinitely. *Exponential growth* (left half of the curve) occurs when resources are not limiting and a population can grow at its *intrinsic rate of increase* (r) or *biotic potential.* Such exponential growth is converted to *logistic growth,* in which the growth rate decreases as the population becomes larger and faces environmental resistance. Over time, the population size stabilizes at or near the *carrying capacity* (K) of its environment, which results in a sigmoid (S-shaped) population growth curve. Depending on resource availability, the size of a population often fluctuates around its carrying capacity, although a population may temporarily exceed its carrying capacity and then suffer a sharp decline or crash in its numbers. *See an animation based on this figure at CengageNOW.* **Question:** What is an example of environmental resistance that humans have not been able to overcome?

New Guinea, and Australia. After World War II, a few of these snakes stowed away on military planes going to the island of Guam. With no enemies or rivals in Guam, they have multiplied exponentially for several decades and have wiped out 8 of Guam's 11 native forest bird species. Their venomous bites have also sent large numbers of people to emergency rooms. Sooner

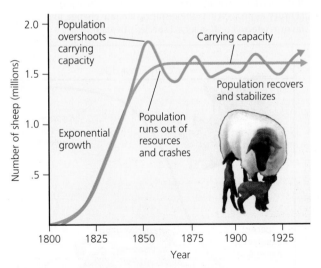

Figure 5-12 *Logistic growth* of a sheep population on the island of Tasmania between 1800 and 1925. After sheep were introduced in 1800, their population grew exponentially, thanks to an ample food supply. By 1855, they had overshot the land's carrying capacity. Their numbers then stabilized and fluctuated around a carrying capacity of about 1.6 million sheep.

or later the brown tree snake will use up most of its food supply in Guam and will then decline in numbers, but meanwhile they are causing serious ecological and economic damage. They may also end up on islands such as those in Hawaii, where they could devastate bird populations.

Changes in the population sizes of keystone species such as the southern sea otter (**Core Case Study**) and the American alligator (Chapter 4 Core Case Study, p. 77) can alter the species composition and biodiversity of an ecosystem. For example, a decline in the population of the southern sea otter caused a decline in the populations of species dependent on them, including the giant kelp. This reduced species diversity of the kelp forest and altered its functional biodiversity by upsetting its food webs and reducing energy flows and nutrient cycling within the forest.

THINKING ABOUT
Southern Sea Otters

Name two species whose populations will likely decline if the population of southern sea otters in kelp beds declines sharply. Name a species whose population would increase if this happened.

When a Population Exceeds Its Habitat's Carrying Capacity, Its Population Can Crash

Some species do not make a smooth transition from exponential growth to logistic growth. Such populations use up their resource supplies and temporarily *overshoot,* or exceed, the carrying capacity of their environment. This occurs because of a *reproductive time lag*—the period needed for the birth rate to fall and the death rate to rise in response to resource overconsumption.

In such cases, the population suffers a *dieback,* or *crash,* unless the excess individuals can switch to new resources or move to an area with more resources. Such a crash occurred when reindeer were introduced onto a small island in the Bering Sea (Figure 5-13, p. 112).

The carrying capacity of an area or volume is not fixed. The carrying capacity of some areas can increase or decrease seasonally and from year to year because of variations in weather and other factors, including an abundance of predators and competitors. For example, a drought can decrease the amount of vegetation growing in an area supporting deer and other herbivores, and this would decrease the normal carrying capacity for those species.

Sometimes when a population exceeds the carrying capacity of an area, it causes damage that reduces the area's carrying capacity. For example, overgrazing by cattle on dry western lands in the United States has reduced grass cover in some areas. This has allowed sagebrush—which cattle cannot eat—to move in, thrive, and replace grasses, reducing the land's carrying capacity for cattle.

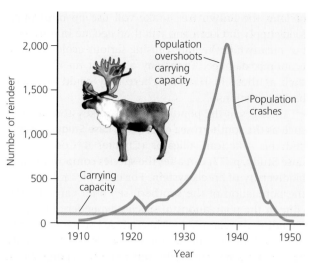

Figure 5-13 Exponential growth, overshoot, and population crash of reindeer introduced to the small Bering Sea island of St. Paul. When 26 reindeer (24 of them female) were introduced in 1910, lichens, mosses, and other food sources were plentiful. By 1935, the herd size had soared to 2,000, overshooting the island's carrying capacity. This led to a population crash, when the herd size plummeted to only 8 reindeer by 1950. **Question:** Why do you think this population grew fast and crashed, unlike the sheep in Figure 5-12?

THINKING ABOUT
Population Overshoot and
Human Ecological Footprints

Humanity's ecological footprint is about 25% larger than the earth's ecological capacity (Figure 1-10, bottom, p. 15) and is growing rapidly. If this keeps up, is the curve for future human population growth more likely to resemble Figure 5-12 or Figure 5-13? Explain.

Species Have Different Reproductive Patterns

Species have different reproductive patterns that can help enhance their survival. Species with a capacity for a high rate of population increase (*r*) are called **r-selected species** (Figure 5-14). These species have many, usually small, offspring and give them little or no parental care or protection. They overcome typically massive losses of offspring by producing so many offspring that a few will likely survive to reproduce many more offspring to begin this reproductive pattern again. Examples include algae, bacteria, rodents, frogs, turtles, annual plants (such as dandelions), and most insects.

Such species tend to be *opportunists*. They reproduce and disperse rapidly when conditions are favorable or when a disturbance opens up a new habitat or niche for invasion. Environmental changes caused by disturbances, such as fires, clear-cutting, and volcanic eruptions, can allow opportunist species to gain a foothold. However, once established, their populations may crash because of unfavorable changes in environmental conditions or invasion by more competitive species. This helps to explain why most opportunist species go

through irregular and unstable boom-and-bust cycles in their population sizes.

At the other extreme are *competitor* or **K-selected species** (Figure 5-14). They tend to reproduce later in life and have a small number of offspring with fairly long life spans. Typically, for K-selected mammals, the offspring develop inside their mothers (where they are safe), are born fairly large, mature slowly, and are cared for and protected by one or both parents, and in some cases by living in herds or groups, until they reach reproductive age. This reproductive pattern results in a few big and strong individuals that can compete for resources and reproduce a few young to begin the cycle again.

Such species are called K-selected species because they tend to do well in competitive conditions when their population size is near the carrying capacity (*K*) of their environment. Their populations typically follow a logistic growth curve (Figure 5-12).

Most large mammals (such as elephants, whales, and humans), birds of prey, and large and long-lived plants (such as the saguaro cactus, and most tropical rain forest trees) are K-selected species. Ocean fish such as orange roughy and swordfish, which are now being depleted by overfishing, are also K-selected. Many of these species—especially those with long times between generations and low reproductive rates like elephants, rhinoceroses, and sharks—are prone to extinction.

Most organisms have reproductive patterns between the extremes of r-selected and K-selected species. In agriculture we raise both r-selected species (crops) and K-selected species (livestock). Individuals of species with different reproductive strategies tend to have different *life expectancies,* or expected lengths of life.

THINKING ABOUT
r-Selected and K-selected Species

If the earth experiences significant warming during this century as projected, is the resulting climate change likely to favor r-selected or K-selected species? Explain.

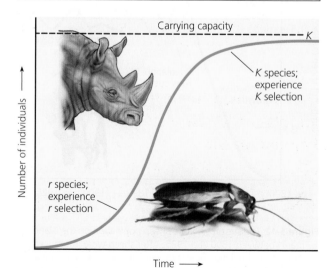

Figure 5-14 Positions of *r-selected* and *K-selected* species on the sigmoid (S-shaped) population growth curve.

Genetic Diversity Can Affect the Size of Small Populations

In most large populations, genetic diversity is fairly constant and the loss or addition of some individuals has little effect on the total gene pool. However, several genetic factors can play a role in the loss of genetic diversity and the survival of small, isolated populations.

One such factor, called the *founder effect*, can occur when a few individuals in a population colonize a new habitat that is geographically isolated from other members of the population (Figure 4-8, p. 87). In such cases, limited genetic diversity or variability may threaten the survival of the colonizing population.

Another factor is a *demographic bottleneck.* It occurs when only a few individuals in a population survive a catastrophe such as a fire or hurricane, as if they had passed through the narrow neck of a bottle. Lack of genetic diversity may limit the ability of these individuals to rebuild the population. Even if the population is able to increase in size, its decreased genetic diversity may lead to an increase in the frequency of harmful genetic diseases.

A third factor is *genetic drift.* It involves random changes in the gene frequencies in a population that can lead to unequal reproductive success. For example, some individuals may breed more than others do and their genes may eventually dominate the gene pool of the population. This change in gene frequency could help or hinder the survival of the population. The founder effect is one cause of genetic drift.

A fourth factor is *inbreeding.* It occurs when individuals in a small population mate with one another. This can occur when a population passes through a demographic bottleneck. This can increase the frequency of defective genes within a population and affect its long-term survival.

Conservation biologists use the concepts of founder effects, demographic bottleneck, genetic drift, inbreeding, and island biogeography (Science Focus, p. 90) to estimate the *minimum viable population size* of rare and endangered species: the number of individuals such populations need for long-term survival.

Under Some Circumstances Population Density Affects Population Size

Population density is the number of individuals in a population found in a particular area or volume. Some factors that limit population growth have a greater effect as a population's density increases. Examples of such *density-dependent population controls* include predation, parasitism, infectious disease, and competition for resources.

Higher population density may help sexually reproducing individuals find mates, but it can also lead to increased competition for mates, food, living space, water, sunlight, and other resources. High population density can help to shield some members from predators, but it can also make large groups such as schools of fish vulnerable to human harvesting methods. In addition, close contact among individuals in dense populations can increase the transmission of parasites and infectious diseases. When population density decreases, the opposite effects occur. Density-dependent factors tend to regulate a population at a fairly constant size, often near the carrying capacity of its environment.

Some factors—mostly abiotic—that can kill members of a population are *density independent.* In other words, their effect is not dependent on the density of the population. For example, a severe freeze in late spring can kill many individuals in a plant population or a population of monarch butterflies (Figure 3-A, left, p. 54), regardless of their density. Other such factors include floods, hurricanes, fire, pollution, and habitat destruction, such as clearing a forest of its trees or filling in a wetland.

Several Different Types of Population Change Occur in Nature

In nature, we find four general patterns of variation in population size: *stable, irruptive, cyclic,* and *irregular.* A species whose population size fluctuates slightly above and below its carrying capacity is said to have a fairly stable population size (Figure 5-12). Such stability is characteristic of many species found in undisturbed tropical rain forests, where average temperature and rainfall vary little from year to year.

For some species, population growth may occasionally explode, or *irrupt,* to a high peak and then crash to a more stable lower level or in some cases to a very low level. Many short-lived, rapidly reproducing species such as algae and many insects have irruptive population cycles that are linked to seasonal changes in weather or nutrient availability. For example, in temperate climates, insect populations grow rapidly during the spring and summer and then crash during the hard frosts of winter.

A third type of fluctuation consists of regular *cyclic fluctuations,* or *boom-and-bust cycles,* of population size over a time period. Examples are lemmings, whose populations rise and fall every 3–4 years, and lynx and snowshoe hare, whose populations generally rise and fall in a 10-year cycle. Ecologists distinguish between *top-down population regulation,* through predation, and *bottom-up population regulation,* in which the size of predator and prey populations is controlled by the scarcity of one or more resources (Figure 5-15, p. 114).

Finally, some populations appear to have *irregular* changes in population size, with no recurring pattern. Some scientists attribute this irregularity to chaos in

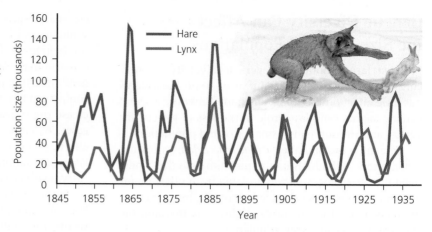

Figure 5-15 Population cycles for the snowshoe hare and Canada lynx. At one time, scientists believed these curves provided circumstantial evidence that these predator and prey populations regulated one another. More recent research suggests that the periodic swings in the hare population are caused by a combination of *top-down population control*—through predation by lynx and other predators—and *bottom-up population control,* in which changes in the availability of the food supply for hares help determine hare population size, which in turn helps determine the lynx population size. (Data from D. A. MacLulich)

such systems. Others scientists contend that it may represent fluctuations in response to periodic catastrophic population crashes due to severe winter weather.

Humans Are Not Exempt from Nature's Population Controls

Humans are not exempt from population overshoot and dieback. Ireland experienced a population crash after a fungus destroyed the potato crop in 1845. About 1 million people died from hunger or diseases related to malnutrition, and 3 million people migrated to other countries, mostly the United States.

During the 14th century the *bubonic plague* spread through densely populated European cities and killed at least 25 million people. The bacterium causing this disease normally lives in rodents. It was transferred to humans by fleas that fed on infected rodents and then bit humans. The disease spread like wildfire through crowded cities, where sanitary conditions were poor and rats were abundant.

Currently, the world is experiencing a global epidemic of eventually fatal AIDS, caused by infection with the human immunodeficiency virus (HIV). Between 1981 and 2007, AIDS killed more than 25 million people (584,000 in the United states) and claims another 2.1 million lives each year—an average of four deaths per minute.

So far, technological, social, and other cultural changes have extended the earth's carrying capacity for the human species. We have increased food production and used large amounts of energy and matter resources to occupy normally uninhabitable areas. As humans spread into larger areas, they interact with and attempt to control the populations of other species such as alligators (Chapter 4 Core Case Study, p. 77) and white-tailed deer in the United States (Case Study, at right), as these two case studies reveal.

Some say we can keep expanding our ecological footprint indefinitely, mostly because of our technological ingenuity. Others say that sooner or later, we will reach the limits that nature always imposes on populations.

HOW WOULD YOU VOTE?

Can we continue to expand the earth's carrying capacity for humans? Cast your vote online at **academic.cengage.com/ biology/miller**.

THINKING ABOUT
The Human Species

If the human species were to suffer a sharp population decline, name three types of species that might move in to occupy part of our ecological niche.

■ CASE STUDY

Exploding White-Tailed Deer Populations in the United States

By 1900, habitat destruction and uncontrolled hunting had reduced the white-tailed deer population in the United States to about 500,000 animals. In the 1920s and 1930s, laws were passed to protect the remaining deer. Hunting was restricted and predators such as wolves and mountain lions that preyed on the deer were nearly eliminated.

These protections worked, and some suburbanites and farmers say perhaps they worked too well. Today there are 25–30 million white-tailed deer in the United States. During the last 50 years, large numbers of Americans have moved into the wooded habitat of deer and provided them with flowers, garden crops, and other plants they like to eat.

Deer like to live in the woods for security and go to nearby fields, orchards, lawns, and gardens for food. Border areas between two ecosystems such as forests and fields are called *edge* habitat, and suburbanization has created an all-you-can-eat edge paradise for deer. Their populations in such areas have soared. In some forests, they are consuming native ground cover vegetation and allowing nonnative weed species to take over. Deer also spread Lyme disease (carried by deer ticks) to humans.

In addition, in deer–vehicle collisions, deer accidentally kill and injure more people each year in the United

States than do any other wild animals. Each year, these 1.5 million accidents injure at least 14,000 people and kill at least 200 (up from 101 deaths in 1993).

There are no easy answers to the deer population problem in the suburbs. Changing hunting regulations to allow killing of more female deer cuts down the overall deer population. But these actions have little effect on deer in suburban areas because it is too dangerous to allow widespread hunting with guns in such populated communities. Some areas have hired experienced and licensed archers who use bows and arrows to help reduce deer numbers. To protect nearby residents the archers hunt from elevated tree stands and shoot their arrows only downward. However, animal activists strongly oppose killing deer on ethical grounds, arguing that this is cruel and inhumane treatment.

Some communities spray the scent of deer predators or rotting deer meat in edge areas to scare off deer. Others use electronic equipment that emits high-frequency sounds, which humans cannot hear, for the same purpose. Some homeowners surround their gardens and yards with a high black plastic mesh fencing that is invisible from a distance. Such deterrents may protect one area, but cause the deer to seek food in someone else's yard or garden.

Deer can also be trapped and moved from one area to another, but this is expensive and must be repeated whenever deer move back into an area. Also, there are questions concerning where to move the deer and how to pay for such programs.

Should we put deer on birth control? Darts loaded with a contraceptive could be shot into female deer to hold down their birth rates. But this is expensive and must be repeated every year. One possibility is an experimental single-shot contraceptive vaccine that causes females to stop producing eggs for several years. Another approach is to trap dominant males and use chemical injections to sterilize them. Both these approaches will require years of testing.

Meanwhile, suburbanites can expect deer to chow down on their shrubs, flowers, and garden plants unless they can protect their properties with high, deer-proof fences or other methods. Deer have to eat every day just as we do. Suburban dwellers might consider not planting their yards with plants that deer like to eat.

THINKING ABOUT
White-Tailed Deer

Some blame the white-tailed deer for invading farms and suburban yards and gardens to eat food that humans have made easily available to them. Others say humans are mostly to blame because they have invaded deer territory, eliminated most of the predators that kept their populations under control, and provided the deer with plenty to eat in their lawns and gardens. Which view do you hold? Do you see a solution to this problem?

5-4 How Do Communities and Ecosystems Respond to Changing Environmental Conditions?

▶ **CONCEPT 5-4** The structure and species composition of communities and ecosystems change in response to changing environmental conditions through a process called ecological succession.

Communities and Ecosystems Change over Time: Ecological Succession

We have seen how changes in environmental conditions (such as loss of habitat) can reduce access to key resources and how invasions by competing species can lead to increases or decreases in species population sizes. Next, we look at how the types and numbers of species in biological communities change in response to changing environmental conditions such as a fires, climate change, and the clearing of forests to plant crops.

Mature forests and other ecosystems do not spring up from bare rock or bare soil. Instead, they undergo changes in their species composition over long periods of time. The gradual change in species composition in a given area is called **ecological succession,** during which, some species colonize an area and their populations become more numerous, while populations of other species decline and may even disappear. In this process, *colonizing* or *pioneer species* arrive first. As environmental conditions change, they are replaced by other species, and later these species may be replaced by another set of species.

Ecologists recognize two main types of ecological succession, depending on the conditions present at the beginning of the process. **Primary succession** involves the gradual establishment of biotic communities in lifeless areas where there is no soil in a terrestrial ecosystem or no bottom sediment in an aquatic ecosystem. The other more common type of ecological succession is called **secondary succession,** in which a series of communities or ecosystems with different species develop in places containing soil or bottom sediment. As part of the earth's natural capital, both types of succession are examples of *natural ecological restoration,* in which various forms life adapt to changes in environmental conditions, resulting in changes to the species composition, population size, and biodiversity in a given area.

Some Ecosystems Start from Scratch: Primary Succession

Primary succession begins with an essentially lifeless area where there is no soil in a terrestrial system (Figure 5-16) or bottom sediment in an aquatic system. Examples include bare rock exposed by a retreating glacier or severe soil erosion, newly cooled lava from a volcanic eruption, an abandoned highway or parking lot, and a newly created shallow pond or reservoir.

Primary succession usually takes a long time because there is no fertile soil to provide the nutrients needed to establish a plant community. Over time, bare rock *weathers* by crumbling into particles and releasing nutrients. Physical weathering occurs when a rock is fragmented, as water in its cracks freezes and expands. Rocks also undergo chemical weathering, reacting with substances in the atmosphere or with precipitation, which can break down the rock's surface material.

The slow process of soil formation begins when *pioneer* or *early successional species* arrive and attach themselves to inhospitable patches of the weathered rock. Examples are lichens and mosses whose seeds or spores

are distributed by the wind and carried by animals. A lichen consists of an alga and a fungus interacting in a mutualistic relationship. The fungi in the lichens provide protection and support for the algae, which, through photosynthesis, provide sugar nutrients for both of the interacting species.

These tough *early successional plant species* start the long process of soil formation by trapping wind-blown soil particles and tiny pieces of detritus, and adding their own wastes and dead bodies. They also secrete mild acids that further fragment and break down the rock. As the lichens spread over the rock, drought-resistant and sun-loving mosses start growing in cracks. As the mosses spread, they form a mat that traps moisture, much like a sponge. When the lichens and mosses die, their decomposing remains add to the growing thin layer of nutrients.

After hundreds to thousands of years, the soil may be deep and fertile enough to store the moisture and nutrients needed to support the growth of *midsuccessional plant species* such as herbs, grasses, and low shrubs. As the shrubs grow and create shade, the lichens and mosses die and decay from lack of sunlight. Next, trees that need lots of sunlight and are adapted to the area's climate and soil usually replace the grasses and shrubs.

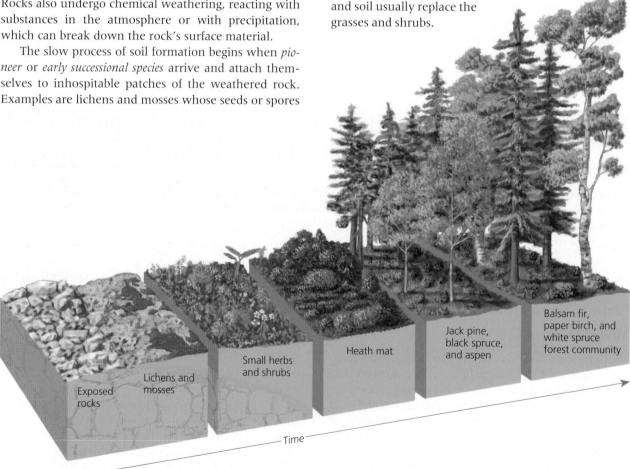

Exposed rocks

Lichens and mosses

Small herbs and shrubs

Heath mat

Jack pine, black spruce, and aspen

Balsam fir, paper birch, and white spruce forest community

Time

Figure 5-16 *Primary ecological succession.* Over almost a thousand years, plant communities developed, starting on bare rock exposed by a retreating glacier on Isle Royal, Michigan (USA) in northern Lake Superior. The details of this process vary from one site to another. **Question:** What are two ways in which lichens, mosses, and plants might get started growing on bare rock?

As these tree species grow and create shade, they are replaced by *late successional plant species* (mostly trees) that can tolerate shade. Unless fire, flooding, severe erosion, tree cutting, climate change, or other natural or human processes disturb the area, what was once bare rock becomes a complex forest community or ecosystem (Figure 5-16).

Primary succession can also take place in a newly created small pond, starting with an influx of sediments and nutrients in runoff from the surrounding land. This sediment can support seeds or spores of plants carried to the pond by winds, birds, or other animals. Over time, this process can transform the pond first into a marsh and eventually to dry land.

Some Ecosystems Do Not Have to Start from Scratch: Secondary Succession

Secondary succession begins in an area where an ecosystem has been disturbed, removed, or destroyed, but some soil or bottom sediment remains. Candidates for secondary succession include abandoned farmland, burned or cut forests, heavily polluted streams, and land that has been flooded. In the soil that remains on disturbed land systems, new vegetation can germinate, usually within a few weeks, from seeds already in the soil and from those imported by wind, birds, and other animals.

In the central, or Piedmont, region of the U.S. state of North Carolina, European settlers cleared many of the mature native oak and hickory forests and planted the land with crops. Later, they abandoned some of this farmland because of erosion and loss of soil nutrients. Figure 5-17 shows one way in which such abandoned farmland has undergone secondary succession over 150–200 years.

CENGAGENOW™ Explore the difference between primary and secondary succession at CengageNOW.

Descriptions of ecological succession usually focus on changes in vegetation. But these changes in turn affect food and shelter for various types of animals. As a consequence, the numbers and types of animals and

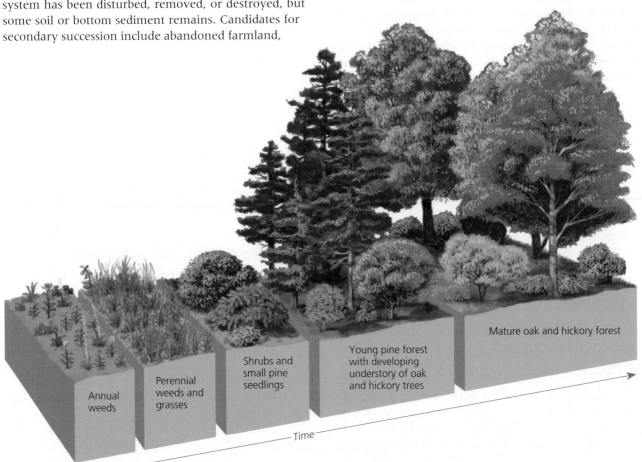

Annual weeds

Perennial weeds and grasses

Shrubs and small pine seedlings

Young pine forest with developing understory of oak and hickory trees

Mature oak and hickory forest

Time

CENGAGENOW™ **Active Figure 5-17** *Natural ecological restoration of disturbed land.* Secondary ecological succession of plant communities on an abandoned farm field in the U.S. state of North Carolina. It took 150–200 years after the farmland was abandoned for the area to become covered with a mature oak and hickory forest. A new disturbance, such as deforestation or fire, would create conditions favoring pioneer species such as annual weeds. In the absence of new disturbances, secondary succession would recur over time, but not necessarily in the same sequence shown here. **Questions:** Do you think the annual weeds (left) would continue to thrive in the mature forest (right)? Why or why not? *See an animation based on this figure at* CengageNOW.

How Do Species Replace One Another in Ecological Succession?

Ecologists have identified three factors that affect how and at what rate succession occurs. One is *facilitation*, in which one set of species makes an area suitable for species with different niche requirements, but less suitable for itself. For example, as lichens and mosses gradually build up soil on a rock in primary succession, herbs and grasses can colonize the site and crowd out the pioneer community of lichens and mosses.

A second factor is *inhibition*, in which some early species hinder the establishment and growth of other species. Inhibition often occurs when plants release toxic chemicals that reduce competition from other plants. For example, pine needles dropped by some species of pines can make the underlying soil acidic and inhospitable to other plant species. Succession then can proceed only when a fire, bulldozer, or other human or natural disturbance removes most of the inhibiting species.

A third factor is *tolerance*, in which late successional plants are largely unaffected by plants at earlier stages of succession because they are not in direct competition with them for key resources. For example, shade tolerant trees and other plants can thrive in the understory of a mature forest (Figure 5-17) because they do not need to compete with the taller species for access to sunlight.

Critical Thinking

Explain how tolerance can increase biodiversity by increasing species diversity and functional diversity (energy flow and chemical cycling) in an ecosystem.

---

decomposers also change. A key point is that primary succession (Figure 5-16) and secondary succession (Figure 5-17) tend to increase biodiversity and thus the sustainability of communities and ecosystems by increasing species richness and interactions among species. Such interactions in turn enhance sustainability by promoting population control and increasing the complexity of food webs for the energy flow and nutrient cycling that make up the functional component of biodiversity (Figure 4-2, p. 79). Ecologists have been conducting research to find out more about the factors involved in ecological succession (Science Focus, above)

During primary or secondary succession, environmental disturbances such as fires, hurricanes, clearcutting of forests, plowing of grasslands, and invasions by nonnative species can interrupt a particular stage of succession, setting it back to an earlier stage. For example, the American alligator (Chapter 4 Core Case Study, p. 77) lives in ponds that would normally become filled in through natural selection. The alligator's movements keep bottom vegetation from growing, thus preventing such succession.

Succession Doesn't Follow a Predictable Path

According to traditional view, succession proceeds in an orderly sequence along an expected path until a certain stable type of *climax community* occupies an area. Such a community is dominated by a few long-lived plant species and is in balance with its environment. This equilibrium model of succession is what ecologists once meant when they talked about the *balance of nature*.

Over the last several decades, many ecologists have changed their views about balance and equilibrium in nature. Under the balance-of-nature view, a large terrestrial community or ecosystem undergoing succession eventually became covered with an expected type of climax vegetation such as a mature forest (Figures 5-16 and 5-17). There is a general tendency for succession to lead to more complex, diverse, and presumably stable ecosystems. But a close look at almost any terrestrial community or ecosystem reveals that it consists of an ever-changing mosaic of patches of vegetation at different stages of succession.

The current view is that we cannot predict the course of a given succession or view it as preordained progress toward an ideally adapted climax plant community or ecosystem. Rather, succession reflects the ongoing struggle by different species for enough light, nutrients, food, and space. Most ecologists now recognize that mature late-successional ecosystems are not in a state of permanent equilibrium, but rather a state of continual disturbance and change.

Living Systems Are Sustained through Constant Change

All living systems from a cell to the biosphere are dynamic systems that are constantly changing in response to changing environmental conditions. Continents move, the climate changes, and disturbances and succession change the composition of communities and ecosystems.

Living systems contain complex networks of positive and negative feedback loops (Figure 2-11, p. 45, and Figure 2-12, p. 45) that interact to provide some degree of stability, or sustainability, over each system's expected life span. This stability, or capacity to withstand external stress and disturbance, is maintained only by constant change in response to changing environmental conditions. For example, in a mature tropical rain forest, some trees die and others take their places. However, unless the forest is cut, burned, or otherwise destroyed, you would still recognize it as a tropical rain forest 50 or 100 years from now.

It is useful to distinguish among two aspects of stability in living systems. One is **inertia,** or **persistence:** the ability of a living system, such as a grassland or a forest, to survive moderate disturbances. A second factor is **resilience:** the ability of a living system to be restored through secondary succession after a moderate disturbance.

Evidence suggests that some ecosystems have one of these properties but not the other. For example, tropical rain forests have high species diversity and high inertia and thus are resistant to significant alteration or destruction. But once a large tract of tropical rain forest is severely damaged, the resilience of the resulting degraded ecosystem may be so low that the forest may not be restored by secondary ecological succession. One reason for this is that most of the nutrients in a tropical rain forest are stored in its vegetation, not in the soil as in most other terrestrial ecosystems. Once the nutrient-rich vegetation is gone, daily rains can remove most of the other nutrients left in the soil and thus prevent a tropical rain forest from regrowing on a large cleared area.

Another reason for why the rain forest cannot recover is that large-scale deforestation can change an area's climate by decreasing the input of water vapor from its trees into the atmosphere. Without such water vapor, rain decreases and the local climate gets warmer. Over many decades, this can allow for the development of a tropical grassland in the cleared area but not for the reestablishment of a tropical rain forest.

By contrast, grasslands are much less diverse than most forests, and consequently they have low inertia and can burn easily. However, because most of their plant matter is stored in underground roots, these ecosystems have high resilience and can recover quickly after a fire as their root systems produce new grasses. Grassland can be destroyed only if its roots are plowed up and something else is planted in its place, or if it is severely overgrazed by livestock or other herbivores.

Variations among species in resilience and inertia are yet another example of biodiversity—one aspect of natural capital that has allowed life on earth to sustain itself for billions of years.

However, there are limits to the stresses that ecosystems and global systems such as climate can take. As a result, such systems can reach a **tipping point,** where any additional stress can cause the system to change in an abrupt and usually irreversible way that often involves collapse. For example, once a certain number of trees have been eliminated from a stable tropical rain forest, it can crash and become a grassland. And continuing to warm the atmosphere by burning fossil fuels that emit CO_2 and cutting down tropical forests that help remove CO_2 could eventually change the global climate system in ways that could last for thousands of years.

Exceeding a tipping point is like falling off a cliff. There is no way back. One of the most urgent scientific research priorities is to identify these and other tipping points and to develop strategies to prevent natural systems from reaching their tipping points.

REVISITING   Southern Sea Otters and Sustainabiliy

Before the arrival of European settlers on the North American west coast, the sea otter population was part of a complex ecosystem made up of bottom-dwelling creatures, kelp, otters, whales, and other species depending on one another for survival. Giant kelp forests served as food and shelter for sea urchins. Otters ate the sea urchins and other kelp eaters. Some species of whales and sharks ate the otters. And detritus from all these species helped to maintain the giant kelp forests. Each of these interacting populations was kept in check by, and helped to sustain, all others.

When humans arrived and began hunting the otters for their pelts, they probably did not know much about the intricate web of life beneath the ocean surface. But with the effects of overhunting, people realized they had done more than simply take otters. They had torn the web, disrupted an entire ecosystem, and triggered a loss of valuable natural resources and services, including biodiversity.

Populations of most plants and animals depend directly or indirectly on solar energy and each population plays a role in the cycling of nutrients in the ecosystems where they live. In addition, the biodiversity found in the variety of species in different terrestrial and aquatic ecosystems provides alternative paths for energy flow and nutrient cycling and better opportunities for natural selection as environmental conditions change. Disrupt these paths and we can decrease the various components of the biodiversity of ecosystems and the sizes of their populations.

In this chapter, we looked more closely at two **principles of sustainability**: *biodiversity promotes sustainability* and *there are always limits to population growth in nature.* Chapter 6 applies the concepts of biodiversity and population dynamics discussed in this chapter to the growth of the human population and its environmental impacts. Chapters 7 and 8 look more closely at biodiversity in the variety of terrestrial ecosystems (such as deserts, grasslands, and forests) and aquatic ecosystems (such as oceans, lakes, rivers, and wetlands) found on the earth.

We cannot command nature
except by obeying her.

SIR FRANCIS BACON

1. Review the Key Questions and Concepts for this chapter on p. 101. Explain how southern sea otters act as a keystone species in kelp beds. Explain why we should care about protecting this species from extinction. Explain why we should help to preserve kelp forests.

2. Define **interspecific competition, predation, parasitism, mutalism,** and **commensalism** and give an example of each. Explain how each of these species interactions can affect the population sizes of species in ecosystems. Distinguish between a **predator** and a **prey** and give an example of each. What is a **predator–prey relationship?** Describe four ways in which prey species can avoid their predators and four ways in which predators can capture these prey.

3. Define and give an example of **coevolution.**

4. Describe and give an example of **resource partitioning** and explain how it can increase species diversity.

5. What is **population dynamics?** Why do most populations live in clumps?

6. Describe four variables that govern changes in population size and write an equation showing how they interact. What is a population's **age structure** and what are three major age group categories? Distinguish among the **biotic potential, intrinsic rate of increase, exponential growth, environmental resistance, carrying capacity,** and **logistic growth** of a population, and use these concepts to explain why there are always limits to

population growth in nature. Why are southern sea otters making a slow comeback and what factors can threaten this recovery? Define and give an example of a **population crash.** Explain why humans are not exempt from nature's population controls.

7. Distinguish between **r-selected species** and **K-selected species** and give an example of each type. Define **population density** and explain how it can affect the size of some but not all populations.

8. Describe the exploding white-tailed deer population problem in the United States and discuss options for dealing with it.

9. What is **ecological succession?** Distinguish between **primary ecological succession** and **secondary ecological succession** and give an example of each. Explain why succession does not follow a predictable path. In terms of stability, distinguish between **inertia (persistence)** and **resilience.** Explain how living systems achieve some degree of stability or sustainability by undergoing constant change in response to changing environmental conditions.

10. Explain how the role of the southern sea otter in its ecosystem (**Core Case Study**) illustrates the population control **principle of sustainability.**

Note: Key Terms are in **bold** type.

1. What difference would it make if the southern sea otter (**Core Case Study**) became prematurely extinct because of human activities? What are three things we could do to help prevent the premature extinction of this species?

2. Use the second law of thermodynamics (p. 43) to explain why predators are generally less abundant than their prey.

3. Explain why most species with a high capacity for population growth (high biotic potential) tend to have small individuals (such as bacteria and flies) while those with a low capacity for population growth tend to have large individuals (such as humans, elephants, and whales).

4. List three factors that have limited human population growth in the past that we have overcome. Describe how we overcame each of these factors. List two factors that may limit human population growth in the future.

5. Why are pest species likely to be extreme examples of r-selected species? Why are many endangered species

likely to be extreme examples of K-selected species?

6. Given current environmental conditions, if you had a choice, would you rather be an r-strategist or a K-strategist? Explain your answer.

7. Is the southern sea otter an r-strategist or a K-strategist species? Explain. How does this affect our efforts to protect this species from premature extinction?

8. How would you reply to someone who argues that we should not worry about our effects on natural systems because natural succession will heal the wounds of human activities and restore the balance of nature?

9. In your own words, restate this chapter's closing quotation by Sir Francis Bacon. Do you agree with this notion? Why or why not?

10. List two questions that you would like to have answered as a result of reading this chapter.

Note: See Supplement 13 (p. S78) for a list of Projects related to this chapter.

The graph below shows the changes in size of an Emperor Penguin population in terms of breeding pairs on Terre Adelie in the Antarctic. Use the graph to answer the questions below.

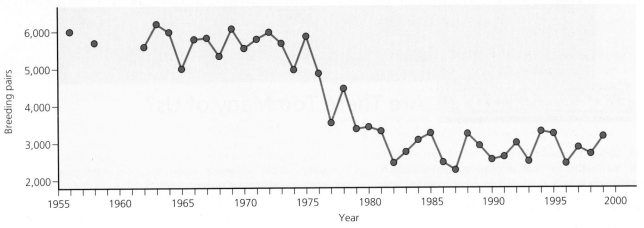

Source: Data from *Nature,* May 10, 2001.

1. What was the approximate carrying capacity of the penguin population on the island from 1960 to 1975? (*Hint:* See Figure 5-11, p. 111.) What was the approximate carrying capacity of the penguin population on the island from 1980 to 1999?

2. What is the percentage decline in the penguin population from 1975 to 1999?

Log on to the Student Companion Site for this book at **academic.cengage.com/biology/miller**, and choose Chapter 5 for many study aids and ideas for further reading and research. These include flash cards, practice quizzing, Weblinks, information on Green Careers, and InfoTrac® College Edition articles.

6 The Human Population and Its Impact

Are There Too Many of Us?

Each week, about 1.6 million people are added to the world's population. As a result, the number of people on the earth is projected to increase from 6.7 to 9.3 billion or more between 2008 and 2050, with most of this growth occurring in the world's developing countries (Figure 6-1). This raises an important question: *Can the world provide an adequate standard of living for a projected 2.6 billion more people by 2050 without causing widespread environmental damage?* There is disagreement over the answer to this question.

According to one view, the planet already has too many people collectively degrading the earth's natural capital. To some analysts, the problem is the sheer number of people in developing countries with 82% of the world's population. To others, it is high per capita resource consumption rates in developed countries—and to an increasing extent in rapidly developing countries such as China and India—that magnify the environmental impact, or ecological footprint, of each person (Figure 1-10, p. 15).

Many argue that both population growth and resource consumption per person are important causes of the environmental problems we face (**Concept 1-5A**, p. 17).

CONCEPT LINK

Another view is that technological advances have allowed us to overcome the environmental resistance that all populations face (Figure 5-11, p. 111) and to increase the earth's carrying capacity for our species. Some analysts argue there is no reason we cannot continue doing so, and they believe that the planet can support billions more people. They also see a growing population as our most valuable resource for solving environmental and other problems and for stimulating economic growth by increasing the number of consumers. As a result, they see no need to control the world's population growth.

Some people view any form of population regulation as a violation of their religious or moral beliefs. Others see it as an intrusion into their privacy and their freedom to have as many children as they want. These people also would argue against any form of population control.

Proponents of slowing and eventually stopping population growth have a different view. They point out that we are not providing the basic necessities for about one of every five people—a total of some 1.4 billion. They ask how we will be able to do so for the projected 2.6 billion more people by 2050.

They also warn of two serious consequences we will face if we do not sharply lower birth rates. *First,* death rates may increase because of declining health and environmental conditions in some areas, as is already happening in parts of Africa. *Second,* resource use and environmental degradation may intensify as more consumers increase their already large ecological footprints in developed countries and in rapidly developing countries, such as China and India (Figure 1-10, p. 15). This could increase environmental stresses such as infectious disease, biodiversity losses, water shortages, traffic congestion, pollution of the seas, and climate change.

This debate over interactions among population growth, economic growth, politics, and moral beliefs is one of the most important and controversial issues in environmental science.

Figure 6-1 Crowded street in China. Together, China and India have 36% of the world's population and the resource use per person in these countries is projected to grow rapidly as they become more modernized (Case Study, p. 15).

L. Young/UNEP/Peter Arnold, Inc.

6-1 How many people can the earth support?

CONCEPT 6-1 We do not know how long we can continue increasing the earth's carrying capacity for humans without seriously degrading the life-support system for humans and many other species.

6-2 What factors influence the size of the human population?

CONCEPT 6-2A Population size increases because of births and immigration and decreases through deaths and emigration.

CONCEPT 6-2B The average number of children born to women in a population (*total fertility rate*) is the key factor that determines population size.

6-3 How does a population's age structure affect its growth or decline?

CONCEPT 6-3 The numbers of males and females in young, middle, and older age groups determine how fast a population grows or declines.

6-4 How can we slow human population growth?

CONCEPT 6-4 Experience indicates that the most effective ways to slow human population growth are to encourage family planning, to reduce poverty, and to elevate the status of women.

Note: Supplements 2 (p. S4), 3 (p. S10), 4 (p. S20), and 13 (p. S78) can be used with this chapter.

The problems to be faced are vast and complex, but come down to this:
6.7 billion people are breeding exponentially.
The process of fulfilling their wants and needs is stripping earth
of its biotic capacity to support life;
a climactic burst of consumption by a single species
is overwhelming the skies, earth, waters, and fauna.

PAUL HAWKEN

6-1 How Many People Can the Earth Support?

▶ **CONCEPT 6-1** We do not know how long we can continue increasing the earth's carrying capacity for humans without seriously degrading the life-support system for humans and many other species.

Human Population Growth Continues but It Is Unevenly Distributed

For most of history, the human population grew slowly (Figure 1-1, p. 5, left part of curve). But for the past 200 years, the human population has experienced rapid exponential growth reflected in the characteristic J-curve (Figure 1-1, right part of curve).

Three major factors account for this population increase. *First,* humans developed the ability to expand into diverse new habitats and different climate zones. *Second,* the emergence of early and modern agriculture allowed more people to be fed for each unit of land area farmed. *Third,* the development of sanitation systems, antibiotics, and vaccines helped control infectious disease agents. As a result, death rates dropped sharply below birth rates and population size grew rapidly.

About 10,000 years ago when agriculture began, there were about 5 million humans on the planet; now there are 6.7 billion of us. It took from the time we arrived until about 1927 to add the first 2 billion people to the planet; less than 50 years to add the next 2 billion (by 1974); and just 25 years to add the next 2 billion (by 1999)—an illustration of the awesome power of exponential growth (Chapter 1 Core Case Study, p. 5). By 2012 we will be trying to support 7 billion people and perhaps 9.3 billion by 2050. Such growth raises the question of whether the earth is over-populated (**Core Case Study**). (See Figure 4, p. S12, in Supplement 3 for a timeline of key events related to human population growth.)

The rate of population growth has slowed, but the world's population is still growing exponentially at a rate of 1.22% a year. This means that 82 million people were added to the world's population during 2008—an average of nearly 225,000 more people each day, or 2.4 more people every time your heart beats. (See *The Habitable Planet,* Video 5, at **www.learner.org/resources/series209.html** for a discussion of how demographers measure population size and growth.)

Geographically, this growth is unevenly distributed. About 1.2 million of these people were added to the

Links: ⬧ **CORE CASE STUDY** refers to the Core Case Study. ❀ **SUSTAINABILITY** refers to the book's sustainability theme. ⬧ **CONCEPT LINK** indicates links to key concepts in earlier chapters.

123

SCIENCE FOCUS

How Long Can the Human Population Keep Growing?

To survive and provide resources for growing numbers of people, humans have modified, cultivated, built on, or degraded a large and increasing portion of the earth's natural systems. Our activities have directly affected, to some degree, about 83% of the earth's land surface, excluding Antarctica (Figure 3, pp. S24–25, in Supplement 4), as our ecological footprints have spread across the globe (**Concept 1-3**, p. 12, and Figure 1-10, p. 15). In other words, human activities have degraded the various components of earth's biodiversity (Figure 4-2, p. 79) and such threats are expected to increase.

We have used technology to alter much of the rest of nature to meet our growing needs and wants in eight major ways (Figure 6-A).

Scientific studies of populations of other species tell us that *no population can continue growing indefinitely* (**Concept 5-3**, p.108), which is one of the four **scientific principles of sustainability** (see back cover). How long can we continue increasing the earth's carrying capacity for our species by sidestepping many of the factors that sooner or later limit the growth of any population?

The debate over this important question has been going on since 1798 when Thomas Malthus, a British economist, hypothesized that the human population tends to increase exponentially, while food supplies tend to increase more slowly at a linear rate. So far, Malthus has been proven wrong. Food pro-

duction has grown at an exponential rate instead of at a linear rate because of genetic and technological advances in industrialized food production.

No one knows how close we are to the environmental limits that sooner or later will control the size of the human population, but mounting evidence indicates that we are steadily degrading the natural capital, which keeps us and other species alive and supports our economies (**Concept 6-1**).

Critical Thinking

How close do you think we are to the environmental limits of human population growth?

NATURAL CAPITAL DEGRADATION

Altering Nature to Meet Our Needs

Reduction of biodiversity

Increasing use of the earth's net primary productivity

Increasing genetic resistance of pest species and disease-causing bacteria

Elimination of many natural predators

Introduction of potentially harmful species into communities

Using some renewable resources faster than they can be replenished

Interfering with the earth's chemical cycling and energy flow processes

Relying mostly on polluting and climate-changing fossil fuels

CENGAGENOW Active Figure 6-A
Major ways in which humans have altered the rest of nature to meet our growing population's resource needs and wants. *See an animation based on this figure at CengageNOW.* **Questions:** Which three of these items do you believe have been the most harmful? Explain. How does your lifestyle contribute directly or indirectly to each of these three items?

world's developed countries, growing at 0.1% a year. About 80.8 million were added to developing countries, growing 15 times faster at 1.5% a year. In other words, most of the world's population growth takes place in already heavily populated parts of world most of which are the least equipped to deal with the pressures of such rapid growth. In our demographically divided world, roughly 1 billion people live in countries with essentially a stable population size while another billion or so live in countries whose populations are projected to at least double between 2008 and 2050.

How many of us are likely to be here in 2050? Answer: 7.8–10.7 billion people, depending mostly on projections about the average number of babies women are likely to have. The medium projection is 9.3 billion people (Figure 6-2). About 97% of this growth is projected to take place in developing countries, where acute poverty is a way of life for about 1.4 billion people.

The prospects for stabilization of the human population in the near future are nil. However, during this century, the human population may level off as it moves from a J-shaped curve of exponential growth to an S-shaped curve of logistic growth because of various factors that can limit human population growth (Figure 5-11, p. 111).

── **HOW WOULD YOU VOTE?** ☑ ──

Should the population of the country where you live be stabilized as soon as possible? Cast your vote online at **academic.cengage.com/biology/miller**.

This raises the question posed in the **Core Case Study** at the beginning of this chapter: How many people can the earth support indefinitely? Some say about 2 billion. Others say as many as 30 billion. This issue has long been a topic of scientific debate (Science Focus, at left).

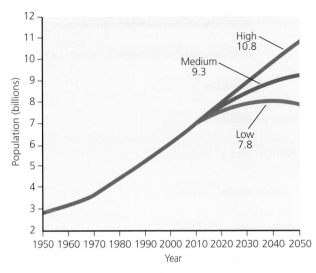

Figure 6-2 *Global connections:* UN world population projections, assuming that by 2050 women will have an average of 2.5 children (high), 2.0 children (medium), or 1.5 children (low). The most likely projection is the medium one—9.3 billion by 2050. (Data from United Nations).

Some analysts believe this is the wrong question. Instead, they say, we should ask what the *optimum sustainable population* of the earth might be, based on the planet's **cultural carrying capacity.** This would be an optimum level that would allow most people to live in reasonable comfort and freedom without impairing the ability of the planet to sustain future generations. (See the Guest Essay by Garrett Hardin at CengageNOW™.)

── **RESEARCH FRONTIER** ──

Determining the optimum sustainable population size for the earth and for various regions. See **academic.cengage.com/biology/miller**.

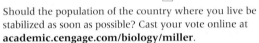

6-2 What Factors Influence the Size of the Human Population?

▶ **CONCEPT 6-2A** Population size increases because of births and immigration and decreases through deaths and emigration.

▶ **CONCEPT 6-2B** The average number of children born to women in a population *(total fertility rate)* is the key factor that determines population size.

The Human Population Can Grow, Decline, or Remain Fairly Stable

On a global basis, if there are more births than deaths during a given period of time, the earth's population increases, and when the reverse is true, it decreases.

When births equal deaths during a particular time period population size does not change.

Human populations of countries and cities grow or decline through the interplay of three factors: *births* (fertility), *deaths* (mortality), and *migration*. We can calculate **population change** of an area by subtracting the

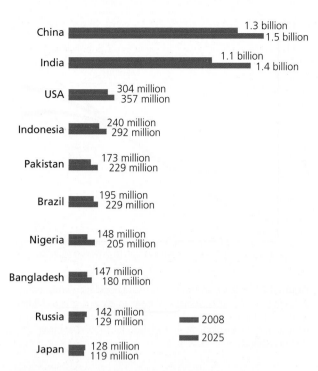

China 1.3 billion / 1.5 billion
India 1.1 billion / 1.4 billion
USA 304 million / 357 million
Indonesia 240 million / 292 million
Pakistan 173 million / 229 million
Brazil 195 million / 229 million
Nigeria 148 million / 205 million
Bangladesh 147 million / 180 million
Russia 142 million / 129 million — 2008 / 2025
Japan 128 million / 119 million

Figure 6-3 *Global connections:* the world's 10 most populous countries in 2008, with projections of their population sizes in 2025. (Data from World Bank and Population Reference Bureau)

number of people leaving a population (through death and emigration) from the number entering it (through birth and immigration) during a specified period of time (usually one year) (**Concept 6-2A**). See Figures 5 and 6, p. S13, in Supplement 3.

$$\text{Population change} = (\text{Births} + \text{Immigration}) - (\text{Deaths} + \text{Emigration})$$

When births plus immigration exceed deaths plus emigration, population increases; when the reverse is true, population declines.

Instead of using the total numbers of births and deaths per year, population experts (demographers) use the **birth rate,** or **crude birth rate** (the number of live births per 1,000 people in a population in a given year), and the **death rate,** or **crude death rate** (the number of deaths per 1,000 people in a population in a given year).

What five countries had the largest numbers of people in 2008? Number 1 was China with 1.3 billion people, or one of every five people in the world (Figures 6-1 and 6-3). Number 2 was India with 1.1 billion people, or one of every six people. Together, China and India have 36% of the world's population. The United States, with 304 million people in 2008—had the world's third largest population but only 4.5% of the world's people. Can you guess the next two most populous countries? What three countries are expected to have the most people in 2025? Look at Figure 6-3 to see if your answers are correct.

Women Are Having Fewer Babies but Not Few Enough to Stabilize the World's Population

Another measurement used in population studies is **fertility rate,** the number of children born to a woman during her lifetime. Two types of fertility rates affect a country's population size and growth rate. The first type, called the **replacement-level fertility rate,** is the average number of children that couples in a population must bear to replace themselves. It is slightly higher than two children per couple (2.1 in developed countries and as high as 2.5 in some developing countries), mostly because some children die before reaching their reproductive years.

Does reaching replacement-level fertility bring an immediate halt to population growth? No, because so many *future* parents are alive. If each of today's couples had an average of 2.1 children, they would not be contributing to population growth. But if all of today's girl children also have 2.1 children, the world's population will continue to grow for 50 years or more (assuming death rates do not rise).

The second type of fertility rate, the **total fertility rate (TFR),** is the average number of children born to women in a population during their reproductive years. This factor plays a key role in determining population size (**Concept 6-2B**). The average fertility rate has been declining. In 2008, the average global TFR was 2.6 children per woman: 1.6 in developed countries (down from 2.5 in 1950) and 2.8 in developing countries (down from 6.5 in 1950). Although the decline in TFR in developing countries is impressive, the TFR remains far above the replacement level of 2.1, not low enough to stabilize the world's population in the near future. See Figures 7 and 8, p. S14, in Supplement 3.

THINKING ABOUT
The Slower Rate of Population Growth

Why has the world's exponential rate of population growth slowed down in the last few decades? What would have to happen for the world's population to stop growing?

■ CASE STUDY
The U.S. Population Is Growing Rapidly

The population of the United States grew from 76 million in 1900 to 304 million in 2008, despite oscillations in the country's TFR (Figure 6-4) and birth rates (Figure 6-5). It took the country 139 years to add its

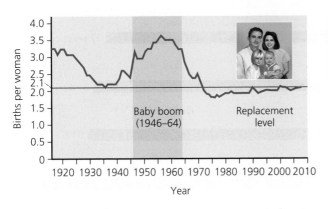

Figure 6-4 Total fertility rates for the United States between 1917 and 2008. **Question:** The U.S. fertility rate has declined and remained at or below replacement levels since 1972, so why is the population of the United States still increasing? (Data from Population Reference Bureau and U.S. Census Bureau)

Figure 6-5 Birth rates in the United States, 1910–2008. Use this figure to trace changes in crude birth rates during your lifetime. (Data from U.S. Bureau of Census and U.S. Commerce Department)

first 100 million people, 52 years to add another 100 million by 1967, and only 39 years to add the third 100 million in 2006. The period of high birth rates between 1946 and 1964 is known as the *baby boom,* when 79 million people were added to the U.S. population. In 1957, the peak of the baby boom, the TFR reached 3.7 children per woman. Since then, it has generally declined, remaining at or below replacement level since 1972. In 2008, the TFR was 2.1 children per woman, compared to 1.6 in China.

The drop in the TFR has slowed the rate of population growth in the United States. But the country's population is still growing faster than that of any other developed country, and of China, and is not close to leveling off. About 2.9 million people (one person every 11 seconds) were added to the U.S. population in 2008. About 66% (1.9 million) of this growth occurred because births outnumbered deaths and 34% (1 million) came from legal and illegal immigration (with someone migrating to the U.S. every 32 seconds).

In addition to the almost fourfold increase in population growth since 1900, some amazing changes in lifestyles took place in the United States during the 20th century (Figure 6-6, p. 128), which led to dramatic increases in per capita resource use and a much larger U.S. ecological footprint (**Concept 1-3,** p. 12, and Figure 1-10, top, p. 15). **CONCEPT LINK**

Here are a few more changes that occurred during the last century. In 1907, the three leading causes of death in the United States were pneumonia, tuberculosis, and diarrhea; 90% of U.S. doctors had no college education; one out of five adults could not read or write; only 6% of Americans graduated from high school; the average U.S. worker earned $200–400 per year and the average daily wage was 22 cents per hour; there were only 9,000 cars in the U.S., and only 232 kilometers (144 miles) of paved roads; a 3-minute phone call from Denver, Colorado, to New York city cost $11; only 30 people lived in Las Vegas, Nevada; most women washed their hair only once a month; marijuana, heroin, and morphine were available over the counter at local drugstores; and there were only 230 reported murders in the entire country.

According to U.S. Census Bureau, the U.S. population is likely to increase from 304 million in 2008 to 438 million by 2050 and then to 571 million by 2100. In contrast, population growth has slowed in other major developed countries since 1950, most of which are expected to have declining populations after 2010. Because of a high per capita rate of resource use and the resulting waste and pollution, each addition to the U.S. population has an enormous environmental impact (Figure 1-9, bottom, p. 14, Figure 1-10, p. 15, and Figure 7 on pp. S28–S29 in Supplement 4).

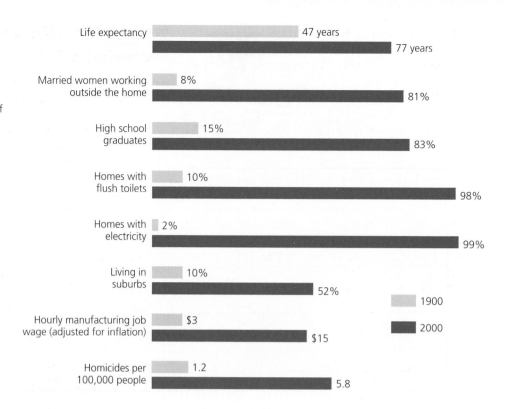

Figure 6-6 Some major changes that took place in the United States between 1900 and 2000. **Question:** Which two of these changes do you think were the most important? (Data from U.S. Census Bureau and Department of Commerce)

Life expectancy — 47 years / 77 years

Married women working outside the home — 8% / 81%

High school graduates — 15% / 83%

Homes with flush toilets — 10% / 98%

Homes with electricity — 2% / 99%

Living in suburbs — 10% / 52%

Hourly manufacturing job wage (adjusted for inflation) — $3 / $15

Homicides per 100,000 people — 1.2 / 5.8

1900 / 2000

Several Factors Affect Birth Rates and Fertility Rates

Many factors affect a country's average birth rate and TFR. One is the *importance of children as a part of the labor force*. Proportions of children working tend to be higher in developing countries.

Another economic factor is the *cost of raising and educating children*. Birth and fertility rates tend to be lower in developed countries, where raising children is much more costly because they do not enter the labor force until they are in their late teens or twenties. In the United States, it costs about $290,000 to raise a middle-class child from birth to age 18. By contrast, many children in poor countries have to work to help their families survive.

The *availability of private and public pension systems* can influence the decision for some couples on how many children to have, especially the poor in developing countries. Pensions reduce a couple's need to have many children to help support them in old age.

Urbanization plays a role. People living in urban areas usually have better access to family planning services and tend to have fewer children than do those living in rural areas (especially in developing countries) where children are often needed to help raise crops and carry daily water and fuelwood supplies.

Another important factor is the *educational and employment opportunities available for women*. TFRs tend to be low when women have access to education and paid employment outside the home. In developing countries, a woman with no education typically has two more children than does a woman with a high school edu-

cation. In nearly all societies, better-educated women tend to marry later and have fewer children.

Another factor is the **infant mortality rate**—the number of children per 1,000 live births who die before one year of age. In areas with low infant mortality rates, people tend to have fewer children because fewer children die at an early age.

Average age at marriage (or, more precisely, the average age at which a woman has her first child) also plays a role. Women normally have fewer children when their average age at marriage is 25 or older.

Birth rates and TFRs are also affected by the *availability of legal abortions*. Each year about 190 million women become pregnant. The United Nations and the World Bank estimate that 46 million of these women get abortions—26 million of them legal and 20 million illegal (and often unsafe). Also, the *availability of reliable birth control methods* allows women to control the number and spacing of the children they have.

Religious beliefs, traditions, and cultural norms also play a role. In some countries, these factors favor large families and strongly oppose abortion and some forms of birth control.

Several Factors Affect Death Rates

The rapid growth of the world's population over the past 100 years is not primarily the result of a rise in the crude birth rate. Instead, it has been caused largely by a decline in crude death rates, especially in developing countries.

More people started living longer and fewer infants died because of increased food supplies and distribution, better nutrition, medical advances such as immunizations and antibiotics, improved sanitation, and safer water supplies (which curtailed the spread of many infectious diseases).

Two useful indicators of the overall health of people in a country or region are **life expectancy** (the average number of years a newborn infant can expect to live) and the **infant mortality rate** (the number of babies out of every 1,000 born who die before their first birthday). Between 1955 and 2008, the global life expectancy increased from 48 years to 68 years (77 years in developed countries and 67 years in developing countries) and is projected to reach 74 by 2050. Between 1900 and 2008, life expectancy in the United States increased from 47 to 78 years and, by 2050, is projected to reach 82 years. In the world's poorest countries, however, life expectancy is 49 years or less and may fall further in some countries because of more deaths from AIDS.

Even though more is spent on health care per person in the United States than in any other country, people in 41 other countries including Canada, Japan, Singapore, and a number of European countries have longer life expectancies than do Americans. Analysts cite two major reasons for this. First, 45 million Americans lack health care insurance, while Canada and many European countries have universal health care insurance. Second, adults in the United States have one of the world's highest obesity rates.

Infant mortality is viewed as one of the best measures of a society's quality of life because it reflects a country's general level of nutrition and health care. A high infant mortality rate usually indicates insufficient food (undernutrition), poor nutrition (malnutrition), and a high incidence of infectious disease (usually from drinking contaminated water and having weakened disease resistance due to undernutrition and malnutrition).

Between 1965 and 2008, the world's infant mortality rate dropped from 20 to 6.3 in developed countries and from 118 to 59 in developing countries (see Figures 9 and 10, p. S15, in Supplement 3). This is good news, but annually, more than 4 million infants (most in developing countries) die of preventable causes during their first year of life—an average of 11,000 mostly unnecessary infant deaths per day. This is equivalent to 55 jet airliners, each loaded with 200 infants younger than age 1, crashing *each day* with no survivors!

The U.S. infant mortality rate declined from 165 in 1900 to 6.6 in 2008. This sharp decline was a major factor in the marked increase in U.S. average life expectancy during this period. Still, some 40 countries, including Taiwan, Cuba, and most of Europe had lower infant mortality rates than the United States had in 2008. Three factors helped keep the U.S. infant mortality rate high: *inadequate health care for poor women during pregnancy and for their babies after birth, drug addiction among pregnant women,* and *a high birth rate among teenagers* (although this rate dropped by almost half between 1991 and 2006).

Migration Affects an Area's Population Size

The third factor in population change is **migration:** the movement of people into (*immigration*) and out of (*emigration*) specific geographic areas.

Most people migrating from one area or country to another seek jobs and economic improvement. But some are driven by religious persecution, ethnic conflicts, political oppression, wars, and environmental degradations such as water and food shortages and soil erosion. According to a U.N. study, there were about 25 million *environmental refugees* in 2005 and the number could reach 50 million by 2010. Environmental scientist Norman Myers warns that, in a warmer world, the number of such refugees could soar to 250 million or more before the end of this century. (See more on this in the Guest Essay by Norman Myers at CengageNOW.)

■ CASE STUDY

The United States: A Nation of Immigrants

Since 1820, the United States has admitted almost twice as many immigrants and refugees as all other countries combined. The number of legal immigrants (including refugees) has varied during different periods because of changes in immigration laws and rates of economic growth (Figure 6-7). Currently, legal and illegal immigration account for about 40% of the country's annual population growth.

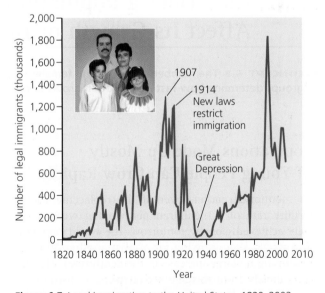

Figure 6-7 Legal immigration to the United States, 1820–2003 (the last year for which data are available). The large increase in immigration since 1989 resulted mostly from the Immigration Reform and Control Act of 1986, which granted legal status to illegal immigrants who could show they had been living in the country for several years. (Data from U.S. Immigration and Naturalization Service and the Pew Hispanic Center)

Between 1820 and 1960, most legal immigrants to the United States came from Europe. Since 1960, most have come from Latin America (53%) and Asia (25%), followed by Europe (14%). In 2007, Latinos (67% of them from Mexico) made up 15% of the U.S. population, and by 2050, are projected to make up 25% of the population. According to the Pew Hispanic Center, 53% of the 100 million Americans that were added to the population between 1967 and 2007 were either immigrants or their children.

There is controversy over whether to reduce legal immigration to the United States. Some analysts would accept new entrants only if they can support themselves, arguing that providing legal immigrants with public services makes the United States a magnet for the world's poor. Proponents of reducing legal immigration argue that it would allow the United States to stabilize its population sooner and help reduce the country's enormous environmental impact from its huge ecological footprint (Figure 1-10, p. 15; and **Concept 1-3**, p. 12). CONCEPT LINK

Polls show that almost 60% of the U.S. public strongly supports reducing legal immigration. There is also intense political controversy over what to do about illegal immigration. In 2007, there were an estimated 11.3 million illegal immigrants in the United States, with about 58% of them from Mexico and 22% from other Latin American countries.

Those opposed to reducing current levels of legal immigration argue that it would diminish the histori-cal role of the United States as a place of opportunity for the world's poor and oppressed and as a source of cultural diversity that has been a hallmark of American culture since its beginnings. In addition, according to several studies, including a 2006 study by the Pew Hispanic Center, immigrants and their descendants pay taxes, take many menial and low-paying jobs that most other Americans shun, start new businesses, create jobs, add cultural vitality, and help the United States succeed in the global economy. Also, according to the U.S. Census Bureau, after 2020, higher immigration levels will be needed to supply enough workers as baby boomers retire.

According to a recent study by the U.N. Population Division, if the United States wants to maintain its current ratio of workers to retirees, it will need to absorb an average of 10.8 million immigrants each year—more than 13 times the current immigration level—through 2050. At that point, the U.S. population would total 1.1 billion people, 73% of them fairly recent immigrants or their descendants. Housing this influx of almost 11 million immigrants per year would require the equivalent of building another New York City every 10 months.

— **HOW WOULD YOU VOTE?** ☑ —

Should legal immigration into the United States be reduced? Cast your vote online at **academic.cengage .com/biology/miller**.

6-3 How Does a Population's Age Structure Affect Its Growth or Decline?

▶ **CONCEPT 6-3** The numbers of males and females in young, middle, and older age groups determine how fast a population grows or declines.

Populations Made Up Mostly of Young People Can Grow Rapidly

As mentioned earlier, even if the replacement-level fertility rate of 2.1 children per woman were magically achieved globally tomorrow, the world's population would keep growing for at least another 50 years (assuming no large increase in the death rate). This results mostly from the **age structure:** the distribution of males and females among age groups in a population—in this case, the world population (**Concept 6-3**).

Population experts construct a population *age-structure diagram* by plotting the percentages or numbers of males and females in the total population in each of three age categories: *prereproductive* (ages 0–14), *reproductive* (ages 15–44), and *postreproductive* (ages 45 and older). Figure 6-8 presents generalized age-structure diagrams for countries with rapid, slow, zero, and negative population growth rates.

Any country with many people younger than age 15 (represented by a wide base in Figure 6-8, far left) has a powerful built-in momentum to increase its population size unless death rates rise sharply. The number of births will rise even if women have only one or two children, because a large number of girls will soon be moving into their reproductive years.

What is one of the world's most important population statistics? *Nearly 28% of the people on the planet were under 15 years old in 2008.* These 1.9 billion young people are poised to move into their prime reproductive

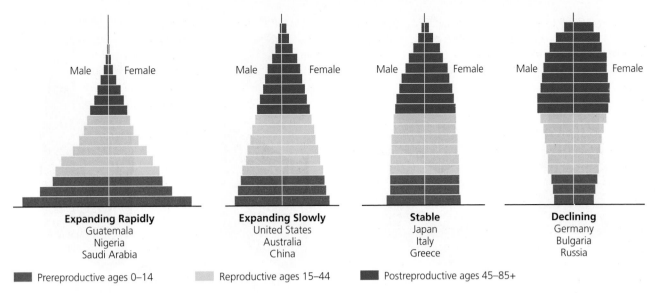

Expanding Rapidly
Guatemala
Nigeria
Saudi Arabia

Expanding Slowly
United States
Australia
China

Stable
Japan
Italy
Greece

Declining
Germany
Bulgaria
Russia

■ Prereproductive ages 0–14   ▢ Reproductive ages 15–44   ■ Postreproductive ages 45–85+

CENGAGENOW™ **Active Figure 6-8** Generalized population age structure diagrams for countries with rapid (1.5–3%), slow (0.3–1.4%), zero (0–0.2%), and negative (declining) population growth rates. A population with a large proportion of its people in the prereproductive age group (far left) has a large potential for rapid population growth. *See an animation based on this figure at* CengageNOW. **Question:** Which of these figures best represents the country where you live? (Data from Population Reference Bureau)

years. In developing countries, the percentage is even higher: 30% on average (41% in Africa) compared with 17% in developed countries (20% in the United States and 16% in Europe). These differences in population age structure between developed and developing countries are dramatic, as Figure 6-9 reveals. This figure also shows why almost all of future human population growth will be in developing countries.

We Can Use Age-Structure Information to Make Population and Economic Projections

Changes in the distribution of a country's age groups have long-lasting economic and social impacts. Between 1946 and 1964, the United States had a *baby boom,* which added 79 million people to its population. Over time, this group looks like a bulge moving up through the country's age structure, as shown in Figure 6-10, p. 132.

Baby boomers now make up almost half of all adult Americans. As a result, they dominate the population's demand for goods and services and play increasingly important roles in deciding who gets elected and what laws are passed. Baby boomers who created the youth market in their teens and twenties are now creating the 50-something market and will soon move on to create a 60-something market. After 2011, when the first baby boomers will turn 65, the number of Americans older than age 65 will grow sharply through 2029 in what

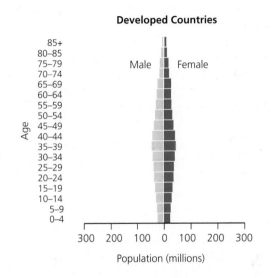

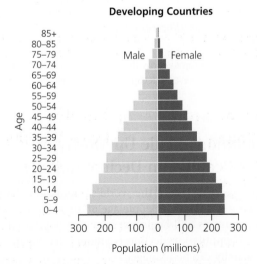

Figure 6-9 *Global outlook:* population structure by age and sex in developing countries and developed countries, 2006. **Question:** If all girls under 15 had only one child during their lifetimes, how do you think these structures would change over time? (Data from United Nations Population Division and Population Reference Bureau)

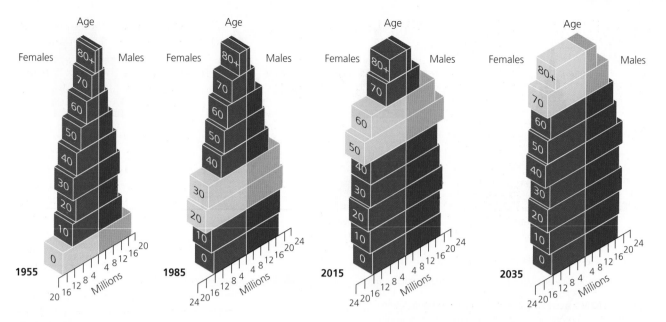

CENGAGENOW™ **Active Figure 6-10** Tracking the baby-boom generation in the United States. U.S. population by age and sex, 1955, 1985, 2015, and 2035 (projected). *See an animation based on this figure at* CengageNOW. (Data from U.S. Census Bureau)

has been called the *graying of America*. In 2008, about 13% of Americans were 65 or older, but that number is projected to increase to about 25% by 2043.

According to some analysts, the retirement of baby boomers is likely to create a shortage of workers in the United States unless immigrant workers or various forms of automation replace some of them. Retired baby boomers may use their political clout to have the smaller number of people in the baby-bust generation that followed them pay higher income, health-care, and social security taxes. However, the rapidly increasing number of immigrants and their descendants may dilute their political power. Their power may also be weakened by the rise of members of the echo baby boom generation (Figure 6-5).

CENGAGENOW™ Examine how the baby boom affects the U.S. age structure over several decades at CengageNOW.

Populations Made Up Mostly of Older People Can Decline Rapidly

As the age structure of the world's population changes and the percentage of people age 60 or older increases, more countries will begin experiencing population declines. If population decline is gradual, its harmful effects usually can be managed.

Japan has the world's highest proportion of elderly people and the lowest proportion of young people. Its population, at 128 million in 2008, is projected to decline to about 96 million by 2050.

Rapid population decline can lead to severe economic and social problems. A country that experiences a fairly rapid "baby bust" or a "birth dearth" when its TFR falls below 1.5 children per couple for a prolonged period sees a sharp rise in the proportion of older people. This puts severe strains on government budgets because these individuals consume an increasingly larger share of medical care, social security funds, and other costly public services, which are funded by a decreasing number of working taxpayers. Such countries can also face labor shortages unless they rely more heavily on automation or massive immigration of foreign workers. In the next two to three decades, countries such as the United States and many European nations with rapidly aging populations will face shortages of health workers. For example, the United Nations will need half a million more nurses by 2020 and twice as many doctors specializing in health care for the elderly (geriatrics).

Figure 6-11 lists some of the problems associated with rapid population decline. Countries faced with a rapidly declining population in the future include Japan, Russia, Germany, Bulgaria, the Czech Republic, Hungary, Poland, Ukraine, Greece, Italy, and Spain.

Populations Can Decline from a Rising Death Rate: The AIDS Tragedy

A large number of deaths from AIDS can disrupt a country's social and economic structure by removing significant numbers of young adults from its age

Some Problems with Rapid Population Decline

Can threaten economic growth

Labor shortages

Less government revenues with fewer workers

Less entrepreneurship and new business formation

Less likelihood for new technology development

Increasing public deficits to fund higher pension and health-care costs

Pensions may be cut and retirement age increased

Figure 6-11 Some problems with rapid population decline. **Question:** Which three of these problems do you think are the most important?

structure. According to the World Health Organization, AIDS had killed 25 million people by 2008. Unlike hunger and malnutrition, which kill mostly infants and children, AIDS kills many young adults.

This change in the young-adult age structure of a country has a number of harmful effects. One is a sharp drop in average life expectancy. In 8 African countries, where 16–39% of the adult population is infected with HIV, life expectancy could drop to 34–40 years.

Another effect is a loss of a country's most productive young adult workers and trained personnel such as scientists, farmers, engineers, teachers, and government, business, and health-care workers. This causes a sharp drop in the number of productive adults available to support the young and the elderly and to grow food and provide essential services. Within a decade, countries such as Zimbabwe and Botswana in sub-Saharan Africa could lose more than a fifth of their adult populations.

Analysts call for the international community—especially developed countries—to create and fund a massive program to help countries ravaged by AIDS in Africa and elsewhere. This program would have two major goals. *First,* reduce the spread of HIV through a combination of improved education and health care. *Second,* provide financial assistance for education and health care as well as volunteer teachers and health-care and social workers to help compensate for the missing young-adult generation.

6-4 How Can We Slow Human Population Growth?

▶ **CONCEPT 6-4** Experience indicates that the most effective ways to slow human population growth are to encourage family planning, to reduce poverty, and to elevate the status of women.

As Countries Develop, Their Populations Tend to Grow More Slowly

Demographers examining birth and death rates of western European countries that became industrialized during the 19th century developed a hypothesis of population change known as the **demographic transition:** as countries become industrialized, first their death rates and then their birth rates decline. According to the hypothesis, based on such data, this transition takes place in four distinct stages (Figure 6-12, p. 134).

Some analysts believe that most of the world's developing countries will make a demographic transi-

tion over the next few decades mostly because modern technology can bring economic development and family planning to such countries. Others fear that the still-rapid population growth in some developing countries might outstrip economic growth and overwhelm some local life-support systems. As a consequence, some of these countries could become caught in a *demographic trap* at stage 2. This is now happening as death rates rise in a number of developing countries, especially in Africa. Indeed, countries in Africa being ravaged by the HIV/AIDS epidemic are falling back to stage 1.

Other factors that could hinder the demographic transition in some developing countries are shortages of scientists and engineers (94% of them work in the industrialized world), shortages of skilled workers,

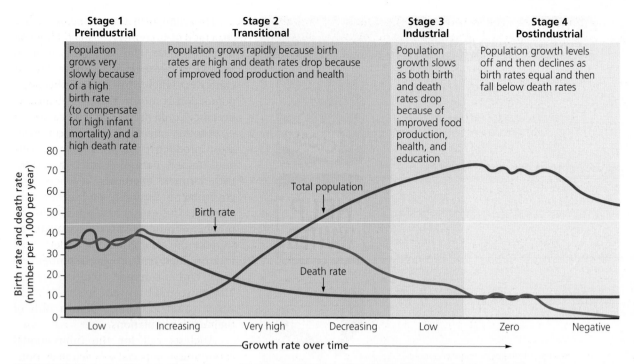

	Stage 1 Preindustrial	Stage 2 Transitional	Stage 3 Industrial	Stage 4 Postindustrial
	Population grows very slowly because of a high birth rate (to compensate for high infant mortality) and a high death rate	Population grows rapidly because birth rates are high and death rates drop because of improved food production and health	Population growth slows as both birth and death rates drop because of improved food production, health, and education	Population growth levels off and then declines as birth rates equal and then fall below death rates

Total population

Birth rate

Death rate

Birth rate and death rate (number per 1,000 per year)

80 70 60 50 40 30 20 10 0

Low — Increasing — Very high — Decreasing — Low — Zero — Negative

Growth rate over time

CENGAGENOW™ **Active Figure 6-12** Four stages of the *demographic transition,* which the population of a country can experience when it becomes industrialized. There is uncertainty about whether this model will apply to some of today's developing countries. *See an animation based on this figure at* CengageNOW. **Question:** At what stage is the country where you live?

insufficient financial capital, large debts to developed countries, and a drop in economic assistance from developed countries since 1985.

CENGAGENOW™ Explore the effects of economic development on birth and death rates and population growth at CengageNOW.

Planning for Babies Works

Family planning provides educational and clinical services that help couples choose how many children to have and when to have them. Such programs vary from culture to culture, but most provide information on birth spacing, birth control, and health care for pregnant women and infants.

Family planning has been a major factor in reducing the number of births throughout most of the world, mostly because of increased knowledge and availability of contraceptives. According to the U.N. Population Division, 58% of married women ages 15–45 in developed countries and 54% in developing countries used modern contraception in 2008. Family planning has also reduced the number of legal and illegal abortions performed each year and decreased the number of deaths of mothers and fetuses during pregnancy.

Studies by the U.N. Population Division and other population agencies indicate that family planning is responsible for at least 55% of the drop in total fertility

rates (TFRs) in developing countries, from 6.0 in 1960 to 3.0 in 2008. Between 1971 and 2008, for example, Thailand used family planning to cut its annual population growth rate from 3.2% to 0.5% and its TFR from 6.4 to 1.6 children per family. Another family planning success involves Iran, which between 1989 and 2000, cut its population growth rate from 2.5% to 1.4%.

Despite such successes, two problems remain. *First,* according to the U.N. Population Fund, 42% of all pregnancies in developing countries are unplanned and 26% end with abortion. *Second,* an estimated 201 million couples in developing countries want to limit the number and determine the spacing of their children, but they lack access to family planning services. According to a recent study by the U.N. Population Fund and the Alan Guttmacher Institute, meeting women's current unmet needs for family planning and contraception could *each year* prevent 52 million unwanted pregnancies, 22 million induced abortions, 1.4 million infant deaths, and 142,000 pregnancy-related deaths.

Some analysts call for expanding family planning programs to include teenagers and sexually active unmarried women, who are excluded from many existing programs. Another suggestion is to develop programs that educate men about the importance of having fewer children and taking more responsibility for raising them. Proponents also call for greatly increased research on developing more effective and more acceptable birth control methods for men.

In 1994, the United Nations held its third Conference on Population and Development in Cairo, Egypt.

One of the conference's goals was to encourage action to stabilize the world's population at 7.8 billion by 2050 instead of the projected 9.2 billion.

The experiences of countries such as Japan, Thailand, South Korea, Taiwan, Iran, and China show that a country can achieve or come close to replacement-level fertility within a decade or two. Such experiences also suggest that the best ways to slow and stabilize population growth are through *investing in family planning, reducing poverty,* and *elevating the social and economic status of women* (**Concept 6-4**).

Empowering Women Can Slow Population Growth

Studies show that women tend to have fewer children if they are educated, hold a paying job outside the home, and live in societies where their human rights are not suppressed. Although women make up roughly half of the world's population, in most societies they do not have the same rights and educational and economic opportunities as men do.

Women do almost all of the world's domestic work and child care for little or no pay and provide more unpaid health care than all of the world's organized health services combined. They also do 60–80% of the work associated with growing food, gathering and hauling wood (Figure 6-13) and animal dung for use as fuel, and hauling water in rural areas of Africa, Latin America, and Asia. As one Brazilian woman put it, "For poor women the only holiday is when you are asleep."

Globally, women account for two-thirds of all hours worked but receive only 10% of the world's income,

and they own less than 2% of the world's land. Also, about 70% of the world's poor and 64% of all 800 million illiterate adults are women.

Because sons are more valued than daughters in many societies, girls are often kept at home to work instead of being sent to school. Globally, some 900 million girls—three times the entire U.S. population—do not attend elementary school. Teaching women to read has a major impact on fertility rates and population growth. Poor women who cannot read often have five to seven children, compared to two or fewer in societies where almost all women can read.

According to Thorya Obaid, executive director of the U.N. Population Fund, "Many women in the developing world are trapped in poverty by illiteracy, poor health, and unwanted high fertility. All of these contribute to environmental degradation and tighten the grip of poverty."

An increasing number of women in developing countries are taking charge of their lives and reproductive behavior. As it expands, such bottom-up change by individual women will play an important role in stabilizing population and reducing environmental degradation.

■ CASE STUDY

Slowing Population Growth in China: The One–Child Policy

China has made impressive efforts to feed its people, bring its population growth under control, and encourage economic growth. Between 1972 and 2008, the country cut its crude birth rate in half and trimmed its

Figure 6-13 Women from a village in the West African country of Burkina Faso returning with fuelwood. Typically they spend 2 hours a day two or three times a week searching for and hauling fuelwood.

Mark Edwards/Peter Arnold, Inc.

TFR from 5.7 to 1.6 children per woman, compared to 2.1 in the United States. Despite such drops China is the world's most populous country and in 2008 added about 6.8 million people to its population (compared to 2.9 million in the United States and 18 million in India). If current trends continue, China's population is expected to peak by about 2033 at around 1.46 billion and then to begin a slow decline.

Since 1980, China has moved 350 million people (an amount greater than the entire U.S. population) from extreme poverty to its consumer middle class and is likely to double that number by 2010. However, about 47% of its people were struggling to live on less than $2 (U.S.) a day in 2006. China also has a literacy rate of 91% and has boosted life expectancy to 73 years. By 2020, some economists project that China could become the world's leading economic power.

In the 1960s, Chinese government officials concluded that the only alternative to mass starvation was strict population control. To achieve a sharp drop in fertility, China established the most extensive, intrusive, and strict family planning and population control program in the world. It discourages premarital sex and urges people to delay marriage and limit their families to one child each. Married couples who pledge to have no more than one child receive more food, larger pensions, better housing, free health care, salary bonuses, free school tuition for their child, and preferential employment opportunities for their child. Couples who break their pledge lose such benefits.

The government also provides married couples with free sterilization, contraceptives, and abortion. However, reports of forced abortions and other coercive actions have brought condemnation from the United States and other national governments.

In China, there is a strong preference for male children, because unlike sons, daughters are likely to marry and leave their parents. A folk saying goes, "Rear a son, and protect yourself in old age." Some pregnant Chinese women use ultrasound to determine the gender of their fetuses, and some get an abortion if it is female. The result: a rapidly growing *gender imbalance* or "bride shortage" in China's population, with a projected 30–40 million surplus of men expected by 2020. Because of this skewed sex ratio, teen-age girls in some parts of rural China are being kidnapped and sold as brides for single men in other parts of the country.

With fewer children, the average age of China's population is increasing rapidly. By 2020, 31% of the Chinese population will be over 60 years old, compared to 8% in 2008. This graying of the Chinese population could lead to a declining work force, higher wages for younger workers, a shortage of funding for continuing economic development, and fewer children and grandchildren to care for the growing number of elderly people. These concerns and other factors may slow economic growth and lead to some relaxation of China's one-child population control policy in the future.

China also faces serious resource and environmental problems that could limit its economic growth. It has 19% of the world's population, but only 7% of the world's freshwater and cropland, 4% of its forests, and 2% of its oil. In 2002, only 15% of China's land area was protected on paper (compared to 23% in the United States, 51% in Japan, and 63% in Venezuela) and only 29% of its rural population had access to adequate sanitation.

In 2005, China's deputy minister of the environment summarized the country's environmental problems: "Our raw materials are scarce, we don't have enough land, and our population is constantly growing. Half of the water in our seven largest rivers is completely useless. One-third of the urban population is breathing polluted air."

China's economy is growing at one of the world's highest rates as the country undergoes rapid industrialization. More middle class Chinese (Case Study, p. 15) will consume more resources per person, increasing China's ecological footprint (Figure 1-10, p. 15) within its own borders and in other parts of the world that provide it with resources (**Concept 1-3**, p. 12). This will put a strain on the earth's natural capital unless China steers a course toward more sustainable economic development.

CONCEPT LINK

■ CASE STUDY
Slowing Population Growth in India

For more than 5 decades, India has tried to control its population growth with only modest success. The world's first national family planning program began in India in 1952, when its population was nearly 400 million. By 2008, after 56 years of population control efforts, India had 1.1 billion people.

In 1952, India added 5 million people to its population. In 2008, it added 18 million—more than any other country. By 2015, India is projected to be the world's most populous country, with its population projected to reach 1.76 billion by 2050.

India faces a number of serious poverty, malnutrition, and environmental problems that could worsen as its population continues to grow rapidly. India has a thriving and rapidly growing middle class of more than 300 million people—roughly equal to the entire U.S. population—many of them highly skilled software developers and entrepreneurs.

By global standards, however, one of every four people in India is poor, despite the fact that since 2004 it has had the world's second fastest growing economy, and by 2007, was the world's fourth largest economy. Such prosperity and progress have not touched many of the nearly 650,000 villages where more than two-thirds of India's population lives. In 2002, only 18% of its rural population had access to adequate sanitation. In 2006, nearly half of the country's labor force was unemployed or underemployed and 80% of its people were struggling to live on less than $2 (U.S.) day (see Photo 2 in the Detailed Contents).

Although India currently is self-sufficient in food grain production, about 40% of its population and more than half of its children suffer from malnutrition, mostly because of poverty. In 2002, only 5% of the country's land was protected on paper.

The Indian government has provided information about the advantages of small families for years and has also made family planning available throughout the country. Even so, Indian women have an average of 2.8 children. Most poor couples still believe they need many children to work and care for them in old age. As in China, the strong cultural preference for male children also means some couples keep having children until they produce one or more boys. The result: even though 90% of Indian couples know of at least one modern birth control method, only 48% actually use one.

Like China, India also faces critical resource and environmental problems. With 17% of the world's people, India has just 2.3% of the world's land resources and 2% of the forests. About half the country's cropland is degraded as a result of soil erosion and over-grazing. In addition, more than two-thirds of its water is seriously polluted, sanitation services often are inadequate, and many of its major cities suffer from serious air pollution.

India is undergoing rapid economic growth, which is expected to accelerate. As members of its huge and growing middle class increase their resource use per person, India's ecological footprint (**Concept 1-3**) (Figure 1-10, p. 15) will expand and increase the pressure on the country's and the earth's natural capital.

On the other hand, economic growth may help to slow population growth by accelerating India's demographic transition. By 2050, India—the largest democracy the world has ever seen—could become the world's leading economic power.

THINKING ABOUT

China, India, the United States, and Overpopulation

 CORE CASE STUDY

Based on population size and resource use per person (Figure 1-10, p. 15) is the United States more overpopulated than China? Explain. Answer the same question for the United States versus India.

REVISITING   Population Growth and Sustainability

 CONCEPT LINK   CORE CASE STUDY   SUSTAINABILITY

This chapter began with a discussion of whether the world is overpopulated (**Core Case Study**). As we have noted, some experts say this is the wrong question to be asking. Instead, they believe we ought to ask, "What is the optimal level of human population that the planet can support *sustainably?*" In other words, "What is the maximum number of people that can live comfortably without seriously degrading the earth's biodiversity and other forms of natural capital and jeopardizing the earth's ability to provide the same comforts for future generations?"

In the first six chapters of this book, you have learned how ecosystems and species have been sustained throughout history in keeping with four **scientific principles of sustainability**—relying on solar energy, biodiversity, population control, and nutrient recycling (see back cover and **Concept 1-6**, p. 23). In this chapter, you may have gained a sense of the need for humans to apply these sustainability principles to their lifestyles and economies, especially with regard to human population growth, globally and in particular countries.

In the next five chapters, you will learn how various principles of ecology and these four **scientific principles of sustainability** can be applied to help preserve the earth's biodiversity.

Our numbers expand but Earth's natural systems do not.

LESTER R. BROWN

REVIEW

1. Review the Key Questions and Concepts in this chapter on p. 123. Do you think the world is overpopulated? Explain.

2. List three factors that account for the rapid growth of the world's human population over the past 200 years. Describe eight ways in which we have used technology to alter nature to meet our growing needs and wants. How many of us are likely to be here in 2050?

3. What is the **cultural carrying capacity** of a population? How do some analysts apply this concept in considering the question of whether the earth is overpopulated?

4. List four variables that affect the **population change** of an area and write an equation showing how they are related. Distinguish between **crude birth rate** and **crude death rate.** What five countries had the largest numbers of people in 2008?

5. What is **fertility rate?** Distinguish between **replacement-level fertility rate** and **total fertility rate (TFR).** Explain why reaching the replacement-level fertility rate will not stop global population growth until about 50 years have passed (assuming that death rates do not rise).

6. Describe population growth in the United States and explain why it is high compared to those of most other developed countries and China. Is the United States overpopulated? Explain.

7. List ten factors that can affect the birth rate and fertility rate of a country. Distinguish between **life expectancy** and **infant mortality rate** and explain how they affect the population size of a country. Why does the United States have a lower life expectancy and higher infant mortality rate than a number of other countries? What is **migration?** Describe immigration into the United States and the issues it raises.

8. What is the **age structure** of a population. Explain how it affects population growth and economic growth. What

are some problems related to rapid population decline from an aging population?

9. What is the **demographic transition** and what are its four stages? What factors could hinder some developing countries from making this transition? What is **family planning?** Describe the roles of family planning, reducing poverty, and elevating the status of women in slowing population growth. Describe China's and India's efforts to control their population growth.

10. How has human population growth (**Core Case Study**) interfered with natural processes related to three of the **scientific principles of sustainability?** Name the three principles, and for each one, describe the effects of rapid human population growth.

Note: Key Terms are in **bold** type.

CRITICAL THINKING

1. List three ways in which you could apply some of what you learned in this chapter to making your lifestyle more environmentally sustainable.

2. Which of the three major environmental worldviews summarized on p. 20 do you believe underlie the two major positions on whether the world is overpopulated (**Core Case Study**)?

3. Identify a major local, national, or global environmental problem, and describe the role of population growth in this problem.

4. Is it rational for a poor couple in a developing country such as India to have four or five children? Explain.

5. Do you believe that the population is too high in **(a)** the world (**Core Case Study**), **(b)** your own country, and **(c)** the area where you live? Explain.

6. Should everyone have the right to have as many children as they want? Explain. Is your belief on this issue consistent with your environmental worldview?

7. Some people have proposed that the earth could solve its population problem by shipping people off to space colonies, each containing about 10,000 people. Assuming we could build such large-scale, self-sustaining space stations (a big assumption), how many of them would we have to build to provide living spaces for the 82 million people added to the earth's population this year? If space shuttles could each carry 100 passengers, how many shuttles would have to be launched each day for a year to

offset the 82 million people added to the population this year? According to your calculations, determine whether this proposal is a logical solution to the earth's population problem. What effect might the daily launching of these shuttles have on global warming? Explain.

8. Some people believe our most important goal should be to sharply reduce the rate of population growth in developing countries where 97% of the world's population growth is expected to take place. Others argue that the most serious environmental problems stem from high levels of resource consumption per person in developed countries, which use 88% of the world's resources and have much larger ecological footprints per person (Figure 1-10, p. 15) than do developing countries. What is your view on this issue? Explain.

9. Congratulations! You are in charge of the world. List the three most important features of your population policy.

10. List two questions that you would like to have answered as a result of reading this chapter.

Note: See Supplement 13 (p. S78) for a list of Projects related to this chapter.

ECOLOGICAL FOOTPRINT ANALYSIS

The chart below shows selected population data for two different countries A and B.

	Country A	Country B
Population (millions)	144	82
Crude birth rate	43	8
Crude death rate	18	10
Infant mortality rate	100	3.8
Total fertility rate	5.9	1.3
Percentage of population under 15 years old	45	14
Percentage of population older than 65 years	3.0	19
Average life expectancy at birth	47	79
Percentage urban	44	75

Source: Data from Population Reference Bureau 2007, *World Population Data Sheet.*

1. Calculate the rates of natural increase for the populations of Country A and Country B. From the rates of natural increase and the data in the table, suggest whether A and B are developed or developing countries, and explain the reasons for your answers.

2. Describe where each of the two countries may be in terms of their stage in the demographic transition (Figure 6-12, p. 134). Discuss factors that could hinder Country A from progressing to later stages in the demographic transition.

3. Explain how the percentage of people under 15 years of age in Country A and in Country B could affect the per capita and total ecological footprints of each country.

LEARNING ONLINE

Log on to the Student Companion Site for this book at **academic.cengage.com/biology/miller**, and choose Chapter 6 for many study aids and ideas for further reading and research. These include flash cards, practice quizzing, Weblinks, information on Green Careers, and InfoTrac® College Edition articles.

7 Climate and Terrestrial Biodiversity

Blowing in the Wind: Connections between Wind, Climate, and Biomes

Terrestrial biomes such as deserts, grasslands, and forests make up one of the components of the earth's biodiversity (Figure 4-2, p. 79). Why is one area of the earth's land surface a desert, another a grassland, and another a forest? The general answer lies in differences in *climate*, resulting mostly from long-term differences in average temperature and precipitation caused by global air circulation, which we discuss in this chapter.

Figure 7-1 Some of the dust blown from West Africa, shown here, can end up as soil nutrients in Amazonian rain forests and toxic air pollutants in the U.S. state of Florida and in the Caribbean. It may also help to suppress hurricanes in the western Atlantic.

Seawifs Project/Nasa/GSFC, Nasa

Wind—an indirect form of solar energy—is an important factor in the earth's climate. It is part of the planet's circulatory system for heat, moisture, plant nutrients, soil particles, and long-lived air pollutants. Without wind, the tropics would be unbearably hot and most of the rest of the planet would freeze.

Winds transport nutrients from one place to another. For example, winds carry dust that is rich in phosphates and iron across the Atlantic Ocean from the Sahara Desert in West Africa (Figure 7-1). These deposits help to build agricultural soils in the Bahamas and to supply nutrients for plants in the upper canopies of rain forests in Brazil. A 2007 study indicated that in 2006, such dust might have helped to reduce hurricane frequency in the southwestern North Atlantic and Caribbean by blocking some energizing sunlight. Dust blown from China's Gobi Desert deposits iron into the Pacific Ocean between Hawaii and Alaska. The iron stimulates the growth of phytoplankton, the minute producers that support ocean food webs. Wind is also a rapidly growing source of renewable energy, as discussed in Chapter 16.

Winds also have a downside. They transport harmful substances. Particles of reddish-brown soil and pesticides banned in the United States are blown from Africa's deserts and eroding farmlands into the sky over the U.S. state of Florida. Some types of fungi in this dust may play a role in degrading or killing coral reefs in the Florida Keys and in the Caribbean.

Particles of iron-rich dust from Africa also enhance the productivity of algae, and have been linked to outbreaks of toxic algal blooms—referred to as *red tides*—in Florida's coastal waters. People who eat shellfish contaminated by a toxin produced in red tides can become paralyzed or can even die. These red tides can also cause fish kills.

Dust, soot, and other long-lived air pollutants from rapidly industrializing China and central Asia are blown across the Pacific Ocean and degrade air quality over parts of the western United States. Asian pollution makes up as much as 10% of West Coast smog—a problem that is expected to get worse as China continues to industrialize. A 2007 study, led by Renyl Zhang, linked such Asian air pollution to intensified storms over the North Pacific Ocean and to increased warming in the polar regions.

The ecological lesson: *Everything we do affects some other part of the biosphere* because *everything is connected.* In this chapter, we examine the key role that climate, including winds, plays in the formation and location of the deserts, grasslands, and forests that make up an important part of the earth's terrestrial biodiversity.

7-1 What factors influence climate?

CONCEPT 7-1 An area's climate is determined mostly by solar radiation, the earth's rotation, global patterns of air and water movement, gases in the atmosphere, and the earth's surface features.

7-2 How does climate affect the nature and locations of biomes?

CONCEPT 7-2 Differences in average annual precipitation and temperature lead to the formation of tropical, temperate, and cold deserts, grasslands, and forests, and largely determine their locations.

7-3 How have we affected the world's terrestrial ecosystems?

CONCEPT 7-3 In many areas, human activities are impairing ecological and economic services provided by the earth's deserts, grasslands, forests, and mountains.

Note: Supplements 2 (p. S4), 4 (p. S20), 5 (p. S31), 8 (p. 47), 10 (p. S59), and 13 (p. S78) can be used with this chapter.

To do science is to search for repeated patterns,
not simply to accumulate facts,
and to do the science of geographical ecology
is to search for patterns of plant and animal life
that can be put on a map.

ROBERT H. MACARTHUR

7-1 What Factors Influence Climate?

▶ **CONCEPT 7-1** An area's climate is determined mostly by solar radiation, the earth's rotation, global patterns of air and water movement, gases in the atmosphere, and the earth's surface features.

The Earth Has Many Different Climates

Weather is a local area's short-term temperature, precipitation, humidity, wind speed, cloud cover, and other physical conditions of the lower atmosphere as measured over hours or days. (Supplement 8, p. S47, introduces you to weather basics.) **Climate** is an area's general pattern of atmospheric or weather conditions measured over long periods of time ranging from decades to thousands of years. As American writer and humorist Mark Twain once said, "Climate is what we expect, weather is what we get." Figure 7-2 (p. 142), depicts the earth's major climate zones, an important component of the earth's natural capital (Figure 1-3, p. 8).

Climate varies in different parts of the earth mostly because patterns of global air circulation and ocean currents distribute heat and precipitation unevenly (Figure 7-3, p. 142). Three major factors determine how air circulates in the lower atmosphere, which helps to distribute heat and moisture from the tropics to other parts of the world:

- *Uneven heating of the earth's surface by the sun.* Air is heated much more at the equator, where the sun's rays strike directly, than at the poles, where sunlight strikes at a slanted angle and spreads out over a much greater area (Figure 7-3, right). These differences in the distribution of incoming solar energy help to explain why tropical regions near the equator are hot, why polar regions are cold, and why temperate regions in between generally have intermediate average temperatures.

- *Rotation of the earth on its axis.* As the earth rotates around its axis, its equator spins faster than its polar regions. As a result, heated air masses rising above the equator and moving north and south to cooler areas are deflected to the west or east over different parts of the planet's surface (Figure 7-3). The atmosphere over these different areas is divided into huge regions called *cells*, distinguished by direction of air movement. And the differing directions of air movement are called *prevailing winds*—major surface winds that blow almost continuously

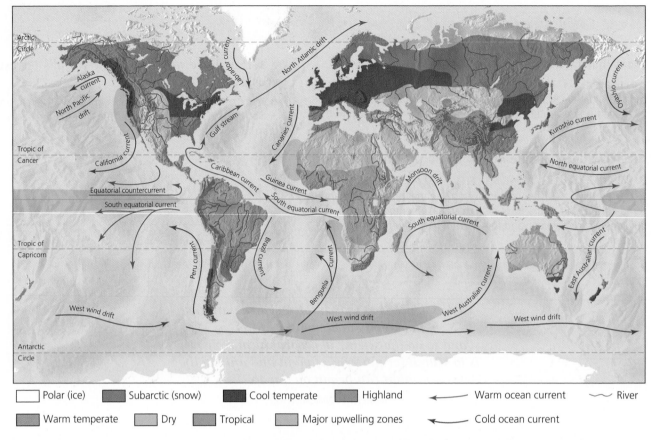

Polar (ice) | Subarctic (snow) | Cool temperate | Highland | ⟵ Warm ocean current | ⟿ River
Warm temperate | Dry | Tropical | Major upwelling zones | ⟵ Cold ocean current

CENGAGENOW™ **Active Figure 7-2 Natural capital:** generalized map of the earth's current climate zones, showing the major contributing ocean currents and drifts and upwelling areas (where currents bring nutrients from the ocean bottom to the surface). Winds play an important role in distributing heat and moisture in the atmosphere, which leads to such climate zones. Winds also cause currents that help distribute heat throughout the world's oceans. *See an animation based on this figure at* CengageNOW™. **Question:** Based on this map what is the general type of climate where you live?

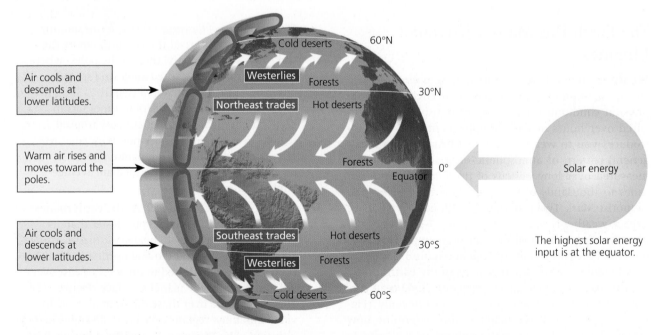

Figure 7-3 *Global air circulation.* The largest input of solar energy occurs at the equator. As this air is heated it rises and moves toward the poles. However, the earth's rotation deflects the movement of the air over different parts of the earth. This creates global patterns of prevailing winds that help distribute heat and moisture in the atmosphere.

and help distribute air, heat, moisture, and dust over the earth's surface (**Core Case Study**).

CORE CASE STUDY

- *Properties of air, water, and land.* Heat from the sun evaporates ocean water and transfers heat from the oceans to the atmosphere, especially near the hot equator. This evaporation of water creates giant cyclical convection cells that circulate air, heat, and moisture both vertically and from place to place in the atmosphere, as shown in Figure 7-4.

Prevailing winds (Figure 7-3) blowing over the oceans produce mass movements of surface water called **currents.** Driven by prevailing winds and the earth's rotation, the earth's major ocean currents (Figure 7-2) redistribute heat from the sun from place to place, thereby influencing climate and vegetation, especially near coastal areas.

The oceans absorb heat from the earth's air circulation patterns; most of this heat is absorbed in tropical waters, which receive most of the sun's heat. This heat and differences in water *density* (mass per unit volume) create warm and cold ocean currents. Prevailing winds and irregularly shaped continents interrupt these cur-

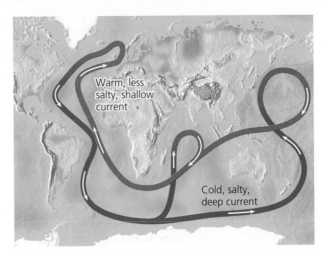

Figure 7-5 *Connected deep and shallow ocean currents.* A connected loop of shallow and deep ocean currents transports warm and cool water to various parts of the earth. This loop, which rises in some areas and falls in others, results when ocean water in the North Atlantic near Iceland is dense enough (because of its salt content and cold temperature) to sink to the ocean bottom, flow southward, and then move eastward to well up in the warmer Pacific. A shallower return current aided by winds then brings warmer, less salty—and thus less dense—water to the Atlantic. This water can cool and sink to begin this extremely slow cycle again. **Question:** How do you think this loop affects the climates of the coastal areas around it?

rents and cause them to flow in roughly circular patterns between the continents, clockwise in the northern hemisphere and counterclockwise in the southern hemisphere.

Heat is also distributed to the different parts of the ocean and the world when ocean water mixes vertically in shallow and deep ocean currents, mostly as a result of differences in the density of seawater. Because it has a higher density, colder seawater sinks and flows beneath warmer and less dense seawater. This creates a connected loop of deep and shallow ocean currents, which act like a giant conveyer belt that moves heat to and from the deep sea and transfers warm and cold water between the tropics and the poles (Figure 7-5).

The ocean and the atmosphere are strongly linked in two ways: ocean currents are affected by winds in the atmosphere (**Core Case Study**), and heat from the ocean affects atmospheric circulation (Figure 7-4). One example of the interactions between the ocean and the atmosphere is the *El Niño–Southern Oscillation,* or *ENSO,* as discussed on pp. S48–S49 in Supplement 8 and in *The Habitable Planet,* Video 3, at **www.learner.org/resources/series209.html**. This large-scale weather phenomenon occurs every few years when prevailing winds in the tropical Pacific Ocean weaken and change direction. The resulting above-average warming of Pacific waters can affect populations of marine species by changing the distribution of plant nutrients. It also alters the weather of at least two-thirds of the earth for one or two years (see Figure 5, p. S49, in Supplement 8).

CORE CASE STUDY

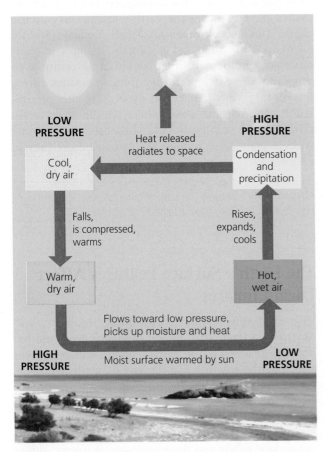

Figure 7-4 Energy transfer by convection in the atmosphere. *Convection* occurs when hot and wet warm air rises, cools, and releases heat and moisture as precipitation (right side). Then the denser cool, dry air sinks, gets warmer, and picks up moisture as it flows across the earth's surface to begin the cycle again.

CENGAGENOW™ Learn more about how oceans affect air movements where you live and all over the world at CengageNOW™.

The earth's air circulation patterns, prevailing winds, and configuration of continents and oceans result in six giant convection cells (like the one shown in Figure 7-4) in which warm, moist air rises and cools, and cool, dry air sinks. Three of these cells are found north of the equator and three are south of the equator. These cells lead to an irregular distribution of climates and deserts, grasslands, and forests, as shown in Figure 7-6 (**Concept 7-1**).

CENGAGENOW™ Watch the formation of six giant convection cells and learn more about how they affect climates at CengageNOW.

THINKING ABOUT
Winds and Biomes

CORE
CASE
STUDY

How might the distribution of the world's forests, grasslands, and deserts shown in Figure 7-6 differ if the prevailing winds shown in Figure 7-3 did not exist?

Greenhouse Gases Warm the Lower Atmosphere

Figure 3-8 (p. 56) shows how energy flows to and from the earth. Small amounts of certain gases, including water vapor (H_2O), carbon dioxide (CO_2), methane

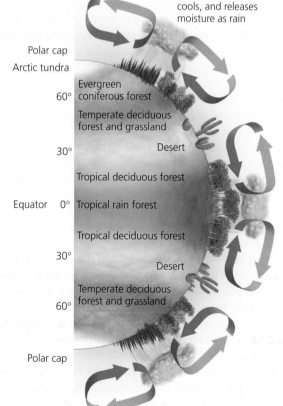

Figure 7-6 *Global air circulation, ocean currents, and biomes.* Heat and moisture are distributed over the earth's surface via six giant convection cells (like the one in Figure 7-4) at different latitudes. The resulting uneven distribution of heat and moisture over the planet's surface leads to the forests, grasslands, and deserts that make up the earth's terrestrial biomes.

Moist air rises, cools, and releases moisture as rain

Polar cap
Arctic tundra
60° Evergreen coniferous forest
Temperate deciduous forest and grassland
30° Desert
Tropical deciduous forest
Equator 0° Tropical rain forest
Tropical deciduous forest
30° Desert
Temperate deciduous forest and grassland
60°
Polar cap

(CH_4), and nitrous oxide (N_2O), in the atmosphere play a role in determining the earth's average temperatures and its climates. These **greenhouse gases** allow mostly visible light and some infrared radiation and ultraviolet (UV) radiation from the sun to pass through the atmosphere. The earth's surface absorbs much of this solar energy and transforms it to longer-wavelength infrared radiation (heat), which then rises into the lower atmosphere.

Some of this heat escapes into space, but some is absorbed by molecules of greenhouse gases and emitted into the lower atmosphere as even longer-wavelength infrared radiation. Some of this released energy radiates into space, and some warms the lower atmosphere and the earth's surface. This natural warming effect of the troposphere is called the **greenhouse effect** (see Figure 3-8, p. 56, and *The Habitable Planet,* Video 2, at **www.learner.org/resources/series209.html**). Without the warming caused by these greenhouse gases, the earth would be a cold and mostly lifeless planet.

Human activities such as burning fossil fuels, clearing forests, and growing crops release carbon dioxide, methane, and nitrous oxide into the atmosphere. Considerable evidence and climate models indicate that there is a 90–99% chance that the large inputs of greenhouse gases into the atmosphere from human activities are enhancing the earth's natural greenhouse effect. This *human-enhanced global warming* (Science Focus, p. 33) could cause climate changes in various places on the earth that could last for centuries to thousands of years. As this warming intensifies during this century, climate scientists expect it to alter precipitation patterns, shift areas where we can grow crops, raise average sea levels, and shift habitats for some types of plants and animals, as discussed more fully in Chapter 19.

CENGAGENOW™ Witness the natural greenhouse effect and see how human activities have affected it at CengageNOW.

The Earth's Surface Features Affect Local Climates

Heat is absorbed and released more slowly by water than by land. This difference creates land and sea breezes. As a result, the world's oceans and large lakes moderate the weather and climates of nearby lands.

Various topographic features of the earth's surface create local and regional weather and climatic conditions that differ from the general climate of a region. For example, mountains interrupt the flow of prevailing surface winds and the movement of storms. When moist air blowing inland from an ocean reaches a mountain range, it is forced upward. As it rises, it cools and expands and then loses most of its moisture as rain and snow on the windward slope of the mountain (the side from which the wind is blowing).

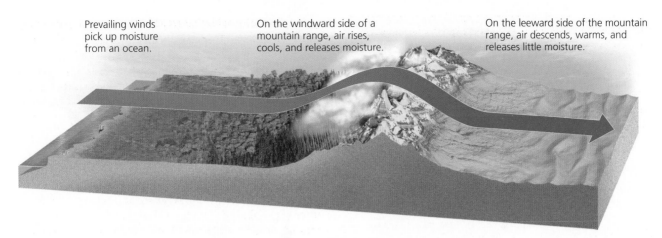

Prevailing winds pick up moisture from an ocean.

On the windward side of a mountain range, air rises, cools, and releases moisture.

On the leeward side of the mountain range, air descends, warms, and releases little moisture.

Figure 7-7 The *rain shadow effect* is a reduction of rainfall and loss of moisture from the landscape on the side of a mountain facing away from prevailing surface winds. Warm, moist air in onshore winds loses most of its moisture as rain and snow on the windward slopes of a mountain range. This leads to semiarid and arid conditions on the leeward side of the mountain range and the land beyond. The Mojave Desert in the U.S. state of California and Asia's Gobi Desert are both produced by this effect.

As the drier air mass passes over the mountaintops it flows down the leeward (away from the wind) slopes, warms up (which increases its ability to hold moisture), and sucks up moisture from the plants and soil below. The loss of moisture from the landscape and the resulting semiarid or arid conditions on the leeward side of high mountains create the **rain shadow effect** (Figure 7-7). Sometimes this leads to the formation of deserts such as Death Valley in the United States, which is in the rain shadow of Mount Whitney, the highest mountain in the Sierra Nevadas. In this way, winds (**Core Case Study**) play a key role in forming some of the earth's deserts.

Cities also create distinct microclimates. Bricks, concrete, asphalt, and other building materials absorb and hold heat, and buildings block wind flow. Motor vehi-

cles and the climate control systems of buildings release large quantities of heat and pollutants. As a result, cities tend to have more haze and smog, higher temperatures, and lower wind speeds than the surrounding countryside.

THINKING ABOUT

Winds and Your Life

What are three changes in your lifestyle that would take place if there were no winds where you live?

CORE CASE STUDY

RESEARCH FRONTIER

Modeling and other research to learn more about how human activities affect climate. See **academic.cengage.com/ biology/miller**.

7-2 How Does Climate Affect the Nature and Locations of Biomes?

▶ **CONCEPT 7-2** Differences in average annual precipitation and temperature lead to the formation of tropical, temperate, and cold deserts, grasslands, and forests, and largely determine their locations.

Climate Affects Where Organisms Can Live

Different climates (Figure 7-2) explain why one area of the earth's land surface is a desert, another a grassland, and another a forest (Figure 7-6) and why global air circulation (Figure 7-3) accounts for different types of deserts, grasslands, and forests (**Concept 7-2**).

Figure 7-8 (p. 146) shows how scientists have divided the world into several major **biomes**—large ter-

restrial regions characterized by similar climate, soil, plants, and animals, regardless of where they are found in the world. The variety of terrestrial biomes and aquatic systems is one of the four components of the earth's biodiversity (Figure 4-2, p. 79, and **Concept 4-1A**, p. 78)—a vital part of the earth's natural capital.

By comparing Figure 7-8 with Figure 7-2 and Figure 1 on pp. S20–S21 in Supplement 4, you can see how the world's major biomes vary with climate.

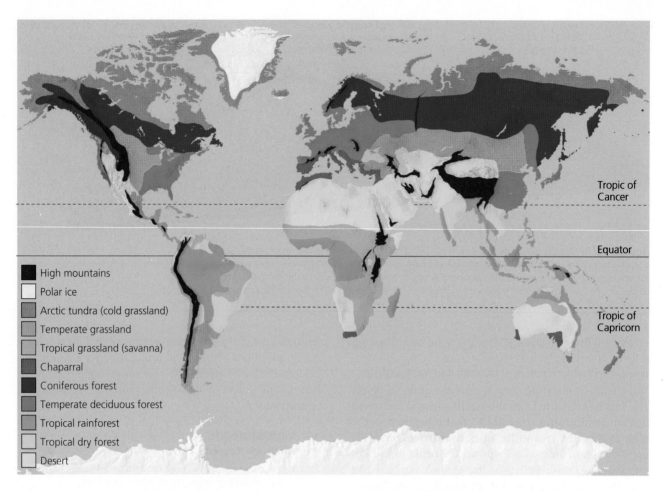

■	High mountains
□	Polar ice
■	Arctic tundra (cold grassland)
■	Temperate grassland
□	Tropical grassland (savanna)
■	Chaparral
■	Coniferous forest
■	Temperate deciduous forest
■	Tropical rainforest
□	Tropical dry forest
□	Desert

CENGAGENOW **Active Figure 7-8** **Natural capital:** the earth's major *biomes*—the main types of natural vegetation in various undisturbed land areas—result primarily from differences in climate. Each biome contains many ecosystems whose communities have adapted to differences in climate, soil, and other environmental factors. Figure 5 on p. S27 in Supplement 4 shows the major biomes of North America. Human activities have removed or altered much of the natural vegetation in some areas for farming, livestock grazing, lumber and fuelwood, mining, and construction of towns and cities (see Figure 3, pp. S24–S25, and Figure 7, pp. S28–S29, in Supplement 4). *See an animation based on this figure at* CengageNOW. **Question:** If you factor out human influences such as farming and urban areas, what kind of biome do you live in?

Figure 3-7 (p. 55) shows how major biomes in the United States (Figure 5, p. S27, in Supplement 4) are related to its different climates.

On maps such as the one in Figure 7-8, biomes are shown with sharp boundaries, each being covered with one general type of vegetation. In reality, *biomes are not uniform.* They consist of a *mosaic of patches,* each with somewhat different biological communities but with similarities typical of the biome. These patches occur mostly because the resources that plants and animals need are not uniformly distributed and because human activities remove and alter the natural vegetation in many areas.

Figure 7-9 shows how climate and vegetation vary with *latitude* and *elevation.* If you climb a tall mountain from its base to its summit, you can observe changes in plant life similar to those you would encounter in traveling from the equator to one of the earth's poles. For example, if you hike up a tall Andes mountain in

Ecuador, your trek can begin in tropical rain forest and end up on a glacier at the summit.

> **THINKING ABOUT**
> **Biomes, Climate, and Human Activities**
>
> Use Figure 7-2 to determine the general type of climate where you live and Figure 7-8 to determine the general type of biome that should exist where you live. Then use Figure 3, pp. S24–S25, and Figure 7, pp. S28–S29, in Supplement 4 to determine how human ecological footprints have affected the general type of biome where you live.

Differences in climate, mostly from average annual precipitation and temperature, lead to the formation of tropical (hot), temperate (moderate), and polar (cold) deserts, grasslands, and forests (Figure 7-10)—another important component of the earth's natural capital (**Concept 7-2**).

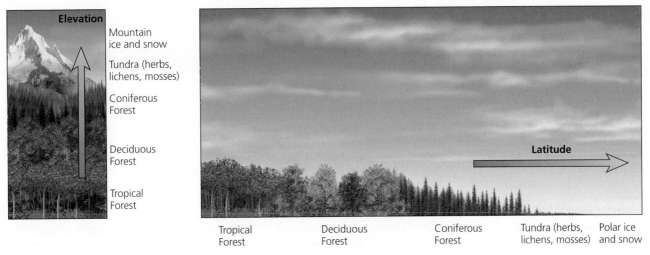

Figure 7-9 Generalized effects of elevation (left) and latitude (right) on climate and biomes. Parallel changes in vegetation type occur when we travel from the equator to the poles or from lowlands to mountaintops. **Question:** How might the components of the left diagram change as the earth warms during this century? Explain.

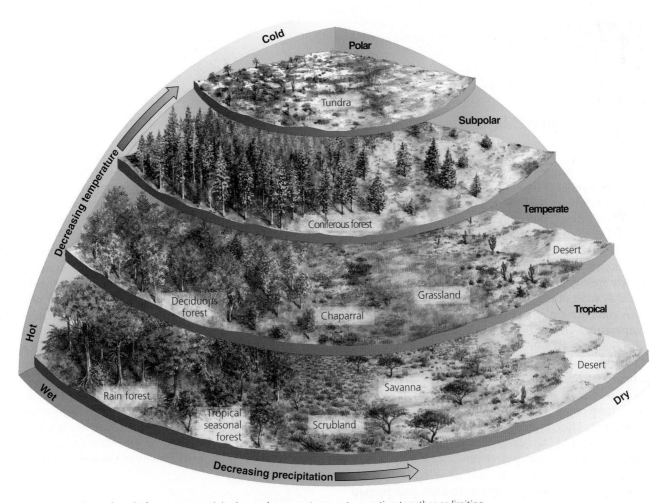

Figure 7-10 Natural capital: average precipitation and average temperature, acting together as limiting factors over a long time, help to determine the type of desert, grassland, or forest biome in a particular area. Although each actual situation is much more complex, this simplified diagram explains how climate helps to determine the types and amounts of natural vegetation found in an area left undisturbed by human activities. (Used by permission of Macmillan Publishing Company, from Derek Elsom, *The Earth,* New York: Macmillan, 1992. Copyright ©1992 by Marshall Editions Developments Limited).

SCIENCE FOCUS

Staying Alive in the Desert

Adaptations for survival in the desert have two themes: *beat the heat,* and *every drop of water counts.*

Desert plants have evolved a number of strategies for doing this. During long hot and dry spells, plants such as mesquite and creosote drop their leaves to survive in a dormant state. *Succulent* (fleshy) *plants,* such as the saguaro ("sah-WAH-ro") cactus (Figure 7-11, middle photo), have three adaptations: they have no leaves, which can lose water by evapotranspiration; they store water and synthesize food in their expandable, fleshy tissue; and they reduce water loss by opening their pores to take up carbon dioxide (CO_2) only at night. The spines of these and many other desert plants guard them from being eaten by herbivores seeking the precious water they hold.

Some desert plants use deep roots to tap into groundwater. Others such as prickly pear and saguaro cacti use widely spread, shallow roots to collect water after brief showers and store it in their spongy tissue.

Evergreen plants conserve water by having wax-coated leaves that reduce water loss. Others, such as annual wildflowers and grasses, store much of their biomass in seeds that remain inactive, sometimes for years, until they receive enough water to germinate. Shortly after a rain, these seeds germinate, grow, and carpet some deserts with dazzling arrays of colorful flowers that last for a few weeks.

Most desert animals are small. Some beat the heat by hiding in cool burrows or rocky crevices by day and coming out at night or in the early morning. Others become dormant during periods of extreme heat or drought. Some larger animals such as camels can drink massive quantities of water when it is available and store it in their fat for use as needed. The camel is also covered with dense hair and does not sweat, which reduces heat gain and water loss through evaporation. Kangaroo rats never drink water. They get the water they need by breaking down fats in seeds that they consume.

Insects and reptiles (such as rattlesnakes and Gila monsters) have thick outer coverings to minimize water loss through evaporation, and their wastes are dry feces and a dried concentrate of urine. Many spiders and insects get their water from dew or from the food they eat.

Critical Thinking

What are three things you would do to survive in the open desert?

There Are Three Major Types of Deserts

In a **desert**, annual precipitation is low and often scattered unevenly throughout the year. During the day, the baking sun warms the ground and causes evaporation of moisture from plant leaves and soil. But at night, most of the heat stored in the ground radiates quickly into the atmosphere. Desert soils have little vegetation and moisture to help store the heat, and the skies above deserts are usually clear. This explains why, in a desert, you may roast during the day but shiver at night.

A combination of low rainfall and different average temperatures creates tropical, temperate, and cold deserts (Figures 7-10 and 7-11).

Tropical deserts (Figure 7-11, top photo), such as the Sahara and Namib of Africa, are hot and dry most of the year (Figure 7-11, top graph). They have few plants and a hard, windblown surface strewn with rocks and some sand. They are the deserts we often see in the movies. Wind-blown dust storms (**Core Case Study**) in the Sahara Desert have increased tenfold since 1950 mostly because of overgrazing and drought due to climate change and human population growth. Another reason is the *SUV connection.* Increasing numbers of four-wheel vehicles speeding over the sand (Figure 7-11, top photo) break the desert's surface crust. Wind storms can then blow the dusty material into the atmosphere.

In *temperate deserts* (Figure 7-11, center photo), such as the Mojave in the southern part of the U. S. state of California, daytime temperatures are high in summer and low in winter and there is more precipitation than in tropical deserts (Figure 7-11, center graph). The sparse vegetation consists mostly of widely dispersed, drought-resistant shrubs and cacti or other succulents adapted to the lack of water and temperature variations. (Figure 1, p. S53, in Supplement 9 shows some components and food web interactions in a temperate desert ecosystem.)

In *cold deserts,* such as the Gobi Desert in Mongolia, vegetation is sparse (Figure 7-11, bottom photo). Winters are cold, summers are warm or hot, and precipitation is low (Figure 7-11, bottom graph). Desert plants and animals have adaptations that help them to stay cool and to get enough water to survive (Science Focus, above).

Desert ecosystems are fragile. Their soils take decades to hundreds of years to recover from disturbances such as off-road vehicles. This is because of their slow plant growth, low species diversity, slow nutrient cycling (due to low bacterial activity in the soils), and lack of water.

THINKING ABOUT
Winds and Deserts

What roles do winds (**Core Case Study**) play in creating and sustaining deserts?

CORE CASE STUDY

Tropical desert

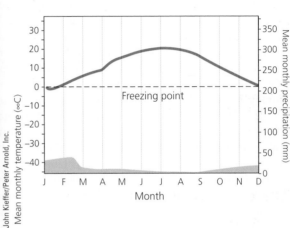

Temperate desert

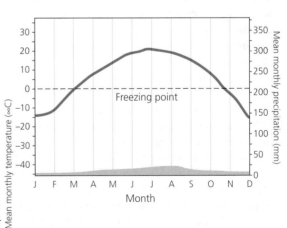

Cold desert

Figure 7-11 Climate graphs showing typical variations in annual temperature (red) and precipitation (blue) in tropical, temperate, and cold deserts. Top photo: a popular (but destructive) SUV rodeo in United Arab Emirates (tropical desert). Center photo: saguaro cactus in the U.S. state of Arizona (temperate desert). Bottom photo: a Bactrian camel in Mongolia's Gobi Desert (cold desert). **Question:** What month of the year has the highest temperature and the lowest rainfall for each of the three types of deserts?

There Are Three Major Types of Grasslands

Grasslands occur mostly in the interiors of continents in areas too moist for deserts and too dry for forests (Figure 7-8). Grasslands persist because of a combination of seasonal drought, grazing by large herbivores, and occasional fires—all of which keep large numbers of shrubs and trees from growing.

The three main types of grassland—tropical, temperate, and cold (arctic tundra)—result from combinations of low average precipitation and various average temperatures (Figures 7-10 and 7-12).

One type of tropical grassland, called a *savanna*, contains widely scattered clumps of trees such as acacia, (Figure 7-12, top photo), which are covered with thorns that help to keep herbivores away. This biome usually has warm temperatures year-round and alternating dry and wet seasons (Figure 7-12, top graph).

Tropical savannas in East Africa have herds of *grazing* (grass- and herb-eating) and *browsing* (twig- and leaf-nibbling) hoofed animals, including wildebeests (Figure 7-12, top photo), gazelles, zebras, giraffes, and antelopes and their predators such as lions, hyenas, and humans. Herds of these grazing and browsing animals migrate to find water and food in response to seasonal and year-to-year variations in rainfall (Figure 7-12, blue region in top graph) and food availability.

In their niches, these and other large herbivores have evolved specialized eating habits that minimize competition among species for the vegetation found on the savanna. For example, giraffes eat leaves and shoots from the tops of trees, elephants eat leaves and branches farther down, wildebeests prefer short grasses, and zebras graze on longer grasses and stems. Savanna plants, like desert plants, are adapted to survive drought and extreme heat. Many have deep roots that can tap into groundwater.

In a *temperate grassland*, winters are bitterly cold, summers are hot and dry, and annual precipitation is fairly sparse and falls unevenly through the year (Figure 7-12, center graph). Because the aboveground parts of most of the grasses die and decompose each year, organic matter accumulates to produce a deep, fertile soil. This soil is held in place by a thick network of intertwined roots of drought-tolerant grasses (unless the topsoil is plowed up, which exposes it to be blown away by high winds found in these biomes). The natural grasses are also adapted to fires, which burn the plant parts above the ground but do not harm the roots, from which new grass can grow.

Two types of temperate grasslands are *tall-grass prairies* and *short-grass prairies* (Figure 7-12, center photo), such as those of the Midwestern and western United States and Canada. Short-grass prairies typically get about 25 centimeters (10 inches) of rain a year, and the grasses have short roots. Tall-grass prairies can get up to 88 centimeters (35 inches) of rain per year, and the grasses have deep roots. Mixed or middle-grass prairies get annual rainfall between these two extremes.

In all prairies, winds blow almost continuously and evaporation is rapid, often leading to fires in the summer and fall. This combination of winds and fires helps to maintain such grasslands by hindering tree growth. (Figure 2, p. S54, in Supplement 9 shows some components and food-web interactions in a temperate tall-grass prairie ecosystem in North America.)

Many of the world's natural temperate grasslands have disappeared because their fertile soils are useful for growing crops (Figure 7-13, p. 152) and grazing cattle.

Cold grasslands, or *arctic tundra* (Russian for "marshy plain"), lie south of the arctic polar ice cap (Figure 7-8). During most of the year, these treeless plains are bitterly cold (Figure 7-12, bottom graph), swept by frigid winds, and covered by ice and snow. Winters are long and dark, and scant precipitation falls mostly as snow.

Under the snow, this biome is carpeted with a thick, spongy mat of low-growing plants, primarily grasses, mosses, lichens, and dwarf shrubs (Figure 7-12, bottom photo, and Figure 1-17, p. 23). Trees and tall plants cannot survive in the cold and windy tundra because they would lose too much of their heat. Most of the annual growth of the tundra's plants occurs during the 7- to 8-week summer, when the sun shines almost around the clock. (Figure 3, p. S55, in Supplement 9 shows some components and food-web interactions in an arctic tundra ecosystem.)

One outcome of the extreme cold is the formation of **permafrost,** underground soil in which captured water stays frozen for more than 2 consecutive years. During the brief summer, the permafrost layer keeps melted snow and ice from soaking into the ground. As a consequence, many shallow lakes, marshes, bogs, ponds, and other seasonal wetlands form when snow and frozen surface soil melt on the waterlogged tundra (Figure 7-12, bottom photo). Hordes of mosquitoes, black flies, and other insects thrive in these shallow surface pools. They serve as food for large colonies of migratory birds (especially waterfowl) that return from the south to nest and breed in the bogs and ponds.

Animals in this biome survive the intense winter cold through adaptations such as thick coats of fur (arctic wolf, arctic fox, and musk oxen) and feathers (snowy owl) and living underground (arctic lemming). In the summer, caribou migrate to the tundra to graze on its vegetation (Figure 1-17, p. 23).

Global warming is causing some of the permafrost in parts of Canada, Alaska, China, Russia, and Mongolia to melt. This disrupts these ecosystems and releases methane (CH_4) and carbon dioxide (CO_2) from the soil into the atmosphere. These two greenhouse gases can accelerate global warming and cause more permafrost to melt, which can lead to further warming and climate change. The melting permafrost causes the soil to sink

Tropical grassland (savanna)

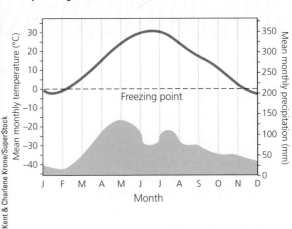

Martin Harvey/Peter Arnold, Inc.

Temperate grassland

Kent & Charlene Krone/SuperStock

Cold grassland (arctic tundra)

M. Leder/Peter Arnold, Inc.

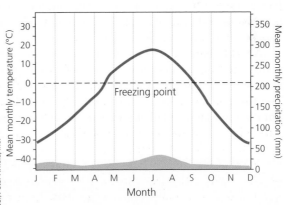

Figure 7-12 Climate graphs showing typical variations in annual temperature (red) and precipitation (blue) in tropical, temperate, and cold (arctic tundra) grassland. Top photo: wildebeests grazing on a savanna in Maasai Mara National Park in Kenya, Africa (tropical grassland). Center photo: wildflowers in bloom on a prairie near East Glacier Park in the U. S. state of Montana (temperate grassland). Bottom photo: arctic tundra (cold grassland) in autumn in front of the Alaska Range, Alaska (USA). **Question:** What month of the year has the highest temperature and the lowest rainfall for each of the three types of grassland?

Figure 7-13 Natural capital degradation: replacement of a biologically diverse temperate grassland with a monoculture crop in the U.S. state of California. When humans remove the tangled root network of natural grasses, the fertile topsoil becomes subject to severe wind erosion unless it is covered with some type of vegetation.

(subside), which can damage buildings, roads, power lines, and other human structures.

Tundra is a fragile biome. Most tundra soils formed about 17,000 years ago when glaciers began retreating after the last Ice Age (Figure 4-7, p. 85). These soils usually are nutrient poor and have little detritus. Because of the short growing season, tundra soil and vegetation recover very slowly from damage or disturbance. Human activities in the arctic tundra—mostly around oil drilling sites, pipelines, mines, and military bases—leave scars that persist for centuries.

Another type of tundra, called *alpine tundra,* occurs above the limit of tree growth but below the permanent snow line on high mountains (Figure 7-9, left). The vegetation is similar to that found in arctic tundra, but it receives more sunlight than arctic vegetation gets.

During the brief summer, alpine tundra can be covered with an array of beautiful wildflowers.

┌─ THINKING ABOUT
Winds and Grassland   CORE CASE STUDY
What roles do winds play in creating and sustaining grasslands?

Temperate Shrubland: Nice Climate, Risky Place to Live

In many coastal regions that border on deserts we find fairly small patches of a biome known as *temperate shrubland* or *chaparral.* Closeness to the sea provides a slightly longer winter rainy season than nearby temperate deserts have, and fogs during the spring and fall reduce evaporation. These biomes are found along coastal areas of southern California in the United States, the Mediterranean Sea, central Chile, southern Australia, and southwestern South Africa (Figure 7-8).

Chaparral consists mostly of dense growths of low-growing evergreen shrubs and occasional small trees with leathery leaves that reduce evaporation (Figure 7-14). The soil is thin and not very fertile. Animal species of the chaparral include mule deer, chipmunks, jackrabbits, lizards, and a variety of birds.

During the long, warm, and dry summers, chaparral vegetation becomes very dry and highly flammable. In the late summer and fall, fires started by lightning or human activities spread with incredible swiftness. Research reveals that chaparral is adapted to and maintained by fires. Many of the shrubs store food reserves in their fire-resistant roots and produce seeds that sprout only after a hot fire. With the first rain, annual grasses and wildflowers spring up and use nutrients released by the fire. New shrubs grow quickly and crowd out the grasses.

People like living in this biome because of its moderate, sunny climate with mild, wet winters and warm, dry summers. As a result, humans have moved in and

Figure 7-14
Chaparral vegetation in the U.S. state of Utah and a typical climate graph.

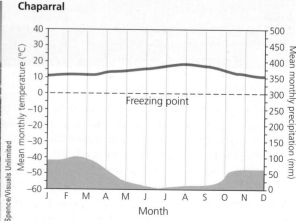

Chaparral

modified this biome considerably (Figure 3, pp. S24–S25, in Supplement 4). The downside of its favorable climate is that people living in chaparral assume the high risk of losing their homes and possibly their lives to frequent fires during the dry season followed by mud slides during rainy seasons.

There Are Three Major Types of Forests

Forest systems are lands dominated by trees. The three main types of forest—*tropical, temperate,* and *cold* (northern coniferous and boreal)—result from combinations of the precipitation level and various average temperatures (Figures 7-10 and 7-15, p. 154).

Tropical rain forests (Figure 7-15, top photo) are found near the equator (Figure 7-8), where hot, moisture-laden air rises and dumps its moisture. These lush forests have year-round, uniformly warm temperatures, high humidity, and heavy rainfall almost daily (Figure 7-15, top graph). This fairly constant warm and wet climate is ideal for a wide variety of plants and animals.

Figure 7-16 (p. 155) shows some of the components and food web interactions in these extremely diverse ecosystems. Tropical rain forests are dominated by *broadleaf evergreen plants,* which keep most of their leaves year-round. The tops of the trees form a dense canopy, which blocks most light from reaching the forest floor, illuminating it with a dim greenish light.

The ground level in such a forest has little vegetation, except near stream banks or where a fallen tree has opened up the canopy and let in sunlight. Many of the plants that do live at the ground level have enormous leaves to capture what little sunlight filters through to the dimly lit forest floor.

Some trees are draped with vines (called lianas) that reach for the treetops to gain access to sunlight. Once in the canopy, the vines grow from one tree to another, providing walkways for many species living there. When a large tree is cut down, its lianas can pull down other trees.

Tropical rain forests have a very high net primary productivity (Figure 3-16, p. 64); they are teeming with life and boast incredible biological diversity. Although tropical rain forests cover only about 2% of the earth's land surface, ecologists estimate that they contain at least half of the earth's known terrestrial plant and animal species. For example, a single tree in a rain forest may support several thousand different insect species. Plants from tropical rain forests are a source of chemicals used as blueprints for making most of the world's prescription drugs. Thus, the plant biodiversity found in this biome saves many human lives.

Tropical rain forest life forms occupy a variety of specialized niches in distinct layers. For example, vegetation layers are structured mostly according to the plants' needs for sunlight, as shown in Figure 7-17 (p. 156). Stratification of specialized plant and animal niches in a tropical rain forest enables the coexistence of a great variety of species (high species richness; see Photo 3 in the Detailed Contents). Much of the animal life, particularly insects, bats, and birds, lives in the sunny *canopy* layer, with its abundant shelter and supplies of leaves, flowers, and fruits. To study life in the canopy, ecologists climb trees, use tall construction cranes, and build platforms and boardwalks in the upper canopy. See *The Habitable Planet,* Videos 4 and 9, at **www.learner.org/resources/series209.html** for information on how scientists gather information about tropical rain forests and the effects of human activities on such forests.

CENGAGENOW™ Learn more about how plants and animals in a rain forest are connected in a food web at CengageNOW.

Because of the dense vegetation, there is little wind in these forests to spread seeds and pollen. Consequently, most rain forest plant species depend on bats, butterflies, birds, bees, and other species to pollinate their flowers and to spread seeds in their droppings.

Dropped leaves, fallen trees, and dead animals decompose quickly because of the warm, moist conditions and the hordes of decomposers. This rapid recycling of scarce soil nutrients explains why there is so little plant litter on the ground. Instead of being stored in the soil, about 90% of plant nutrients released by decomposition are quickly taken up and stored by trees, vines, and other plants. This is in sharp contrast to temperate forests, where most plant nutrients are found in the soil.

Because of these ecological processes and the almost daily rainfall, which leaches nutrients from the soil, the soils in most rain tropical forests contain few plant nutrients. This helps explain why rain forests are not good places to clear and grow crops or graze cattle on a sustainable basis. Despite this ecological limitation, many of these forests are being cleared or degraded for logging, growing crops, grazing cattle, and mineral extraction (Chapter 3 Core Case Study, p. 50).

So far, at least half of these forests have been destroyed or disturbed by human activities and the pace of destruction and degradation of these centers of terrestrial biodiversity is increasing (Figure 3-1, p. 50). Ecologists warn that without strong conservation measures, most of these forests will probably be gone within your lifetime, and with them perhaps a quarter of the world's species. This will reduce the earth's biodiversity and help to accelerate global warming and the resulting climate change by eliminating large areas of trees that remove carbon dioxide from the atmosphere (see *The Habitable Planet,* Video 9, at **www.learner.org/resources/series209.html**).

Temperate deciduous forests (Figure 7-15, center photo) grow in areas with moderate average temperatures that change significantly with the season. These areas have long, warm summers, cold but not too severe winters,

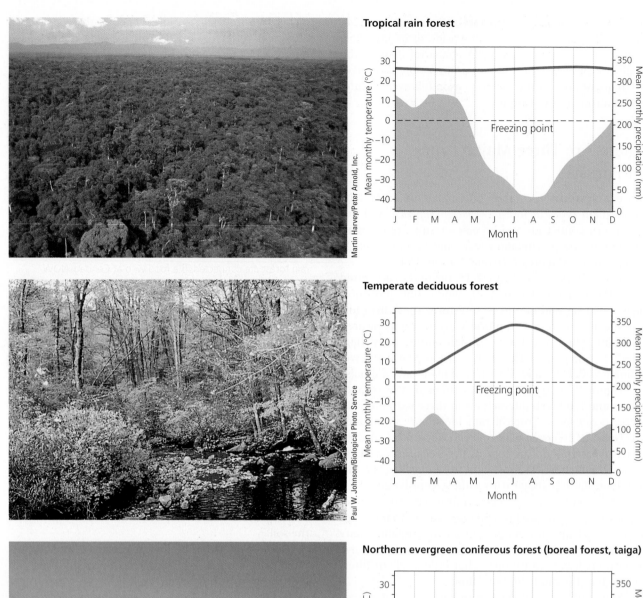

Tropical rain forest

Temperate deciduous forest

Northern evergreen coniferous forest (boreal forest, taiga)

Figure 7-15 Climate graphs showing typical variations in annual temperature (red) and precipitation (blue) in tropical, temperate, and cold (northern coniferous and boreal) forests. Top photo: the closed canopy of a tropical rain forest in the western Congo Basin of Gabon, Africa. Middle photo: a temperate deciduous forest in the U.S. state of Rhode Island during the fall. (Photo 4 in the Detailed Contents shows this same area of forest during winter.) Bottom photo: a northern coniferous forest in the Malheur National Forest and Strawberry Mountain Wilderness in the U.S. state of Oregon. **Question:** What month of the year has the highest temperature and the lowest rainfall for each of the three types of forest?

CENGAGENOW™ **Active Figure 7-16** Some components and interactions in a *tropical rain forest ecosystem*. When these organisms die, decomposers break down their organic matter into minerals that plants use. Colored arrows indicate transfers of matter and energy between producers; primary consumers (herbivores); secondary, or higher-level, consumers (carnivores); and decomposers. Organisms are not drawn to scale. *See an animation based on this figure at CengageNOW.*

Blue and gold macaw

Harpy eagle

Ocelot

Squirrel monkeys

Climbing monstera palm

Katydid

Slaty-tailed trogon

Green tree snake

Tree frog

Ants

Bacteria

Bromeliad

Fungi

Producer to primary consumer

Primary to secondary consumer

Secondary to higher-level consumer

All producers and consumers to decomposers

and abundant precipitation, often spread fairly evenly throughout the year (Figure 7-15, center graph).

This biome is dominated by a few species of *broad-leaf deciduous trees* such as oak, hickory, maple, poplar, and beech. They survive cold winters by dropping their leaves in the fall and becoming dormant through the winter (see Photo 4 in the Detailed Contents). Each spring, they grow new leaves whose colors change in the fall into an array of reds and golds before the leaves drop.

Because of a slow rate of decomposition, these forests accumulate a thick layer of slowly decaying leaf litter, which is a storehouse of nutrients. (Figure 4, p. S56, in Supplement 9, shows some components and food web interactions in a temperate deciduous forest ecosystem.) On a global basis, this biome has been disturbed by human activity more than any other terrestrial biome. Many forests have been cleared for growing crops or developing urban areas. However, within

100–200 years, abandoned cropland can return to a deciduous forest through secondary ecological succession (Figure 5-17, p. 117).

The temperate deciduous forests of the eastern United States were once home to such large predators as bears, wolves, foxes, wildcats, and mountain lions (pumas). Today, most of the predators have been killed or displaced, and the dominant mammal species often is the white-tailed deer (Case Study, p. 114), along with smaller mammals such as squirrels, rabbits, opossums, raccoons, and mice.

Warblers, robins, and other bird species migrate to these forests during the summer to feed and breed. Many of these species are declining in numbers because of loss or fragmentation of their summer and winter habitats.

Evergreen coniferous forests (Figure 7-15, bottom photo) are also called *boreal forests* and *taigas* ("TIE-guhs"). These cold forests are found just south of the

Figure 7-17 Stratification of specialized plant and animal niches in a *tropical rain forest*. Filling such specialized niches enables species to avoid or minimize competition for resources and results in the coexistence of a great variety of species.

arctic tundra in northern regions across North America, Asia, and Europe (Figure 7-8) and above certain altitudes in the High Sierra and Rocky Mountains of the United States. In this subarctic climate, winters are long, dry, and extremely cold; in the northernmost taigas, winter sunlight is available only 6–8 hours per day. Summers are short, with cool to warm temperatures (Figure 7-15, bottom graph), and the sun shines up to 19 hours a day.

Most boreal forests are dominated by a few species of *coniferous* (cone-bearing) *evergreen trees* such as spruce, fir, cedar, hemlock, and pine that keep most of their narrow-pointed leaves (needles) year-round (Figure 7-15, bottom photo). The small, needle-shaped, waxy-coated leaves of these trees can withstand the intense cold and drought of winter, when snow blankets the ground. Such trees are ready to take advantage of the brief summers in these areas without taking time to grow new needles. Plant diversity is low because few species can survive the winters when soil moisture is frozen.

Beneath the stands of these trees is a deep layer of partially decomposed conifer needles. Decomposition is slow because of the low temperatures, waxy coating on conifer needles, and high soil acidity. The decomposing needles make the thin, nutrient-poor soil acidic, which prevents most other plants (except certain shrubs) from growing on the forest floor.

This biome contains a variety of wildlife. Year-round residents include bears, wolves, moose, lynx, and many burrowing rodent species. Caribou spend the winter in taiga and the summer in arctic tundra. During the brief summer, warblers and other insect-eating birds feed on hordes of flies, mosquitoes, and caterpillars. (Figure 5, p. S57, in Supplement 9 shows some components and food web interactions in an evergreen coniferous forest ecosystem.)

Coastal coniferous forests or *temperate rain forests* (Figure 7-18) are found in scattered coastal temperate areas that have ample rainfall or moisture from dense ocean fogs. Dense stands of large conifers such as Sitka spruce, Douglas fir, and redwoods once dominated undisturbed areas of this biome along the coast of North America, from Canada to northern California in the United States.

Figure 7-18 Temperate rain forest in Olympic National Park in the U.S. state of Washington.

THINKING ABOUT
Winds and Forests

CORE CASE STUDY

What roles do winds play in creating temperate and coniferous forests?

Mountains Play Important Ecological Roles

Some of the world's most spectacular environments are high on *mountains* (Figure 7-19), steep or high lands which cover about one-fourth of the earth's land surface (Figure 7-8). Mountains are places where dramatic changes in altitude, slope, climate, soil, and vegetation take place over a very short distance (Figure 7-9, left).

About 1.2 billion people (18% of the world's population) live in mountain ranges or on their edges and 4 billion people (59% of the world's population) depend on mountain systems for all or some of their water. Because of the steep slopes, mountain soils are easily eroded when the vegetation holding them in place is removed by natural disturbances, such as landslides and avalanches, or human activities, such as timber cutting and agriculture. Many freestanding mountains are *islands of biodiversity* surrounded by a sea of lower-elevation landscapes transformed by human activities.

Mountains play important ecological roles. They contain the majority of the world's forests, which are habitats for much of the planet's terrestrial biodiversity. They often provide habitats for endemic species found nowhere else on earth. They also serve as sanctuaries for animal species driven to migrate from lowland areas to higher altitudes.

Mountains also help to regulate the earth's climate. Mountaintops covered with ice and snow affect climate by reflecting solar radiation back into space. This helps to cool the earth and offset global warming. However, many of the world's mountain glaciers are melting, mostly because of global warming. While glaciers reflect solar energy, the darker rocks exposed by melting glaciers absorb that energy. This helps to increase global warming, which melts more glaciers and warms the atmosphere more—an example of a runaway positive feedback loop.

Mountains can affect sea levels by storing and releasing water in glacial ice. As the earth gets warmer, mountaintop glaciers and other land-based glaciers can melt, adding water to the oceans and helping to raise sea levels.

Finally, mountains play a critical role in the hydrologic cycle by serving as major storehouses of water. In the warmer weather of spring and summer, much of their snow and ice melts and is released to streams for use by wildlife and by humans for drinking and irrigating crops. As the earth warms, mountaintop snowpacks and glaciers melt earlier each year. This can lower food production if water needed to irrigate crops during the summer has already been released.

Despite their ecological, economic, and cultural importance, the fate of mountain ecosystems has not been a high priority for governments or for many environmental organizations.

Figure 7-19 Mountains such as these in Mount Rainier National Park in the U.S. state of Washington play important ecological roles.

7-3 How Have We Affected the World's Terrestrial Ecosystems?

▶ **CONCEPT 7-3** In many areas, human activities are impairing ecological and economic services provided by the earth's deserts, grasslands, forests, and mountains.

Humans Have Disturbed Most of the Earth's Land

The human species dominates most of the planet. In many areas, human activities are impairing some of the ecological and economic services provided by the world's deserts, grasslands, forests, and mountains (**Concept 7-3**).

According to the 2005 Millennium Ecosystem Assessment, about 62% of the world's major terrestrial ecosystems are being degraded or used unsustainably (see Figure 3, pp. S24–S25, and Figure 7, pp. S28–S29, in Supplement 4), as the human ecological footprint intensifies and spreads across the globe (Figure 1-10, p. 15, and **Concept 1-3**, p. 12). This environmental destruction and degradation is increas-

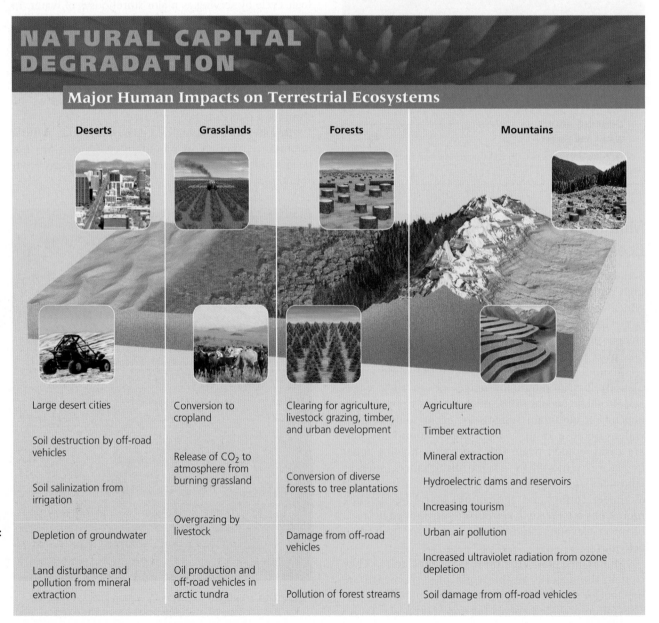

NATURAL CAPITAL DEGRADATION

Major Human Impacts on Terrestrial Ecosystems

Deserts	Grasslands	Forests	Mountains
Large desert cities	Conversion to cropland	Clearing for agriculture, livestock grazing, timber, and urban development	Agriculture
Soil destruction by off-road vehicles	Release of CO₂ to atmosphere from burning grassland		Timber extraction
Soil salinization from irrigation		Conversion of diverse forests to tree plantations	Mineral extraction
			Hydroelectric dams and reservoirs
			Increasing tourism
Depletion of groundwater	Overgrazing by livestock	Damage from off-road vehicles	Urban air pollution
			Increased ultraviolet radiation from ozone depletion
Land disturbance and pollution from mineral extraction	Oil production and off-road vehicles in arctic tundra	Pollution of forest streams	Soil damage from off-road vehicles

Figure 7-20 Major human impacts on the world's deserts, grasslands, forests, and mountains. **Question:** Which two of the impacts on each of these biomes do you think are the most harmful?

ing in many parts of the world. Figure 7-20 summarizes some of the human impacts on the world's deserts, grasslands, forests, and mountains.

How long can we keep eating away at these terrestrial forms of natural capital without threatening our economies and the long-term survival of our own and other species? No one knows. But there are increasing signs that we need to come to grips with this vital issue.

This will require protecting the world's remaining wild areas from development. In addition, many of the land areas we have degraded need to be restored. However, such efforts to achieve a balance between exploitation and conservation are highly controversial because of timber, mineral, fossil fuel, and other re-

sources found in or under many of the earth's biomes. These issues are discussed in Chapter 10.

┌─ **RESEARCH FRONTIER** ──────────────────┐
Better understanding of the effects of human activities on terrestrial biomes and how we can reduce these impacts. See **academic.cengage.com/biology/miller**.
└───┘

┌─ **THINKING ABOUT** ─────────────────────┐
Sustainability
Develop four guidelines for preserving the earth's terrestrial biodiversity based on the four **scientific principles of sustainability** (see back cover).
└───┘

REVISITING — Winds and Sustainability

CORE CASE STUDY · SUSTAINABILITY

This chapter's opening **Core Case Study** described how winds connect parts of the planet to one another. Next time you feel or hear the wind blowing, think about these global connections. As part of the global climate system, winds play important roles in creating and sustaining the world's deserts, grasslands, and forests through the four **scientific principles of sustainability** (see back cover). Winds promote sustainability by helping to distribute solar energy and to recycle the earth's nutrients. In turn, this helps support biodiversity, which in turn affects species interactions that help control population sizes.

Scientists have made a good start in understanding the ecology of the world's terrestrial biomes. One of the major lessons from their research is: in nature, *everything is connected.* According to these scientists, we urgently need more research on the workings of the world's terrestrial biomes and on how they are interconnected. With such information, we will have a clearer picture of how our activities affect the earth's natural capital and of what we can do to help sustain it.

When we try to pick out anything by itself,
we find it hitched to everything else
in the universe.

JOHN MUIR

REVIEW

1. Review the Key Questions and Concepts for this chapter on p. 141. Describe the environmentally beneficial and harmful effects of the earth's winds.

2. Distinguish between **weather** and **climate.** Describe three major factors that determine how air circulates in the lower atmosphere. Describe how the properties of air, water, and land affect global air circulation. How is heat distributed to different parts of the ocean? Explain how global air circulation and ocean currents lead to the forests, grasslands, and deserts that make up the earth's terrestrial biomes.

3. Define and give four examples of a **greenhouse gas.** What is the **greenhouse effect** and why is it important to the earth's life and climate? What is the **rain shadow effect** and how can it lead to the formation of inland deserts? Why do cities tend to have more haze and smog,

higher temperatures, and lower wind speeds than the surrounding countryside?

4. What is a **biome?** Explain why there are three major types of each of the major biomes (deserts, grasslands, and forests). Describe how climate and vegetation vary with latitude and elevation.

5. Describe how the three major types of deserts differ in their climate and vegetation. How do desert plants and animals survive?

6. Describe how the three major types of grasslands differ in their climate and vegetation. What is a savanna? Why have many of the world's temperate grasslands disappeared? What is **permafrost?** Distinguish between arctic tundra and alpine tundra.

7. What is temperate shrubland or chaparral? Why is this biome a desirable place to live? Why is it a risky place to live?

8. What is a **forest system** and what are the three major types of forests? Describe how these three types differ in their climate and vegetation. Why is biodiversity so high in tropical rain forests? Why do most soils in tropical rain forests have few plant nutrients? Describe what happens in temperate deciduous forests in the winter and fall. What are coastal coniferous or temperate rain forests? What important ecological roles do mountains play?

9. Describe how human activities have affected the world's deserts, grasslands, forests, and mountains.

10. Describe the connections between the earth's winds, climates, and biomes (**Core Case Study**) and the four **scientific principles of sustainability** (see back cover).

Note: Key Terms are in **bold** type.

CRITICAL THINKING

1. What would happen to **(a)** the earth's species and **(b)** your lifestyle if the winds stopped blowing (**Core Case Study**)?

2. List a limiting factor for each of the following ecosystems: **(a)** a desert, **(b)** arctic tundra, **(c)** temperate grassland, **(d)** the floor of a tropical rain forest, and **(e)** a temperate deciduous forest.

3. Why do deserts and arctic tundra support a much smaller biomass of animals than do tropical forests?

4. Why do most animals in a tropical rain forest live in its trees?

5. Why do most species living at high latitudes or high altitudes tend to have generalist ecological niches while those living in the tropics tend to have specialist ecological niches (Figure 4-11, p. 91)?

6. Which biomes are best suited for **(a)** raising crops and **(b)** grazing livestock? Use the four **scientific principles of sustainability** (see back cover) to come

up with four guidelines for growing food and grazing livestock in these biomes on a more sustainable basis.

7. What type of biome do you live in? List three ways in which your lifestyle is harming this biome?

8. You are a defense attorney arguing in court for sparing a tropical rain forest. Give your three most important arguments for the defense of this ecosystem.

9. Congratulations! You are in charge of the world. What are the three most important features of your plan to help sustain the earth's terrestrial biodiversity?

10. List two questions that you would like to have answered as a result of reading this chapter.

Note: See Supplement 13 (p. S78) for a list of Projects related to this chapter.

DATA ANALYSIS

In this chapter, you learned how long-term variations in the average temperature and average precipitation play a major role in determining the types of deserts, forests, and grasslands found in different parts of the world. Below are typical annual climate graphs for a tropical grassland (savanna) in Africa and a temperate grassland in the midwestern United States.

Tropical grassland (savanna)

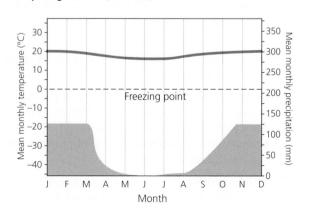

Temperate grassland

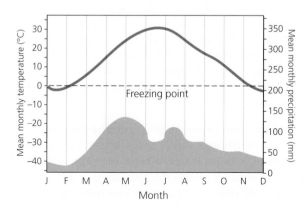

1. In what month (or months) does the most precipitation fall in each of these areas?

2. What are the driest months in each of these areas?

3. What is the coldest month in the tropical grassland?

4. What is the warmest month in the temperate grassland?

LEARNING ONLINE

Log on to the Student Companion Site for this book at **academic.cengage.com/biology/miller**, and choose Chapter 7 for many study aids and ideas for further reading and research. These include flash cards, practice quizzing, Weblinks, information on Green Careers, and InfoTrac® College Edition articles.

Aquatic Biodiversity

Why Should We Care about Coral Reefs?

Coral reefs form in clear, warm coastal waters of the tropics and subtropics (Figure 8-1, left). These stunningly beautiful natural wonders are among the world's oldest, most diverse, and most productive ecosystems. In terms of biodiversity, they are the marine equivalents of tropical rain forests.

Coral reefs are formed by massive colonies of tiny animals called *polyps* (close relatives of jellyfish). They slowly build reefs by secreting a protective crust of limestone (calcium carbonate) around their soft bodies. When the polyps die, their empty crusts remain behind as a platform for more reef growth. The resulting elaborate network of crevices, ledges, and holes serves as calcium carbonate "condominiums" for a variety of marine animals.

Coral reefs are the result of a mutually beneficial relationship between the polyps and tiny single-celled algae called *zooxanthellae* ("zoh-ZAN-thel-ee") that live in the tissues of the polyps. In this example of mutualism (p. 106), the algae provide the polyps with food and oxygen through photosynthesis, and help to produce calcium carbonate, which forms the coral skeleton. Algae also give the reefs their stunning coloration. The polyps, in turn, provide the algae with a well-protected home and some of their nutrients.

Although coral reefs occupy only about 0.2% of the ocean floor, they provide important ecological and economic services. They help moderate atmospheric temperatures by removing CO_2 from the atmosphere, and they act as natural barriers that protect 15% of the world's coastlines from erosion caused by battering waves and storms. And they provide habitats for one-quarter of all marine organisms. Economically, coral reefs produce about one-tenth of the global fish catch—one-fourth of the catch in developing countries—and they provide jobs and building materials for some of the world's poorest countries. Coral reefs also support important fishing and tourism industries.

Finally, these biological treasures give us an underwater world to study and enjoy. Each year, more than 1 million scuba divers and snorkelers visit coral reefs to experience these wonders of aquatic biodiversity.

According to a 2005 report by the World Conservation Union, 15% of the world's coral reefs have been destroyed and another 20% have been damaged by coastal development, pollution, overfishing, warmer ocean temperatures, increasing ocean acidity, and other stresses. And another 25–33% of these centers of aquatic biodiversity could be lost within 20–40 years. One problem is *coral bleaching* (Figure 8-1, upper right). It occurs when stresses such as increased temperature cause the algae, upon which corals depend for food, to die off, leaving behind a white skeleton of calcium carbonate. Another threat is the increasing acidity of ocean water as it absorbs some of the CO_2 produced by the burning of carbon-containing fossils fuels. The CO_2 reacts with ocean water to form a weak acid, which can slowly dissolve the calcium carbonate that makes up the corals.

The degradation and decline of these colorful oceanic sentinels should serve as a warning about threats to the health of the oceans, which provide us with crucial ecological and economic services.

Figure 8-1 *A healthy coral reef* in the Red Sea covered by colorful algae (left) and a bleached coral reef that has lost most of its algae (right) because of changes in the environment (such as cloudy water or high water temperatures). With the colorful algae gone, the white limestone of the coral skeleton becomes visible. If the environmental stress is not removed and no other alga species fill the abandoned niche, the corals die. These diverse and productive ecosystems are being damaged and destroyed at an alarming rate.

Sergio Hanquet/Peter Arnold, Inc

8-1 What is the general nature of aquatic systems?

CONCEPT 8-1A Saltwater and freshwater aquatic life zones cover almost three-fourths of the earth's surface with oceans dominating the planet.

CONCEPT 8-1B The key factors determining biodiversity in aquatic systems are temperature, dissolved oxygen content, availability of food, and availability of light and nutrients necessary for photosynthesis.

8-2 Why are marine aquatic systems important?

CONCEPT 8-2 Saltwater ecosystems are irreplaceable reservoirs of biodiversity and provide major ecological and economic services.

8-3 How have human activities affected marine ecosystems?

CONCEPT 8-3 Human activities threaten aquatic biodiversity and disrupt ecological and economic services provided by saltwater systems.

8-4 Why are freshwater ecosystems important?

CONCEPT 8-4 Freshwater ecosystems provide major ecological and economic services and are irreplaceable reservoirs of biodiversity.

8-5 How have human activities affected freshwater ecosystems?

CONCEPT 8-5 Human activities threaten biodiversity and disrupt ecological and economic services provided by freshwater lakes, rivers, and wetlands.

Note: Supplements 2 (p. S4), 4 (p. S20), 5 (p. S31), 7 (p. S46), 9 (p. S53), and 13 (p. S78) can be used with this chapter.

*If there is magic on this planet,
it is contained in water.*

LOREN EISLEY

8-1   What Is the General Nature of Aquatic Systems?

▶ **CONCEPT 8-1A** Saltwater and freshwater aquatic life zones cover almost three-fourths of the earth's surface with oceans dominating the planet.

▶ **CONCEPT 8-1B** The key factors determining biodiversity in aquatic systems are temperature, dissolved oxygen content, availability of food, and availability of light and nutrients necessary for photosynthesis.

Most of the Earth Is Covered with Water

When viewed from a certain point in outer space, the earth appears to be almost completely covered with water (Figure 8-2). Saltwater covers about 71% of the earth's surface, and freshwater occupies roughly another 2.2%. Yet, in proportion to the entire planet, it all amounts to a thin and precious film of water.

Although the *global ocean* is a single and continuous body of water, geographers divide it into four large areas—the Atlantic, Pacific, Arctic, and Indian Oceans—separated by the continents. The largest ocean is the Pacific, which contains more than half of the earth's water and covers one-third of the earth's surface.

The aquatic equivalents of biomes are called **aquatic life zones.** The distribution of many aquatic organisms is determined largely by the water's *salinity*—the amounts of various salts such as sodium chloride (NaCl) dissolved in a given volume of water. As a

Ocean hemisphere   Land–ocean hemisphere

Figure 8-2 *The ocean planet.* The salty oceans cover 71% of the earth's surface. Almost all of the earth's water is in the interconnected oceans, which cover 90% of the planet's mostly ocean hemisphere (left) and half of its land–ocean hemisphere (right). Freshwater systems cover less than 2.2% of the earth's surface (**Concept 8-1A**).

result, aquatic life zones are classified into two major types: **saltwater** or **marine** (oceans and their accompanying estuaries, coastal wetlands, shorelines, coral reefs, and mangrove forests) and **freshwater** (lakes,

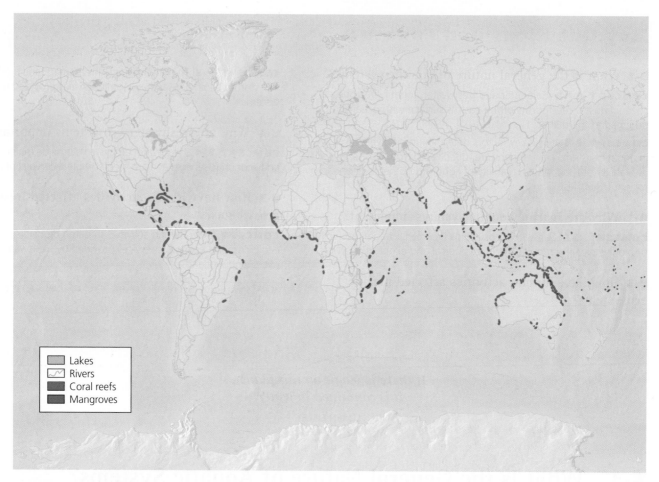

Figure 8-3 Natural capital: distribution of the world's major saltwater oceans, coral reefs, mangroves, and freshwater lakes and rivers. **Question:** Why do you think most coral reefs lie in the southern hemisphere?

rivers, streams, and inland wetlands). Although some systems such as estuaries are a mix of saltwater and freshwater, we classify them as marine systems for purposes of discussion.

Figure 8-3 shows the distribution of the world's major oceans, coral reefs, mangroves, lakes, and rivers. These aquatic systems play vital roles in the earth's biological productivity, climate, biogeochemical cycles, and biodiversity, and they provide us with fish, shellfish, minerals, recreation, transportation routes, and many other economically important goods and services.

Most Aquatic Species Live in Top, Middle, or Bottom Layers of Water

Saltwater and freshwater life zones contain several major types of organisms. One such type consists of weakly swimming, free-floating **plankton,** which can be divided into three groups, the first of which is *phytoplankton* ("FY-toe-plank-ton," Greek for "drifting plants"; see bottom of Figure 3-14, p. 63), which includes many types of algae. They and various rooted plants near shorelines are primary producers that support most

aquatic food webs. See *The Habitable Planet,* Videos 2 and 3, at **www.learner.org/resources/series209.html.**

The second group is *zooplankton* ("ZOH-uh-plank-ton," Greek for "drifting animals"; see bottom of Figure 3-14, p. 63). They consist of primary consumers (herbivores) that feed on phytoplankton and secondary consumers that feed on other zooplankton. They range from single-celled protozoa to large invertebrates such as jellyfish.

A third group consists of huge populations of much smaller plankton called *ultraplankton.* These extremely small photosynthetic bacteria may be responsible for 70% of the primary productivity near the ocean surface.

A second major type of organisms is **nekton,** strongly swimming consumers such as fish, turtles, and whales. The third type, **benthos,** consists of bottom dwellers such as oysters, which anchor themselves to one spot; clams and worms, which burrow into the sand or mud; and lobsters and crabs, which walk about on the sea floor. A fourth major type is **decomposers** (mostly bacteria), which break down organic compounds in the dead bodies and wastes of aquatic organisms into nutrients that can be used by aquatic primary producers.

Most forms of aquatic life are found in the *surface, middle,* and *bottom* layers of saltwater and freshwater systems, which we explore later in this chapter. In most aquatic systems, the key factors determining the types and numbers of organisms found in these layers are *temperature, dissolved oxygen content, availability of food,* and *availability of light and nutrients required for photosynthesis,* such as carbon (as dissolved CO_2 gas), nitrogen (as NO_3^-), and phosphorus (mostly as PO_4^{3-}) (**Concept 8-1B**).

In deep aquatic systems, photosynthesis is largely confined to the upper layer—the *euphotic* or *photic* zone, through which sunlight can penetrate. The depth of the euphotic zone in oceans and deep lakes can be reduced when the water is clouded by excessive algal growth (algal blooms) resulting from nutrient overloads. This cloudiness, called **turbidity,** can occur naturally, such as from algal growth, or can result from disturbances such as clearing of land, which causes silt to flow into bodies of water. This is one of the problems plaguing coral reefs (**Core Case Study**), as excessive turbidity due to silt runoff prevents photosynthesis and causes the corals to die.

CORE CASE STUDY

In shallow systems such as small open streams, lake edges, and ocean shorelines, ample supplies of nutrients for primary producers are usually available. By contrast, in most areas of the open ocean, nitrates, phosphates, iron, and other nutrients are often in short supply, and this limits net primary productivity (NPP) (Figure 3-16, p. 64).

8-2 Why Are Marine Aquatic Systems Important?

▶ **CONCEPT 8-2** Saltwater ecosystems are irreplaceable reservoirs of biodiversity and provide major ecological and economic services.

Oceans Provide Important Ecological and Economic Resources

Oceans provide enormously valuable ecological and economic services (Figure 8-4). One estimate of the combined value of these goods and services from all marine coastal ecosystems is over $12 trillion per year, nearly equal to the annual U.S. gross domestic product.

As land dwellers, we have a distorted and limited view of the blue aquatic wilderness that covers most of the earth's surface. We know more about the surface of the moon than about the oceans. According to aquatic scientists, the scientific investigation of poorly understood marine and freshwater aquatic systems could yield immense ecological and economic benefits.

Marine aquatic systems are huge reservoirs of biodiversity. They include many different ecosystems, which host a great variety of species, genes, and biological and chemical processes, thus sustaining four major components of the earth's biodiversity (Figure 4-2, p. 79). Marine life is found in three major *life zones:* the coastal zone, open sea, and ocean bottom (Figure 8-5, p. 166).

— **RESEARCH FRONTIER** —
Discovering, cataloging, and studying the huge number of unknown aquatic species and their interactions. See **academic .cengage.com/biology/miller**.

The **coastal zone** is the warm, nutrient-rich, shallow water that extends from the high-tide mark on

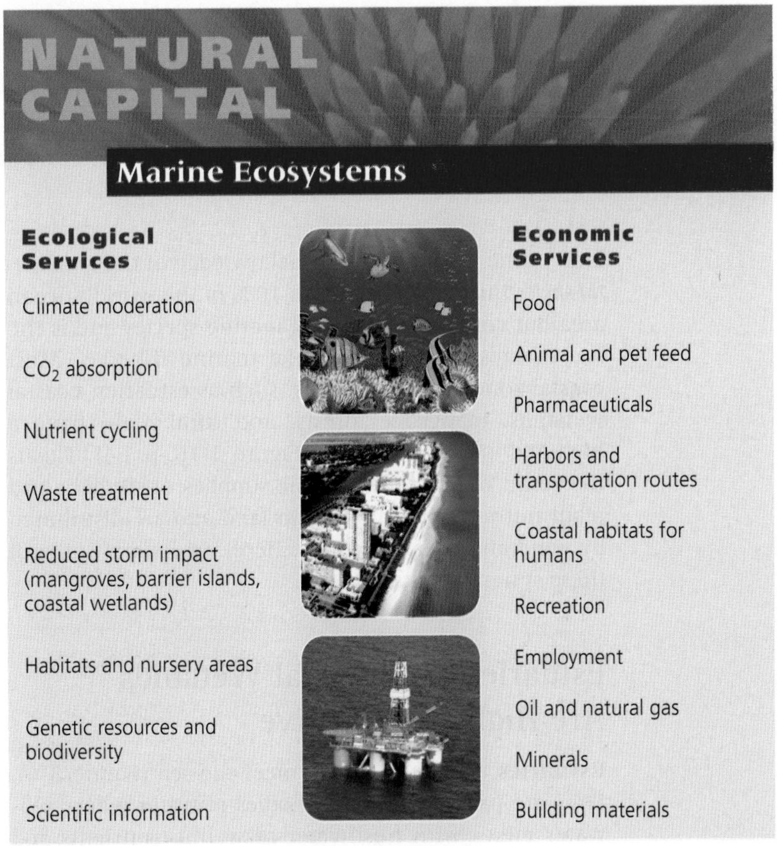

NATURAL CAPITAL

Marine Ecosystems

Ecological Services	Economic Services
Climate moderation	Food
CO_2 absorption	Animal and pet feed
Nutrient cycling	Pharmaceuticals
Waste treatment	Harbors and transportation routes
Reduced storm impact (mangroves, barrier islands, coastal wetlands)	Coastal habitats for humans
	Recreation
Habitats and nursery areas	Employment
	Oil and natural gas
Genetic resources and biodiversity	Minerals
Scientific information	Building materials

Figure 8-4 Major ecological and economic services provided by marine systems (**Concept 8-2**). **Question:** Which two ecological services and which two economic services do you think are the most important? Why?

Figure 8-5
Natural capital: major life zones and vertical zones (not drawn to scale) in an ocean. Actual depths of zones may vary. Available light determines the euphotic, bathyal and abyssal zones. Temperature zones also vary with depth, shown here by the red curve.
Question: How is an ocean like a rain forest? (*Hint:* see Figure 7-17, p. 156.)

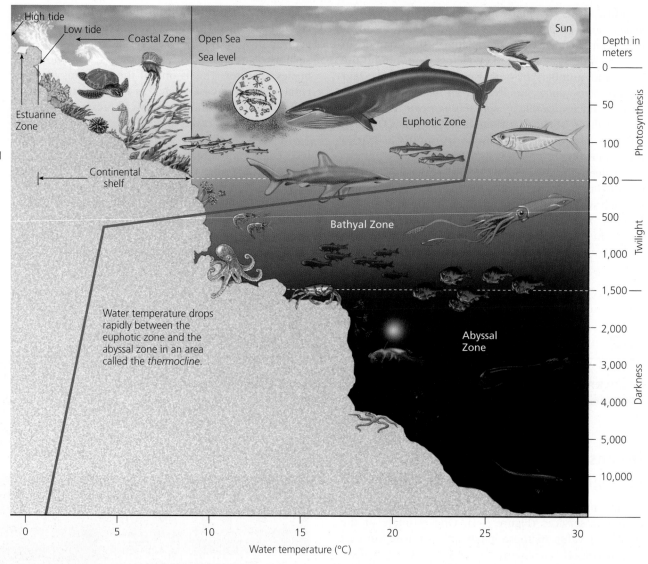

land to the gently sloping, shallow edge of the *continental shelf*. It makes up less than 10% of the world's ocean area but contains 90% of all marine species and is the site of most large commercial marine fisheries. Most coastal zone aquatic systems, such as estuaries, coastal wetlands, mangrove forests, and coral reefs, have a high NPP per unit of area (Figure 3-16, p. 64). This is the result of the zone's ample supplies of sunlight and plant nutrients that flow from land and are distributed by wind and ocean currents. Here, we look at some of these systems in more detail.

Estuaries and Coastal Wetlands Are Highly Productive

Estuaries are where rivers meet the sea (Figure 8-6). They are partially enclosed bodies of water where seawater mixes with freshwater as well as nutrients and pollutants from streams, rivers, and runoff from the land.

Estuaries and their associated **coastal wetlands**—coastal land areas covered with water all or part of the year—include river mouths, inlets, bays, sounds, and salt marshes in temperate zones (Figure 8-7), and mangrove forests in tropical zones (Figure 8-3). They are some of the earth's most productive ecosystems because of high nutrient inputs from rivers and nearby land, rapid circulation of nutrients by tidal flows, and ample sunlight penetrating the shallow waters.

Seagrass beds are another component of coastal marine biodiversity. They consist of at least 60 species of plants that grow underwater in shallow marine and estuarine areas along most continental coastlines. These highly productive and physically complex systems support a variety of marine species. They also help stabilize shorelines and reduce wave impacts.

Life in these coastal ecosystems is harsh. It must adapt to significant daily and seasonal changes in tidal and river flows, water temperatures and salinity, and runoff of eroded soil sediment and other pollutants from the land. Because of these stresses, despite their

Figure 8-6 View of an *estuary* from space. The photo shows the sediment plume (turbidity caused by runoff) at the mouth of Madagascar's Betsiboka River as it flows through the estuary and into the Mozambique Channel. Because of its topography, heavy rainfall, and the clearing of forests for agriculture, Madagascar is the world's most eroded country.

NASA

Herring gulls

Peregrine falcon

Snowy egret

Cordgrass

Short-billed dowitcher

Marsh periwinkle

Phytoplankton

Zooplankton and small crustaceans

Smelt

Soft-shelled clam

Bacteria

Clamworm

 Producer to primary consumer

 Primary to secondary consumer

Secondary to higher-level consumer

All consumers and producers to decomposers

SuperStock

Figure 8-7 Some components and interactions in a *salt marsh ecosystem* in a temperate area such as the United States. When these organisms die, decomposers break down their organic matter into minerals used by plants. Colored arrows indicate transfers of matter and energy between consumers (herbivores), secondary or higher-level consumers (carnivores), and decomposers. Organisms are not drawn to scale. The photo shows a salt marsh in Peru.

Theo Allofs/Visuals Unlimited

productivity, some coastal ecosystems have low plant diversity, composed of a few species that can withstand the daily and seasonal variations.

Mangrove forests are the tropical equivalent of salt marshes. They are found along some 70% of gently sloping sandy and silty coastlines in tropical and subtropical regions, especially Southeast Asia (Figure 8-3). The dominant organisms in these nutrient-rich coastal forests are mangroves—69 different tree species that can grow in salt water. They have extensive root systems that often extend above the water, where they can obtain oxygen and support the trees during periods of changing water levels (Figure 8-8).

These coastal aquatic systems provide important ecological and economic services. They help to maintain water quality in tropical coastal zones by filtering toxic pollutants, excess plant nutrients, and sediments, and by absorbing other pollutants. They provide food, habitats, and nursery sites for a variety of aquatic and terrestrial species. They also reduce storm damage and coastal erosion by absorbing waves and storing excess water produced by storms and tsunamis. Historically, they have sustainably supplied timber and fuelwood to coastal communities.

Loss of mangroves can lead to polluted drinking water, caused by inland intrusion of saltwater into aquifers that are used to supply drinking water. Despite their ecological and economic importance, in 2008, the U.N. Food and Agriculture Organization estimated that between 1980 and 2005 at least one-fifth of the world's mangrove forests were lost mostly because of human coastal development.

Rocky and Sandy Shores Host Different Types of Organisms

The gravitational pull of the moon and sun causes *tides* to rise and fall about every 6 hours in most coastal areas. The area of shoreline between low and high tides is called the **intertidal zone.** Organisms living in this zone must be able to avoid being swept away or crushed by waves, and must deal with being immersed during high tides and left high and dry (and much hotter) at low tides. They must also survive changing levels of salinity when heavy rains dilute saltwater. To deal with such stresses, most intertidal organisms hold on to something, dig in, or hide in protective shells.

On some coasts, steep *rocky shores* are pounded by waves. The numerous pools and other habitats in their intertidal zones contain a great variety of species that occupy different niches in response to daily and seasonal changes in environmental conditions such as temperature, water flows, and salinity (Figure 8-9, top).

Other coasts have gently sloping *barrier beaches,* or *sandy shores,* that support other types of marine organisms (Figure 8-9, bottom). Most of them keep hidden from view and survive by burrowing, digging, and tunneling in the sand. These sandy beaches and their adjoining coastal wetlands are also home to a variety of shorebirds that feed in specialized niches on crustaceans, insects, and other organisms (Figure 4-13, p. 93). Many of these species also live on *barrier islands*—low, narrow, sandy islands that form offshore, parallel to some coastlines.

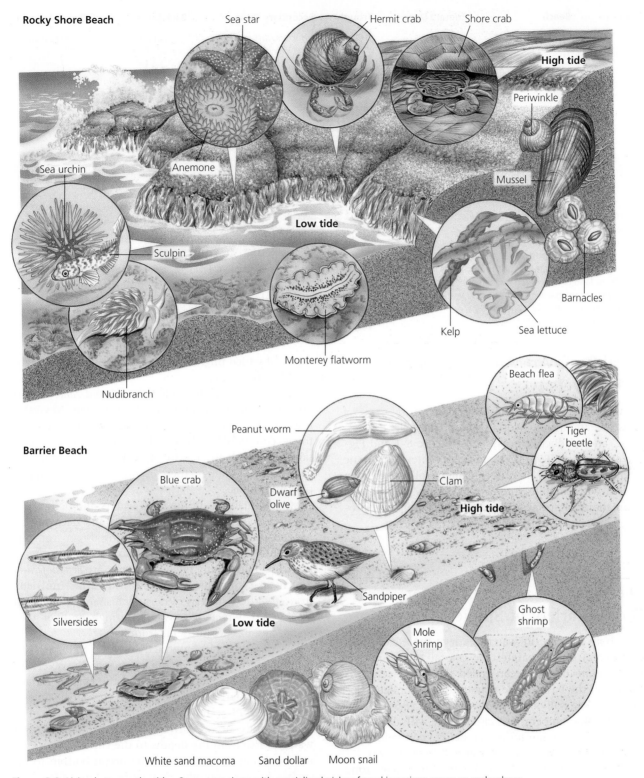

Rocky Shore Beach

Sea star

Hermit crab

Shore crab

High tide

Periwinkle

Sea urchin

Anemone

Mussel

Sculpin

Low tide

Barnacles

Kelp

Sea lettuce

Nudibranch

Monterey flatworm

Barrier Beach

Peanut worm

Beach flea

Tiger beetle

Blue crab

Dwarf olive

Clam

High tide

Silversides

Sandpiper

Low tide

Mole shrimp

Ghost shrimp

White sand macoma

Sand dollar

Moon snail

Figure 8-9 *Living between the tides.* Some organisms with specialized niches found in various zones on rocky shore beaches (top) and barrier or sandy beaches (bottom). Organisms are not drawn to scale.

Undisturbed barrier beaches generally have one or more rows of natural sand dunes in which the sand is held in place by plant roots (Figure 8-10, p. 170). These dunes are the first line of defense against the ravages of the sea. Such real estate is so scarce and valuable that coastal developers frequently remove the protective dunes or build behind the first set of dunes and cover them with buildings and roads. Large storms can then flood and even sweep away seaside buildings and severely erode the sandy beaches. Some people incorrectly call these human-influenced events "natural disasters."

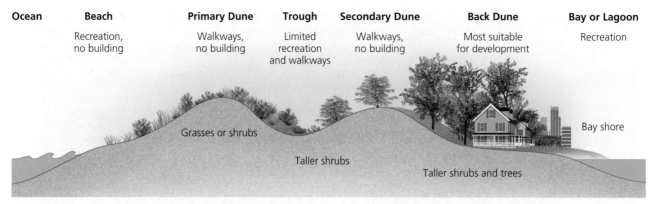

Ocean	Beach	Primary Dune	Trough	Secondary Dune	Back Dune	Bay or Lagoon
	Recreation, no building	Walkways, no building	Limited recreation and walkways	Walkways, no building	Most suitable for development	Recreation

Grasses or shrubs

Taller shrubs

Taller shrubs and trees

Bay shore

Figure 8-10 *Primary and secondary dunes* on gently sloping sandy barrier beaches help protect land from erosion by the sea. The roots of grasses that colonize the dunes hold the sand in place. Ideally, construction is allowed only behind the second strip of dunes, and walkways to the ocean beach are built so as not to damage the dunes. This helps to preserve barrier beaches and to protect buildings from damage by wind, high tides, beach erosion, and flooding from storm surges. Such protection is rare in some coastal areas because the short-term economic value of oceanfront land is considered much higher than its long-term ecological value. Rising sea levels from global warming may put many barrier beaches under water by the end of this century. **Question:** Do you think that the long- and short-term ecological values of oceanfront dunes outweigh the short-term economic value of removing them for coastal development? Explain.

Coral Reefs Are Amazing Centers of Biodiversity

As we noted in the **Core Case Study,** coral reefs are among the world's oldest, most diverse, and productive ecosystems (Figure 8-1 and Figure 4-10, left, p. 89). These amazing centers of aquatic biodiversity are the marine equivalents of tropical rain forests, with complex interactions among their diverse populations of species (Figure 8-11). Coral reefs provide homes for one-fourth of all marine species.

The Open Sea and Ocean Floor Host a Variety of Species

The sharp increase in water depth at the edge of the continental shelf separates the coastal zone from the vast volume of the ocean called the **open sea.** Primarily on the basis of the penetration of sunlight, this deep blue sea is divided into three *vertical zones* (see Figure 8-5). But temperatures also change with depth and can be used to define zones that help to determine species diversity in these layers (Figure 8-5, red curve).

The *euphotic zone* is the brightly lit upper zone where drifting phytoplankton carry out about 40% of the world's photosynthetic activity (see *The Habitable Planet,* Video 3, at **www.learner.org/resources/ series209.html**). Nutrient levels are low (except around upwellings, Figure 7-2, p. 142), and levels of dissolved oxygen are high. Large, fast-swimming predatory fishes such as swordfish, sharks, and bluefin tuna populate this zone.

The *bathyal zone* is the dimly lit middle zone, which, because it gets little sunlight, does not contain photosynthesizing producers. Zooplankton and smaller fishes,

many of which migrate to feed on the surface at night, populate this zone.

The deepest zone, called the *abyssal zone,* is dark and very cold; it has little dissolved oxygen. Nevertheless, the deep ocean floor is teeming with life—enough to be considered a major life zone—because it contains enough nutrients to support a large number of species, even though there is no sunlight to support photosynthesis.

Most organisms of the deep waters and ocean floor get their food from showers of dead and decaying organisms—called *marine snow*—drifting down from upper lighted levels of the ocean. Some of these organisms, including many types of worms, are *deposit feeders,* which take mud into their guts and extract nutrients from it. Others such as oysters, clams, and sponges are *filter feeders,* which pass water through or over their bodies and extract nutrients from it.

Average primary productivity and NPP per unit of area are quite low in the open sea. However, because open sea covers so much of the earth's surface, it makes the largest contribution to the earth's overall NPP. Also, NPP is much higher in some open sea areas where winds, ocean currents, and other factors cause water to rise from the depths to the surface. These *upwellings* bring nutrients from the ocean bottom to the surface for use by producers (Figure 7-2, p. 142).

In 2007, a team of scientists led by J. Craig Venter released a report that dramatically challenged scientists' assumptions about biodiversity in the open sea. After sailing around the world and spending 2 years collecting data, they found that the open sea contains many more bacteria, viruses, and other microbes than scientists had previously assumed.

CENGAGENOW™ Learn about ocean provinces where all ocean life exists at CengageNOW™.

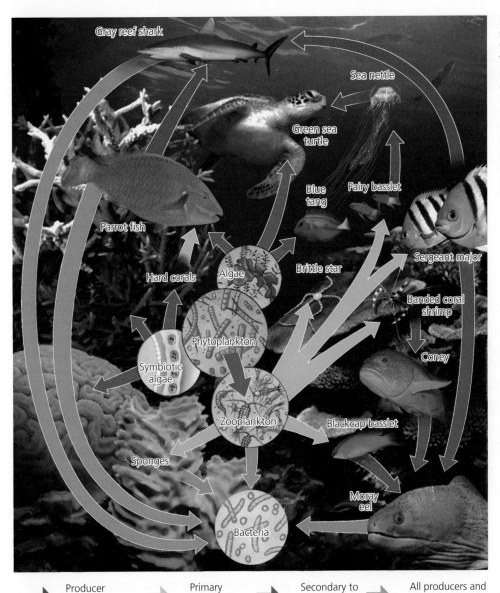

Figure 8-11 Natural capital: some components and interactions in a *coral reef ecosystem*. When these organisms die, decomposers break down their organic matter into minerals used by plants. Colored arrows indicate transfers of matter and energy between producers, primary consumers (herbivores), secondary or higher-level consumers (carnivores), and decomposers. Organisms are not drawn to scale.

Gray reef shark

Sea nettle

Green sea turtle

Blue tang

Fairy basslet

Parrot fish

Hard corals

Algae

Brittle star

Sergeant major

Banded coral shrimp

Symbiotic algae

Phytoplankton

Coney

Zooplankton

Blackcap basslet

Sponges

Bacteria

Moray eel

➤ Producer to primary consumer

➤ Primary to secondary consumer

➤ Secondary to higher-level consumer

➤ All producers and consumers to decomposers

8-3 How Have Human Activities Affected Marine Ecosystems?

CONCEPT 8-3 Human activities threaten aquatic biodiversity and disrupt ecological and economic services provided by saltwater systems.

Human Activities Are Disrupting and Degrading Marine Systems

Human activities are disrupting and degrading some ecological and economic services provided by marine aquatic systems, especially coastal wetlands, shorelines, mangrove forests, and coral reefs (**Concept 8-3**). (See *The Hab-*

itable Planet, Video 9, at **www.learner.org/resources/series209.html**.) Thus, a single largely land-based species—humans—is increasingly threatening the biological diversity and ecosystem services provided by the oceans that cover about 71% of the earth's surface.

In 2008, the U.S. National Center for Ecological Analysis and Synthesis (NCEAS) used computer models

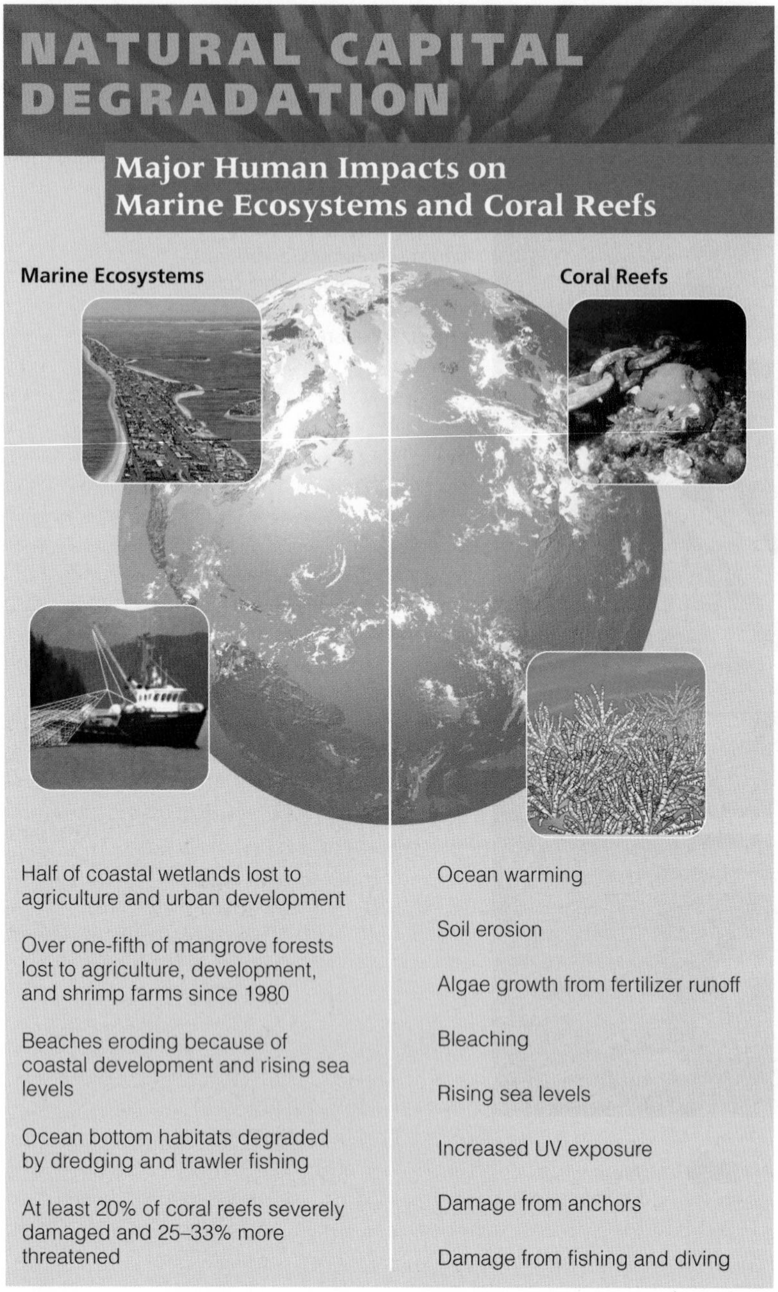

NATURAL CAPITAL DEGRADATION

Major Human Impacts on Marine Ecosystems and Coral Reefs

Marine Ecosystems

Coral Reefs

Half of coastal wetlands lost to agriculture and urban development

Over one-fifth of mangrove forests lost to agriculture, development, and shrimp farms since 1980

Beaches eroding because of coastal development and rising sea levels

Ocean bottom habitats degraded by dredging and trawler fishing

At least 20% of coral reefs severely damaged and 25–33% more threatened

Ocean warming

Soil erosion

Algae growth from fertilizer runoff

Bleaching

Rising sea levels

Increased UV exposure

Damage from anchors

Damage from fishing and diving

Figure 8-12 Major threats to marine ecosystems (left) and particularly coral reefs (right) resulting from human activities (**Concept 8-3**). **Questions:** Which two of the threats to marine ecosystems do you think are the most serious? Why? Which two of the threats to coral reefs do you think are the most serious? Why?

Major threats to marine systems from human activities include:

- Coastal development, which destroys and pollutes coastal habitats (see *The Habitable Planet*, Video 5, at **www.learner.org/resources/series209.html**)

- Overfishing, which depletes populations of commercial fish species

- Runoff of nonpoint source pollution such as fertilizers, pesticides, and livestock wastes from the land (see *The Habitable Planet*, Videos 7 and 8, at **www.learner.org/resources/series209.html**)

- Point source pollution such as sewage from passenger cruise ships and spills from oil tankers

- Habitat destruction from coastal development and trawler fishing boats that drag weighted nets across the ocean bottom

- Invasive species, introduced by humans, that can deplete populations of native aquatic species and cause economic damage

- Climate change, enhanced by human activities, that could cause a rise in sea levels, which could destroy coral reefs and flood coastal marshes and coastal cities (see *The Habitable Planet*, Videos 7 and 8, at **www.learner.org/resources/series209.html**)

- Climate change from burning fossil fuels, which is also threatening marine ecosystems by warming the oceans and making them more acidic

- Pollution and degradation of coastal wetlands and estuaries (Case Study, below)

Figure 8-12 shows some of the effects of such human impacts on marine systems (left) and coral reefs (right). According to a 2007 study by O. Hoegh-Guldberg and 16 other scientists, unless we take action soon to significantly reduce carbon dioxide emissions, the oceans may become too acidic and too warm for most of the world's coral reefs to survive this century, and the important ecological and economic services they provide will be lost. We will examine some of these impacts more closely in Chapter 11.

─ **RESEARCH FRONTIER** ─

Learning more about harmful human impacts on marine ecosystems and how to reduce these impacts. See **academic.cengage.com/biology/miller**.

to analyze and provide the first-ever comprehensive map of the effects of 17 different types of human activities on the world's oceans. In this four-year study, the international team of scientists found that human activity has heavily affected 41% of the world's ocean area, with no area left completely untouched.

In their desire to live near the coast, people are destroying or degrading the natural resources and services (Figure 8-4) that make coastal areas so enjoyable and valuable. In 2006, about 45% of the world's population (including more than half of the U.S. population) lived along or near coasts. By 2040, up to 80% of the world's people are projected to be living in or near coastal zones.

■ CASE STUDY

The Chesapeake Bay—an Estuary in Trouble

Since 1960, the Chesapeake Bay (Figure 8-13)—the largest estuary in the United States—has been in serious trouble from water pollution, mostly because of human activities. One problem is population growth. Between 1940 and 2007, the number of people living

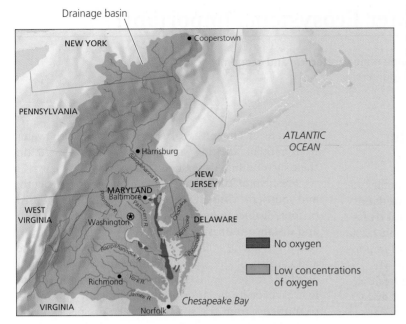

Drainage basin

NEW YORK

• Cooperstown

PENNSYLVANIA

• Harrisburg

Susquehanna R.

MARYLAND
Baltimore •

Potomac R.   *Patuxent R.*

WEST
VIRGINIA

Washington ✪

Rappahannock R.

Choptank   DELAWARE

Nanticoke

Pocomoke

ATLANTIC
OCEAN

NEW
JERSEY

Richmond •   *York R.*

VIRGINIA

James R.

Norfolk •

Chesapeake Bay

■ No oxygen

☐ Low concentrations
of oxygen

Figure 8-13 *Chesapeake Bay,* the largest estuary in the United States, is severely degraded as a result of water pollution from point and nonpoint sources in six states and from the atmospheric deposition of air pollutants.

in the Chesapeake Bay area grew from 3.7 million to 16.6 million. And the rate of population increase in the bay area has increased since 1990. With more than 450 people moving into the watershed each day, the population may soon reach 17 million.

The estuary receives wastes from point and nonpoint sources scattered throughout a huge drainage basin that includes 9 large rivers and 141 smaller streams and creeks in parts of six states (Figure 8-13). The bay has become a huge pollution sink because only 1% of the waste entering it is flushed into the Atlantic Ocean. It is also so shallow that people can wade through much of it.

Phosphate and nitrate levels have risen sharply in many parts of the bay, causing algal blooms and oxygen depletion. Commercial harvests of its once-abundant oysters, crabs, and several important fishes have fallen sharply since 1960 because of a combination of pollution, overfishing, and disease.

Point sources, primarily sewage treatment plants and industrial plants (often in violation of their discharge permits), account for 60% by weight of the phosphates. Nonpoint sources—mostly runoff of fertilizer and animal wastes from urban, suburban, and agricultural land and deposition from the atmosphere—account for 60% by weight of the nitrates. According to a 2004 study by the Chesapeake Bay Foundation, animal manure is the largest source of nitrates and phosphates from agricultural pollution.

In 1983, the United States implemented the Chesapeake Bay Program. In this ambitious attempt at *integrated coastal management,* citizens' groups, communities, state legislatures, and the federal government are working together to reduce pollution inputs into the bay. Strategies include establishing land-use regulations in the bay's six watershed states to reduce agricultural and urban runoff, banning phosphate detergents, upgrading sewage treatment plants, and monitoring

industrial discharges more closely. In addition, wetlands are being restored and large areas of the bay are being replanted with sea grasses to help filter out nutrients and other pollutants.

A century ago, oysters were so abundant that they filtered and cleaned the Chesapeake's entire volume of water every 3 days. This important form of natural capital helped to remove excess nutrients and reduce algal blooms that decreased dissolved oxygen levels. Now the oyster population has been reduced to the point where this filtration process takes a year.

Officials in the states of Maryland and Virginia are evaluating whether to rebuild the Chesapeake's oyster population by introducing an Asian oyster that appears resistant to two parasites that have killed off many of the bay's native oysters. The Asian oysters grow bigger and faster and taste as good as native oysters. But introducing the nonnative Asian oyster is unpredictable and irreversible, and some researchers warn that this nonnative species may not help to clean the water, because it requires clean water in order to flourish.

The hard work on improving the water quality of the Chesapeake Bay has paid off. Between 1985 and 2000, phosphorus levels declined 27%, nitrogen levels dropped 16%, and grasses growing on the bay's floor have made a comeback. This is a significant achievement, given the increasing population in the watershed and the fact that nearly 40% of the nitrogen inputs come from the atmosphere.

There is still a long way to go, and a sharp drop in state and federal funding has slowed progress. During the summer of 2005, more than 40% of the bay had too little dissolved oxygen to support many kinds of aquatic life. And according to a 2006 report by the Chesapeake Bay Foundation, "the bay's health remains dangerously out of balance." Yet despite some setbacks, the Chesapeake Bay Program shows what can be done when diverse groups work together to achieve goals that benefit both wildlife and people.

THINKING ABOUT
The Chesapeake Bay

What are three ways in which Chesapeake Bay area residents could apply the four **scientific principles of sustainability** (see back cover) to try to improve the environmental quality of the bay?

SUSTAINABILITY

8-4 Why Are Freshwater Ecosystems Important?

▶ **CONCEPT 8-4** Freshwater ecosystems provide major ecological and economic services and are irreplaceable reservoirs of biodiversity.

Water Stands in Some Freshwater Systems and Flows in Others

Freshwater life zones include *standing* (lentic) bodies of freshwater, such as lakes, ponds, and inland wetlands, and *flowing* (lotic) systems, such as streams and rivers. Although these freshwater systems cover less than 2.2% of the earth's surface, they provide a number of important ecological and economic services (Figure 8-14).

Lakes are large natural bodies of standing freshwater formed when precipitation, runoff, or groundwater seepage fills depressions in the earth's surface. Causes of such depressions include glaciations (Lake Louise, Alberta, Canada), crustal displacement (Lake Nyasa in East Africa), and volcanic activity (Crater Lake in the U.S. state of Oregon). Lakes are supplied with water from rainfall, melting snow, and streams that drain their surrounding watershed.

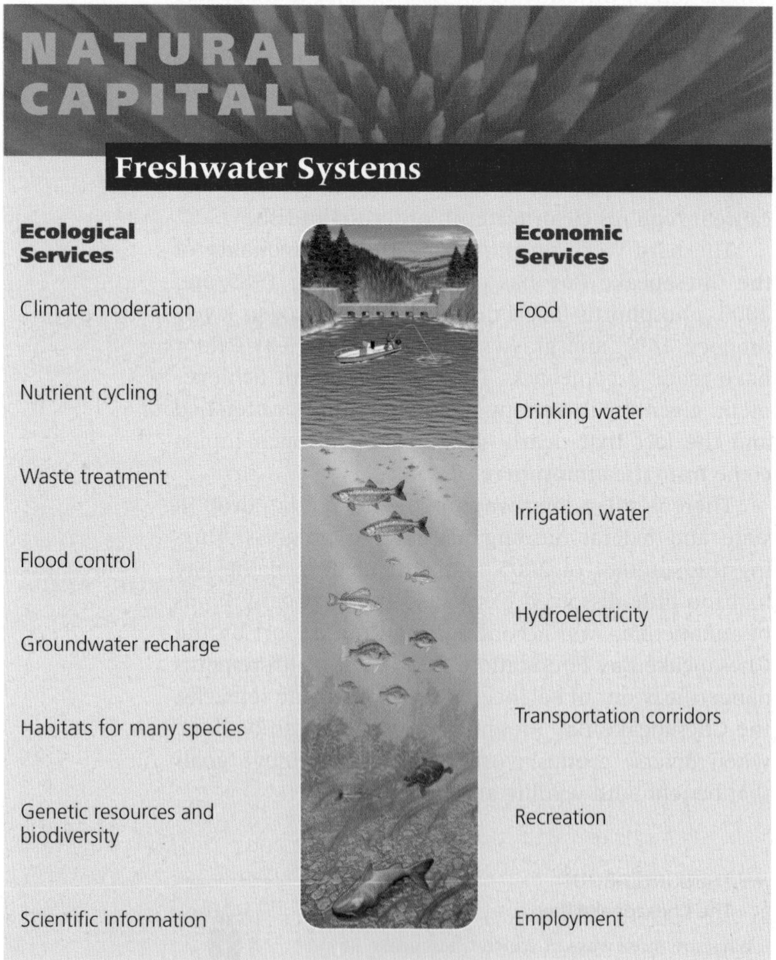

Freshwater lakes vary tremendously in size, depth, and nutrient content. Deep lakes normally consist of four distinct zones that are defined by their depth and distance from shore (Figure 8-15). The top layer, called the *littoral* ("LIT-tore-el") *zone,* is near the shore and consists of the shallow sunlit waters to the depth at which rooted plants stop growing. It has a high biological diversity because of ample sunlight and inputs of nutrients from the surrounding land. Species living in the littoral zone include many rooted plants and animals such as turtles, frogs, crayfish, and many fishes such bass, perch, and carp.

Next is the *limnetic* ("lim-NET-ic") *zone:* the open, sunlit surface layer away from the shore that extends to the depth penetrated by sunlight. The main photosynthetic body of the lake, this zone produces the food and oxygen that support most of the lake's consumers. Its most abundant organisms are microscopic phytoplankton and zooplankton. Some large fishes spend most of their time in this zone, with occasional visits to the littoral zone to feed and reproduce.

Next comes the *profundal* ("pro-FUN-dahl") *zone:* the deep, open water where it is too dark for photosynthesis to occur. Without sunlight and plants, oxygen levels are often low here. Fishes adapted to the lake's cooler and darker water are found in this zone.

The bottom of the lake contains the *benthic* ("BEN-thic") *zone,* inhabited mostly by decomposers, detritus feeders, and some fishes. The benthic zone is nourished mainly by dead matter that falls from the littoral and limnetic zones and by sediment washing into the lake.

Some Lakes Have More Nutrients Than Others

Ecologists classify lakes according to their nutrient content and primary productivity. Lakes that have a small supply of plant nutrients are called **oligotrophic** (poorly nourished) **lakes** (Figure 8-16, left). Often, this type of lake is deep and has steep banks.

Glaciers and mountain streams supply water to many such lakes, bringing little in the way of sediment or microscopic life to cloud the water. These lakes usually have crystal-clear water and small populations of phytoplankton and fishes such as smallmouth bass and trout. Because of their low levels of nutrients, these lakes have a low net primary productivity.

NATURAL CAPITAL

Freshwater Systems

Ecological Services	Economic Services
Climate moderation	Food
Nutrient cycling	Drinking water
Waste treatment	Irrigation water
Flood control	Hydroelectricity
Groundwater recharge	Transportation corridors
Habitats for many species	
Genetic resources and biodiversity	Recreation
Scientific information	Employment

Figure 8-14 Major ecological and economic services provided by freshwater systems (**Concept 8-4**). **Question:** Which two ecological services and which two economic services do you think are the most important? Why?

Sunlight

Painted turtle

Green frog

Blue-winged teal

Muskrat

Pond snail

Littoral zone

Plankton

Diving beetle

Limnetic zone

Profundal zone

Benthic zone

Northern pike

Yellow perch

Bloodworms

CENGAGENOW™ **Active Figure 8-15**
Distinct zones of life in a fairly deep temperate zone lake. *See an animation based on this figure at* CengageNOW.
Question: How are deep lakes like tropical rain forests? (*Hint:* See Figure 7-17, p. 156)

Over time, sediment, organic material, and inorganic nutrients wash into most oligotrophic lakes, and plants grow and decompose to form bottom sediments. A lake with a large supply of nutrients needed by producers is called a **eutrophic** (well-nourished) **lake** (Figure 8-16, right). Such lakes typically are shallow and have murky brown or green water with high turbidity. Because of their high levels of nutrients, these lakes have a high net primary productivity.

Human inputs of nutrients from the atmosphere and from nearby urban and agricultural areas can accelerate the eutrophication of lakes, a process called **cultural eutrophication.** This process often puts excessive nutrients into lakes, which are then described

Jack Carey

Bill Banazewski/Visuals Unlimited

Figure 8-16 The effect of nutrient enrichment on a lake. Crater Lake in the U.S. state of Oregon (left) is an example of an *oligotrophic lake* that is low in nutrients. Because of the low density of plankton, its water is quite clear. The lake on the right, found in western New York State, is a *eutrophic lake.* Because of an excess of plant nutrients, its surface is covered with mats of algae and cyanobacteria.

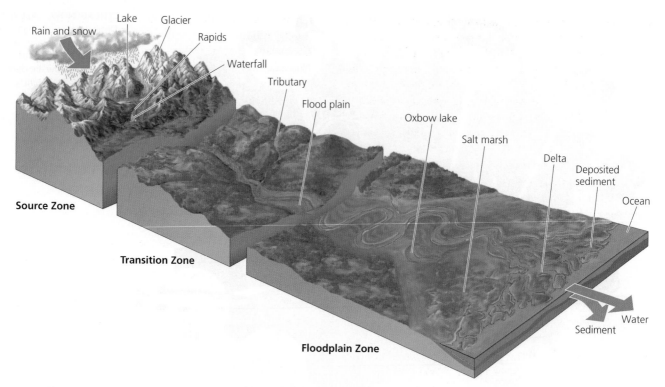

Labels in figure:
Rain and snow — Lake — Glacier — Rapids — Waterfall — Tributary — Flood plain — Oxbow lake — Salt marsh — Delta — Deposited sediment — Ocean — Water — Sediment

Source Zone

Transition Zone

Floodplain Zone

Figure 8-17 Three zones in the downhill flow of water: *source zone* containing mountain (headwater) streams; *transition zone* containing wider, lower-elevation streams; and *floodplain zone* containing rivers, which empty into the ocean.

as **hypereutrophic.** Many lakes fall somewhere between the two extremes of nutrient enrichment. They are called **mesotrophic lakes.**

Freshwater Streams and Rivers Carry Water from the Mountains to the Oceans

Precipitation that does not sink into the ground or evaporate becomes **surface water.** It becomes **runoff** when it flows into streams. A **watershed,** or **drainage basin,** is the land area that delivers runoff, sediment, and dissolved substances to a stream. Small streams join to form rivers, and rivers flow downhill to the ocean (Figure 8-17).

In many areas, streams begin in mountainous or hilly areas that collect and release water falling to the earth's surface as rain or snow that melts during warm seasons. The downward flow of surface water and groundwater from mountain highlands to the sea typically takes place in three aquatic life zones characterized by different environmental conditions: the *source zone,* the *transition zone,* and the *floodplain zone.* Rivers and streams can differ somewhat from this generalized model.

In the first, narrow *source zone* (Figure 8-17, top), headwaters, or mountain highland streams are usually shallow, cold, clear, and swiftly flowing. As this

turbulent water flows and tumbles downward over waterfalls and rapids, it dissolves large amounts of oxygen from the air. Most of these streams are not very productive because of a lack of nutrients and primary producers. Their nutrients come primarily from organic matter (mostly leaves, branches, and the bodies of living and dead insects) that falls into the stream from nearby land.

The source zone is populated by cold-water fishes (such as trout in some areas), which need lots of dissolved oxygen. Many fishes and other animals in fast-flowing headwater streams have compact and flattened bodies that allow them to live under stones. Others have streamlined and muscular bodies that allow them to swim in the rapid and strong currents. Most plants are algae and mosses attached to rocks and other surfaces under water.

In the *transition zone* (Figure 8-17, middle), headwater streams merge to form wider, deeper, and warmer streams that flow down gentler slopes with fewer obstacles. They can be more turbid (from suspended sediment), slower flowing, and have less dissolved oxygen than headwater streams have. The warmer water and other conditions in this zone support more producers and cool-water and warm-water fish species (such as black bass) with slightly lower oxygen requirements.

As streams flow downhill, they shape the land through which they pass. Over millions of years, the friction of moving water may level mountains and cut deep canyons, and rock and soil removed by the water are deposited as sediment in low-lying areas. In these *floodplain zones* (Figure 8-17, bottom), streams join into wider and deeper rivers that flow across broad, flat valleys. Water in this zone usually has higher temperatures and less dissolved oxygen than water in the two higher zones. These slow-moving rivers sometimes support fairly large populations of producers such as algae and cyanobacteria and rooted aquatic plants along the shores.

Because of increased erosion and runoff over a larger area, water in this zone often is muddy and contains high concentrations of suspended particulate matter (silt). The main channels of these slow-moving, wide, and murky rivers support distinctive varieties of fishes (such as carp and catfish), whereas their backwaters support species similar to those present in lakes. At its mouth, a river may divide into many channels as it flows through deltas, built up by deposited sediment, and coastal wetlands and estuaries, where the river water mixes with ocean water (Figure 8-6).

Coastal deltas and wetlands and inland wetlands and floodplains are important parts of the earth's natural capital. They absorb and slow the velocity of floodwaters from coastal storms, hurricanes (see Case Study below), and tsunamis. Deposits of sediments and nutrients at the mouths of rivers build up protective coastal deltas.

Streams receive many of their nutrients from bordering land ecosystems. Such nutrient inputs come from falling leaves, animal feces, insects, and other forms of biomass washed into streams during heavy rainstorms or by melting snow. Thus, the levels and types of nutrients in a stream depend on what is happening in the stream's watershed.

■ CASE STUDY
Dams, Deltas, Wetlands, Hurricanes, and New Orleans

Coastal deltas, mangrove forests, and coastal wetlands provide considerable natural protection against flood damage from coastal storms, hurricanes, typhoons, and tsunamis.

When we remove or degrade these natural speed bumps and sponges, any damage from a natural disaster such as a hurricane or typhoon is intensified. As a result, flooding in places like New Orleans, Louisiana (USA), the U.S. Gulf Coast, and Venice, Italy, are largely self-inflicted unnatural disasters. For example, the U.S. state of Louisiana, which contains about 40% of all coastal wetlands in the lower 48 states, has lost more than a fifth of such wetlands since 1950 to oil and gas wells and other forms of coastal development.

Dams and levees have been built along most of the world's rivers to control water flows and provide electricity (from hydroelectric power plants). This helps to reduce flooding along rivers, but it also traps sediments

that normally are deposited in deltas, which are continually rebuilt by such sediments. As a result, most of the world's river deltas are sinking rather than rising, and their protective coastal wetlands are being flooded. Thus, they no longer provide natural flood protection for coastal communities.

For example, the Mississippi River once delivered huge amounts of sediments to its delta each year. But the multiple dams, levees, and canals in this river system funnel much of this load through the wetlands and out into the Gulf of Mexico. Instead of building up delta lands, this causes them to subside. Other human processes that are increasing such subsidence include extraction of groundwater and oil and natural gas. As freshwater wetlands are lost, saltwater from the Gulf has intruded and killed many plants that depended on river water, further degrading this coastal aquatic system.

This helps to explain why the U.S. city of New Orleans, Louisiana, which was flooded by Hurricane Katrina in 2005 (Figure 8-18), is 3 meters (10 feet)

Figure 8-18 Much of the U.S. city of New Orleans, Louisiana, was flooded by the storm surge that accompanied Hurricane Katrina, which made landfall just east of the city on August 29, 2005. When the surging water rushed through the Mississippi River Gulf Outlet, a dredged waterway on the edge of the city, it breached a floodwall, and parts of New Orleans were flooded with 2 meters (6.5 feet) of water within a few minutes. Within a day, floodwaters reached a depth of 6 meters (nearly 20 feet) in some places; 80% of the city was under water at one point. The hurricane killed more than 1,800 people, and caused more than $100 billion in damages, making it the costliest and deadliest hurricane in the U.S. history. In addition, a variety of toxic chemicals from flooded industrial and hazardous waste sites, as well as oil and gasoline from more than 350,000 ruined cars and other vehicles, were released into the stagnant floodwaters. After the water receded, parts of New Orleans were covered with a thick oily sludge.

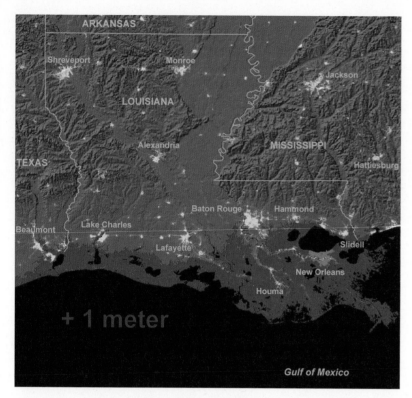

Figure 8-19 Projection of how a 1-meter (3.3-foot) rise in sea level from global warming by the end of this century would put New Orleans and much of Louisiana's current coast under water. (Used by permission from Jonathan Overpeck and Jeremy Weiss, University of Arizona)

The good news is that we now understand some of the connections between dams, deltas, wetlands, barrier islands, sea level rise, and hurricanes. The question is whether we will use such ecological and geological wisdom to change our ways or suffer the increasingly severe consequences of our own actions.

THINKING ABOUT
New Orleans

Do you think that a sinking city such as New Orleans, Louisiana (USA), should be rebuilt and protected with higher levees, or should the lower parts of the city be allowed to revert to wetlands that would help to protect nearby coastal areas? Explain.

---

below sea level and, in the not-too-distant future, will probably be 6 meters (20 feet) below sea level. Add to this the reduction of the protective effects of coastal and inland wetlands and barrier islands and you have a recipe for a major *unnatural* disaster.

To make matters worse, global sea levels have risen almost 0.3 meters (1 foot) since 1900 and are projected to rise 0.3–0.9 meter (1–3 feet) by the end of this century. Figure 8-19 shows a projection of how such a rise in sea level would put New Orleans and other current areas of the Louisiana coast under water. Most of this projected rise is due to the expansion of water and melting ice caused by global warming—another unnatural disaster helped along mostly by our burning of fossil fuels and clearing of large areas of the world's tropical forests.

Governments can spend hundreds of billions of dollars building or rebuilding higher levees around cities such as New Orleans. But sooner or later increasingly stronger hurricanes and rising sea levels will overwhelm these defenses and cause even greater damage and loss of life.

For example, much of New Orleans is a 3-meter (10-foot)-deep bathtub or bowl, with parts of the city 0.9–3 meters (3–9 feet) below sea level. According to engineers, even if we build levees high enough to make it a 6-meter (20-foot)-deep bathtub, a Category 5 hurricane and rising sea levels will eventually overwhelm such defenses and lead to a much more serious unnatural disaster.

Freshwater Inland Wetlands Are Vital Sponges

Inland wetlands are lands covered with freshwater all or part of the time (excluding lakes, reservoirs, and streams) and located away from coastal areas. They include *marshes* (dominated by grasses and reeds with few trees), *swamps* (dominated by trees and shrubs), and *prairie potholes* (depressions carved out by ancient glaciers). Other examples are *floodplains,* which receive excess water during heavy rains and floods, and the wet *arctic tundra* in summer (Figure 7-12, bottom photo, p. 151). Some wetlands are huge; others are small.

Some wetlands are covered with water year-round. Others, called *seasonal wetlands,* remain under water or are soggy for only a short time each year. The latter include prairie potholes, floodplain wetlands, and bottomland hardwood swamps. Some stay dry for years before water covers them again. In such cases, scientists must use the composition of the soil or the presence of certain plants (such as cattails, bulrushes, or red maples) to determine that a particular area is a wetland. Wetland plants are highly productive because of an abundance of nutrients. Many of these wetlands are important habitats for game fishes, muskrats, otters, beavers, migratory waterfowl, and many other bird species.

Inland wetlands provide a number of other free ecological and economic services, which include:

- Filtering and degrading toxic wastes and pollutants

- Reducing flooding and erosion by absorbing storm water and releasing it slowly and by absorbing overflows from streams and lakes

- Helping to replenish stream flows during dry periods

- Helping to recharge groundwater aquifers

- Helping to maintain biodiversity by providing habitats for a variety of species

- Supplying valuable products such as fishes and shellfish, blueberries, cranberries, wild rice, and timber

- Providing recreation for birdwatchers, nature photographers, boaters, anglers, and waterfowl hunters

8-5 How Have Human Activities Affected Freshwater Ecosystems?

▶ **CONCEPT 8-5** Human activities threaten biodiversity and disrupt ecological and economic services provided by freshwater lakes, rivers, and wetlands.

Human Activities Are Disrupting and Degrading Freshwater Systems

Human activities are disrupting and degrading many of the ecological and economic services provided by freshwater rivers, lakes, and wetlands (**Concept 8-5**). Such activities affect freshwater systems in four major ways. *First,* dams, and canals fragment about 40% of the world's 237 large rivers. They alter and destroy terrestrial and aquatic wildlife habitats along rivers and in coastal deltas and estuaries by reducing water flow and increasing damage from coastal storms (Case Study, p. 177). *Second,* flood control levees and dikes built along rivers disconnect the rivers from their floodplains, destroy aquatic habitats, and alter or reduce the functions of nearby wetlands.

For example, the 2,341-mile Missouri River in the west central United States has been harnessed by levees and a series of six dams, which have disrupted the seasonal variations in the flow of the river and changed water temperatures on some stretches. This hinders the growth of insect populations and interferes with spawning cycles of fishes and the feeding habits of shorebirds, and thus severely disrupts entire food webs. The dams and levees also have interrupted sediment flow and distribution, degrading shoreline habitats maintained by sediments. As a result of these disturbances, dozens of native species have declined, and the biodiversity of the ecosystem is threatened.

A *third* major human impact on freshwater systems comes from cities and farms, which add pollutants and excess plant nutrients to nearby streams, rivers, and lakes. For example, runoff of nutrients into a lake (cultural eutrophication, Figure 8-16, right) causes explosions in the populations of algae and cyanobacteria, which deplete the lake's dissolved oxygen. When these organisms die and sink to the lake bottom, decomposers go to work and further deplete the oxygen in deeper waters. Fishes and other species can then die off, causing a major loss in biodiversity.

Fourth, many inland wetlands have been drained or filled to grow crops or have been covered with concrete, asphalt, and buildings. See the following Case Study.

■ CASE STUDY

Inland Wetland Losses in the United States

About 95% of the wetlands in the United States contain freshwater and are found inland. The remaining 5% are saltwater or coastal wetlands. Alaska has more of the nation's inland wetlands than the other 49 states put together.

More than half of the inland wetlands estimated to have existed in the continental United States during the 1600s no longer exist. About 80% of lost wetlands were destroyed to grow crops. The rest were lost to mining, forestry, oil and gas extraction, highways, and urban development. The heavily farmed U.S. state of Iowa has lost about 99% of its original inland wetlands.

This loss of natural capital has been an important factor in increased flood and drought damage in the United States—more examples of unnatural disasters. Many other countries have suffered similar losses. For example, 80% of all wetlands in Germany and France have been destroyed.

We look further into human impacts on aquatic systems in Chapter 11. There, we also explore possible solutions to environmental problems that result from these impacts, as well as ways to sustain aquatic biodiversity.

Coral Reefs and Sustainability

This chapter's opening case study pointed out the ecological and economic importance of the world's incredibly diverse coral reefs. They are living examples of the four **scientific principles of sustainability** in action. They thrive on solar energy, participate in the cycling of carbon and other chemicals, are a prime example of aquatic biodiversity, and have a network of interactions among species that helps to maintain sustainable population sizes.

In this chapter, we have seen that coral reefs and other aquatic systems are being severely stressed by a variety of human activities. Research shows when such harmful human activities are reduced, coral reefs and other stressed aquatic systems can recover fairly quickly.

In other words, from a scientific standpoint, we know what to do. Whether or not we act is primarily a political and ethical problem. This requires educating leaders and citizens about the ecological and economic importance of the earth's aquatic ecosystems and about the need to include the economic values of such ecosystem services in the prices of goods and services generated with the use of these resources. Solving these problems also requires individual citizens to put pressure on elected officials and business leaders to change the ways we treat these important repositories of natural capital.

. . . the sea, once it casts its spell, holds one in its net of wonders forever.

JACQUES-YVES COUSTEAU

REVIEW

1. Review the Key Questions and Concepts for this chapter on p. 163. What is a **coral reef** and why should we care about coral reefs? What is coral bleaching?

2. What percentage of the earth's surface is covered with water? What is an **aquatic life zone?** Distinguish between a **saltwater (marine)** life zone and a **freshwater** life zone. What major types of organisms live in the top, middle, and bottom layers of aquatic life zones? Define **plankton** and describe three types of plankton. Distinguish among **nekton, benthos,** and **decomposers** and give an example of each. What five factors determine the types and numbers of organisms found in the three layers of aquatic life zones? What is **turbidity,** and how does it occur? Describe one of its harmful impacts.

3. What major ecological and economic services are provided by marine systems? What are the three major life zones in an ocean? Distinguish between the **coastal zone** and the **open sea.** Distinguish between an **estuary** and a **coastal wetland** and explain why they have high net primary productivities. What is a **mangrove forest** and what is its ecological and economic importance? What is the **intertidal zone?** Distinguish between rocky and sandy shores. Why does the open sea have a low net primary productivity?

4. What human activities pose major threats to marine systems and to coral reefs?

5. Explain why the Chesapeake Bay is an estuary in trouble. What is being done about some of its problems?

6. What major ecological and economic services do freshwater systems provide? What is a **lake?** What four zones are found in most lakes? Distinguish among **oligotrophic, eutrophic, hypereutrophic,** and **mesotrophic lakes.** What is **cultural eutrophication?**

7. Define **surface water, runoff,** and **watershed (drainage basin).** Describe the three zones that a stream passes through as it flows from mountains to the sea. Describe the relationships between dams, deltas, wetlands, hurricanes, and flooding in New Orleans, Louisiana (USA).

8. Give three examples of **inland wetlands** and explain the ecological importance of such wetlands.

9. What are four ways in which human activities are disrupting and degrading freshwater systems? Describe inland wetlands in the United States in terms of the area of wetlands lost and the resulting loss of ecological and economic services.

10. How is the degradation of many of the earth's coral reefs (**Core Case Study**) a reflection of our failure to follow the four **scientific principles of sustainability?** Describe this connection for each principle.

Note: Key Terms are in **bold** type.

1. List three ways in which you could apply **Concepts 8-3** and **8-5** to make your lifestyle more environmentally sustainable.

2. What are three steps governments and industries could take to protect remaining coral reefs (**Core Case Study**)? What are three ways in which individuals can help to protect those reefs?

3. Why do aquatic plants such as phytoplankton tend to be very small, whereas most terrestrial plants such as trees tend to be larger and have more specialized structures such as stems and leaves for growth?

4. Why are some aquatic animals, especially marine mammals such as whales, extremely large compared with terrestrial animals?

5. How would you respond to someone who proposes that we use the deep portions of the world's oceans to deposit our radioactive and other hazardous wastes because the deep oceans are vast and are located far away from human habitats? Give reasons for your response.

6. Suppose a developer builds a housing complex overlooking a coastal salt marsh and the result is pollution and degradation of the marsh. Describe the effects of such a development on the wildlife in the marsh, assuming at least one species is eliminated as a result. (See Figure 8-7.)

7. How does a levee built on a river affect species such as deer and hawks living in a forest overlooking the river?

8. Suppose you have a friend who owns property that includes a freshwater wetland, and the friend tells you he is planning to fill the wetland to make more room for his lawn and garden. What would you say to this friend?

9. Congratulations! You are in charge of the world. What are the three most important features of your plan to help sustain the earth's aquatic biodiversity?

10. List two questions that you would like to have answered as a result of reading this chapter.

Note: See Supplement 13 (p. S78) for a list of Projects related to this chapter.

At least 25% of the world's coral reefs have been severely damaged. A number of factors have played a role in this serious loss of aquatic biodiversity (Figure 8-12, right, p. 172), including ocean warming, sediment from coastal soil erosion, excessive algae growth from fertilizer runoff, coral bleaching, rising sea levels, overfishing, and damage from hurricanes.

In 2005, scientists Nadia Bood, Melanie McField, and Rich Aronson conducted research to evaluate the recovery of coral reefs in Belize from the combined effects of mass bleaching and Hurricane Mitch in 1998. Some of these reefs are in protected waters where no fishing is allowed. The researchers speculated that reefs in waters where no fishing is allowed should recover faster than reefs in water where fishing is allowed. The graphs below show some of the data they collected from three highly protected (no fishing) sites and three unprotected (fishing) sites to evaluate their hypothesis.

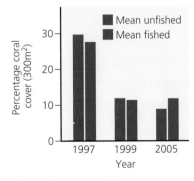
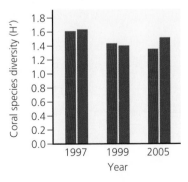

Effects of restricting fishing on the recovery of unfished and fished coral reefs damaged by the combined effects of mass bleaching and Hurricane Mitch in 1998. (Data from Melanie McField, et al., *Status of Caribbean Coral Reefs After Bleaching and Hurricanes in 2005,* NOAA, 2008. (Report available at **www.coris.noaa.gov/activities/caribbean_rpt/**)

1. By about what percentage did the mean coral cover drop in the protected (unfished) reefs between 1997 and 1999?

2. By about what percentage did the mean coral cover drop in the protected (unfished) reefs between 1997 and 2005?

3. By about what percentage did the coral cover drop in the unprotected (fished) reefs between 1997 and 1999?

4. By about what percentage did the coral cover change in the unprotected (fished) reefs between 1997 and 2005?

5. Do these data support the hypothesis that coral reef recovery should occur faster in areas where fishing is prohibited? Explain.

LEARNING ONLINE

Log on to the Student Companion Site for this book at **academic.cengage.com/biology/miller**, and choose Chapter 8 for many study aids and ideas for further reading and research. These include flash cards, practice quizzing, Weblinks, information on Green Careers, and InfoTrac® College Edition articles.

Sustaining Biodiversity: The Species Approach

The Passenger Pigeon: Gone Forever

In 1813, bird expert John James Audubon saw a single huge flock of passenger pigeons that took three days to fly past him and was so dense that it darkened the skies.

By 1900, North America's passenger pigeon (Figure 9-1), once the most numerous bird species on earth, had disappeared from the wild because of a combination of uncontrolled commercial hunting and habitat loss as forests were cleared to make room for farms and cities. These birds were good to eat, their feathers made good pillows, and their bones were widely used for fertilizer. They were easy to kill because they flew in gigantic flocks and nested in long, narrow, densely packed colonies.

Commercial hunters would capture one pigeon alive, sew its eyes shut, and tie it to a perch called a stool. Soon a curious flock would land beside this "stool pigeon"—a term we now use to describe someone who turns in another person for breaking the law. Then the birds would be shot or ensnared by nets that could trap more than 1,000 of them at once.

Beginning in 1858, passenger pigeon hunting became a big business. Shotguns, traps, artillery, and even dynamite were used. People burned grass or sulfur below their roosts to suffocate the birds. Shooting galleries used live birds as targets. In 1878, one professional pigeon trapper made $60,000 by killing 3 million birds at their nesting grounds near Petoskey, Michigan.

By the early 1880s, only a few thousand birds remained. At that point, recovery of the species was doomed because the females laid only one egg per nest each year. On March 24, 1900, a young boy in the U.S. state of Ohio shot the last known wild passenger pigeon.

Eventually all species become extinct or evolve into new species. The archeological record reveals five *mass extinctions* since life on the earth began—each a massive impoverishment of life on the earth. These mass extinctions were caused by natural phenomena, such as major climate change or large asteroids hitting the earth, which drastically altered the earth's environmental conditions.

There is considerable evidence that we are now in the early stage of a sixth great extinction. Evidence indicates that we humans are causing this mass extinction as our population grows and as we consume more resources, disturb more land and aquatic systems, use more of the earth's net primary productivity, and cause changes to the earth's climate.

Scientists project that during this century, human activities, especially those that cause habitat destruction and climate change, will lead to the premature extinction of one-fourth to one-half of the world's plant and animal species—an incredibly rapid rate of extinction. And there will be no way to restore what we have lost, because species extinction is forever. If we keep impoverishing the earth's biodiversity, eventually, our species will also become impoverished.

Michael Sewell/Peter Arnold, Inc.

Figure 9-1 Lost natural capital: passenger pigeons have been extinct in the wild since 1900 because of human activities. The last known passenger pigeon died in the U.S. state of Ohio's Cincinnati Zoo in 1914.

9-1 What role do humans play in the premature extinction of species?

CONCEPT 9-1A We are degrading and destroying biodiversity in many parts of the world, and these threats are increasing.

CONCEPT 9-1B Species are becoming extinct 100 to 1,000 times faster than they were before modern humans arrived on the earth (the *background rate*), and by the end of this century, the extinction rate is expected to be 10,000 times the background rate.

9-2 Why should we care about preventing premature species extinction?

CONCEPT 9-2 We should prevent the premature extinction of wild species because of the economic and ecological services they provide and because they have a right to exist regardless of their usefulness to us.

9-3 How do humans accelerate species extinction?

CONCEPT 9-3 The greatest threats to any species are (in order) loss or degradation of its habitat, harmful invasive species, human population growth, pollution, climate change, and overexploitation.

9-4 How can we protect wild species from extinction resulting from our activities?

CONCEPT 9-4A We can use existing environmental laws and treaties and work to enact new laws designed to prevent premature species extinction and protect overall biodiversity.

CONCEPT 9-4B We can help to prevent premature species extinction by creating and maintaining wildlife refuges, gene banks, botanical gardens, zoos, and aquariums.

CONCEPT 9-4C According to the *precautionary principle*, we should take measures to prevent or reduce harm to the environment and to human health, even if some of the cause-and-effect relationships have not been fully established, scientifically.

Note: Supplements 2 (p. S4), 4 (p. S20), 9 (p. S53), and 13 (p. S78) can be used with this chapter.

*The last word in ignorance is the person who says of an animal or plant:
"What good is it?" . . . If the land mechanism as a whole is good,
then every part of it is good, whether we understand it or not. . . .
Harmony with land is like harmony with a friend;
you cannot cherish his right hand and chop off his left.*

ALDO LEOPOLD

9-1 What Role Do Humans Play in the Premature Extinction of Species?

▶ **CONCEPT 9-1A** We are degrading and destroying biodiversity in many parts of the world, and these threats are increasing.

▶ **CONCEPT 9-1B** Species are becoming extinct 100 to 1,000 times faster than they were before modern humans arrived on the earth (the *background rate*), and by the end of this century, the extinction rate is expected to be 10,000 times the background rate.

Human Activities Are Destroying and Degrading Biodiversity

We have depleted and degraded some of the earth's biodiversity, and these threats are expected to increase (**Concept 9-1A**). According to biodiversity expert Edward O. Wilson, "The natural world is everywhere disappearing before our eyes—cut to pieces, mowed down, plowed under, gobbled up, replaced by human artifacts."

According to the 2005 Millennium Ecosystem Assessment and other studies, humans have disturbed, to some extent, at least half and probably about 83% of the earth's land surface (excluding Antarctica and Greenland; see Figure 3, pp. S24–S25, in Supplement 4). Most of this disturbance involves filling in wetlands or converting grasslands and forests to crop fields and urban areas. Such disturbances eliminate large numbers of species by destroying or degrading their habitats, as discussed in more detail in Chapter 10.

Links: CORE CASE STUDY refers to the Core Case Study. SUSTAINABILITY refers to the book's sustainability theme. CONCEPT LINK indicates links to key concepts in earlier chapters.

Human activities are also degrading the earth's *aquatic biodiversity,* as discussed in Chapter 11.

Extinctions Are Natural but Sometimes They Increase Sharply

In due time, all species become extinct. During most of the 3.56 billion years that life has existed on the earth, there has been a continuous, low level of extinction of species known as **background extinction.**

An **extinction rate** is expressed as a percentage or number of species that go extinct within a certain time period such as a year. For example, one extinction per million species per year would be $1/1,000,000 = 0.000001$ species per year. Expressed as a percentage, this is 0.000001×100, or 0.0001%—the estimated *background extinction rate* existing before humans came on the scene.

The balance between formation of new species and extinction of existing species determines the earth's biodiversity (**Concept 4-4A**, p. 86). Overall, the earth's biodiversity has increased for several hundred million years, except during a few periods. The extinction of many species in a relatively short period of geologic time is called a **mass extinction.** Geological and other records indicate that the earth has experienced five mass extinctions when 50–95% of the world's species appear to have become extinct. After each mass extinction, biodiversity eventually returned to equal or higher levels, but each recovery required millions of years.

The causes of past mass extinctions are poorly understood but probably involved global changes in environmental conditions. Examples are major climate change or a large-scale catastrophe such as a collision between the earth and a comet or large asteroid. The last of these mass extinctions took place about 65 million years ago. One hypothesis is that the last mass extinction taking place about 65 million years ago occurred after a large asteroid hit the planet and spewed huge amounts of dust and debris into the atmosphere.

This could have reduced the input of solar energy and cooled the planet enough to wipe out the dinosaurs and many other forms of earth's life at that time.

Biologists distinguish among three levels of species extinction. *Local extinction* occurs when a species is no longer found in an area it once inhabited but is still found elsewhere in the world. Most local extinctions involve losses of one or more populations of species. *Ecological extinction* occurs when so few members of a species are left that it can no longer play its ecological roles in the biological communities where it is found.

In *biological extinction,* a species, such as the passenger pigeon (**Core Case Study**, Figure 9-1) is no longer found anywhere on the earth. Biological extinction is forever and represents an irreversible loss of natural capital.

Some Human Activities Cause Premature Extinctions, and the Pace Is Speeding Up

Although extinction is a natural biological process, it has accelerated as human populations have spread over the globe, consuming large quantities of resources, and creating large ecological footprints (Figure 1-10, p. 15). As a result, human activities are destroying the earth's biodiversity at an unprecedented and accelerating rate. Figure 9-2 shows a few of the many species that have become prematurely extinct mostly because of human activities.

Using the methods described in the Science Focus box (p. 188), scientists from around the world, who published the 2005 Millennium Ecosystem Assessment, estimated that the current annual rate of species extinction is at least 100 to 1,000 times the background rate of about 0.0001%, which existed before modern humans appeared some 150,000 years ago. This amounts to an extinction rate of 0.01% to 0.1% a year. Conservation biologists project that during this century the extinction rate caused by habitat loss, climate change mostly due to global warming, and other

Passenger pigeon   Great auk   Dodo   Golden toad   Aepyornis (Madagascar)

Figure 9-2 Lost natural capital: some animal species that have become prematurely extinct largely because of human activities, mostly habitat destruction and overhunting. **Question:** Why do you think birds top the list of extinct species?

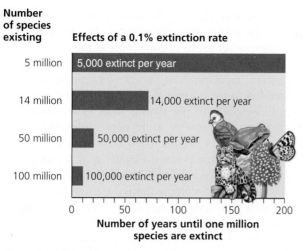

Number of species existing

Effects of a 0.1% extinction rate

5 million — 5,000 extinct per year

14 million — 14,000 extinct per year

50 million — 50,000 extinct per year

100 million — 100,000 extinct per year

0   50   100   150   200

Number of years until one million species are extinct

Figure 9-3 Effects of a 0.1% extinction rate.

human activities will increase to 10,000 times the background rate (**Concept 9-1B**). This will amount to an annual extinction rate of 1% per year.

How many species are we losing prematurely each year? The answer depends on how many species are on the earth and the rate of species extinction. Assuming that the extinction rate is 0.1%, each year we lose 5,000 species if there are 5 million species on earth and 14,000 species if there are 14 million species—the current best scientific estimate. See Figure 9-3 for more examples.

According to researchers Edward O. Wilson and Stuart Pimm, at a 1% extinction rate, at least one-fourth of the world's current animal and plant species could be gone by 2050 and half could vanish by the end of this century. In the words of biodiversity expert Norman Myers, "Within just a few human generations, we shall—in the absence of greatly expanded conservation efforts—impoverish the biosphere to an extent that will persist for at least 200,000 human generations or twenty times longer than the period since humans emerged as a species."

THINKING ABOUT
Extinction

How might your lifestyle change if human activities cause the premature extinction of up to half of the world's species in your lifetime? List three aspects of your lifestyle that contribute to this threat to the earth's natural capital.

Most extinction experts consider extinction rates of 0.01–1% to be conservative estimates (**Concept 9-1B**) for several reasons. *First,* both the rate of species loss and the extent of biodiversity loss are likely to increase during the next 50–100 years because of the projected growth of the world's human population and resource use per person (Figure 1-10, p. 15, and Figure 3, pp. S24–S25, in Supplement 4) and climate change caused mostly by global warming.

Second, current and projected extinction rates are much higher than the global average in parts of the world that are highly endangered centers of biodiver-

sity. Conservation biologists urge us to focus our efforts on slowing the much higher rates of extinction in such *hotspots* as the best and quickest way to protect much of the earth's biodiversity from being lost prematurely. (We discuss this further in Chapter 10.)

Third, we are eliminating, degrading, fragmenting, and simplifying many biologically diverse environments—such as tropical forests, tropical coral reefs, wetlands, and estuaries—that serve as potential colonization sites for the emergence of new species (**Concept 4-4B**, p. 86). Thus, in addition to increasing the rate of extinction, we may be limiting the long-term recovery of biodiversity by reducing the rate of speciation for some species. In other words, we are creating a *speciation crisis.* (See the Guest Essay by Normal Myers on this topic at CengageNOW™.)

CONCEPT LINK

Philip Levin, Donald Levin, and other biologists also argue that the increasing fragmentation and disturbance of habitats throughout the world may increase the speciation rate for rapidly reproducing opportunist species such as weeds, rodents, and cockroaches and other insects. Thus, the real threat to biodiversity from current human activities may be long-term erosion in the earth's variety of species and habitats. Such a loss of biodiversity would reduce the ability of life to adapt to changing conditions by creating new species.

Endangered and Threatened Species Are Ecological Smoke Alarms

Biologists classify species heading toward biological extinction as either *endangered* or *threatened.* An **endangered species** has so few individual survivors that the species could soon become extinct over all or most of its natural range (the area in which it is normally found). Like the passenger pigeon (**Core Case Study,** Figure 9-1) and several other bird species (Figure 9-2), they may soon disappear from the earth. A **threatened species** (also known as a vulnerable species) is still abundant in its natural range but, because of declining numbers, is likely to become endangered in the near future.

CORE CASE STUDY

The International Union for the Conservation of Nature and Natural Resources (IUCN)—also known as the World Conservation Union—is a coalition of the world's leading conservation groups. Since the 1960s, it has published annual *Red Lists,* which have become the world standard for listing the world's threatened species. In 2007, the list included 16,306 plants and animals that are in danger of extinction—60% higher than the number listed in 1995. Those compiling the list say it greatly underestimates the true number of threatened species because only a tiny fraction of 1.8 million known species have been assessed, and of the estimated total of 4–100 million additional species that have not been catalogued or studied. You can examine the Red Lists database online at **www.iucnredlist.org.** Figure 9-4 shows a few of the roughly 1,300 species

Grizzly bear

Kirkland's warbler

Knowlton cactus

Florida manatee

African elephant

Utah prairie dog

Swallowtail butterfly

Humpback chub

Golden lion tamarin

Siberian tiger

Giant panda

Black-footed ferret

Whooping crane

Northern spotted owl

Blue whale

Mountain gorilla

Florida panther

California condor

Hawksbill sea turtle

Black rhinoceros

Figure 9-4 Endangered natural capital: some species that are endangered or threatened with premature extinction largely because of human activities. Almost 30,000 of the world's species and roughly 1,300 of those in the United States are officially listed as being in danger of becoming extinct. Most biologists believe the actual number of species at risk is much larger.

SCIENCE FOCUS

Estimating Extinction Rates Is Not Easy

Conservation biologists who are trying to catalog extinctions, estimate past extinction rates, and project future rates have three problems. *First,* the extinction of a species typically takes such a long time that it is not easy to document. *Second,* we have identified only about 1.8 million of the world's estimated 4 million to 100 million species. *Third,* scientists know little about the nature and ecological roles of most of the species that have been identified.

One approach to estimating future extinction rates is to study records documenting the rates at which mammals and birds (which have been the easiest to observe) have become extinct since humans arrived and to compare this with fossil records of extinctions prior to the arrival of humans. Determining the rates at which minor DNA copying mistakes occur can help scientists to track how long various species typically last before becoming extinct. Such evidence indicates that

under normal circumstances, species survive for 1 million to 10 million years before going extinct.

Another approach is to observe how the number of species present increases with the size of an area. This *species–area relationship* suggests that, on average, a 90% loss of habitat causes the extinction of 50% of the species living in that habitat. This is based on the *theory of island biogeography* (Science Focus, p. 90). Scientists use this model to estimate the number of current and future extinctions in patches or "islands" of shrinking habitat surrounded by degraded habitats or by rapidly growing human developments.

Scientists also use mathematical models to estimate the risk of a particular species becoming endangered or extinct within a certain period of time. These models include factors such as trends in population size, changes in habitat availability, interactions with other species, and genetic factors. Re-

searchers know that their estimates of extinction rates are based on insufficient data and sampling and incomplete models. They are continually striving to get more data and to improve the models used to estimate extinction rates.

At the same time, they point to clear evidence that human activities have accelerated the rate of species extinction and that this rate is still increasing. According to these biologists, arguing over the numbers and waiting to get better data and models should not be used as excuses for inaction. They agree with the advice of Aldo Leopold (Individuals Matter, p. 22) in his thoughts about preventing premature extinction: "To keep every cog and wheel is the first precaution of intelligent tinkering."

Critical Thinking

How would you improve the estimation of extinction rates?

---

officially listed as endangered and protected under the U.S. Endangered Species Act.

Some species have characteristics that make them especially vulnerable to ecological and biological extinction (Figure 9-5). As biodiversity expert Edward O. Wilson puts it, "The first animal species to go are the big, the slow, the tasty, and those with valuable parts such as tusks and skins."

Some species also have *behavioral characteristics* that make them prone to extinction. The passenger pigeon (**Core Case Study,** Figure 9-1) and the Carolina parakeet nested in large flocks that made them easy to kill. Key deer, which live only in the U.S. Florida Keys, are "nicotine addicts" that get killed by cars because they forage for cigarette butts along highways. Some types of species are more threatened with premature extinction from human activities than others are (Figure 9-6).

> **RESEARCH FRONTIER**
>
> Identifying and cataloguing the millions of unknown species and improving models for estimating extinction rates. See **academic.cengage.com/biology/miller**.

Figure 9-5 Characteristics of species that are prone to ecological and biological extinction. **Question:** Which of these characteristics helped lead to the premature extinction of the passenger pigeon within a single human lifetime?

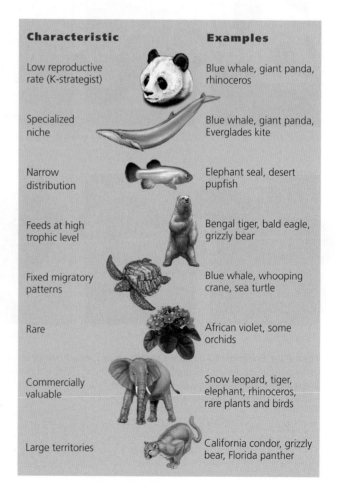

Characteristic	Examples
Low reproductive rate (K-strategist)	Blue whale, giant panda, rhinoceros
Specialized niche	Blue whale, giant panda, Everglades kite
Narrow distribution	Elephant seal, desert pupfish
Feeds at high trophic level	Bengal tiger, bald eagle, grizzly bear
Fixed migratory patterns	Blue whale, whooping crane, sea turtle
Rare	African violet, some orchids
Commercially valuable	Snow leopard, tiger, elephant, rhinoceros, rare plants and birds
Large territories	California condor, grizzly bear, Florida panther

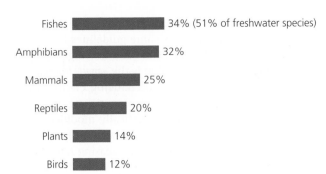

Fishes ████████ 34% (51% of freshwater species)

Amphibians ████████ 32%

Mammals ███████ 25%

Reptiles ██████ 20%

Plants █████ 14%

Birds ████ 12%

Figure 9-6 Endangered natural capital: percentage of various types of species threatened with premature extinction because of human activities (**Concept 9-1A**). **Question:** Why do you think fishes top this list? (Data from World Conservation Union, Conservation International, World Wide Fund for Nature, 2005 Millennium Ecosystem Assessment, and the Intergovernmental Panel on Climate Change)

9-2 Why Should We Care about Preventing Premature Species Extinction?

▶ **CONCEPT 9-2** We should prevent the premature extinction of wild species because of the economic and ecological services they provide and because they have a right to exist regardless of their usefulness to us.

Species Are a Vital Part of the Earth's Natural Capital

So what is all the fuss about? If all species eventually become extinct, why should we worry about premature extinctions? Does it matter that the passenger pigeon (**Core Case Study**) became prematurely extinct because of human activities, or that the remaining orangutans (Figure 9-7) or some unknown plant or insect in a tropical forest might suffer the same fate?

CORE CASE STUDY

New species eventually evolve to take the places of those lost through mass extinctions. So why should we care if we speed up the extinction rate over the next 50–100 years? The answer: because it will take 5–10 million years for natural speciation to rebuild the biodiversity we are likely to destroy during your lifetime.

Biodiversity researchers say we should act now to prevent premature extinction of species partly for their **instrumental value**—their usefulness to us in providing many of the ecological and economic services that

Figure 9-7 Natural capital degradation: endangered orangutans in a tropical forest. In 1900, there were over 315,000 wild orangutans. Now there are less than 20,000 and they are disappearing at a rate of over 2,000 per year because of illegal smuggling and clearing of their forest habitat in Indonesia and Malaysia to make way for oil palm plantations. An illegally smuggled orangutan typically sells for a street price of $10,000. According to 2007 study by the World Wildlife Fund (WWF), projected climate change will further devastate remaining orangutan populations in Indonesia and Malaysia. **Question:** How would you go about trying to set a price on the ecological value of an orangutan?

age fotostock/SuperStock

make up the earth's natural capital (Figure 1-3, p. 8) (**Concept 9-2**).

Instrumental values take two forms. One is *use values,* which benefit us in the form of economic goods and services, ecological services, recreation, scientific information, and the continuation of such uses for future generations. Each year, Americans, as a whole, spend more than three times as many hours watching wildlife—doing nature photography and bird watching, for example—as they spend watching movies or professional sporting events. A diversity of plant species provides economic value in the form of food crops, fuelwood, lumber, paper, drugs, and medicines (Figure 9-8). Bioprospectors search tropical forests and other ecosystems for plants and animals that have chemicals that can be converted into useful medicinal drugs. A 2005 United Nations University report concluded that 62% of all cancer drugs were derived from the discoveries of bioprospectors. **GREEN CAREER:** Bioprospecting

Species diversity also provides economic benefits from wildlife tourism, or *ecotourism,* which generates between $950,000 and $1.8 million per minute in tourist expenditures worldwide. Conservation biologist Michael Soulé estimates that one male lion living to age 7 generates $515,000 in tourist dollars in Kenya, but only $1,000 if killed for its skin. Similarly, over a lifetime of 60 years, a Kenyan elephant is worth about $1 million in ecotourist revenue—many times more than its tusks are worth when they are sold illegally for their ivory (Science Focus, at right). Ecotourism should

not cause ecological damage, but some of it does. The website for this chapter lists some guidelines for evaluating eco-tours. **GREEN CAREER:** Ecotourism guide

Another instrumental value is the *genetic information* that allows species to adapt to changing environmental conditions through evolution. Genetic engineers use this information to produce genetically modified crops and foods. Scientists warn of the alarming loss of genetic diversity resulting from our increasing reliance on a small number of crop plants for feeding the world. Such a loss also results from the premature extinction of wild plants whose genes could be used by genetic engineers to develop improved crop varieties.

One of the tragedies of the current extinction crisis is that we do not know what we are losing, because no one has ever seen or named many of the species that are becoming extinct. Consequently, we know nothing about their genetic makeup, their roles in sustaining ecosystems, or how they might be used to improve human welfare. Carelessly eliminating many of the species that make up the world's vast genetic library is like burning books that we have never read.

The other major form of instrumental value is *nonuse values,* of which there are several types. For example, there is *existence value*—the satisfaction of knowing that a redwood forest, a wilderness, orangutans (Figure 9-7), and wolf packs exist, even if we will never see them or get direct use from them. For many people, biodiversity holds *aesthetic value.* For example, we can appreciate a tree, an orangutan, or a tropical bird (Figure 9-9) for its beauty. A third type is *bequest value,*

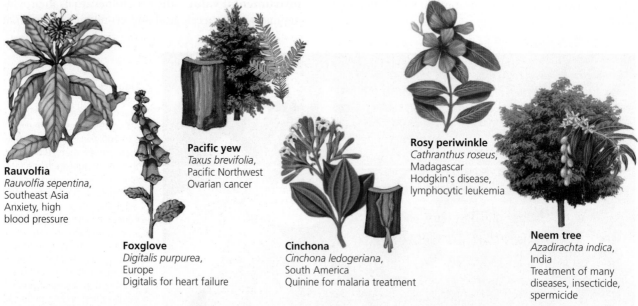

Rauvolfia
Rauvolfia sepentina,
Southeast Asia
Anxiety, high
blood pressure

Foxglove
Digitalis purpurea,
Europe
Digitalis for heart failure

Pacific yew
Taxus brevifolia,
Pacific Northwest
Ovarian cancer

Cinchona
Cinchona ledogeriana,
South America
Quinine for malaria treatment

Rosy periwinkle
Cathranthus roseus,
Madagascar
Hodgkin's disease,
lymphocytic leukemia

Neem tree
Azadirachta indica,
India
Treatment of many
diseases, insecticide,
spermicide

Figure 9-8 Natural capital: *nature's pharmacy.* Parts of these and a number of other plant and animal species (many of them found in tropical forests) are used to treat a variety of human ailments and diseases. Nine of the ten leading prescription drugs originally came from wild organisms. About 2,100 of the 3,000 plants identified by the National Cancer Institute as sources of cancer-fighting chemicals come from tropical forests. Despite their economic and health potential, fewer than 1% of the estimated 125,000 flowering plant species in tropical forests (and a mere 1,100 of the world's 260,000 known plant species) have been examined for their medicinal properties. Once the active ingredients in the plants have been identified, they can usually be produced synthetically. Many of these tropical plant species are likely to become extinct before we can study them.

SCIENCE FOCUS

Using DNA to Reduce Illegal Killing of Elephants for Their Ivory

There are about 400,000 elephants remaining in the wild, most of them in Africa and the rest in Asia. Elephants have long been valued for the ivory in their tusks, but in 1989, an international treaty instituted a ban on the trading of such ivory.

Before the 1989 treaty, poachers slaughtered about 87,000 elephants a year for their ivory. Although this treaty reduced the poaching, it did not stop it. In recent years, the illegal slaughter of elephants for ivory has increased again to about 25,000 a year, most of it in the African countries of Cameroon, Nigeria, and Democratic Republic of the Congo. This poaching has risen sharply since 2004, fueled by sharply rising prices of high-quality ivory, mostly in China. Because of the money to be made, organized crime has

become more heavily involved in this illegal trade.

In 2007, scientists began developing a DNA-based map of elephant populations that allows them to use DNA analysis of seized illegal ivory to determine where the elephants were killed. They hope to use such data to identify poaching hot spots and help international law enforcement authorities to focus their anti-poaching efforts.

On the other hand, elephant populations have exploded in some areas, such as parts of South Africa, and are destroying vegetation and affecting the populations of other species.

A single adult elephant devours up to 300 kilograms (660 pounds) of grass, leaves, and twigs a day. Elephants can also uproot trees and disturb the soil. The South African

government is considering killing (culling) enough elephants each year to keep the population down in selected areas. The culled ivory would be sold in the international marketplace with the proceeds used to benefit the populations of local villagers and to help pay for conservation efforts. DNA analysis could be used to distinguish such culled ivory from poached ivory.

Supporters say that increasing the amount of ivory legally available in the marketplace could help to reduce poaching by lowering market prices. Some animal rights groups oppose elephant culling on ethical grounds.

Critical Thinking

Do you favor culling elephants in areas where large populations are degrading vegetation? Explain.

---

based on the fact that people will pay to protect some forms of natural capital for use by future generations.

Finally, species diversity holds *ecological value*, because it is a vital component of the key ecosystem functions of energy flow, nutrient cycling (Figure 3-12, p. 60), and population control, in keeping with three of the four **scientific principles of sustainability** (see back cover). In other words, the species in ecosystems provide essential *ecosystem* or *natural services*, an important component of the natural capital (Figure 1-3, p. 8, orange items) that supports and sustains the earth's life and economies. Thus, in protecting species from premature extinction and in protecting their vital habitats from environmental degradation (as we discuss in the next chapter), we are helping to sustain our own health and well-being.

Are We Ethically Obligated to Prevent Premature Extinction?

Some scientists and philosophers believe that each wild species has **intrinsic** or **existence value** based on its inherent right to exist and play its ecological roles, regardless of its usefulness to us (**Concept 9-2**). According to this view, we have an ethical responsibility to protect species from becoming prematurely extinct as a result of human activities and to prevent the degradation of the world's ecosystems and its overall biodiversity.

Each species in the encyclopedia of life is a masterpiece of evolution that possesses a unique combination

Figure 9-9 Many species of wildlife, such as this endangered scarlet macaw in Brazil's Amazon rain forest, are a source of beauty and pleasure. These and other colorful species of parrots can become endangered when they are removed from the wild and sold (sometimes illegally) as pets.

SCIENCE FOCUS

Why Should We Care about Bats?

Worldwide there are 950 known species of bats—the only mammals that can fly. But bats have two traits that make them vulnerable to extinction. *First,* they reproduce slowly. *Second,* many bat species live in huge colonies in caves and abandoned mines, which people sometimes close up. This prevents them from leaving to get food, or it can interrupt their hibernation if they have to leave their shelter to escape being trapped.

Bats play important ecological roles. About 70% of all bat species feed on crop-damaging nocturnal insects (Figure 5-3, p. 105) and other insect pest species such as mosquitoes. This makes them the major nighttime SWAT team for such insects.

In some tropical forests and on many tropical islands, *pollen-eating bats* pollinate flowers, and *fruit-eating bats* distribute plants throughout these forests by excreting undigested seeds. As keystone species, such bats are vital for maintaining plant biodiversity and for regenerating large areas of tropical forest that has been cleared by humans. If you enjoy bananas, cashews, dates, figs, avocados, or mangos, you can thank bats.

Many people mistakenly view bats as fearsome, filthy, aggressive, rabies-carrying bloodsuckers. But most bat species are harmless to people, livestock, and crops. In the United States, only 10 people have died of bat-transmitted disease in more than 4 decades of record keeping; more Americans die each year from being hit by falling coconuts.

Because of unwarranted fears of bats and lack of knowledge about their vital ecological roles, several bat species have been driven to extinction. Currently, about one-fourth of the world's bat species are listed as endangered or threatened. And thousands of bats are dying from an unknown illness in the northeastern United States. Because of the important ecological and economic roles they play, conservation biologists urge us to view bats as valuable allies, not as enemies to kill.

Critical Thinking

Has reading this material changed your view of bats? Can you think of two things that could be done to help protect bat species from premature extinction?

---

of genetic traits. These traits allow a species to become adapted to its natural environment and to changing environmental conditions through natural selection. Many analysts believe that we have no right to prematurely erase these unique genetic packages. On this basis, we have an ethical obligation to control our resource consumption to help protect all species, which make up a key component of the earth's biodiversity (Figure 4-2, p. 79), and in the process implement one of the four **scientific principles of sustainability** (see back cover).

Biologist Edward O. Wilson contends that because of the billions of years of biological connections leading to the evolution of the human species, we have an inherent genetic kinship with the natural world. He calls this phenomenon *biophilia* (love of life).

Evidence of this natural and emotional affinity for life is seen in the preference most people have for almost any natural scene over one from an urban environment. Given a choice, most people prefer to live in an area where they can see water and natural landscapes, such as a grassland or a forest. They also have an affinity for parks, wildlife, and pets and enjoy birdwatching, hiking, camping, fishing, and other outdoor activities. More people visit zoos and aquariums than attend all professional sporting events combined.

Some have the opposite feeling—a fear of many forms of wildlife—called *biophobia.* For example, some movies, books, and TV programs condition us to fear or be repelled by certain species such as alligators (Chapter 4 Core Case Study, p. 77), cockroaches (Case Study, p. 92), sharks (Case Study, p. 96), bats (Science Focus above), and bacteria (Science Focus, p. 61). Many people have lived so long in artificial urban settings that they are largely disconnected from wildlife and from outdoor experiences in nature.

Some people distinguish between the survival rights of plants and those of animals, mostly for practical reasons. Poet Alan Watts once said he was a vegetarian "because cows scream louder than carrots."

Other people distinguish among various types of species. For example, they might think little about getting rid of the world's mosquitoes, cockroaches, rats, or disease-causing bacteria, but feel protective of panda bears, elephants, and whales.

THINKING ABOUT

The Passenger Pigeon

In earlier times, many people viewed huge flocks of passenger pigeons (**Core Case Study**) as pests that devoured grain and left massive piles of their waste. Do you think this justified the passenger pigeon's premature extinction? Explain. If you believe that premature extinction of an undesirable species is justified, what would be your three top candidates? What might be some harmful ecological effects of such extinctions?

Some biologists caution us not to focus primarily on protecting relatively large organisms—the plants and animals we can see that are familiar to us. They remind us that the true foundation of the earth's ecosystems and ecological processes is made up of invisible bacteria and the algae, fungi, and other *microorganisms* that decompose the bodies of larger organisms and recycle the nutrients needed by all life.

How Do Humans Accelerate Species Extinction?

Loss of Habitat Is the Single Greatest Threat to Species: Remember HIPPCO

Figure 9-10 shows the underlying and direct causes of the endangerment and premature extinction of wild species. Conservation biologists summarize the most important causes of premature extinction using the acronym **HIPPCO: H**abitat destruction, degradation, and fragmentation; **I**nvasive (nonnative) species; **P**opulation and resource use growth (too many people consuming too many resources); **P**ollution; **C**limate change; and **O**verexploitation (**Concept 9-3**).

According to biodiversity researchers, the greatest threat to wild species is habitat loss (Figure 9-11, p. 194), degradation, and fragmentation. The passenger pigeon (**Core Case Study**, Figure 9-1) is only one of many species whose extinction was hastened by loss of habitat from forest clearing.

Deforestation in tropical areas (Figure 3-1, p. 50) is the greatest eliminator of species, followed by the destruction and degradation of coral reefs and wetlands, plowing of grasslands, and pollution of streams, lakes, and oceans. Globally, temperate biomes have been affected more by habitat loss and degradation than have tropical biomes because of widespread economic development in temperate countries over the past 200 years. Such development is now shifting to many tropical biomes.

Island species—many of them *endemic species* found nowhere else on earth—are especially vulnerable to extinction when their habitats are destroyed, degraded, or fragmented. This is why the collection of islands that make up the U.S. state of Hawaii are America's "extinction capital"—with 63% of its species at risk.

Any habitat surrounded by a different one can be viewed as a *habitat island* for most of the species that live there. Most national parks and other nature reserves are habitat islands, many of them encircled by potentially damaging logging, mining, energy extraction, and industrial activities. Freshwater lakes are also habitat islands that are especially vulnerable to the introduction of nonnative species and pollution.

Habitat fragmentation—by roads, logging, agriculture, and urban development—occurs when a large,

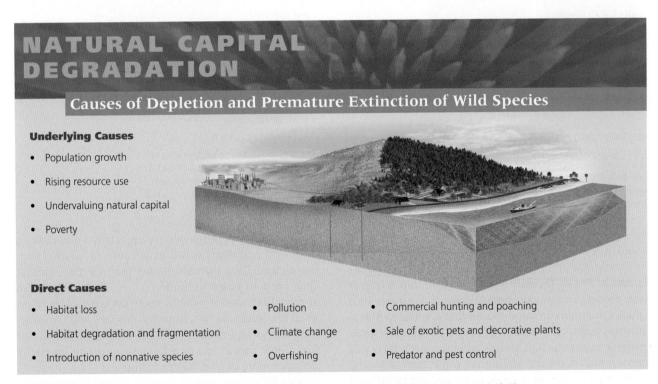

NATURAL CAPITAL DEGRADATION

Causes of Depletion and Premature Extinction of Wild Species

Underlying Causes

- Population growth
- Rising resource use
- Undervaluing natural capital
- Poverty

Direct Causes

- Habitat loss
- Habitat degradation and fragmentation
- Introduction of nonnative species

- Pollution
- Climate change
- Overfishing

- Commercial hunting and poaching
- Sale of exotic pets and decorative plants
- Predator and pest control

Figure 9-10 Underlying and direct causes of depletion and premature extinction of wild species (**Concept 9-3**). The major direct causes of wildlife depletion and premature extinction are habitat loss, degradation, and fragmentation. This is followed by the deliberate or accidental introduction of harmful invasive (nonnative) species into ecosystems. **Question:** What are two direct causes that are related to each of the underlying causes?

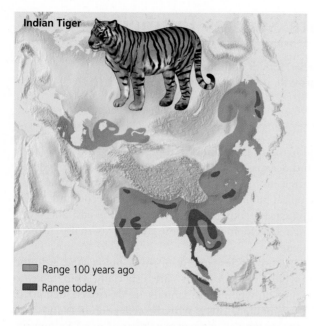

Indian Tiger

◻ Range 100 years ago
◼ Range today

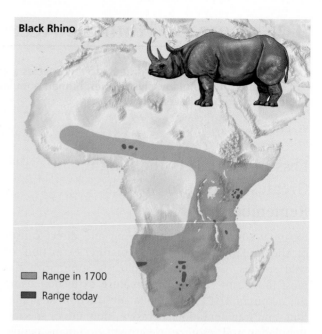

Black Rhino

◻ Range in 1700
◼ Range today

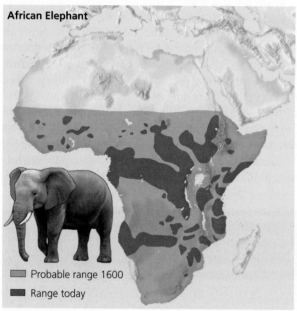

African Elephant

◻ Probable range 1600
◼ Range today

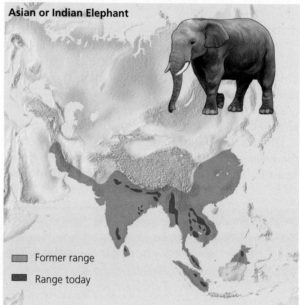

Asian or Indian Elephant

◻ Former range
◼ Range today

CENGAGENOW™ Active Figure 9-11 Natural capital degradation: reductions in the ranges of four wildlife species, mostly as the result of habitat loss and hunting. What will happen to these and millions of other species when the world's human population doubles and per capita resource consumption rises sharply in the next few decades? *See an animation based on this figure at* CengageNOW. **Question:** Would you support expanding these ranges even though this would reduce the land available for people to grow food and live on? Explain. (Data from International Union for the Conservation of Nature and World Wildlife Fund)

contiguous area of habitat is reduced in area and divided into smaller, more scattered, and isolated patches, or habitat islands. This process can decrease tree populations in forests (Science Focus, at right), block migration routes, and divide populations of a species into smaller and more isolated groups that are more vulnerable to predators, competitor species, disease, and catastrophic events such as storms and fires. Also, it creates barriers that limit the abilities of some species to disperse and colonize new areas, to get enough to eat, and to find mates. Migrating species also face dangers from fences, farms, paved areas, skyscrapers, and cell phone towers.

Certain types of species are especially vulnerable to local and regional extinction because of habitat fragmentation. They include species that are rare, species that need to roam unhindered over large areas, and species that cannot rebuild their populations because of a low reproductive capacity. Species with specialized niches and species that are sought by people for furs, food, medicines, or other uses are also especially threatened by habitat fragmentation.

Scientists use the theory of island biogeography (Science Focus, p. 90) to help them understand the effects of fragmentation on species extinction and to develop ways to help prevent such extinction.

SCIENCE FOCUS

Studying the Effects of Forest Fragmentation on Old-Growth Trees

Tropical rain forests typically consist of large numbers of different tree species with only a few members of each species in an area. Thus, most of the old-growth tree species in an area are rare and vulnerable to local or regional extinction when the forest is disturbed.

Tropical biologist Bill Laurance and his colleagues have been studying the nature of topical rain forests and how they are affected by human activities for over 25 years. One of his research interests is the effect of increasing fragmentation of tropical rain forests as people establish more roads, crop plantations, settlements, and cattle grazing areas.

The edges of forest fragments are often invaded by sun-loving species such as vines, which can gradually take over and cause the loss of a fragment's rare old-growth trees. Laurance is trying to determine how large the undisturbed inner core of a fragment must be to afford protection to its rare old-growth tree species.

His research team studies this by looking at the trees and other plant species in designated plots near the edge of a forest fragment and in the fragment's interior. In this example of muddy-boots ecology, they identify the tree species present in each plot. Then they measure the height and diameter of each tree to calculate its biomass and thus the total biomass of the trees in each plot. Such measurements are repeated every two years for two decades or more to determine changes in the species composition of the study plots.

Such painstaking research reveals that within 100 meters (330 feet) of the edge of a forest fragment, typically up to 36% of the biomass of old-growth trees is lost within 10-17 years after fragmentation. Plots in a fragment's interior show little loss of their tree biomass. Scientists can use such data to estimate how large a fragment must be in order to prevent the loss of rare trees within its protected core habitat. For more details on this research see *The Habitable Planet*, Video 9, at **www.learner.org/resources/series209.html**.

Critical Thinking

What are two ways to reduce the fragmentation of tropical rain forests?

CENGAGENOW See how serious the habitat fragmentation problem is for elephants, tigers, and rhinos at CengageNOW.

■ CASE STUDY

A Disturbing Message from the Birds

Approximately 70% of the world's nearly 10,000 known bird species are declining in numbers, and roughly one of every eight (12%) of these bird species is threatened with extinction, mostly because of habitat loss, degradation, and fragmentation. About three-fourths of the threatened bird species live in forests, many of which are being cleared at a rapid rate, especially in the tropical areas in Asia and Latin America (Figure 9-12).

Some 40% of Indonesia's moist, tropical forests, particularly in Borneo and Sumatra, has been cleared for lumber and palm plantations. The harvested palm oil is used as biofuel, mostly in European nations. As a result, 75% of the bird species in Sumatra's lowland forests are on the verge of extinction.

In Brazil, 115 bird species are threatened, mostly because of the burning and clearing of Amazon forests for farms and ranches. Other threats to Brazil's bird species are the loss of 93% of Brazil's Atlantic coastal rain forest and, most recently, the clearing of the country's savannah-like *cerrado* area to establish soybean plantations.

A 2007 joint study by the National Audubon Society and the American Bird Conservacy found that 30% of

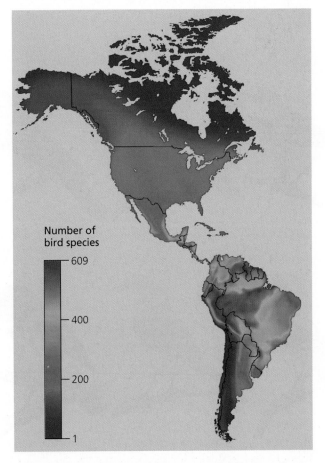

Figure 9-12 Distribution of bird species in North America and Latin America. **Question:** Why do you think more bird species are found in Latin America than in North America? (Data from The Nature Conservancy, Conservation International, World Wildlife Fund, and Environment Canada).

all North American bird species (25% of those living in the United States) and 70% of those living in grasslands are declining in numbers or are at risk of disappearing. A 2007 study by the American Bird Conservancy found that Hawaii's forests are the most threatened bird habitat. The government lists 30 of 71 remaining bird species in the Hawaiian Islands as endangered or threatened. Three other seriously threatened U.S. bird habitats are stream watersheds in the Southwest, tall-grass prairies in the Midwest, and beaches and marsh areas along the country's coastlines.

The numbers and distribution of North American bird species that can prosper around humans, such as robins, blackbirds, and starlings, have increased over the last 35 years. But populations of many forest songbirds have declined sharply. The greatest decline has occurred among long-distance migrant species such as tanagers, orioles, thrushes, vireos, and warblers that nest deep in North American woods in the summer and spend their winters in Central or South America or the Caribbean Islands. Figure 9-13 shows the 10 most threatened U.S. songbird species.

The primary culprit for these declines appears to be habitat loss and fragmentation of the birds' breeding habitats. In North America, woodlands are being cleared and broken up by roads and developments. In Central and South America, tropical forest habitats, mangroves, and wetland forests are suffering the same fate.

After habitat loss, the intentional or accidental introduction of nonnative species such as bird-eating cats, rats, snakes, and mongooses is the second greatest danger, affecting about 28% of the world's threatened birds. Fifty-two of the world's 388 parrot species (Figure 9-9) are threatened by a combination of habitat loss and capture for the pet trade (often illegal), especially in Europe and the United States.

At least 23 species of seabirds face extinction. Many seabirds drown after becoming hooked on one of the many baited lines put out by fishing boats. And populations of 40% of the world's water birds are in decline because of the global loss of wetlands.

Millions of migrating birds are killed each year when they collide with power lines, communications towers, and skyscrapers that have been erected in the middle of their migration routes. While U.S. hunters kill about 121 million birds a year, as many as 1 billion birds in the United States die each year when they fly into glass windows, especially those in tall city buildings that are lit up at night—the number one cause of U.S. bird mortality. Other threats to birds are oil spills, exposure to pesticides, herbicides that destroy their habi-

Cerulean warbler   Sprague's pipit   Bichnell's thrush   Black-capped vireo   Golden-cheeked warbler

Florida scrub jay   California gnatcatcher   Kirtland's warbler   Henslow's sparrow   Bachman's warbler

Figure 9-13 The 10 most threatened species of U.S. songbirds. Most of these species are vulnerable because of habitat loss and fragmentation from human activities. An estimated 12% of the world's known bird species may face premature extinction due mostly to human activities during this century. (Data from National Audubon Society)

SCIENCE FOCUS

Vultures, Wild Dogs, and Rabies: Some Unexpected Scientific Connections

In 2004, the World Conservation Union placed three species of vultures found in India and South Asia on the critically endangered list. During the early 1990s, there were more than 40 million of these carcass-eating vultures. But within a few years their populations had fallen by more than 97%.

This is an interesting scientific mystery, but should anyone care if various vulture species disappear? The answer is yes.

Scientists were puzzled, but they eventually discovered that the vultures were being poisoned by *diclofenac*. This anti-inflammatory drug reduces pain in cows and in humans and is used to increase milk production in cows.

But it causes kidney failure in vultures that feed on the carcasses of these cows.

As the vultures died off, huge numbers of cow carcasses, normally a source of food for the vultures, were consumed by wild dogs and rats whose populations the vultures helped control by reducing their food supply. As wild dog populations exploded due to a greatly increased food supply, the number of dogs with rabies also increased. This increased the risks to people bitten by rabid dogs. In 1997 alone, more than 30,000 people in India died of rabies—more than half the world's total number of rabies deaths that year.

Thus, protecting these vulture species from extinction can end up protecting millions of

people from a life-threatening disease. Unraveling often unexpected ecological connections in nature is not only fascinating but also vital to our own lives and health.

Some who argue against protecting species and ecosystems from harmful human activities frame the issue as a choice between protecting people or wildlife. Conservation biologists reject this as a misleading conclusion. To them, the goal is to protect both wildlife and people because their fates and well-being are interconnected.

Critical Thinking

What would happen to your life and lifestyle if most of the world's vultures disappeared?

---

tats, and ingestion of toxic lead shotgun pellets, which fall into wetlands, and lead sinkers left by anglers.

The greatest new threat to birds is climate change. A 2006 review, done for the World Wildlife Fund (WWF), of more than 200 scientific articles found that climate change is causing declines of bird populations in every part of the world. And climate change is expected to increase sharply during this century.

Migratory, mountain, island, wetland, Antarctic, Arctic, and sea birds are especially at risk from climate change. The researchers have warned that protecting many current areas with high bird diversity will not help, because climate change will force many bird species to shift to unprotected zones. Island and mountain birds may simply have nowhere to go.

Conservation biologists view this decline of bird species with alarm. One reason is that birds are excellent *environmental indicators* because they live in every climate and biome, respond quickly to environmental changes in their habitats, and are relatively easy to track and count.

Furthermore, birds perform a number of important economic and ecological services in ecosystems throughout the world. They help control populations of rodents and insects (which decimate many tree species), remove dead animal carcasses (a food source for some birds), and spread plants throughout their habitats by helping with pollination and by consuming and excreting plant seeds.

Extinctions of birds that play key and specialized roles in pollination and seed dispersal, especially in tropical areas, may lead to extinctions of plants dependent on these ecological services. Then some specialized animals that feed on these plants may also become

extinct. This cascade of extinctions, in turn, can affect our own food supplies and well-being. Protecting birds and their habitats is not only a conservation issue. It is an important issue for human health as well (Science Focus, above).

Biodiversity scientists urge us to listen more carefully to what birds are telling us about the state of the environment, for the birds' sake, as well as for ours.

THINKING ABOUT
Bird Extinctions

How does your lifestyle directly or indirectly contribute to the premature extinction of some bird species? What are three things that you think should be done to reduce the premature extinction of birds?

RESEARCH FRONTIER

Learning more about why birds are declining, what it implies for the biosphere, and what can be done about it. See **academic.cengage.com/biology/miller**.

Some Deliberately Introduced Species Can Disrupt Ecosystems

After habitat loss and degradation, the biggest cause of premature animal and plant extinctions is the deliberate or accidental introduction of harmful invasive species into ecosystems (**Concept 9-3**).

Most species introductions are beneficial to us, although they often displace native species. We depend heavily on introduced species for ecosystem services, food, shelter, medicine, and aesthetic enjoyment.

According to a 2000 study by ecologist David Pimentel, introduced species such as corn, wheat, rice, and other food crops, and cattle, poultry, and other livestock provide more than 98% of the U.S. food supply. Similarly, nonnative tree species are grown in about 85% of the world's tree plantations. Some deliberately introduced species have also helped to control pests.

The problem is that some introduced species have no natural predators, competitors, parasites, or pathogens to help control their numbers in their new habitats. Such nonnative species can reduce or wipe out populations of many native species, trigger ecological disruptions, cause human health problems, and lead to economic losses.

In 1988, for example, a giant African land snail was imported into Brazil as a cheap substitute for conventional escargot (snails) used as a source of food. It grows to the size of a human fist and weighs 1 kilogram (2.2 pounds) or more. When export prices for escargot fell, breeders dumped the snails into the wilds. Now it has spread to 23 of Brazil's states and devours everything from lettuce to mouse droppings. It also can carry rat lungworm, a parasite that burrows into the human brain and causes meningitis, and another parasite that can rupture the intestines. Authorities eventually banned the snail, but it was too late. So far, the snail has been unstoppable.

Figure 9-14 shows some of the estimated 7,100 harmful invasive species that, after being deliberately or accidentally introduced into the United States, have caused ecological and economic harm. Nonnative species threaten almost half of the roughly 1,300 endangered and threatened species in the United States and 95% of those in the state of Hawaii, according to the U.S. Fish and Wildlife Service. According to biologist Thomas Lovejoy, harmful invader species cost the U.S. public an average of $261,000 per minute! The situation in China is much worse. Biologist David Pimentel estimates that, globally, damage to watersheds, soils, and wildlife by bioinvaders may be costing as much $44,400 per second. And the damages are rising rapidly.

Some deliberately introduced species are plants such as kudzu (Case Study, at right). Deliberately introduced animal species have also caused ecological and economic damage. Consider the estimated 1 million European wild (feral) boars (Figure 9-14) found in parts of Florida and other U.S. states. They compete for food with endangered animals, root up farm fields, and cause traffic accidents. Game and wildlife officials have failed to control their numbers through hunting and trapping and say there is no way to stop them. Another example is the estimated 30 million feral cats and 41 million outdoor pet cats found in the United States. Most were introduced to the environment when they were abandoned by their owners and left to breed in the wild; they kill about 568 million birds per year. Because of pet overpopulation and abandonment, pet shelters in the United States are forced to kill over 14 million cats

and dogs a year. Shelter officials urge owners to spay or neuter their cats and keep them indoors.

■ CASE STUDY
The Kudzu Vine

An example of a deliberately introduced plant species is the *kudzu* ("CUD-zoo") *vine,* which, in the 1930s, was imported from Japan and planted in the southeastern United States in an attempt to control soil erosion. Kudzu does control erosion. But it is so prolific and difficult to kill that it engulfs hillsides, gardens, trees, abandoned houses and cars, stream banks, patches of forest, and anything else in its path (Figure 9-15, p. 200).

This plant, which is sometimes called "the vine that ate the South," has spread throughout much of the southeastern United States. It could spread as far north as the Great Lakes by 2040 if climate change caused by global warming occurs as projected.

Kudzu is considered a menace in the United States but Asians use a powdered kudzu starch in beverages, gourmet confections, and herbal remedies for a range of diseases. A Japanese firm has built a large kudzu farm and processing plant in the U.S. state of Alabama and ships the extracted starch to Japan. And almost every part of the kudzu plant is edible. Its deep-fried leaves are delicious and contain high levels of vitamins A and C. Stuffed kudzu leaves, anyone?

Although kudzu can engulf and kill trees, it might eventually save some trees from loggers. Researchers at the Georgia Institute of Technology indicate that it could be used in place of trees as a source of fiber for making paper. And a preliminary 2005 study indicated that kudzu powder could be used to reduce alcoholism and binge drinking. Ingesting small amounts of the powder can lessen one's desire for alcohol.

Some Accidentally Introduced Species Can Also Disrupt Ecosystems

Welcome to one of the downsides of global trade, travel, and tourism. Many unwanted nonnative invaders arrive from other continents as stowaways on aircraft, in the ballast water of tankers and cargo ships, and as hitchhikers on imported products such as wooden packing crates. Cars and trucks can also spread the seeds of nonnative plant species embedded in their tire treads. Many tourists return home with living plants that can multiply and become invasive. These plants might also harbor insects that can escape, multiply rapidly, and threaten crops.

In the 1930s, the extremely aggressive Argentina fire ant (Figures 9-14 and 9-16, p. 200) was introduced accidentally into the United States in Mobile, Alabama.

Deliberately Introduced Species

Purple loosestrife

European starling

African honeybee ("Killer bee")

Nutria

Salt cedar (Tamarisk)

Marine toad (Giant toad)

Water hyacinth

Japanese beetle

Hydrilla

European wild boar (Feral pig)

Accidentally Introduced Species

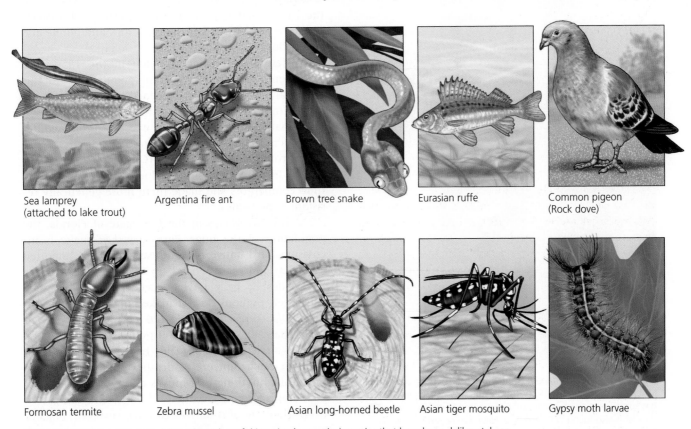

Sea lamprey (attached to lake trout)

Argentina fire ant

Brown tree snake

Eurasian ruffe

Common pigeon (Rock dove)

Formosan termite

Zebra mussel

Asian long-horned beetle

Asian tiger mosquito

Gypsy moth larvae

Figure 9-14 Some of the more than 7,100 harmful invasive (nonnative) species that have been deliberately or accidentally introduced into the United States.

Figure 9-15 Kudzu taking over an abandoned house in the U.S. state of Mississippi. This vine, which can grow 5 centimeters (2 inches) per hour, was deliberately introduced into the United States for erosion control. Digging it up or burning it cannot stop it. Grazing by goats and repeated doses of herbicides can destroy it, but goats and herbicides also destroy other plants, and herbicides can contaminate water supplies. Scientists have found a common fungus that can kill kudzu within a few hours, apparently without harming other plants. Stay tuned.

The ants may have arrived on shiploads of lumber or coffee imported from South America. Without natural predators, fire ants have spread rapidly by land and water (they can float) throughout the South, from Texas to Florida and as far north as Tennessee and Virginia.

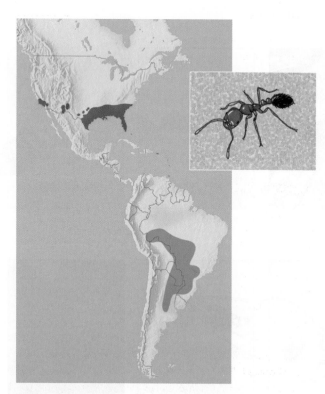

Figure 9-16 The *Argentina fire ant,* introduced accidentally into Mobile, Alabama, in the 1930s from South America (green area), has spread over much of the southern United States (red area). This invader is also found in Puerto Rico, New Mexico, and California.
Question: How might this accidental introduction of fire ants have been prevented? (Data from S. D. Porter, Agricultural Research Service, U.S. Department of Agriculture)

When these ants invade an area, they can wipe out as much as 90% of native ant populations. Mounds containing fire ant colonies cover many farm fields and invade people's yards. Walk on one of these mounds, and as many as 100,000 ants may swarm out of their nest to attack you with painful and burning stings. They have killed deer fawns, birds, livestock, pets, and at least 80 people who were allergic to their venom. In the United States, they also do an estimated $68,000 of economic damage per hour to crops and phone and power lines.

Widespread pesticide spraying in the 1950s and 1960s temporarily reduced fire ant populations. But this chemical warfare actually hastened the advance of the rapidly multiplying fire ants by reducing populations of many native ant species. Even worse, it promoted development of genetic resistance to pesticides in the fire ants through natural selection (**Concept 4-2B**, p. 80). In other words, we helped wipe out their competitors and made them more genetically resistant to pesticides.

CONCEPT LINK

In the Everglades in the U.S. state of Florida, the population of the huge *Burmese python* snake is increasing. This native of Southeast Asia was imported as a pet, and many ended up being dumped in the Everglades by people who learned that, when they get larger, pythons do not make good pets. They can live 25 years, reach 6 meters (20 feet) in length, weigh more than 90 kilograms (200 pounds), and have the girth of a telephone pole. They have razor-sharp teeth and can catch, squeeze to death, and swallow whole practically anything that moves and is warm-blooded, including raccoons, a variety of birds, and full-grown deer. They are also known to have survived alligator attacks. They are slowly spreading to other areas and, by 2100, could be found in most of the southern half of the continental United States.

Prevention Is the Best Way to Reduce Threats from Invasive Species

Once a harmful nonnative species becomes established in an ecosystem, its removal is almost impossible—somewhat like trying to get smoke back into a chimney. Clearly, the best way to limit the harmful impacts of nonnative species is to prevent them from being introduced and becoming established.

Scientists suggest several ways to do this:

- Fund a massive research program to identify the major characteristics of successful invader species and the types of ecosystems that are vulnerable to invaders (Figure 9-17).

- Greatly increase ground surveys and satellite observations to detect and monitor species invasions and develop better models for predicting how they will spread.

- Step up inspection of imported goods and goods carried by travelers that are likely to contain invader species.

- Identify major harmful invader species and establish international treaties banning their transfer from one country to another, as is now done for endangered species.

- Require cargo ships to discharge their ballast water and replace it with saltwater at sea before entering ports, or require them to sterilize such water or pump nitrogen into the water to displace dissolved oxygen and kill most invader organisms.

- Increase research to find and introduce natural predators, parasites, bacteria, and viruses to control populations of established invaders.

Characteristics of Successful Invader Species	Characteristics of Ecosystems Vulnerable to Invader Species
■ High reproductive rate, short generation time (r-selected species)	■ Climate similar to habitat of invader
■ Pioneer species	■ Absence of predators on invading species
■ Long lived	■ Early successional systems
■ High dispersal rate	■ Low diversity of native species
■ Generalists	■ Absence of fire
■ High genetic variability	■ Disturbed by human activities

Figure 9-17 Some general characteristics of successful invader species and ecosystems vulnerable to invading species. **Question:** Which, if any, of the characteristics on the right-hand side could humans influence?

WHAT CAN YOU DO?
Controlling Invasive Species

- ■ Do not capture or buy wild plants and animals

- ■ Do not remove wild plants from their natural areas

- ■ Do not dump the contents of an aquarium into waterways, wetlands, or storm drains

- ■ When camping, use wood found near your camp site instead of bringing firewood from somewhere else

- ■ Do not dump unused bait into any waterways

- ■ After dogs visit woods or the water, brush them before taking them home

- ■ After each use, clean your mountain bike, canoe, boat, hiking boots, and other gear before heading for home

Figure 9-18 Individuals Matter: ways to prevent or slow the spread of harmful invasive species. **Questions:** Which two of these actions do you think are the most important? Why? Which of these actions do you plan to take?

RESEARCH FRONTIER

Learning more about invasive species, why they thrive, and how to control them. See **academic.cengage.com/biology/miller**.

Figure 9-18 shows some of the things you can do to help prevent or slow the spread of these harmful invaders.

Population Growth, Overconsumption, Pollution, and Climate Change Can Cause Species Extinctions

Human population growth and excessive and wasteful consumption of resources have greatly expanded the human ecological footprint (Figure 1-10, p. 15; and Figure 3, pp. S24–S25, in Supplement 4), which has eliminated vast areas of wildlife habitat (Figure 9-11). Acting together, these factors have caused premature extinction of many species (**Concept 9-3**). (See *The Habitable Planet,* Video 13, at **www.learner.org/resources/series209.html**).

Pollution also threatens some species with extinction (**Concept 9-3**), as has been shown by the unintended effects of certain pesticides. According to the U.S. Fish and Wildlife Service, each year pesticides kill about one-fifth of the beneficial honeybee colonies in United States (Case Study, p. 202), more than 67 million birds, and 6–14 million fish. They also threaten about one-fifth of the country's endangered and threatened species.

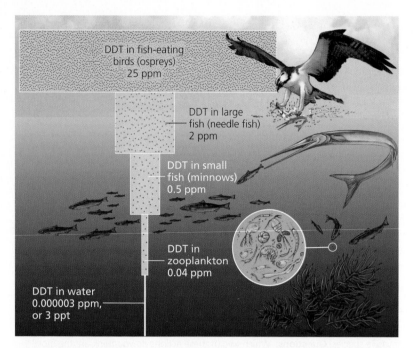

Figure 9-19 *Bioaccumulation* and *biomagnification*. DDT is a fat-soluble chemical that can accumulate in the fatty tissues of animals. In a food chain or web, the accumulated DDT can be biologically magnified in the bodies of animals at each higher trophic level. The concentration of DDT in the fatty tissues of organisms was biomagnified about 10 million times in this food chain in an estuary near Long Island Sound in the U.S. state of New York. If each phytoplankton organism takes up from the water and retains one unit of DDT, a small fish eating thousands of zooplankton (which feed on the phytoplankton) will store thousands of units of DDT in its fatty tissue. Each large fish that eats 10 of the smaller fish will ingest and store tens of thousands of units, and each bird (or human) that eats several large fish will ingest hundreds of thousands of units. Dots represent DDT. **Question:** How does this story demonstrate the value of pollution prevention?

During the 1950s and 1960s, populations of fish-eating birds such as ospreys, brown pelicans (see Photo 1 in the Detailed Contents), and bald eagles plummeted. A chemical derived from the pesticide DDT, when biologically magnified in food webs (Figure 9-19), made the birds' eggshells so fragile that they could not reproduce successfully. Also hard hit were such predatory birds as the prairie falcon, sparrow hawk, and peregrine falcon, which help to control rabbits, ground squirrels, and other crop eaters.

Since the U.S. ban on DDT in 1972, most of these species have made a comeback. For example, after eliminating DDT and after crackdowns on hunting and habitat destruction, the American bald eagle has rebounded from only 417 breeding pairs in the lower 48 states in 1963 to almost 10,000 breeding pairs in 2007. This was enough to have it removed from the endangered species list. The comeback of this species from the brink of extinction is one of the greatest wildlife protection successes in U.S. history.

A 2004 study by Conservation International predicted that climate change caused mostly by global warming (Science Focus, p. 33) could drive more than a quarter of all land animals and plants to extinction by the end of this century. Some scientific studies indicate that polar bears (Case Study, p. 203) and 10 of the world's 17 penguin species are already threatened because of higher temperatures and melting sea ice in their polar habitats.

■ **CASE STUDY**

Where Have All the Honeybees Gone?

Three-quarters of all flowering plants in North America—including most fruit and vegetable crops—rely on pollinators such as bees, bats (Science Focus, p. 192), butterflies, and hummingbirds for fertilization.

According to a 2006 report by the U.S. National Academy of Sciences (NAS), populations of such vital pollinators are declining across North America, and honeybee populations have been in increasing trouble for over 2 decades. The report warns that continued decreases in wild populations of such pollinators could disrupt food production and ecosystems.

The report includes a specific warning on the decline of the honeybee, which pollinates more than 110 commercially grown crops that are vital to U.S. agriculture, including up to one-third of U.S. fruit and vegetable crops. Globally, about one-third of the human diet comes from insect-pollinated plants, and the honeybee is responsible for 80% of that pollination, according to the U.S. Department of Agriculture. Adult honeybees live on honey that they make from nectar they collect from flowering plants, and they feed their young with protein-rich pollen.

Honeybees are big business. In the United States, honeybee colonies are managed by beekeepers who rent the bees out for their pollination services, especially for major crops such as almonds, apples, and blueberries. By 1994, such colonies had replaced an estimated 98% of the wild, free-range honeybees in the United States.

According to the NAS report, there has been a 30% drop in U.S. honeybee populations since the 1980s. Causes include pesticide exposure (the wax in beehives absorbs these and other airborne toxins), attacks by parasitic mites that can wipe out a colony in hours, and invasion by African honeybees (killer bees, p. 93).

Since 2006, a growing number of bee colonies in 27 states have suffered what researchers call "bee colony collapse disorder," or what French scientists call "mad bee disease," in which the worker bees in a colony vanish without a trace. When beekeepers inspect what were once healthy and strong colonies, they find all of the adult worker bees gone and an abandoned queen bee.

More than a quarter of the country's 2.4 million honeybee colonies (each with 30,000 to 100,000 individual bees) have been lost. Nobody knows where the missing bees went or what is causing them to leave the hives. Possible causes include parasites, fungi, bacteria, and pesticides. Scientists also suspect a virus that paralyzes bees, which may have come from Israel or from bees imported from Australia to help replace declining U.S. honeybee populations. Another problem is poor nutrition and stress caused when colonies of bees are fed an artificial diet while being trucked around the country and rented out for pollination.

Scientists are also finding sharp declines in some species of bumblebees found in the United States. These bees are responsible for pollinating an estimated 15% of the crops grown in the United States, especially those that are raised in greenhouses, such as tomatoes, peppers, and strawberries. Bumblebees collect pollen and nectar to feed their young, but make very little honey.

Declining bee populations have also been reported in Brazil, Taiwan, Guatemala, and parts of Europe. China, where some argue that pesticides are overused, gives us a glimpse of a future without honeybees. Apple orchards in China are now largely hand-fertilized by humans in the absence of bees. Some scientists warn that if it continues and grows, *bee colony collapse disorder* could lead to *agricultural collapse disorder* in parts of the world.

┌─ **THINKING ABOUT** ─────────────
Bees

What difference would it make to your lifestyle if most of the honeybees or bumblebees disappeared? What two things would you do to help reduce the loss of honeybees?
└──────────────────────────────────

■ CASE STUDY
Polar Bears and Global Warming

The world's 20,000–25,000 polar bears are found in 19 populations distributed across the frozen Arctic. About 60% of them are in Canada, and the rest are found in arctic areas in Denmark, Norway, Russia, and the U.S. state of Alaska.

Throughout the winter, the bears hunt for seals on floating sea ice (Figure 9-20), which expands southward each winter and contracts as the temperature rises each summer. Normally the bears swim from one patch of sea ice to another to hunt and eat seals during winter as their body fat accumulates. In the summer and fall as sea ice breaks up, the animals fast and live off their body fat for several months until hunting resumes when the ice again expands.

Evidence shows that the Arctic is warming twice as fast as the rest of the world and that the average annual area of floating summer sea ice in the Arctic is declining and is breaking up earlier each year. Scientists project that summer sea ice could be gone by 2030 and perhaps as soon as 2012.

This means that polar bears have less time to feed and to store the fat they need in order to survive their summer and fall months of fasting. As a result, they must fast longer, which weakens them. As females become weaker, their ability to reproduce and keep their young cubs alive declines. And as bears grow hungrier, they are more likely to go to human settlements looking for food. The resulting increase in bear sightings gives people the false impression that their populations are increasing.

Polar bears are strong swimmers, but ice shrinkage has forced them to swim longer distances to find enough food and to spend more time during winter hunting on land where prey is nearly impossible to find. Several studies link global warming and diminished sea ice to polar bears drowning or starving while in search of prey and, in some cases, to cannibalism among the bears.

According to a 2006 study by the IUCN–World Conservation Union, the world's total polar bear population is likely to decline by 30–35% by 2050, and by the end of this century, the bears may be found only in zoos.

Another threat to the bears in some areas is the buildup of toxic PCBs, DDT (Figure 9-19), and other pesticides in their fatty tissue, which can adversely affect their development, behavior, and reproduction. And in 2007, the U.S. Fish and Wildlife Service estimated that Russian poachers are killing 100–250 polar bears a year.

In 2007, the IUCN listed polar bears as threatened in their annual red list of endangered species, and in 2008, the U.S. government listed the polar bear as threatened under the Endangered Species Act.

Figure 9-20 Polar bear with seal prey on floating ice in Svalbard, Norway. Polar bears in the Arctic are likely to become extinct sometime during this century because global warming is melting the floating sea ice on which they hunt seals.

THINKING ABOUT

Polar Bears

What difference would it make if all of the world's polar bears disappeared? List two things you would do to help protect the world's remaining polar bears from premature extinction.

Illegal Killing, Capturing, and Selling of Wild Species Threatens Biodiversity

Some protected species are illegally killed for their valuable parts or are sold live to collectors (**Concept 9-3**). Such *poaching* endangers many larger animals and some rare plants. Globally, this illegal trade in wildlife earns smugglers at least $10 billion a year—an average of $19,000 a minute. Organized crime has moved into illegal wildlife smuggling because of the huge profits involved—surpassed only by the illegal international trade in drugs and weapons. Rapidly growing wildlife smuggling is a high-profit, low-risk business because few of the smugglers are caught or punished. At least two-thirds of all live animals smuggled around the world die in transit.

To poachers, a live mountain gorilla is worth $150,000, a giant panda pelt $100,000, a chimpanzee $50,000, an Imperial Amazon macaw $30,000, and a Komodo dragon reptile from Indonesia $30,000. A poached rhinoceros horn (Figure 9-21) can be worth as much as $55,500 per kilogram ($25,000 per pound). It is used to make dagger handles in the Middle East and as a fever reducer and alleged aphrodisiac in China and other parts of Asia.

According to a 2005 study by the International Fund for Animal Welfare, the Internet has become a key market for the illegal global trade in thousands of live threatened and endangered animals and products made from such animals. For example, U.S. websites offered live chimpanzees dressed as dolls for $60,000–65,000 each and a 2-year-old highly endangered Siberian tiger for $70,000.

An important way to combat the illegal trade in these species is through research and education. Some people are dedicating their time and energy to this problem. For example, scientist Jane Goodall has devoted her life to understanding and protecting chimpanzees (Individuals Matter, at right).

In 1900, an estimated 100,000 tigers roamed free in the world. Despite international protection, only about 3,500 tigers remain in the wild, on an ever shrinking range (Figure 9-11, top left), according to a 2006 study by the World Conservation Union. Between 1900 and 2007, the estimated number of tigers in India plunged from 40,000 to about 1,400. In 2007, Eric Dinerstein and his colleagues estimated that tigers have 41% less habitat than they had in 1997, mostly because of deforestation and forest fragmentation, and now live in

Martin Harvey/Peter Arnold, Inc

Figure 9-21 White rhinoceros killed by a poacher for its horn in South Africa. **Question:** What would you say if you could talk to the poacher of this animal?

just 7% of their historic range around the world. Today all five tiger subspecies are endangered in the wild, although at least 11,000 captive tigers of mixed ancestry exist behind bars.

Tigers are also threatened because they are killed for their coats and body parts. The Bengal or Indian tiger is at risk because a coat made from its fur can sell for as much as $100,000 in Tokyo. Wealthy collectors have paid $10,000 to $20,000 for a Bengal tiger rug. With the body parts of a single tiger worth as much as $25,000, and because few of the poachers are caught or punished, it is not surprising that illegal hunting has skyrocketed. According to a 2006 study by tiger experts, without emergency action to curtail poaching and preserve their habitat, few if any tigers may be left in the wild within 20 years.

THINKING ABOUT

Tigers

What difference would it make if all the world's tigers disappeared? What two things would you do to help protect the world's remaining tigers from premature extinction?

The global legal and illegal trade in wild species for use as pets is also a huge and very profitable business. Many owners of wild pets do not know that, for every live animal captured and sold in the pet market, an estimated 50 others are killed or die in transit. Most are also unaware that some imported exotic animals can carry dangerous infectious diseases.

About 25 million U.S. households have exotic birds as pets, 85% of them imported. More than 60 bird species, mostly parrots, (Figure 9-9), are endangered or threatened because of this wild bird trade. Ironically, keeping birds as pets can also be dangerous for people. A 1992 study suggested that keeping a pet bird indoors for more than 10 years doubles a person's chances of getting lung cancer from inhaling tiny particles of bird dander.

Other wild species whose populations are depleted because of the pet trade include amphibians, reptiles, mammals, and tropical fishes (taken mostly from the coral reefs of Indonesia and the Philippines). Divers catch tropical fish by using plastic squeeze bottles of poisonous cyanide to stun them. For each fish caught alive, many more die. In addition, the cyanide solution kills the coral animals that create the reef.

Some exotic plants, especially orchids and cacti, are endangered because they are gathered (often illegally) and sold to collectors to decorate houses, offices, and landscapes. A collector may pay $5,000 for a single rare orchid. A mature crested saguaro cactus can earn cactus rustlers as much as $15,000.

THINKING ABOUT
Collecting Wild Species

Some people believe it is unethical to collect wild animals and plants for display and personal pleasure. They believe we should leave most exotic wild species in the wild. Explain why you agree or disagree with this view.

As commercially valuable species become endangered, their black market demand soars. This positive feedback loop increases their chances of premature extinction from poaching. Most poachers are not caught and the money they can make far outweighs the small risk of being caught, fined, or imprisoned.

On the other hand, species also hold great value by surviving in the wild. According to the U.S. Fish and Wildlife Service, collectors of exotic birds may pay $10,000 for a threatened hyacinth macaw smuggled out of Brazil. But during its lifetime, a single macaw left in the wild might yield as much as $165,000 in tourist revenues.

In Thailand, biologist Pilai Poonswad decided to do something about poachers taking Great Indian hornbills—large, beautiful, and rare birds—from a rain forest. She visited the poachers in their villages and showed them why the birds are worth more alive than dead. Today, some former poachers earn money by taking ecotourists into the forest to see these magnificent birds. Because of their vested financial interest in preserving the hornbills, they now help to protect the birds from poachers. Individuals matter.

Jane Goodall

Primatologist and anthropologist Jane Goodall (Figure 9-A) spent 45 years studying chimpanzee social and family life in Gombe National Park in the African country of Tanzania. One of her major scientific contributions was the discovery that chimpanzees have tool-making skills. She observed that some chimpanzees modified twigs or blades of grass and then poked them into termite mounds. When the termites latched onto these primitive tools, the chimpanzees would pull them out and eat the termites.

In 1977, she established the Jane Goodall Institute, which supports the research at Gombe National Park and, with 19 offices around the world, works to protect chimpanzees and their habitats. Dr. Goodall spends nearly 300 days a year traveling and educating people throughout the world about chimpanzees and the need to protect the environment.

Figure 9-A Jane Goodall with a young chimpanzee living in Tanzania's Gombe National Park.

M. Gunther/BIOS/Peter Arnold, Inc.

Goodall is also president of Advocates for Animals, an animal rights organization based in Edinburgh, Scotland, that campaigns against the use of animals for medical research, zoos, farming, and sport hunting. She has received many awards and prizes for her scientific and conservation contributions. She has also written 23 books for adults and children and has produced 14 films about the lives and importance of chimpanzees.

Rising Demand for Bush Meat Threatens Some African Species

Indigenous people in much of West and Central Africa have sustainably hunted wildlife for *bush meat,* a source of food, for centuries. But in the last two decades bush meat hunting in some areas has skyrocketed as local people try to provide food for rapidly growing populations or seek to make a living by supplying restaurants with exotic meat (Figure 9-22, p. 206). Logging roads have enabled miners, ranchers, and settlers to move into once inaccessible forests, which has made it easier to hunt animals for bush meat. And a 2004 study showed that people living in coastal areas of West Africa have increased bush meat hunting because local fish harvests have declined due to overfishing by heavily subsidized European Union fishing fleets.

So what is the big deal? After all, people have to eat. For most of our existence, humans have survived by hunting and gathering wild species.

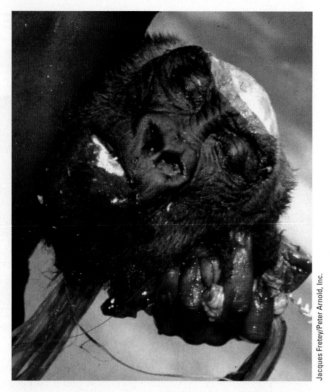

Figure 9-22 *Bush meat,* such as this severed head of a lowland gorilla in the Congo, is consumed as a source of protein by local people in parts of West Africa and sold in the national and international marketplace. You can find bush meat on the menu in Cameroon and the Congo in West Africa as well as in Paris, London, Toronto, New York, and Washington, D.C. It is often supplied by poaching. Wealthy patrons of some restaurants regard gorilla meat as a source of status and power. **Question:** How, if at all, is this different from killing a cow for food?

One problem today is that bush meat hunting has led to the local extinction of many wild animals in parts of West Africa and has driven one species—Miss Waldron's red colobus monkey—to complete extinction. It is also a factor in reducing gorilla, orangutan (Figure 9-7), chimpanzee, elephant, and hippopotamus populations. This practice also threatens forest carnivores, such as crowned eagles and leopards, by depleting their main prey species.

Some conservationists fear that within 1 or 2 decades, the Congo basin's rain forest—the world's second largest remaining tropical forest—will contain few large mammals, and most of Africa's great apes will be extinct. Another problem is that butchering and eating some forms of bush meat has helped to spread fatal diseases such as HIV/AIDS and the Ebola virus to humans.

The U.S. Agency for International Development is trying to reduce unsustainable hunting for bush meat in some areas by introducing alternative sources of food, such as fish farms. They are also showing villagers how to breed large rodents such as cane rats as a source of food.

THINKING ABOUT

The Passenger Pigeon and Humans

Humans exterminated the passenger pigeon (Figure 9-1) within a single human lifetime because it was considered a pest and because of its economic value. Suppose a species superior to us arrived and began taking over the earth with the goal of using the planet more sustainably. The first thing they might do is to exterminate us. Do you think such an action would be justified? Explain.

9-4 How Can We Protect Wild Species from Extinction Resulting from Our Activities?

▶ **CONCEPT 9-4A** We can use existing environmental laws and treaties and work to enact new laws designed to prevent premature species extinction and protect overall biodiversity.

▶ **CONCEPT 9-4B** We can help to prevent premature species extinction by creating and maintaining wildlife refuges, gene banks, botanical gardens, zoos, and aquariums.

▶ **CONCEPT 9-4C** According to the *precautionary principle,* we should take measures to prevent or reduce harm to the environment and to human health, even if some of the cause-and-effect relationships have not been fully established, scientifically.

International Treaties Help to Protect Species

Several international treaties and conventions help to protect endangered and threatened wild species (**Concept 9-4A**). One of the most far reaching is the 1975 *Convention on International Trade in Endangered Species* (CITES). This treaty, now signed by 172 countries, bans hunting, capturing, and selling of threatened or endan-

gered species. It lists some 900 species that cannot be commercially traded as live specimens or wildlife products because they are in danger of extinction. It also restricts international trade of roughly 5,000 species of animals and 28,000 plants species that are at risk of becoming threatened.

CITES has helped reduce international trade in many threatened animals, including elephants, crocodiles, cheetahs, and chimpanzees. But the effects of this treaty are limited because enforcement varies from country to country, and convicted violators often pay only small fines. Also, member countries can exempt themselves from protecting any listed species, and much of the highly profitable illegal trade in wildlife and wildlife products goes on in countries that have not signed the treaty.

The *Convention on Biological Diversity* (CBD), ratified by 190 countries (but not by the United States), legally commits participating governments to reversing the global decline of biodiversity and to equitably sharing the benefits from use of the world's genetic resources. This includes efforts to prevent or control the spread of ecologically harmful invasive species.

This convention is a landmark in international law because it focuses on ecosystems rather than on individual species and it links biodiversity protection to issues such as the traditional rights of indigenous peoples. However, because some key countries including the United States have not ratified it, implementation has been slow. Also, the law contains no severe penalties or other enforcement mechanisms.

■ CASE STUDY
The U.S. Endangered Species Act

The *Endangered Species Act of 1973* (ESA; amended in 1982, 1985, and 1988) was designed to identify and protect endangered species in the United States and abroad (**Concept 9-4A**). This act is probably the most far-reaching environmental law ever adopted by any nation, which has made it controversial. Canada and a number of other countries have similar laws.

Under the ESA, the National Marine Fisheries Service (NMFS) is responsible for identifying and listing endangered and threatened ocean species, while the U.S. Fish and Wildlife Services (USFWS) is to identify and list all other endangered and threatened species. Any decision by either agency to add a species to, or remove one from, the list must be based on biological factors alone, without consideration of economic or political factors. However, economic factors can be used in deciding whether and how to protect endangered habitat and in developing recovery plans for listed species.

The ESA also forbids federal agencies (except the Defense Department) to carry out, fund, or authorize projects that would jeopardize an endangered or threatened species or destroy or modify its critical habitat. For offenses committed on private lands, fines as high as $100,000 and 1 year in prison can be imposed to ensure protection of the habitats of endangered species. This part of the act has been controversial because at least 90% of the listed species live totally or partially on private land. The ESA also makes it illegal for Americans to sell or buy any product made from an endangered or threatened species or to hunt, kill, collect, or injure such species in the United States.

Between 1973 and 2007, the number of U.S. species on the official endangered and threatened lists increased from 92 to about 1,350 species—55% of them plants and 45% animals. According to a 2000 study by The Nature Conservancy, one-third of the country's species are at risk of extinction, and 15% of all species are at high risk—far more than the roughly 1,350 species on the country's endangered species list. The study also found that many of the country's rarest and most imperiled species are concentrated in a few areas, called *hotspots*. To conservation biologists, protecting such areas should be a top priority.

For each species listed, the USFWS or the NMFS is supposed to prepare a recovery plan that includes designation and protection of critical habitat. Examples of successful recovery plans include those for the American alligator (Chapter 4 Core Case Study, p. 77), the gray wolf, the peregrine falcon, and the bald eagle.

The ESA also requires that all commercial shipments of wildlife and wildlife products enter or leave the country through one of nine designated ports. Only 120 full-time USFWS inspectors examine shipments of wild animals that enter the United States through these ports, airports, and border crossings. They can inspect only a small fraction of the more than 200 million wild animals brought legally into the United States each year. Also, tens of millions of such animals are brought in illegally, but few illegal shipments of endangered or threatened animals or plants are confiscated (Figure 9-23, p. 208). Even when they are caught, many violators are not prosecuted, and convicted violators often pay only a small fine.

In addition, people who smuggle or buy imported exotic animals are rarely aware that many of them carry dangerous infectious diseases, such as hantavirus, Ebola virus, Asian bird flu, herpes B virus (carried by most adult macaques), and salmonella (from pets such as hamsters, turtles, and iguanas), which can spread from pets to humans. The country's small number of wildlife inspectors does not have the capability or budget to detect such diseases.

Since 1982, the ESA has been amended to give private landowners economic incentives to help save endangered species living on their lands. The goal is to strike a compromise between the interests of private landowners and those of endangered and threatened species.

With one of these approaches, called a *habitat conservation plan* (HCP), a landowner, developer, or logger is allowed to destroy some critical habitat in exchange for taking steps to protect members of the species. Such measures might include setting aside a part of the species' habitat as a protected area, paying to relocate the species to another suitable habitat, or contributing money to have the government buy suitable habitat elsewhere. Once the plan is approved, it cannot be changed, even if new data show that the plan is inadequate to protect a species and help it recover. The ESA has also been used to protect endangered and threatened marine reptiles,

Figure 9-23 Confiscated products made from endangered species. Because of a scarcity of funds and inspectors, probably no more than one-tenth of the illegal wildlife trade in the United States is discovered. The situation is even worse in most other countries.

Steve Hillebrand/U.S. Fish and Wildlife Service.

Other critics would go further and do away with this act entirely. But because this step is politically unpopular with the American public, most efforts are designed to weaken the act and reduce its meager funding. In 2007, the USFWS issued a new interpretation of the ESA that would allow it to protect plants and animals only in areas where they are struggling to survive, rather than listing a species over its entire range. If this new policy survives court tests, it would remove about 80% of the roughly 1,350 U.S. species listed as endangered or threatened. Many wildlife biologists see this proposed policy as a deceptive way to gut the ESA by weakening existing protections and making it more difficult to list new species.

Most conservation biologists and wildlife scientists agree that the ESA needs to be simplified and streamlined. But they contend that it has not been a failure (Science Focus, at right).

such as turtles, and mammals, especially whales, seals, and sea lions, as discussed in Chapter 11.

Some believe that the Endangered Species Act should be weakened or repealed, and others believe it should be strengthened and modified to focus on protecting ecosystems. Opponents of the act contend that it puts the rights and welfare of endangered plants and animals above those of people. They argue that it has not been effective in protecting endangered species and has caused severe economic losses by hindering development on private lands. Since 1995, efforts to weaken the ESA have included the following suggested changes:

- Make protection of endangered species on private land voluntary.

- Have the government compensate landowners if they are forced to stop using part of their land to protect endangered species.

- Make it harder and more expensive to list newly endangered species by requiring government wildlife officials to navigate through a series of hearings and peer-review panels and requiring hard data instead of computer-based models.

- Eliminate the need to designate critical habitats.

- Allow the secretary of the interior to permit a listed species to become extinct without trying to save it.

- Allow the secretary of the interior to give any state, county, or landowner a permanent exemption from the law, with no requirement for public notification or comment.

We Can Establish Wildlife Refuges and Other Protected Areas

In 1903, President Theodore Roosevelt established the first U.S. federal wildlife refuge at Pelican Island, Florida, to help protect birds such as the brown pelican (see Photo 1 in the Detailed Contents) from extinction. Since then, the National Wildlife Refuge System has grown to include 547 refuges. Each year, more than 35 million Americans visit these refuges to hunt, fish, hike, and watch birds and other wildlife.

More than three-fourths of the refuges serve as wetland sanctuaries vital for protecting migratory waterfowl. One-fifth of U.S. endangered and threatened species have habitats in the refuge system, and some refuges have been set aside for specific endangered species (**Concept 9-5B**). These areas have helped Florida's key deer, the brown pelican, and the trumpeter swan to recover. According to a General Accounting Office study, however, activities considered harmful to wildlife occur in nearly 60% of the nation's wildlife refuges.

Conservation biologists call for setting aside more refuges for endangered plants. They also urge Congress and state legislatures to allow abandoned military lands that contain significant wildlife habitat to become national or state wildlife refuges.

Gene Banks, Botanical Gardens, and Wildlife Farms Can Help Protect Species

Gene or *seed banks* preserve genetic information and endangered plant species by storing their seeds in refrigerated, low-humidity environments (**Concept 9-4B**). More than 100 seed banks around the world collectively hold about 3 million samples.

SCIENCE FOCUS

Accomplishments of the Endangered Species Act

Critics of the ESA call it an expensive failure because only 37 species have been removed from the endangered list. Most biologists insist that it has not been a failure, for four reasons.

First, species are listed only when they face serious danger of extinction. This is like setting up a poorly funded hospital emergency room that takes only the most desperate cases, often with little hope for recovery, and saying it should be shut down because it has not saved enough patients.

Second, it takes decades for most species to become endangered or threatened. Not surprisingly, it also takes decades to bring a species in critical condition back to the point where it can be removed from the critical list. Expecting the ESA—which has been in existence only since 1973—to quickly repair the biological depletion of many decades is unrealistic.

Third, according to federal data, the conditions of more than half of the listed species are stable or improving, and 99% of the protected species are still surviving. A hospital emergency room taking only the most desperate cases and then stabilizing or improving the conditions of more than half of its patients and keeping 99% of them alive would be considered an astounding success.

Fourth, the ESA budget included only $58 million in 2005—about what the Department of Defense spends in a little more than an hour—or 20¢ per year per U.S. citizen. To its supporters, it is amazing that the ESA, on such a small budget, has managed to stabilize or improve the conditions of more than half of the listed species.

Its supporters would agree that the act can be improved and that federal regulators have sometimes been too heavy handed in enforcing it. But instead of gutting or doing away with the ESA, biologists call for it to be strengthened and modified to help protect ecosystems and the nation's overall biodiversity.

A study by the U.S. National Academy of Sciences recommended three major changes to make the ESA more scientifically sound and effective:

- Greatly increase the meager funding for implementing the act.

- Develop recovery plans more quickly. A 2006 study by the Government Accountability Office, found that species with recovery plans have a better chance of getting off the endangered list, and it recommended that any efforts to reform the law should continue to require recovery plans.

- When a species is first listed, establish a core of its survival habitat as critical, as a temporary emergency measure that could support the species for 25–50 years.

Most biologists and wildlife conservationists believe that the United States needs a new law that emphasizes protecting and sustaining biological diversity and ecosystem functioning (**Concept 9-4A**) rather than focusing mostly on saving individual species. We discuss this idea further in Chapter 10.

Critical Thinking

Should the U.S. Endangered Species Act be modified to more effectively protect and sustain the nation's overall biodiversity? Explain.

Scientists urge the establishment of many more such banks, especially in developing countries. But some species cannot be preserved in gene banks. The banks are also expensive to operate and can be destroyed by fires and other mishaps.

The world's 1,600 *botanical gardens* and *arboreta* contain living plants representing almost one-third of the world's known plant species. However, they contain only about 3% of the world's rare and threatened plant species and have too little space and funding to preserve most of those species.

We can take pressure off some endangered or threatened species by raising individuals on *farms* for commercial sale. Farms in Florida raise alligators for their meat and hides. Butterfly farms flourish in Papua New Guinea, where many butterfly species are threatened by development activities.

Zoos and Aquariums Can Protect Some Species

Zoos, aquariums, game parks, and animal research centers are being used to preserve some individuals of critically endangered animal species, with the long-term goal of reintroducing the species into protected wild habitats (**Concept 9-4B**).

Two techniques for preserving endangered terrestrial species are egg pulling and captive breeding. *Egg pulling* involves collecting wild eggs laid by critically endangered birds and then hatching them in zoos or research centers. In *captive breeding,* some or all of the wild individuals of a critically endangered species are captured for breeding in captivity, with the aim of reintroducing the offspring into the wild. Captive breeding has been used to save the peregrine falcon and the California condor (Case Study, p. 210).

Other techniques for increasing the populations of captive species include artificial insemination, embryo transfer (surgical implantation of eggs of one species into a surrogate mother of another species), use of incubators, and cross-fostering (in which the young of a rare species are raised by parents of a similar species). Scientists also use computer databases, which hold information on family lineages of species in zoos, and DNA analysis to match individuals for mating—a computer dating service for zoo animals—and to prevent genetic erosion through inbreeding.

The ultimate goal of captive breeding programs is to build up populations to a level where they can be reintroduced into the wild. But after more than two decades of captive breeding efforts, only a handful of endangered species have been returned to the wild. Examples shown in Figure 9-3 include the black-footed ferret, the California condor (Case Study, right), and the golden lion tamarin. Most reintroductions fail because of lack of suitable habitat, inability of individuals bred in captivity to survive in the wild, renewed overhunting, or poaching of some of the returned individuals.

Lack of space and money limits efforts to maintain breeding populations of endangered animal species in zoos and research centers. The captive population of each species must number 100–500 individuals to avoid extinction through accident, disease, or loss of genetic diversity through inbreeding. Recent genetic research indicates that 10,000 or more individuals are needed for an endangered species to maintain its capacity for biological evolution.

Public aquariums that exhibit unusual and attractive fish and some marine animals such as seals and dolphins help to educate the public about the need to protect such species. But mostly because of limited funds, public aquariums have not served as effective gene banks for endangered marine species, especially marine mammals that need large volumes of water.

Instead of seeing zoos and aquariums as sanctuaries, some critics claim that most of them imprison once-wild animals. They also contend that zoos and aquariums can foster the false notion that we do not need to preserve large numbers of wild species in their natural habitats. Proponents counter that these facilities play an important role in educating the public about wildlife and the need to protect biodiversity.

Regardless of their benefits and drawbacks, zoos, aquariums, and botanical gardens are not biologically or economically feasible solutions for the growing problem of premature extinction of species. Figure 9-24 lists some things you can do to deal with this problem.

■ CASE STUDY

Trying to Save the California Condor

At one time the California condor (Figure 9-4), North America's largest bird, was nearly extinct with only 22 birds remaining in the wild. To save the species, one approach was to capture the remaining birds and breed them in captivity at zoos.

The captured birds were isolated from human contact as much as possible, and to reduce genetic defects, closely related individuals were prevented from breeding. As of 2007, 135 condors had been released back into the wild throughout the southwestern United States.

A major threat to these birds is lead poisoning resulting when they ingest lead pellets from ammunition in animal carcasses or gut piles left behind by hunters. A lead-poisoned condor quickly becomes weak and mentally impaired and dies of starvation, or is killed by predators.

A coalition of conservationist and health organizations is lobbying state game commissions and legislatures to ban the use of lead in ammunition and to require use of less harmful substitutes. They also urge people who hunt in condor ranges to remove all killed animals or to hide carcasses and gut piles by burying them, covering them with brush or rocks, or putting them in inaccessible areas.

WHAT CAN YOU DO?

Protecting Species

- Do not buy furs, ivory products, or other items made from endangered or threatened animal species
- Do not buy wood or paper products produced by cutting old-growth forests in the tropics
- Do not buy birds, snakes, turtles, tropical fish, and other animals that are taken from the wild
- Do not buy orchids, cacti, or other plants that are taken from the wild
- Spread the word. Talk to your friends and relatives about this problem and what they can do about it

Figure 9-24 Individuals matter: ways to help prevent premature extinction of species. **Question:** Which two of these suggestions do you believe are the most important? Why?

THINKING ABOUT

The California Condor's Comeback

CORE CASE STUDY

What are some differences between the stories of the condor and the passenger pigeon (**Core Case Study**) that might give the condor a better chance of avoiding premature extinction than the passenger pigeon had?

The Precautionary Principle

Some might argue that, because we have identified fewer than 2 million of the estimated 5–100 million species on the earth, it makes little sense to take drastic measures to preserve them.

Conservation biologists disagree. They remind us that the earth's species are the primary components of its biodiversity, which should not be degraded because of the economic and ecological services it provides. They call for us to use great caution in making potentially harmful changes to communities and ecosystems and to take precautionary action to help *prevent* potentially serious environmental problems, including premature extinctions.

This approach is based on the **precautionary principle:** When substantial preliminary evidence indicates that an activity can harm human health or the environment, we should take precautionary measures to prevent or reduce such harm, even if some of the cause-and-effect relationships have not been fully established, scientifically (**Concept 9-4C**). It is based on the commonsense idea behind many adages such as "Better safe than sorry" and "Look before you leap."

Scientists use the precautionary principle to argue for preservation of species, and also for preserving entire ecosystems, which is the focus of the next two chapters.

Passenger Pigeons and Sustainability

The disappearance of the passenger pigeon (**Core Case Study**) in a short time was a blatant example of the effects of activities undertaken by uninformed or uncaring people. Since the passenger pigeon became extinct, we have learned a lot about how to protect birds and other species from premature extinction resulting from our activities. We have also learned much about the importance of wild species as key components of the earth's biodiversity and of the natural capital that supports all life and economies.

Yet, despite these efforts, there is overwhelming evidence that we are in the midst of wiping out as many as half of the world's wild species within your lifetime. Ecological ignorance accounts for some of the failure to deal with this problem. But to many, the real cause of this failure is political. They would argue that we lack the will to act on our scientific knowledge and our ethical judgments.

In keeping with the four **scientific principles of sustainability** (see back cover), acting to prevent the premature extinction of species helps to preserve the earth's biodiversity and to maintain species interactions that help control population sizes, energy flow, and matter cycling in ecosystems. Thus it is not only for the species that we ought to act, but also for the overall long-term health of the biosphere on which we all depend, and for the health and well-being of our own species. Protecting wild species and their habitats is a way of protecting ourselves and our descendants.

The problem is complex, and so are the solutions. Protecting biodiversity is no longer simply a matter of passing and enforcing endangered species laws and setting aside parks and preserves. It will also require slowing climate change, which will affect many species and their habitats. And it will require reducing the size and impact of our ecological footprints (Figure 1-10, p. 15). Although the solutions to biodiversity loss and degradation are complex, they are well within our reach.

The great challenge of the twenty-first century is to raise people everywhere to a decent standard of living while preserving as much of the rest of life as possible.

EDWARD O. WILSON

REVIEW

1. Review the Key Questions and Concepts for this chapter on p. 184. What factors led to the premature extinction of the passenger pigeon in the United States?

2. Distinguish between **background extinction** and **mass extinction.** What is the **extinction rate** of a species? Describe how scientists estimate extinction rates. Give four reasons why many extinction experts believe that human activities are now causing a sixth mass extinction. Distinguish between **endangered species** and **threatened species.** List some characteristics that make some species especially vulnerable to extinction.

3. What are two reasons for trying to prevent the premature extinction of wild species? What is the **instrumental value** of a species? List six types of instrumental values provided by wild species. How are scientists using DNA analysis to reduce the illegal killing of elephants? What is the **intrinsic (existence) value** of a species?

4. What is biophilia? Why should we care about bats?

5. What is **HIPPCO?** In order, what are the six largest causes of premature extinction of species resulting from human activities? Why are island species especially vulnerable to extinction? What is habitat fragmentation, and how does it threaten many species?

6. Describe the threats to bird species in the world and in the United States. List three reasons why we should be alarmed by the decline of bird species.

7. Give two examples of the harmful effects of nonnative species that have been introduced **(a)** deliberately and **(b)** accidentally. List ways to limit the harmful impacts of nonnative species. Describe the roles of population growth, overconsumption, pollution, and climate change in the premature extinction of wild species. Describe what is happening to many of the honeybees in the United States and what economic and ecological roles they play. Explain how pesticides such as DDT can be biomagnified in food chains and webs. Explain how global warming is threatening polar bears.

8. Describe the poaching of wild species and give three examples of species that are threatened by this illegal activity. Describe the work of Jane Goodall in protecting wild primates. Why are tigers likely to disappear from the wild by the end of this century? Describe the threat to some forms of wildlife from increased hunting for bush meat.

9. Describe two international treaties that are used to help protect species. Describe the U.S. Endangered Species Act, how successful it has been, and the controversy over this act. Describe the roles of wildlife refuges, gene banks, botanical gardens, wildlife farms, zoos, and aquariums in protecting some species.

10. Describe how protecting wild species from premature extinction (**Core Case Study**) is in keeping with the four **scientific principles of sustainability**.

Note: Key Terms are in **bold** type.

CRITICAL THINKING

1. List three ways in which you could apply **Concept 9-3** to make your lifestyle more environmentally sustainable.

2. What are three aspects of your lifestyle that directly or indirectly contribute to the premature extinction of some bird species (Case Study, p. 195)? What are three things that you think should be done to reduce the premature extinction of birds?

3. Discuss your gut-level reaction to the following statement: "Eventually, all species become extinct. Thus, it does not really matter that the passenger pigeon (**Core Case Study**) is extinct, and that the whooping crane and the world's remaining tiger species are endangered mostly because of human activities." Be honest about your reaction, and give arguments for your position.

4. Do you accept the ethical position that each species has the inherent right to survive without human interference, regardless of whether it serves any useful purpose for humans? Explain. Would you extend this right to the *Anopheles* mosquito, which transmits malaria, and to infectious bacteria? Explain.

5. Wildlife ecologist and environmental philosopher Aldo Leopold wrote, "To keep every cog and wheel is the first precaution of intelligent tinkering." Explain how this statement relates to the material in this chapter. Explain how protecting wild species and their habitats is an important way to protect the health and well-being of people.

6. What would you do if **(a)** your yard and house were invaded by fire ants, **(b)** you found bats flying around your yard at night, and **(c)** deer invaded your yard and ate your shrubs, flowers, and vegetables?

7. Which of the following statements best describes your feelings toward wildlife?
   a. As long as it stays in its space, wildlife is okay.
   b. As long as I do not need its space, wildlife is okay.
   c. I have the right to use wildlife habitat to meet my own needs.
   d. When you have seen one redwood tree, elephant, or some other form of wildlife, you have seen them all, so lock up a few of each species in a zoo or wildlife park and do not worry about protecting the rest.
   e. Wildlife should be protected.

8. Environmental groups in a heavily forested state want to restrict logging in some areas to save the habitat of an endangered squirrel. Timber company officials argue that the well-being of one type of squirrel is not as important as the well-being of the many families who will be affected if the restriction causes the company to lay off hundreds of workers. If you had the power to decide this issue, what would you do and why? Can you come up with a compromise?

9. Congratulations! You are in charge of preventing the premature extinction, caused by human activities, of the world's existing species. What are three things you would do to accomplish this goal?

10. List two questions that you would like to have answered as a result of reading this chapter.

Note: See Supplement 13 (p. S78) for a list of Projects related to this chapter.

DATA ANALYSIS

Examine these data released by the World Resources Institute and answer the following questions.

Country	Total Land Area in Square Kilometers (square miles)	Protected Area as Percent of Total Land Area (2003)	Total Number of Known Breeding Bird Species (1992–2002)	Number of Threatened Breeding Bird Species (2002)	Threatened Breeding Bird Species as Percent of Total Number of Known Breeding Bird Species
Afghanistan	647,668 (250,000)	0.3	181	11	
Cambodia	181,088 (69,900)	23.7	183	19	
China	9,599,445 (3,705,386)	7.8	218	74	
Costa Rica	50,110 (19,730)	23.4	279	13	
Haiti	27,756 (10,714)	0.3	62	14	
India	3.287,570 (1,269,388)	5.2	458	72	
Rwanda	26,344 (10,169)	7.7	200	9	
United States	9,633,915 (3,781,691)	15.8	508	55	

Source: Data from **earthtrends.wri.org/country_profiles/index.php?theme=7**.
World Resources Institute, *Earth Trends, Biodiversity and Protected Areas, Country Profiles*

1. For each of the eight countries, complete the table by filling in the last column.

2. Arrange the countries from largest to smallest according to Total Land Area. Does there appear to be any correlation between the size of country and the percentage of Threatened Breeding Bird Species? Explain your reasoning.

LEARNING ONLINE

Log on to the Student Companion Site for this book at **academic.cengage.com/biology/miller**, and choose Chapter 9 for many study aids and ideas for further reading and research. These include flash cards, practice quizzing, Weblinks, information on Green Careers, and InfoTrac® College Edition articles.

10 Sustaining Terrestrial Biodiversity: The Ecosystem Approach

Reintroducing Gray Wolves to Yellowstone

Around 1800 at least 350,000 gray wolves (Figure 10-1), roamed over the lower 48 states, especially in the West, and survived mostly by preying on bison, elk, caribou, and mule deer. But between 1850 and 1900, most of them were shot, trapped, and poisoned by ranchers, hunters, and government employees. When Congress passed the U.S. Endangered Species Act in 1973, only a few hundred gray wolves remained outside of Alaska, primarily in Minnesota and Michigan. In 1974, the gray wolf was listed as an endangered species in the lower 48 states.

Ecologists recognize the important role this keystone predator species once played in parts of the West , especially in the northern Rocky Mountain states of Montana, Wyoming, and Idaho where Yellowstone National Park is located. The wolves culled herds of bison, elk, caribou, and mule deer, and kept down coyote populations. They also provided uneaten meat for scavengers such as ravens, bald eagles, ermines, grizzly bears, and foxes. When wolves declined, herds of plant-browsing elk, moose, and mule deer expanded and devastated vegetation such as willow and aspen trees often found growing near streams and rivers. This increased soil erosion and threatened habitats of other wildlife species such as beavers, which, as foundation species (p. 96), helped to maintain wetlands.

In 1987, the U.S. Fish and Wildlife Service (USFWS) proposed reintroducing gray wolves into the Yellowstone National Park ecosystem to help restore and sustain its biodiversity. The proposal brought angry protests, some from area ranchers who feared the wolves would leave the park and attack their cattle and sheep. Other objections came from hunters who feared the wolves would kill too many big-game animals, and from mining and logging companies fearing that the government would halt their operations on wolf-populated federal lands.

In 1995 and 1996, federal wildlife officials caught gray wolves in Canada and relocated 41 of them in Yellowstone National Park. Scientists estimate that the long-term carrying capacity of the park is 110 to 150 gray wolves. In 2007, the park had 171 gray wolves. Overall, this experiment in ecosystem restoration has helped to re-establish and sustain some of the biodiversity that the Yellowstone ecosystem once had, as discussed later in this chapter.

In 2008, the USFWS decided to remove the gray wolf from protection under the Endangered Species Act in the states of Montana, Wyoming, and Utah. Several conservation groups filed suits to have the courts overturn this decision. The wolves in the park will remain protected. But 6 of the park's 11 wolf packs travel outside of the park boundaries during part of every year. If the courts allow removing the wolves from the endangered species list, it will be legal to kill any of these packs' individuals found outside the park.

Biologists warn that human population growth, economic development, and poverty are exerting increasing pressure on ecosystems and the services they provide to sustain biodiversity. This chapter is devoted to helping us understand and sustain the earth's forests, grasslands, and other storehouses of terrestrial biodiversity.

Tom Kitchin/Tom Stack & Associates

Figure 10-1 Natural capital restoration: the *gray wolf.* After becoming almost extinct in much of the western United States, in 1974 the gray wolf was listed and protected as an endangered species. Despite intense opposition by ranchers, hunters, miners, and loggers 41 members of this keystone species were reintroduced to their former habitat in the Yellowstone National Park and central Idaho in 1995 and 1996. By 2007, there were about 171 gray wolves in the park.

10-1 What are the major threats to forest ecosystems?

CONCEPT 10-1A Forest ecosystems provide ecological services far greater in value than the value of raw materials obtained from forests.

CONCEPT 10-1B Unsustainable cutting and burning of forests, along with diseases and insects, made worse by global warming, are the chief threats to forest ecosystems.

CONCEPT 10-1C Tropical deforestation is a potentially catastrophic problem because of the vital ecological services at risk, the high rate of tropical deforestation, and its growing contribution to global warming.

10-2 How should we manage and sustain forests?

CONCEPT 10-2 We can sustain forests by emphasizing the economic value of their ecological services, protecting old-growth forests, harvesting trees no faster than they are replenished, and using sustainable substitute resources.

10-3 How should we manage and sustain grasslands?

CONCEPT 10-3 We can sustain the productivity of grasslands by controlling the number and distribution of grazing livestock and by restoring degraded grasslands.

10-4 How should we manage and sustain parks and nature reserves?

CONCEPT 10-4 Sustaining biodiversity will require protecting much more of the earth's remaining undisturbed land area as parks and nature reserves.

10-5 What is the ecosystem approach to sustaining biodiversity?

CONCEPT 10-5A We can help to sustain biodiversity by identifying severely threatened areas and protecting those with high plant diversity (biodiversity hotspots) and those where ecosystem services are being impaired.

CONCEPT 10-5B Sustaining biodiversity will require a global effort to rehabilitate and restore damaged ecosystems.

CONCEPT 10-5C Humans dominate most of the earth's land, and preserving biodiversity will require sharing as much of it as possible with other species.

Note: Supplements 2 (p. S4), 4 (p. S20), 5 (p. S31), 9 (p. S53), and 13 (p. S78) can be used with this chapter.

There is no solution, I assure you,
to save Earth's biodiversity other than preservation
of natural environments in reserves large enough
to maintain wild populations sustainably.

EDWARD O. WILSON

10-1 What Are the Major Threats to Forest Ecosystems

▶ **CONCEPT 10-1A** Forest ecosystems provide ecological services far greater in value than the value of raw materials obtained from forests.

▶ **CONCEPT 10-1B** Unsustainable cutting and burning of forests, along with diseases and insects, made worse by global warming, are the chief threats to forest ecosystems.

▶ **CONCEPT 10-1C** Tropical deforestation is a potentially catastrophic problem because of the vital ecological services at risk, the high rate of tropical deforestation, and its growing contribution to global warming.

Forests Vary in Their Make-Up, Age, and Origins

Natural and planted forests occupy about 30% of the earth's land surface (excluding Greenland and Antarctica). Figure 7-8 (p. 146) shows the distribution of the world's boreal, temperate, and tropical forests. Tropical forests (Figure 7-15, top, p. 154) account for more than half of the world's forest area, and boreal (northern coniferous) forests (Figure 7-15, bottom) account for one quarter.

Forest managers and ecologists classify natural forests into two major types based on their age and structure. The first type is an **old-growth forest:** an uncut or regenerated primary forest that has not been seriously disturbed by human activities or natural disasters for

---

Figure 10-2 **Natural capital:** an old-growth forest in the U.S. state of Washington's Olympic National Forest (left) and an old-growth tropical rain forest in Queensland, Australia (right).

200 years or more (Figure 10-2). Old-growth or primary forests are reservoirs of biodiversity because they provide ecological niches for a multitude of wildlife species (Figure 7-16, p. 155, and Figure 7-17, p. 156).

The second type is a **second-growth forest:** a stand of trees resulting from secondary ecological succession (Figure 5-17, p. 117, and Figure 7-15, center photo, p. 154). These forests develop after the trees in

an area have been removed by human activities, such as clear-cutting for timber or cropland, or by natural forces, such as fire, hurricanes, or volcanic eruption.

A **tree plantation,** also called a **tree farm** or **commercial forest** (Figure 10-3), is a managed tract with uniformly aged trees of one or two genetically uniform species that usually are harvested by clear-cutting as soon as they become commercially valuable. The land

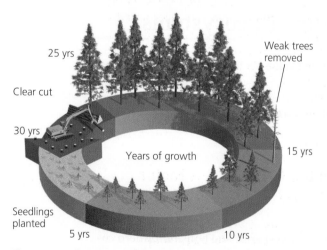

Figure 10-3 Short (25- to 30-year) rotation cycle of cutting and regrowth of a monoculture tree plantation used in modern industrial forestry. In tropical countries, where trees can grow more rapidly year-round, the rotation cycle can be 6–10 years. Old-growth or second-growth forests are clear-cut to provide land for growing most tree plantations (see photo, right). **Question:** What are two ways in which this process can degrade an ecosystem?

is then replanted and clear-cut again in a regular cycle. When managed carefully, such plantations can produce wood at a fast rate and thus increase their owners' profits. Most of this wood goes to paper mills and to mills that produce composites used as a substitute for natural wood.

But tree plantations with only one or two tree species are much less biologically diverse and probably less sustainable than old-growth and second-growth forests because they violate nature's biodiversity **principle of sustainability** (see back cover). And repeated cycles of cutting and replanting can eventually deplete the soil of nutrients and lead to an irreversible ecological tipping point that can hinder the growth of any type of forest on the land. There is also controversy over the increased use of genetically engineered tree species whose seeds could spread to other areas and threaten the diversity of second- and old-growth forests.

According to 2007 estimates by the FAO, about 60% of the world's forests are second-growth forests, 36% are old-growth or primary forests, and 4% are tree plantations (6% in the United States). In order, five countries—Russia, Canada, Brazil, Indonesia, and Papua New Guinea—have more than three-fourths of the world's remaining old-growth forests. In order, China (which has little original forest left), India, the United States, Russia, Canada, and Sweden account for about 60% of the world's tree plantations. Some conservation biologists urge establishing tree plantations only on land that has already been cleared or degraded instead of putting them in place of existing old-growth or secondary forests. One day, tree plantations may supply most of the world's demand for industrial wood, and this will help to protect the world's remaining forests.

Forests Provide Important Economic and Ecological Services

Forests provide highly valuable ecological and economic services (Figure 10-4 and **Concept 10-1A**). For example, through photosynthesis, forests remove CO_2 from the atmosphere and store it in organic compounds (biomass). By performing this ecological service, forests help to stabilize the earth's temperature and to slow global warming as a part of the global carbon cycle (Figure 3-18, p. 68). Scientists have attempted to estimate the economic value of the ecological services provided by the world's forests and other ecosystems (Science Focus, p. 218).

— **RESEARCH FRONTIER** —

Refining estimates of the economic values of ecological services provided by forests and other major ecosystems. See **academic.cengage.com/biology/miller**.

NATURAL CAPITAL

Forests

Ecological Services	Economic Services
Support energy flow and chemical cycling	Fuelwood
Reduce soil erosion	Lumber
Absorb and release water	Pulp to make paper
Purify water and air	Mining
Influence local and regional climate	Livestock grazing
Store atmospheric carbon	Recreation
Provide numerous wildlife habitats	Jobs

Figure 10-4 Major ecological and economic services provided by forests (**Concept 10-1A**). **Question:** Which two ecological services and which two economic services do you think are the most important?

Most biologists believe that the clearing and degrading of the world's remaining old-growth forests is a serious global environmental threat because of the important ecological and economic services they provide (**Concept 10-1A**). For example, traditional medicines, used by 80% of the world's people, are derived mostly from natural plants in forests, and chemicals found in tropical forest plants are used as blueprints for making most of the world's prescription drugs (Figure 9-8, p. 190). Forests are also habitats for about two-thirds of the earth's terrestrial species. In addition, they are home to more than 300 million people, and one of every four people depend on forests for their livelihoods.

Unsustainable Logging Is a Major Threat to Forest Ecosystems

Along with highly valuable ecological services, forests provide us with raw materials, especially wood.

The first step in harvesting trees is to build roads for access and timber removal. Even carefully designed logging roads have a number of harmful effects (Figure 10-5, p. 218)—namely, increased erosion and sediment runoff into waterways, habitat fragmentation (see Science Focus, p. 195, and *The Habitable Planet,*

SCIENCE FOCUS

Putting a Price Tag on Nature's Ecological Services

The long-term health of an economy cannot be separated from the health of the natural systems that support it. Currently, forests and other ecosystems are valued mostly for their economic services (Figure 10-4, right). But suppose we took into account the monetary value of the ecological services provided by forests (Figure 10-4, left).

In 1997, a team of ecologists, economists, and geographers—led by ecological economist Robert Costanza of the University of Vermont—estimated the monetary worth of the earth's ecological services and the biological income they provide. They estimated the latter to be at least $33.2 trillion per year—close to the economic value of all of the goods and services produced annually throughout the world. The amount of money required to provide such interest income, and thus the estimated value of the world's natural capital, would have to be at least $500 trillion—an average of about $73,500 for each person on earth!

According to this study, the world's forests provide us with ecological services worth at least $4.7 trillion per year—hundreds of times more than their economic value. And these are very conservative estimates.

Costanza's team examined many studies and a variety of methods used to estimate the values of ecosystems. For example, some researchers estimated people's *willingness to pay* for ecosystem services that are not marketed, such as natural flood control and carbon storage. These estimates were added to the known values of marketed goods like timber to arrive at a total value for an ecosystem.

The researchers estimated total global areas of 16 major categories of ecosystems, including forests, grasslands, and other terrestrial and aquatic systems. They multiplied those areas by the values per hectare of various ecosystem services to get the estimated economic values of these forms of natural capital. Some of the results for forests are shown in Figure 10-A. Note that the collective value of these ecosystem services is much greater than the value of timber and other raw materials extracted from forests (**Concept 10-1A**).

These researchers hope their estimates will alert people to three important facts: the earth's ecosystem services are essential for all humans and their economies; their economic value is huge; and they are an ongoing source of ecological income as long as they are used sustainably.

However, unless such estimates are included in the market prices of goods and services—through market tools such as regulations and taxes that discourage biodiversity degradation and through subsidies that protect biodiversity—the world's forests and other ecosystems will continue to be degraded. For example, the governments of countries such as Brazil and Indonesia provide subsidies that encourage the clearing or burning of tropical forests to plant vast soybean and oil palm plantations. This sends the powerful message that it makes more economic sense to destroy or degrade these centers of biodiversity than it does to leave them intact.

Critical Thinking

Some analysts believe that we should not try to put economic values on the world's irreplaceable ecological services because their value is infinite. Do you agree with this view? Explain. What is the alternative?

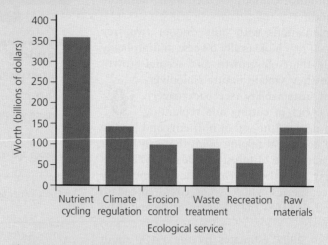

Figure 10-A Estimated annual global economic values of some ecological services provided by forests compared to the raw materials they produce (in billions of dollars).

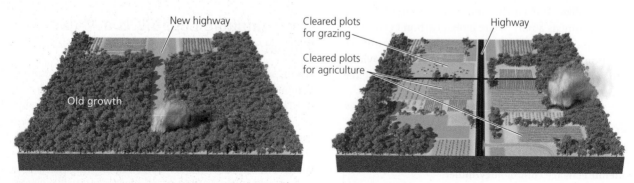

Figure 10-5 Natural capital degradation: Building roads into previously inaccessible forests paves the way to fragmentation, destruction, and degradation.

(a) Selective cutting

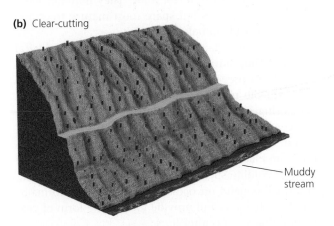

Clear stream

(b) Clear-cutting

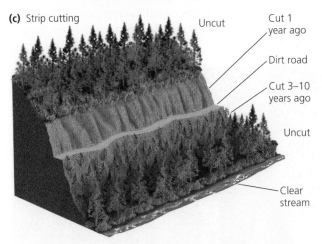

Muddy stream

(c) Strip cutting

Uncut

Cut 1 year ago

Dirt road

Cut 3–10 years ago

Uncut

Clear stream

Figure 10-6 Major tree harvesting methods. **Question:** If you were cutting trees in a forest you owned, which method would you choose and why?

Video 9, at **www.learner.org/resources/series209. html**), and loss of biodiversity. Logging roads also expose forests to invasion by nonnative pests, diseases, and wildlife species. And they open once-inaccessible forests to farmers, miners, ranchers, hunters, and off-road vehicle users.

Once loggers reach a forest area, they use a variety of methods to harvest the trees (Figure 10-6). With *selective cutting*, intermediate-aged or mature trees in an uneven-aged forest are cut singly or in small groups

(Figure 10-6a). But often, loggers remove all the trees from an area in what is called a *clear-cut* (Figures 10-6b and 10-7). Clear-cutting is the most efficient way for a logging operation to harvest trees, but it can do considerable harm to an ecosystem.

For example, scientists found that removing all the tree cover from a watershed greatly increases water runoff and loss of soil nutrients (Chapter 2 Core Case Study, p. 28). This increases soil erosion, which in turn causes more vegetation to die, leaving barren ground that can be eroded further, an example of a harmful positive feedback loop (Figure 2-11, p. 45, and **Concept 2-5A**, p. 44). More erosion also means more pollution of streams in the watershed. ◀ CONCEPT LINK And loss of vegetation destroys habitat and degrades biodiversity. Figure 10-8 (p. 220) summarizes some advantages and disadvantages of clear-cutting.

CENGAGENOW™ Learn more about how deforestation can affect the drainage of a watershed and disturb its ecosystem at CengageNOW™.

A variation of clear-cutting that allows a more sustainable timber yield without widespread destruction is *strip cutting* (Figure 10-6c). It involves clear-cutting a strip of trees along the contour of the land within a corridor narrow enough to allow natural regeneration within a few years. After regeneration, loggers cut another strip next to the first, and so on.

Figure 10-7 Clear-cut logging in the U.S. state of Washington.

TRADE-OFFS

Clear-Cutting Forests

Advantages	Disadvantages
Higher timber yields	Reduces biodiversity
Maximum profits in shortest time	Destroys and fragments wildlife habitats
Can reforest with fast-growing trees	Increases water pollution, flooding, and erosion on steep slopes
Good for tree species needing full or moderate sunlight	Eliminates most recreational value

Figure 10-8 Advantages and disadvantages of clear-cutting forests. **Question:** Which single advantage and which single disadvantage do you think are the most important? Why?

Biodiversity experts are alarmed at the growing practice of illegal, uncontrolled, and unsustainable logging taking place in 70 countries, especially in Africa and Southeast Asia (**Concept 10-1B**). Such logging has ravaged 37 of the 41 national parks in the African country of Kenya and now makes up 73–80% of all logging in Indonesia.

Complicating this issue is global trade in timber and wood products. For example, China, which has cut most of its own natural forests, imports more tropical rain forest timber than any other nation. Much of this timber is harvested illegally and unsustainably and is used to make furniture, plywood, flooring, and other products that are sold in the global marketplace.

Fire, Insects, and Climate Change Can Threaten Forest Ecosystems

Two types of fires can affect forest ecosystems. *Surface fires* (Figure 10-9, left) usually burn only undergrowth and leaf litter on the forest floor. They may kill seedlings and small trees but spare most mature trees and allow most wild animals to escape.

Occasional surface fires have a number of ecological benefits. They burn away flammable ground material and help to prevent more destructive fires. They also free valuable mineral nutrients tied up in slowly decomposing litter and undergrowth, release seeds from the cones of lodgepole pines, stimulate the germination of certain tree seeds such as those of the giant sequoia and jack pine, and help control tree diseases and insects. Wildlife species such as deer, moose, muskrat, and quail depend on occasional surface fires to maintain their habitats and provide food in the form of vegetation that sprouts after fires.

Another type of fire, called a *crown fire* (Figure 10-9, right), is an extremely hot fire that leaps from treetop

Figure 10-9 Surface fires (left) usually burn undergrowth and leaf litter on a forest floor. They can help to prevent more destructive crown fires (right) by removing flammable ground material. In fact, carefully controlled surface fires are deliberately set sometimes to prevent buildup of flammable ground material in forests. They also recycle nutrients and thus help to maintain the productivity of a variety of forest ecosystems. **Question:** What is another way in which a surface fire might benefit a forest?

White pine blister rust

Pine shoot beetle

Beech bark disease

Sudden oak death

Hemlock woolly adelgid

Figure 10-10 Natural capital degradation: some of the nonnative insect species and disease organisms that have invaded U.S. forests and are causing billions of dollars in damages and tree loss. The light green and orange colors in the map show areas where green or red overlap with yellow. (Data from U.S. Forest Service)

to treetop, burning whole trees. Crown fires usually occur in forests that have not experienced surface fires for several decades, a situation that allows dead wood, leaves, and other flammable ground litter to accumulate. These rapidly burning fires can destroy most vegetation, kill wildlife, increase soil erosion, and burn or damage human structures in their paths.

As part of a natural cycle, forest fires are not a major threat to forest ecosystems. But they are serious threats in parts of the world where people intentionally burn forests to clear the land, mostly to make way for crop plantations (**Concept 10-1B**). This can result in dramatic habitat losses, air pollution, and increases in atmospheric CO_2.

Accidental or deliberate introductions of foreign diseases and insects are a major threat to forests in the United States and elsewhere. Figure 10-10 shows some nonnative species of pests and disease organisms that are causing serious damage to certain tree species in parts of the United States.

There are several ways to reduce the harmful impacts of tree diseases and insect pests on forests. One is to ban imported timber that might introduce harmful new diseases or insects; another is to remove or clearcut infected and infested trees. We can also develop tree species that are genetically resistant to common tree diseases. Another approach is to control insect pests by applying conventional pesticides. Scientists also use biological control (bugs that eat harmful bugs) combined with very small amounts of conventional pesticides.

On top of these threats, projected climate change from global warming could harm many forests. For example, sugar maples are sensitive to heat, and in the U.S. region of New England, rising temperatures could kill these trees and, consequently, a productive maple syrup industry. Rising temperatures would also make many forest areas more suitable for insect pests and increase the size of pest populations. The resulting combination of drier forests and more dead trees could increase the incidence and intensity of forest fires (**Concept 10-1B**). This would add more of the greenhouse gas CO_2 to the atmosphere, which would further increase atmospheric temperatures and cause even more forest fires in a runaway positive feedback loop (Figure 2-11, p. 45).

We Have Cut Down Almost Half of the World's Forests

Deforestation is the temporary or permanent removal of large expanses of forest for agriculture, settlements, or other uses. Surveys by the World Resources Institute (WRI) indicate that over the past 8,000 years, human activities have reduced the earth's original forest cover by about 46%, with most of this loss occurring in the last 60 years.

Deforestation continues at a rapid rate in many parts of the world. The U.N. Food and Agricultural Organization (FAO) and World Resources Institute (WRI) surveys indicate that the global rate of forest cover loss between 1990 and 2005 was between 0.2%

and 0.5% per year, and that at least another 0.1–0.3% of the world's forests were degraded every year, mostly to grow crops and graze cattle. If these estimates are correct, the world's forests are being cleared or degraded exponentially at a rate of 0.3–0.8% per year, with much higher rates in some areas.

These losses are concentrated in developing countries, especially those in the tropical areas of Latin America, Indonesia, and Africa (Figure 3, pp. S24–S25, in Supplement 4). In its 2007 *State of the World's Forests* report, U.N. Food and Agricultural Organization estimated that about 130,000 square kilometers (50,000 square miles) of tropical forests are cleared each year (Figure 10-11)—equivalent to the total area of Greece or the U.S. state of Mississippi. We examine tropical forest losses further in the next subsection.

In addition to losses of tropical forests, scientists are concerned about the increased clearing of the northern boreal forests of Alaska, Canada, Scandinavia, and Russia, which together make up about one-fourth of the world's forested area. These vast coniferous forests (Figure 7-15, bottom photo, p. 154) are the world's greatest terrestrial storehouse of organic carbon and play a major role in the carbon cycle (Figure 3-18, p. 68) and in climate regulation for the entire planet. They also contain more than 70,000 plant and animal species. Surveys indicate that the total area of boreal forests lost every year is about twice the total area of Brazil's vast rain forests. In 2007, a group of 1,500 scientists from around the world signed a letter calling for the Canadian government to protect half of Canada's threatened boreal forests (of which only 10% are protected now) from logging, mining, and oil and gas extraction.

Figure 10-11 Natural capital degradation: extreme tropical deforestation in Chiang Mai, Thailand. The clearing of trees that absorb carbon dioxide increases global warming. It also dehydrates the soil by exposing it to sunlight. The dry topsoil blows away, which prevents the reestablishment of a forest in this area.

S. Channanrith-UNEP/Peter Arnold, Inc.

According to the WRI, if current deforestation rates continue, about 40% of the world's remaining intact forests will have been logged or converted to other uses within 2 decades, if not sooner. Clearing large areas of forests, especially old-growth forests, has important short-term economic benefits (Figure 10-4, right), but it also has a number of harmful environmental effects (Figure 10-12).

— HOW WOULD YOU VOTE? ☑

Should there be a global effort to sharply reduce the cutting of old-growth forests? Cast your vote online at **academic .cengage.com/biology/miller**.

In some countries, there is encouraging news about forest use. In 2007, the FAO reported that the net total forest cover in several countries, including the United States (see Case Study below), changed very little or increased between 2000 and 2005. Some of the increases resulted from natural reforestation by secondary ecological succession on cleared forest areas and abandoned croplands. But such increases were also due to the spread of commercial tree plantations.

■ CASE STUDY

Many Cleared Forests in the United States Have Grown Back

Forests that cover about 30% of the U.S. land area provide habitats for more than 80% of the country's wildlife species and supply about two-thirds of the nation's surface water. Old-growth forests once covered more than half of the nation's land area. But between 1620 when Europeans first arrived and 1920, the old-growth forests of the eastern United States were decimated.

Today, forests (including tree plantations) cover more area in the United States than they did in 1920. Many of the old-growth forests that were cleared or partially cleared between 1620 and 1920 have grown back naturally through secondary ecological succession (Figure 5-17, p. 117). There are fairly diverse second-growth (and in some cases third-growth) forests in every region of the United States, except much of the West. In 1995, environmental writer Bill McKibben cited forest regrowth in the United States—especially in the East—as "the great environmental story of the United States, and in some ways, the whole world."

Every year, more wood is grown in the United States than is cut and the total area planted with trees increases. Protected forests make up about 40% of the country's total forest area, mostly in the *National Forest System,* which consists of 155 national forests managed by the U.S. Forest Service (USFS).

On the other hand, since the mid-1960s, an increasing area of the nation's remaining old-growth and fairly diverse second-growth forests has been cut down and replaced with biologically simplified tree planta-

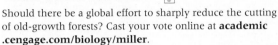

NATURAL CAPITAL DEGRADATION

Deforestation

■ Decreased soil fertility from erosion

■ Runoff of eroded soil into aquatic systems

■ Premature extinction of species with specialized niches

■ Loss of habitat for native species and migratory species such as birds and butterflies

■ Regional climate change from extensive clearing

■ Release of CO_2 into atmosphere

■ Acceleration of flooding

Figure 10-12 Harmful environmental effects of deforestation, which can reduce biodiversity and the ecological services provided by forests (Figure 10-4, left). **Question:** What are three products you have used recently that might have come from old-growth forests?

tions. According to biodiversity researchers, this reduces overall forest biodiversity and disrupts ecosystem processes such as energy flow and chemical cycling. And if such plantations are harvested too frequently, it could also deplete forest soils of key nutrients. Many biodiversity researchers favor establishing tree plantations only on land that has already been degraded instead of cutting old-growth and second-growth forests in order to replace them with tree plantations.

Tropical Forests Are Disappearing Rapidly

Tropical forests (Figure 7-15, top photo, p. 154) cover about 6% of the earth's land area—roughly the area of the lower 48 U.S. states. Climatic and biological data suggest that mature tropical forests once covered at least twice as much area as they do today; the majority of tropical forest loss has taken place since 1950 (Chapter 3 Core Case Study, p. 50).

Satellite scans and ground-level surveys indicate that large areas of tropical rain forests and tropical dry forests are being cut rapidly in parts of Africa, Southeast Asia (Figure 10-11), and South America (Figure 3-1, p. 50, and Figure 10-13, p. 224). A 2006 study by the U.S. National Academy of Sciences found that between 1990 and 2005, Brazil and Indonesia led the world in tropical forest loss. Illegal tree felling in 37 of 41 of Indonesia's supposedly protected parks account for three-quarters of the country's logging. According to the United Nations, Indonesia, which currently has the world's most diverse combination of plants, animals, and marine life, has already lost an estimated

Figure 10-13
Satellite images
of Amazon
deforestation
in the state
of Rondônia,
Brazil, between
1975 and 2001.

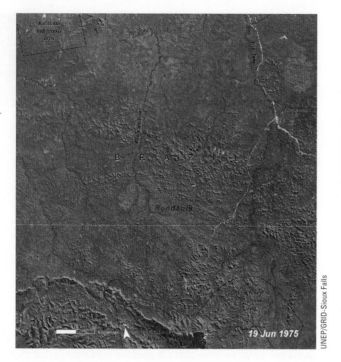

72% of its original intact forest, and 98% of its remaining forests will be gone by 2022.

Studies indicate that at least half of the world's known species of terrestrial plants and animals live in tropical rain forests (Figure 10-14). Because of their specialized niches (Figure 7-17, p. 156, and **Concept 4-6A**, p. 91) these species are highly vulnerable to extinction when their forest habitats are destroyed or degraded. Tropical deforestation is the main reason that more than 8,000 tree species—10% of the world's total—are threatened with extinction.

Brazil has more than 30% of the world's remaining tropical rain forest and an estimated 30% of the world's terrestrial plant and animal species (Figure 10-14, left)

Figure 10-14 *Species diversity:* two species found in tropical forests are part of the earth's biodiversity. On the left is an endangered *white ukari* in a Brazilian tropical forest. On the right is the world's largest flower, the *flesh flower* (Rafflesia) growing in a tropical rain forest of West Sumatra, Indonesia. The flower of this leafless plant can be as large 1 meter (3.3 feet) in diameter and weigh 7 kilograms (15 pounds). The plant gives off a smell like rotting meat, presumably to attract flies and beetles that pollinate the flower. After blossoming once a year for a few weeks, the blood red flower dissolves into a slimy black mass.

in its vast Amazon basin, which covers an area larger than India.

According to Brazil's government and forest experts, the percentage of its Amazon basin that had been deforested or degraded increased from 1% in 1970 to 16–20% in 2005 (Figure 10-13). Between 2005 and 2007, this rate increased sharply, with people cutting forests mostly to make way for cattle ranching and large plantations of crops such as soybeans used for cattle feed. In 2004, researchers at the Smithsonian Tropical Research Institute estimated that loggers, ranchers, and farmers in Brazil were clearing and burning an area equivalent to a loss of 11 football fields a minute! (**Concept 10-1C**)

Because of difficulties in estimating tropical forest loss, yearly estimates of global tropical deforestation vary widely from 50,000 square kilometers (19,300 square miles)—roughly the size of Costa Rica or the U.S. state of West Virginia—to 170,000 square kilometers (65,600 square miles)—about the size of the South American country of Uruguay or the U.S. state of Florida. At such rates, half of the world's remaining tropical forests will be gone in 35–117 years, resulting in a dramatic loss and degradation of biodiversity and the ecosystem services it provides.

RESEARCH FRONTIER

Improving estimates of rates of tropical deforestation. See **academic.cengage.com/biology/miller**.

Causes of Tropical Deforestation Are Varied and Complex

Tropical deforestation results from a number of interconnected basic and secondary causes (Figure 10-15). Population growth and poverty combine to drive subsistence farmers and the landless poor to tropical forests, where they try to grow enough food to survive. Government subsidies can accelerate deforestation by reducing the costs of timber harvesting, cattle grazing,

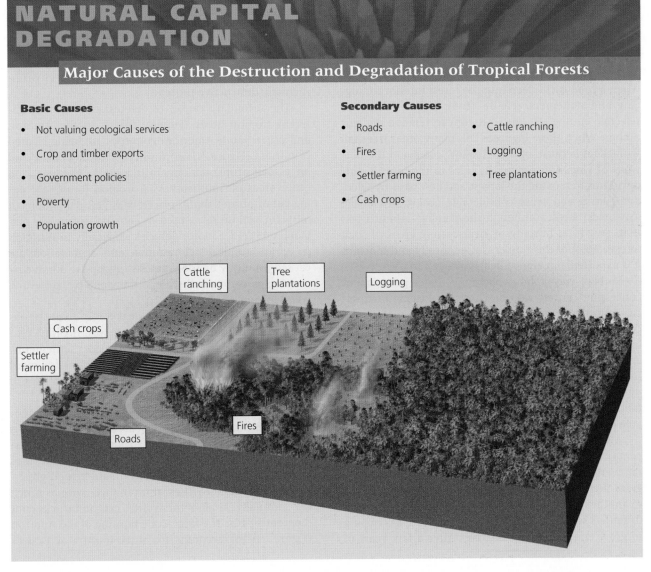

NATURAL CAPITAL DEGRADATION

Major Causes of the Destruction and Degradation of Tropical Forests

Basic Causes

- Not valuing ecological services
- Crop and timber exports
- Government policies
- Poverty
- Population growth

Secondary Causes

- Roads
- Fires
- Settler farming
- Cash crops
- Cattle ranching
- Logging
- Tree plantations

Cattle ranching · Tree plantations · Logging · Cash crops · Settler farming · Fires · Roads

Figure 10-15
Major interconnected causes of the destruction and degradation of tropical forests. The importance of specific secondary causes varies in different parts of the world. **Question:** If we could eliminate the basic causes, which if any of the secondary causes might automatically be eliminated?

Figure 10-16 Natural capital degradation: large areas of tropical forest in Brazil's Amazon basin are burned each year to make way for cattle ranches, small-scale farms, and plantation crops such as soybeans.
Questions: What are three ways in which your lifestyle probably contributes to this process? How, in turn, might this process affect your life?

Herbert Giradet/Peter Arnold, Inc.

and establishing vast plantations of crops such as soybeans and oil palm.

The degradation of a tropical forest usually begins when a road is cut deep into the forest interior for logging and settlement (Figure 10-5). Loggers then use selective cutting (Figure 10-6, top) to remove the best timber. When these big trees fall, many other trees fall with them because of their shallow roots and the network of vines connecting the trees in the forest's canopy. Thus, removing the largest and best trees by selective cutting can cause considerable ecological damage in tropical forests, although the damage is much less than that from burning or clear-cutting areas of such forests. Most of this timber is used locally and much of it is exported, but a great deal is also left to rot.

Foreign corporations operating under government concession contracts do much of the logging in tropical countries. Once a country's forests are gone, the companies move on to another country, leaving ecological devastation behind. For example, the Philippines and Nigeria have lost most of their once-abundant tropical hardwood forests and now are net importers of forest products. Several other tropical countries are following this ecologically and economically unsustainable path.

After the best timber has been removed, timber companies or the government often sell the land to ranchers. Within a few years, the cattle typically overgraze the land and the ranchers move their operations to another forest area. Then they sell the degraded land to settlers who have migrated to tropical forests hoping to grow enough food to survive. After a few years of

crop growing and erosion from rain, the nutrient-poor tropical soil is depleted of nutrients. Then the settlers move on to newly cleared land to repeat this environmentally destructive process.

The secondary causes of deforestation vary in different tropical areas. Tropical forests in the Amazon and other South American countries are being cleared or burned (Figure 10-16) mostly for cattle grazing and large soybean plantations. In Indonesia, Malaysia, and other areas of Southeast Asia, tropical forests are being cut or burned and replaced with vast plantations of oil palm, whose oil is used in cooking, cosmetics, and biodiesel fuel for motor vehicles (especially in Europe). In Africa, tropical deforestation and degradation are caused primarily by individuals struggling to survive by clearing plots for small-scale farming and by harvesting wood for fuel.

Burning is widely used to clear forest areas for agriculture, settlement, and other purposes. Healthy rain forests do not burn naturally. But roads, settlements, and farming, grazing, and logging operations fragment them (Science Focus, p. 195). The resulting patches of forest dry out and readily ignite.

The burning of tropical forests is a major component of human-enhanced global warming, which is projected to change the global climate at an increasing rate. Scientists estimate that globally, these fires account for at least 20% of all human-created greenhouse gas emissions. They also produce twice as much CO_2, annually, as all of the world's cars and trucks emit (**Concept 10-1C**). The large-scale burning of the Amazon (Figure 10-16) accounts for three-fourths of Brazil's greenhouse gas emissions, making Brazil the world's fourth largest emitter of such gases.

A 2005 study by forest scientists found that widespread fires in the Amazon are changing weather patterns by raising temperatures and reducing rainfall. Resulting droughts dry out the forests and make them more likely to burn—another example of a runaway positive feedback loop **Concept 2-5A**, p. 44, and Figure 2-11, p. 45). This process is converting deforested areas of tropical forests to tropical grassland (savanna)—an example of reaching an irreversible ecological *tipping point*. When the forests disappear, rainfall declines and yields of the crops planted on the land drop

CONCEPT
LINK

sharply. Models project that if current burning and deforestation rates continue, 20–30% of the Amazon will be turned into savanna in the next 50 years, and perhaps all of it could be so converted by 2080.

> **THINKING ABOUT**
> **Tropical Forests**
>
> Why should you care whether most of the world's remaining tropical forests are burned or cleared and converted to savanna within your lifetime? What are three things you could do to help slow the rate of this depletion and degradation of the earth's natural capital?

10-2 How Should We Manage and Sustain Forests?

▶ **CONCEPT 10-2** We can sustain forests by emphasizing the economic value of their ecological services, protecting old-growth forests, harvesting trees no faster than they are replenished, and using sustainable substitute resources.

We Can Manage Forests More Sustainably

Biodiversity researchers and a growing number of foresters have called for more sustainable forest management. Figure 10-17 lists ways to achieve this goal (**Concept 10-2**). Certification of sustainably grown timber and of sustainably produced forest products can help people consume forest products more sustainably (Science Focus, p. 228). **GREEN CAREER:** Sustainable forestry

SOLUTIONS

Sustainable Forestry

- Identify and protect forest areas high in biodiversity

- Rely more on selective cutting and strip cutting

- No clear-cutting on steep slopes

- No logging of old-growth forests

- Sharply reduce road building into uncut forest areas

- Leave most standing dead trees and fallen timber for wildlife habitat and nutrient recycling

- Plant tree plantations primarily on deforested and degraded land

- Certify timber grown by sustainable methods

- Include ecological services of forests in estimating their economic value

Figure 10-17 Ways to grow and harvest trees more sustainably (**Concept 10-2**). **Question:** Which three of these solutions do you think are the most important? Why?

We Can Improve the Management of Forest Fires

In the United States, the Smokey Bear educational campaign undertaken by the Forest Service and the National Advertising Council has prevented countless forest fires. It has also saved many lives and prevented billions of dollars in losses of trees, wildlife, and human structures.

At the same time, this educational program has convinced much of the public that all forest fires are bad and should be prevented or put out. Ecologists warn that trying to prevent all forest fires increases the likelihood of destructive crown fires (Figure 10-9, right) by allowing accumulation of highly flammable underbrush and smaller trees in some forests.

According to the U.S. Forest Service, severe fires could threaten 40% of all federal forest lands, mainly because of fuel buildup resulting from past rigorous fire protection programs (the Smokey Bear era), increased logging in the 1980s that left behind highly flammable logging debris (called *slash*), and greater public use of federal forest lands.

Ecologists and forest fire experts have proposed several strategies for reducing fire-related harm to forests and people. One approach is to set small, contained surface fires to remove flammable small trees

Certifying Sustainably Grown Timber

Collins Pine owns and manages a large area of productive timberland in the northeastern part of the U.S. state of California. Since 1940, the company has used selective cutting to help maintain the ecological and economic sustainability of its timberland.

Since 1993, Scientific Certification Systems (SCS) has evaluated the company's timber production. SCS, which is part of the nonprofit Forest Stewardship Council (FSC), was formed to develop a list of environmentally sound practices for use in certifying timber and products made from such timber.

Each year, SCS evaluates Collins Pine's landholdings and has consistently found that: their cutting of trees has not exceeded long-term forest regeneration; roads and harvesting systems have not caused unreasonable ecological damage; soils are not damaged; and downed wood (boles) and standing dead trees (snags) are left to provide wildlife habitat. As a result, SCS judges the company to be a good employer and a good steward of its land and water resources.

According to the FSC, between 1995 and 2007, the area of the world's forests in 76 countries that meets its international certification standards grew more than 16-fold. The countries with the largest areas of FSC-certified forests are, in order, Canada, Russia, Sweden, the United States, Poland, and Brazil. Despite this progress, by 2007, less than 10% of the world's forested area was certified. FSC also certifies 5,400 manufacturers and distributors of wood products.

Critical Thinking

Should governments provide tax breaks for sustainably grown timber to encourage this practice? Explain.

and underbrush in the highest-risk forest areas. Such *prescribed fires* require careful planning and monitoring to try to keep them from getting out of control. As an alternative to prescribed burns, local officials in populated parts of fire-prone California use herds of goats (kept in moveable pens) to eat away underbrush.

A second strategy is to allow many fires on public lands to burn, thereby removing flammable underbrush and smaller trees, as long as the fires do not threaten human structures and life. A third approach is to protect houses and other buildings in fire-prone areas by thinning a zone of about 60 meters (200 feet) around them and eliminating the use of flammable materials such as wooden roofs.

A fourth approach is to thin forest areas vulnerable to fire by clearing away small fire-prone trees and underbrush under careful environmental controls. Many forest fire scientists warn that such thinning should not involve removing economically valuable medium-size and large trees for two reasons. First, these are the most fire-resistant trees. Second, their removal encourages dense growth of more flammable young trees and underbrush and leaves behind highly flammable slash. Many of the worst fires in U.S. history—including some of those during the 1990s—burned through cleared forest areas containing slash. A 2006 study by U.S. Forest Service researchers found that thinning forests without using prescribed burning to remove accumulated brush and deadwood can greatly increase rather than decrease fire damage.

Despite such warnings from forest scientists, the U.S. Congress under lobbying pressure from timber companies passed the 2003 *Healthy Forests Restoration Act*. It allows timber companies to cut down economically valuable medium-size and large trees in 71% of the country's national forests in return for clearing away smaller, more fire-prone trees and underbrush.

However, the companies are not required to conduct prescribed burns after completing the thinning process.

This law also exempts most thinning projects from environmental reviews, which are currently required by forest protection laws in the national forests. According to many biologists and forest fire scientists, this law is likely to *increase* the chances of severe forest fires because it ignores the four strategies scientists have suggested for better management of forest fires. Critics of the Healthy Forests Restoration Act of 2003 say that healthier forests could be maintained at a much lower cost to taxpayers by giving communities in fire prone areas grants to implement these recommendations.

HOW WOULD YOU VOTE? ✓

Do you support repealing or modifying the Healthy Forests Restoration Act of 2003? Cast your vote online at **academic .cengage.com/biology/miller**.

We Can Reduce the Demand for Harvested Trees

One way to reduce the pressure on forest ecosystems is to improve the efficiency of wood use. According to the Worldwatch Institute and forestry analysts, *up to 60% of the wood consumed in the United States is wasted unnecessarily.* This results from inefficient use of construction materials, excess packaging, overuse of junk mail, inadequate paper recycling, and failure to reuse wooden shipping containers.

One reason for cutting trees is to provide pulp for making paper, but paper can be made out of fiber that does not come from trees. China uses rice straw and other agricultural residues to make much of its paper.

Figure 10-18 Solutions: pressure to cut trees to make paper could be greatly reduced by planting and harvesting a fast-growing plant known as kenaf. According to the USDA, kenaf is "the best option for tree-free papermaking in the United States" and could replace wood-based paper within 20–30 years. **Question:** Would you invest in a kenaf plantation? Explain.

Most of the small amount of tree-free paper produced in the United States is made from the fibers of a rapidly growing woody annual plant called *kenaf* (pronounced "kuh-NAHF"; Figure 10-18). Kenaf and other nontree fibers such as hemp yield more paper pulp per hectare than tree farms and require fewer pesticides and herbicides. It is estimated, that within 2 to 3 decades we could essentially eliminate the need to use trees to make paper. However, while timber companies successfully lobby for government subsidies to grow and harvest trees to make paper, there are no major lobbying efforts or subsidies for producing paper from kenaf or kudzu (Figure 9-15, p. 200).

■ **CASE STUDY**
Deforestation and the Fuelwood Crisis

Another major strain on forests, especially in tropical areas, is the practice of cutting of trees for fuelwood. About half of the wood harvested each year and three-fourths of that in developing countries is used for fuel.

Fuelwood and charcoal made from wood are used for heating and cooking by more than 2 billion people in developing countries (Figure 6-13, p. 135). As the demand for fuelwood in urban areas exceeds the sustainable yield of nearby forests, expanding rings of deforested land encircle such cities. By 2050, the demand for fuelwood could easily be 50% greater than the amount that can be sustainably supplied.

Haiti, a country with 9 million people, was once a tropical paradise covered largely with forests. Now it is an ecological disaster. Largely because its trees were cut for fuelwood, only about 2% of its land is forested. With the trees gone, soils have eroded away, making it much more difficult to grow crops. This unsustainable use of natural capital has led to a downward spiral of environmental degradation, poverty, disease, social injustice, crime, and violence. As a result, Haiti is classified as one of the world's leading *failing states* (Figure 17, p. S19, Supplement 3).

One way to reduce the severity of the fuelwood crisis in developing countries is to establish small plantations of fast-growing fuelwood trees and shrubs around farms and in community woodlots. Another approach to this problem is to burn wood more efficiently by providing villagers with cheap, more fuel-efficient, and less-polluting wood stoves, household biogas units that run on methane produced from crop and animal wastes, solar ovens, and electric hotplates powered by solar- or wind-generated electricity. This will also greatly reduce premature deaths from indoor air pollution caused by open fires and poorly designed stoves.

In addition, villagers can switch to burning the renewable sun-dried roots of various gourds and squash plants. Scientists are also looking for ways to produce charcoal for heating and cooking without cutting down trees. For example, Professor Amy Smith, of MIT in Cambridge, Massachusetts (USA), is developing a way to make charcoal from the fibers in a waste product called bagasse, which is left over from sugar cane processing in Haiti. Because sugarcane charcoal burns cleaner than wood charcoal, using it could help Haitians reduce indoor air pollution.

Countries such as South Korea, China, Nepal, and Senegal, have used such methods to reduce fuelwood shortages, sustain biodiversity through reforestation, and reduce soil erosion. Indeed, the mountainous country of South Korea is a global model for its successful reforestation following severe deforestation during the war between North and South Korea, which ended in 1953. Today, forests cover almost two-thirds of the country, and tree plantations near villages supply fuelwood on a sustainable basis. However, most countries suffering from fuelwood shortages are cutting trees for fuelwood and forest products 10–20 times faster than new trees are being planted. Shifting government subsidies from the building of logging roads to the planting of trees would help to increase forest cover worldwide.

INDIVIDUALS MATTER

Wangari Maathai and Kenya's Green Belt Movement

In the mid-1970s, Wangari Maathai (Figure 10-B) took stock of environmental conditions in her native Kenya. Tree-lined streams she had known as a child had dried up. Farms and plantations that were draining the watersheds and degrading the soil had replaced vast areas of forest. The Sahara Desert was encroaching from the north.

Something inside her told Maathai she had to do something about this degradation. Starting with a small tree nursery in her backyard, she founded the Green Belt Movement in 1977. The main goal of this highly regarded women's self-help group is to organize poor women in rural Kenya to plant and protect millions of trees in order to combat deforestation and provide fuelwood. By 2004, the 50,000 members of this grassroots group had established 6,000 village nurseries and planted and protected more than 30 million trees.

The women are paid a small amount for each seedling they plant that survives. This gives them an income to help break the cycle of poverty. It also improves the environment because trees reduce soil erosion and provide fruits, fuel, building materials, fodder for livestock, shade, and beauty. Having more trees also reduces the distances women and children have to walk to get fuelwood for cooking and heating. The success of this project has sparked the creation of similar programs in more than 30 other African countries.

Charlotte Thege/Peter Arnold, Inc.

In 2004, Maathai became the first African woman and the first environmentalist to be awarded the Nobel Peace Prize for her lifelong efforts. Within an hour of learning that she had won the prize, Maathai planted a tree, telling onlookers it was "the best way to celebrate." In her speech accepting the award, she said the purpose of the Green Belt program was to help people "make the connections between their own personal actions and the problems they witness in their environment and society." She urged everyone in the world to plant a tree as a symbol of commitment and hope.

Figure 10-B Wangari Maathai was the first Kenyan woman to earn a Ph.D. and to head an academic department at the University of Nairobi. In 1977, she organized the internationally acclaimed Green Belt Movement. For her work in protecting the environment, she has received many honors, including the Goldman Prize, the Right Livelihood Award, the U.N. Africa Prize for Leadership, and the 2004 Nobel Peace Prize. After years of being harassed, beaten, and jailed for opposing government policies, she was elected to Kenya's parliament as a member of the Green Party in 2002. In 2003, she was appointed Assistant Minister for Environment, Natural Resources, and Wildlife.

In 2006, she launched a project to plant a billion trees worldwide in 2007 to help fight poverty and climate change. The project greatly exceeded expectations with the planting of 2 billlion trees in 55 countries. In 2008, the UNEP set a goal of planting an additional 5 billion trees.

Wangari tells her story in her book *The Green Belt Movement: Sharing the Approach and the Experience,* published by Lantern Books in 2003.

This in turn would help to slow global warming, as more trees would remove more of the carbon dioxide that we are adding to the atmosphere.

Governments and Individuals Can Act to Reduce Tropical Deforestation

In addition to reducing fuelwood demand, analysts have suggested other ways to protect tropical forests and use them more sustainably. One way is to help new settlers in tropical forests to learn how to practice small-scale sustainable agriculture and forestry. Another is to harvest some of the renewable resources such as fruits and nuts in rain forests on a sustainable basis. And strip cutting (Figure 10-6c) can be used to harvest tropical trees for lumber.

In Africa's northern Congo Republic, some nomadic forest-dwelling pygmies go into the forests carrying hand-held satellite tracking devices in addition to their traditional spears and bows. They use these Global Positioning System (GPS) devices to identify their hunting grounds, burial grounds, water holes, sacred areas, and areas rich in medicinal plants. They then download such information on computers to provide a map of areas that need to be protected from logging, mining, and other destructive activities.

Debt-for-nature swaps can make it financially attractive for countries to protect their tropical forests. In such swaps, participating countries act as custodians of protected forest reserves in return for foreign aid or debt relief. In a similar strategy called *conservation concessions,* governments or private conservation organizations pay nations for concessions to preserve their natural resources.

Loggers can also use tropical forests more sustainably by using gentler methods for harvesting trees. For example, cutting canopy vines (lianas) before felling a tree and using the least obstructed paths to remove the logs can sharply reduce damage to neighboring trees. In addition, governments and individuals can mount efforts to reforest and rehabilitate degraded tropical forests and watersheds (see Individuals Matter, at left) and clamp down on illegal logging.

Finally, each of us as consumers can reduce the demand that fuels illegal and unsustainable logging in tropical forests. For building projects, we can use substitutes for wood such as bamboo and recycled plastic building materials (**Concept 10-2**). Recycled waste lumber is another alternative, now marketed by companies such as TerraMai and EcoTimber.

We can also buy only lumber and wood products that are certified as sustainably produced (Science Focus, p. 228). Growing awareness of tropical deforestation and the resulting consumer pressure caused the giant retail company Home Depot to take action. It reported in 2007 that 80% of the wood it carries meets such certification standards.

These and other ways to protect tropical forests are summarized in (Figure 10-19).

SOLUTIONS

Sustaining Tropical Forests

Prevention

Protect the most diverse and endangered areas

Educate settlers about sustainable agriculture and forestry

Subsidize only sustainable forest use

Protect forests with debt-for-nature swaps and conservation concessions

Certify sustainably grown timber

Reduce poverty

Slow population growth

Restoration

Encourage regrowth through secondary succession

Rehabilitate degraded areas

Concentrate farming and ranching in already-cleared areas

Figure 10-19 Ways to protect tropical forests and use them more sustainably (**Concept 10-2**). **Question:** Which three of these solutions do you think are the most important? Why?

10-3 How Should We Manage and Sustain Grasslands?

▶ **CONCEPT 10-3** We can sustain the productivity of grasslands by controlling the number and distribution of grazing livestock and by restoring degraded grasslands.

Some Rangelands Are Overgrazed

Grasslands provide many important ecological services, including soil formation, erosion control, nutrient cycling, storage of atmospheric carbon dioxide in biomass, and maintenance of biodiversity.

After forests, the ecosystems most widely used and altered by human activities are grasslands. **Rangelands** are unfenced grasslands in temperate and tropical climates that supply *forage,* or vegetation, for grazing (grass-eating) and browsing (shrub-eating) animals. Cattle, sheep, and goats graze on about 42% of the world's grassland. The 2005 Millennium Ecosystem Assessment estimated that continuing on our present course will increase that percentage to 70% by 2050. Livestock also graze in **pastures**—managed grasslands

or enclosed meadows usually planted with domesticated grasses or other forage.

Blades of rangeland grass grow from the base, not at the tip. So as long as only the upper half of the blade is eaten and its lower half remains, rangeland grass is a renewable resource that can be grazed again and again.

Moderate levels of grazing are healthy for grasslands, because removal of mature vegetation stimulates rapid regrowth and encourages greater plant diversity. The key is to prevent both overgrazing and undergrazing by domesticated livestock and wild herbivores. **Overgrazing** occurs when too many animals graze for too long and exceed the carrying capacity of a rangeland area (Figure 10-20, left, p. 232). It reduces grass cover, exposes the soil to erosion by water and wind, and compacts the soil (which diminishes its capacity

Figure 10-20 Natural capital degradation: overgrazed (left) and lightly grazed (right) rangeland.

to hold water). Overgrazing also enhances invasion by species such as sagebrush, mesquite, cactus, and cheatgrass, which cattle will not eat.

Scientists have also learned that, before settlers made them into rangeland, natural grassland ecosystems were maintained partially by periodic wildfires sparked by lightning. Fires were important because they burned away mesquite and other invasive shrubs, keeping the land open for grasses. Ecologists have studied grasslands of the Malpai Borderlands—an area on the border between the southwestern U.S. states of Arizona and New Mexico—where ranchers, with the help of the federal government, not only allowed overgrazing for more than a century, but also fought back fires and kept the grasslands from burning. Consequently, trees and shrubs replaced grasses, the soil was badly eroded, and the area lost most of its value for grazing.

Since 1993, ranchers, scientists, environmentalists, and government agencies have joined forces to restore the native grasses and animal species to the Malpai Borderlands. Land managers conduct periodic controlled burns on the grasslands, and the ecosystem has now been largely reestablished. What was once a classic example of unsustainable resource management became a valuable scientific learning experience and a management success story.

About 200 years ago, grass may have covered nearly half the land in the southwestern United States. Today, it covers only about 20%, mostly because of a combination of prolonged droughts and overgrazing, which created footholds for invader species that now cover many former grasslands.

Limited data from FAO surveys in various countries indicate that overgrazing by livestock has caused as much as a fifth of the world's rangeland to lose productivity. Some grasslands suffer from **undergrazing,** where absence of grazing for long periods (at least 5 years) can reduce the net primary productivity of grassland vegetation and grass cover.

We Can Manage Rangelands More Sustainably

The most widely used method for more sustainable management of rangeland is to control the number of grazing animals and the duration of their grazing in a given area so that the carrying capacity of the area is not exceeded (**Concept 10-3**). One way of doing this is *rotational grazing* in which cattle are confined by portable fencing to one area for a short time (often only 1–2 days) and then moved to a new location.

Livestock tend to aggregate around natural water sources, especially thin strips of lush vegetation along streams or rivers known as *riparian zones,* and around ponds established to provide water for livestock. Overgrazing by cattle can destroy the vegetation in such areas (Figure 10-21, left). Protecting overgrazed land from further grazing by moving livestock around and by fencing off these areas can eventually lead to its natural ecological restoration (Figure 10-21, right). Ranchers can also move cattle around by providing supplemental feed at selected sites and by strategically locating water holes and tanks and salt blocks.

A more expensive and less widely used method of rangeland management is to suppress the growth of unwanted invader plants by use of herbicides, mechanical

Figure 10-21 Natural capital restoration: in the mid-1980s, cattle had degraded the vegetation and soil on this stream bank along the San Pedro River in the U.S. state of Arizona (left). Within 10 years, the area was restored through natural regeneration after the banning of grazing and off-road vehicles (right) (**Concept 10-3**).

removal, or controlled burning. A cheaper way to discourage unwanted vegetation in some areas is through controlled, short-term trampling by large numbers of livestock.

Replanting barren areas with native grass seeds and applying fertilizer can increase growth of desirable vegetation and reduce soil erosion. But this is an expensive way to restore severely degraded rangeland. The better option is to prevent degradation by using the methods described above and in the following case study.

■ CASE STUDY

Grazing and Urban Development in the American West—Cows or Condos?

The landscape is changing in ranch country. Since 1980, millions of people have moved to parts of the southwestern United States, and a growing number of ranchers have sold their land to developers. Housing developments, condos, and small "ranchettes" are creeping out from the edges of many southwestern cities and towns. Most people moving to the southwestern states value the landscape for its scenery and recreational opportunities, but uncontrolled urban development can degrade these very qualities.

For decades some environmental scientists and environmentalists have sought to reduce overgrazing on these lands and, in particular, to reduce or eliminate livestock grazing permits on public lands. They have not had the support of ranchers or of the government. They have also pushed for decreased timber cutting and increased recreational opportunities in the national forests and grasslands. These efforts have made private tracts of land, especially near protected public lands, more desirable and valuable to people who enjoy outdoor activities and can afford to live in scenic areas.

Now, because of this population surge, ranchers, ecologists, and environmentalists are joining together to help preserve cattle ranches as the best hope for sustaining the key remaining grasslands and the habitats they provide for native species. They are working together to identify areas that are best for sustainable grazing, areas best for sustainable urban development, and areas that should be neither grazed nor developed. One strategy involves land trust groups, which pay ranchers for *conservation easements*—deed restrictions that bar future owners from developing the land. These groups are also pressuring local governments to zone the land in order to prevent large-scale development in ecologically fragile rangeland areas.

Some ranchers are also reducing the harmful environmental impacts of their herds. They rotate their cattle away from riparian areas (Figure 10-21), use far less fertilizer and pesticides, and consult with range and wildlife scientists about ways to make their ranch operations more economically and ecologically sustainable.

10-4 How Should We Manage and Sustain Parks and Nature Reserves?

▶ **CONCEPT 10-4** Sustaining biodiversity will require protecting much more of the earth's remaining undisturbed land area as parks and nature reserves.

National Parks Face Many Environmental Threats

Today, more than 1,100 major national parks are located in more than 120 countries (see Figure 7-12, top, p. 151; Figure 7-18, p. 157; Figure 7-19, p. 157; and Figure 8-8, p. 168). However, most of these national parks are too small to sustain a lot of large animal species. And many parks suffer from invasions by nonnative species that compete with and reduce the populations of native species and worsen ecological disruption.

Parks in developing countries possess the greatest biodiversity of all parks, but only about 1% of these parklands are protected. Local people in many of these countries enter the parks illegally in search of wood, cropland, game animals, and other natural products for their daily survival. Loggers and miners operate illegally in many of these parks, as do wildlife poachers who kill animals to obtain and sell items such as rhino horns, elephant tusks, and furs. Park services in most developing countries have too little money and too few personnel to fight these invasions, either by force or through education.

■ CASE STUDY

Stresses on U.S. Public Parks

The U.S. national park system, established in 1912, includes 58 major national parks, sometimes called the country's crown jewels. States, counties, and cities also operate public parks.

Popularity is one of the biggest problems for many parks. Between 1960 and 2007, the number of visitors to U.S. national parks more than tripled, reaching 273 million. The Great Smoky Mountains National Park in the states of Tennessee and North Carolina, the country's most frequently visited national park, hosts about 9 million visitors each year. Many state parks are located near urban areas and receive about twice as many visitors per year as do the national parks. Visitors often expect parks to have grocery stores, laundries, bars, and other such conveniences.

During the summer, users entering the most popular parks face long backups and experience noise, congestion, eroded trails, and stress instead of peaceful solitude. In some parks and other public lands, noisy and polluting dirt bikes, dune buggies, jet skis, snowmobiles, and other off-road vehicles degrade the aesthetic experience for many visitors, destroy or damage fragile vegetation (Figure 10-22), and disturb wildlife. There is controversy over whether these machines should be allowed in national parks.

Photo courtesy of Kevin Walker

Figure 10-22 Natural capital degradation: damage from off-road vehicles in a proposed wilderness area near the U.S. city of Moab, Utah. Such vehicles pollute the air, damage soils and vegetation, threaten wildlife, and degrade wetlands and streams.

THINKING ABOUT
National Parks and Off-Road Vehicles

Do you support allowing off-road vehicles in national parks? Explain. If you do, what restrictions, if any, would you put on their use?

Parks also suffer damage from the migration or deliberate introduction of nonnative species. European wild boars (imported to the state of North Carolina in 1912 for hunting) threaten vegetation in parts of the Great Smoky Mountains National Park. Nonnative mountain goats in Washington State's Olympic National Park trample native vegetation and accelerate soil erosion. Nonnative species of plants, insects, and worms entering the parks on vehicle tires and hikers' gear also degrade the biodiversity of parklands.

Effects of Reintroducing the Gray Wolf to Yellowstone National Park

For over a decade, wildlife ecologist Robert Crabtree and a number of other scientists have been studying the effects of reintroducing the gray wolf into the Yellowstone National Park (**Core Case Study**). They have put radio-collars on most of the wolves to gather data and track their movements. They have also studied changes in vegetation and the populations of various plant and animal species. Results of this research have suggested that the return of the gray wolf, a keystone predator species, has sent ecological ripples through the park's ecosystem.

Elk, the main herbivores in the Yellowstone system, are the primary food source for the wolves, but wolves also kill some moose, mule deer, and bison. Not surprisingly, elk populations have declined with the return of wolves. However, drought, grizzly bears (which kill elk calves), and a severe winter in 1997 have contributed to this decline. Leftovers of elk killed by wolves provide an important food source for grizzly bears

and other scavengers such as bald eagles and ravens.

Before the wolves returned, elk had been browsing on willow shoots and other vegetation near the banks of streams and rivers. With the return of wolves, the elk retreated to higher ground. This has spurred the regrowth of aspen, cottonwoods, and willow trees in these riparian areas and increased populations of riparian songbirds.

This regrowth of trees has in turn helped to stabilize and shade stream banks, which lowered the water temperature and made it better habitat for trout. Beavers seeking willow and aspen for food and dam construction have returned. The beaver dams established wetlands and created more favorable habitat for aspens.

The wolves have also cut in half the population of coyotes—the top predators in the absence of wolves. This has resulted in fewer coyote attacks on cattle and sheep on surrounding ranches and has increased populations of red fox and smaller animals such as

ground squirrels, mice, and gophers hunted by coyotes, eagles, and hawks.

Elk in the park are hunted in limited numbers, but wolves, as a protected species, are not hunted. However, wolves kill one another in clashes between packs, and a few have been killed by cars. The wolves also face threats from dogs that visitors bring to the park. The dogs carry parvovirus, which can kill wolf pups.

Wolves are an important factor in the Yellowstone ecosystem. But there are many other interacting factors involved in the structure and functioning of this complex ecosystem. Decades of research will be needed to unravel and better understand these interactions. For more information, see *The Habitable Planet*, Video 4, at **www.learner.org/resources/series209.html**.

Critical Thinking

Do you approve or disapprove of the reintroduction of the gray wolf into the Yellowstone National Park system? Explain.

At the same time, native species—some of them threatened or endangered—are killed or removed illegally in almost half of all U.S. national parks. This is what happened with the gray wolf until it was successfully reintroduced into Yellowstone National Park after a half century's absence (Science Focus, above). Not all park visitors understand the rules that protect species, and rangers have to spend an increasing amount of their time on law enforcement and crowd control instead of on conservation management and education.

Many U.S. national parks have become threatened islands of biodiversity surrounded by a sea of commercial development. Nearby human activities that threaten wildlife and recreational values in many national parks include mining, logging, livestock grazing, use of coal-burning power plants, oil drilling, water diversion, and urban development.

Polluted air, drifting hundreds of kilometers from cities, kills ancient trees in California's Sequoia National Park and often degrades the awesome views at Arizona's Grand Canyon. The Great Smoky Mountains, named for the natural haze emitted by their lush vegetation, ironically have air quality similar to that of Los Angeles, California, and vegetation on their highest peaks has been damaged by acid rain. According to the National Park Service, air pollution, mostly from coal-fired power plants and dense vehicle traffic, degrades scenic views in U.S. national parks more than 90% of the time.

Another problem, reported by the U.S. General Accounting Office, is that the national parks need at least $6 billion for long overdue repairs of trails, buildings, and other infrastructure. Some analysts say more of these funds could come from private concessionaires who provide campgrounds, restaurants, hotels, and other services for park visitors. They pay franchise fees averaging only about 6–7% of their gross receipts, and many large concessionaires with long-term contracts pay as little as 0.75%. Analysts say these percentages could reasonably be increased to around 20%.

Figure 10-23 (p. 236) lists other suggestions made by various analysts for sustaining and expanding the national park system in the United States.

Nature Reserves Occupy Only a Small Part of the Earth's Land

Most ecologists and conservation biologists believe the best way to preserve biodiversity is to create a worldwide network of protected areas. (See the chapter opening quote on p. 215.) Currently, only 12% of the earth's land area is protected strictly or partially in nature reserves, parks, wildlife refuges, wilderness, and other areas. This 12% figure is misleading because no more than 5% of the earth's land is strictly protected from potentially harmful human activities. In other words, *we have*

SOLUTIONS

National Parks

- Integrate plans for managing parks and nearby federal lands
- Add new parkland near threatened parks
- Buy private land inside parks
- Locate visitor parking outside parks and provide shuttle buses for people touring heavily used parks
- Increase federal funds for park maintenance and repairs
- Raise entry fees for visitors and use resulting funds for park management and maintenance
- Seek private donations for park maintenance and repairs
- Limit the number of visitors in crowded park areas
- Increase the number of park rangers and their pay
- Encourage volunteers to give visitor lectures and tours

Figure 10-23 Suggestions for sustaining and expanding the national park system in the United States. **Question:** Which two of these solutions do you think are the most important? Why? (Data from Wilderness Society and National Parks and Conservation Association).

reserved 95% of the earth's land for human use, and most of the remaining area consists of ice, tundra, or desert—places where most people do not want to live.

Conservation biologists call for full protection of at least 20% of the earth's land area in a global system of biodiversity reserves that would include multiple examples of all the earth's biomes (**Concept 10-4**). But powerful economic and political interests oppose this idea.

Protecting more of the earth's land from unsustainable use will require action and funding by national governments and private groups, bottom-up political pressure by concerned individuals, and cooperative ventures involving governments, businesses, and private conservation organizations. Such groups play an important role in establishing wildlife refuges and other reserves to protect biological diversity.

For example, since its founding by a group of professional ecologists in 1951, The Nature Conservancy—with more than 1 million members worldwide—has created the world's largest system of private natural areas and wildlife sanctuaries in 30 countries. In the United States, efforts by The Nature Conservancy and private landowners have protected land, waterways, and wetlands in local and state trusts totaling roughly the area of the U.S. state of Georgia.

Eco-philanthropists are using some of their wealth to buy up wilderness areas in South America, and they are donating the preserved land to the governments of various countries. For example, Douglas and Kris Tompkins have created 11 wilderness parks in Latin America. In 2005, they donated two new national parks to Chile and Argentina.

In the United States, private, nonprofit *land trust groups* have protected large areas of land. Members pool their financial resources and accept tax-deductible donations to buy and protect farmlands, grasslands, woodlands, and urban green spaces.

Some governments are also making progress. By 2007, the Brazilian government had officially protected 23% of the Amazon—an area the size of France—from development. However, many of these areas are protected only on paper and are not always secure from illegal resource removal and degradation.

Most developers and resource extractors oppose protecting even the current 12% of the earth's remaining undisturbed ecosystems. They contend that these areas might contain valuable resources that would add to economic growth. Ecologists and conservation biologists disagree. They view protected areas as islands of biodiversity and natural capital that help to sustain all life and economies and serve as centers of future evolution. See Norman Myer's Guest Essay on this topic at CengageNOW.

HOW WOULD YOU VOTE?

Should at least 20% of the earth's land area be strictly protected from economic development? Cast your vote online at **academic.cengage.com/biology/miller**.

Designing and Connecting Nature Reserves

Large reserves sustain more species and provide greater habitat diversity than do small reserves. They also minimize exposure to natural disturbances (such as fires and hurricanes), invading species, and human disturbances from nearby developed areas.

In 2007, scientists reported on the world's largest and longest running study of forest fragmentation, which took place in the Amazon. They found that conservation of large reserves in the Amazon was even more important than was previously thought. Because the Amazon rain forest is so diverse, a large expanse of it may contain dozens of ecosystem types, each of which is different enough from the others to support unique species. Therefore, developing just a part of such a large area could result in the elimination of many types of habitats and species.

However, research indicates that in other locales, several well-placed, medium-sized reserves may better protect a wider variety of habitats and preserve more biodiversity than would a single large reserve of the same total area. When deciding on whether to recommend large- or medium-sized reserves in a particular area, conservation biologists must carefully consider its various ecosystems.

Whenever possible, conservation biologists call for using the *buffer zone concept* to design and manage nature

Biosphere Reserve

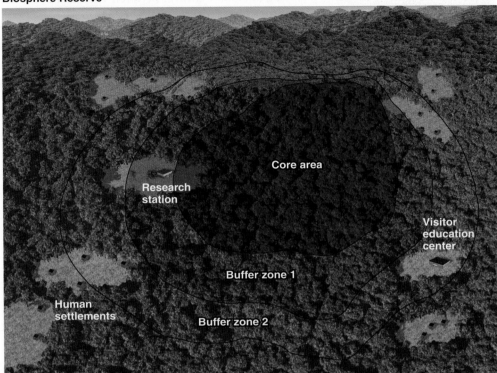

Research
station

Core area

Visitor
education
center

Buffer zone 1

Human
settlements

Buffer zone 2

Figure 10-24 Solutions: a model *biosphere reserve.* Each reserve contains a protected inner core surrounded by two buffer zones that local and indigenous people can use for sustainable logging, growing limited crops, grazing cattle, hunting, fishing, and ecotourism. **Question:** Do you think some of these reserves should be free of all human activity, including ecotourism? Why or why not?

reserves. This means protecting an inner core of a reserve by usually establishing two buffer zones in which local people can extract resources sustainably without harming the inner core. Instead of shutting people out of the protected areas and likely creating enemies, this approach enlists local people as partners in protecting a reserve from unsustainable uses such as illegal logging and poaching.

The United Nations has used this principle in creating its global network of 529 biosphere reserves in 105 countries (Figure 10-24). According to Craig Leisher, an economist for The Nature Conservancy, "Local people are often the best people in developing countries to manage these conservation areas, because they want them to survive in the long term as well."

So far, most biosphere reserves fall short of these ideals and receive too little funding for their protection and management. An international fund to help make up the shortfall would cost about $100 million per year—about the amount spent every 90 minutes on weapons by the world's nations.

Establishing protected *habitat corridors* between isolated reserves helps to support more species and allows migration by vertebrates that need large ranges. Corridors also permit migration of individuals and populations when environmental conditions in a reserve deteriorate, forcing animals to move to a new location, and they support animals that must make seasonal migrations to obtain food. Corridors may also enable some species to shift their ranges if global climate change makes their current ranges uninhabitable.

On the other hand, corridors can threaten isolated populations by allowing movement of pest species, disease, fire, and invasive species between reserves. They also increase exposure of migrating species to natural predators, human hunters, and pollution. In addition, corridors can be costly to acquire, protect, and manage. Nevertheless, an extensive study, reported in 2006, showed that areas connected by corridors host a greater variety of birds, insects, small mammals, and plant species. And in that study, nonnative species did not invade the connected areas.

The creation of large reserves connected by corridors on an eco-regional scale is the grand goal of many conservation biologists. This idea is being put into practice in places such as Costa Rica (see the following Case Study).

RESEARCH FRONTIER

Learning how to design, locate, connect, and manage networks of effective nature preserves. See **academic.cengage.com/biology/miller**.

■ **CASE STUDY**
Costa Rica—A Global Conservation Leader

Tropical forests once completely covered Central America's Costa Rica, which is smaller in area than the U.S. state of West Virginia and about one-tenth the size of France. Between 1963 and 1983, politically powerful ranching families cleared much of the country's forests to graze cattle.

Despite such widespread forest loss, tiny Costa Rica is a superpower of biodiversity, with an estimated 500,000 plant and animal species. A single park in Costa Rica is home to more bird species than are found in all of North America.

In the mid-1970s, Costa Rica established a system of nature reserves and national parks that, by 2006, included about a quarter of its land—6% of it reserved for indigenous peoples. Costa Rica now devotes a larger proportion of its land to biodiversity conservation than does any other country.

The country's parks and reserves are consolidated into eight zoned *megareserves* (Figure 10-25). Each reserve contains a protected inner core surrounded by two buffer zones that local and indigenous people can use for sustainable logging, crop farming, cattle grazing, hunting, fishing, and ecotourism.

Costa Rica's biodiversity conservation strategy has paid off. Today, the country's largest source of income is its $1-billion-a-year tourism business, almost two-thirds of which involves ecotourism.

To reduce deforestation, the government has eliminated subsidies for converting forest to rangeland. It also pays landowners to maintain or restore tree coverage. The goal is to make it profitable to sustain forests. Between 2007 and 2008, the government planted nearly 14 million trees, which helps to preserve the country's biodiversity. As they grow, the trees also remove carbon dioxide from the air and help the country to meet its goal of reducing net CO_2 emissions to zero by 2021.

The strategy has worked: Costa Rica has gone from having one of the world's highest deforestation rates to having one of the lowest. Between 1986 and 2006, the country's forest cover grew from 26% to 51%.

Protecting Wilderness Is an Important Way to Preserve Biodiversity

One way to protect undeveloped lands from human exploitation is by legally setting them aside as large areas of undeveloped land called **wilderness** (**Concept 10-4**). Theodore Roosevelt (Figure 4, p. S34, in Supplement 5), the first U.S. president to set aside protected areas, summarized what we should do with wilderness: "Leave it as it is. You cannot improve it."

Wilderness protection is not without controversy (see the following Case Study). Some critics oppose protecting large areas for their scenic and recreational value for a relatively small number of people. They believe this is an outmoded ideal that keeps some areas of the planet from being economically useful to people here today. But to most biologists, the most important reasons for protecting wilderness and other areas from exploitation and degradation involve long-term needs. One such need is to *preserve biodiversity* as a vital part of the earth's natural capital. Another is to protect wilderness areas *as centers for evolution* in response to mostly unpredictable changes in environmental conditions. In other words, wilderness serves as a biodiversity bank and an eco-insurance policy.

■ CASE STUDY

Controversy over Wilderness Protection in the United States

In the United States, conservationists have been trying to save wild areas from development since 1900. Overall, they have fought a losing battle. Not until 1964 did Congress pass the Wilderness Act (Figure 6, p. S35, in Supplement 5). It allowed the government to protect undeveloped tracts of public land from development as part of the National Wilderness Preservation System.

The area of protected wilderness in the United States increased tenfold between 1970 and 2007. Even so, only about 4.6% of U.S. land is protected as wilderness—almost three-fourths of it in Alaska. Only 1.8% of the land area of the lower 48 states is protected, most of it in the West. In other words, Americans have reserved at least 98% of the continental United States to be used as they see fit and have protected less than 2% as wilderness. According to a 1999 study by the World Conservation Union, the United States ranks 42nd among nations in terms of terrestrial area protected as wilderness, and Canada is in 36th place.

In addition, only 4 of the 413 wilderness areas in the lower 48 states are large enough to sustain the species they contain. The system includes only 81 of the country's 233 distinct ecosystems. Most wilderness ar-

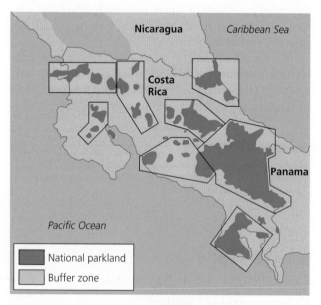

Figure 10-25 Solutions: Costa Rica has consolidated its parks and reserves into eight zoned *megareserves* designed to sustain about 80% of the country's rich biodiversity. Green areas are protected reserves and yellow areas are nearby buffer zones, which can be used for sustainable forms of forestry, agriculture, hydropower, hunting, and other human activities.

eas in the lower 48 states are threatened habitat islands in a sea of development.

Scattered blocks of public lands with a total area roughly equal to that of the U.S. state of Montana could qualify for designation as wilderness. About 60% of such land in the national forests. For more than 20 years, these areas were temporarily protected under the Roadless Rule—a federal regulation that put undeveloped areas of national forests off-limits to road building and logging while they were evaluated for wilderness protection.

For decades, politically powerful oil, gas, mining, and timber industries have sought entry to these areas—which are owned jointly by all citizens of the United States—to develop resources there. Their efforts paid off in 2005 when the secretary of the interior ended protection of roadless areas within the national forest system that were being considered for classification as wilderness. The secretary also began allowing states to classify old cow paths and off-road vehicle trails as roads (Figure 10-22), which would disqualify their surrounding areas from protection as wilderness.

THINKING ABOUT
Protecting Wolves and Wild Lands

CORE CASE STUDY

How do you think protecting wolves, in part by reintroducing them to areas such as Yellowstone National Park (**Core Case Study**), helps to protect the forest areas where they live?

10-5 What Is the Ecosystem Approach to Sustaining Biodiversity?

▶ **CONCEPT 10-5A** We can help to sustain biodiversity by identifying severely threatened areas and protecting those with high plant diversity (biodiversity hotspots) and those where ecosystem services are being impaired.

▶ **CONCEPT 10-5B** Sustaining biodiversity will require a global effort to rehabilitate and restore damaged ecosystems.

▶ **CONCEPT 10-5C** Humans dominate most of the earth's land, and preserving biodiversity will require sharing as much of it as possible with other species.

We Can Use a Four-Point Strategy to Protect Ecosystems

Most biologists and wildlife conservationists believe that we must focus more on protecting and sustaining ecosystems, and the biodiversity contained within them, than on saving individual species. Their goals certainly include preventing premature extinction of species, but they argue the best way to do that is to protect threatened habitats and ecosystem services. This *ecosystems approach* generally would employ the following four-point plan:

- Map global ecosystems and create an inventory of the species contained in each of them and the ecosystem services they provide.

- Locate and protect the most endangered ecosystems and species, with emphasis on protecting plant biodiversity and ecosystem services.

- Seek to restore as many degraded ecosystems as possible.

- Make development *biodiversity-friendly* by providing significant financial incentives (such as tax breaks and write-offs) and technical help to private landowners who agree to help protect endangered ecosystems.

Some scientists have argued that we need new laws to embody this strategy. In the United States, for example, there is support for amending the Endangered Species Act, or possibly even replacing it with a new law focused on protection of ecosystems and biodiversity.

Protecting Global Biodiversity Hotspots Is an Urgent Priority

In reality, few countries are physically, politically, or financially able to set aside and protect large biodiversity reserves. To protect as much of the earth's remaining biodiversity as possible, some conservation biologists urge adoption of an *emergency action* strategy to identify and quickly protect **biodiversity hotspots** (**Concept 10-5A**)—an idea first proposed in 1988 by environmental scientist Norman Myers. (See his Guest Essay on this topic at CengageNOW.) These "ecological arks" are areas especially rich in plant species that are found nowhere else and are in great danger of extinction. These areas suffer serious ecological disruption, mostly because of rapid human population growth and the resulting pressure on natural resources. (See Case Study p. 240.) Myers and his colleagues at Conservation International relied primarily on the diversity of plant species to identify biodiversity hotspot areas because data on plant diversity was more readily available and was also thought to be an indicator of animal diversity.

Figure 10-26 shows 34 global terrestrial biodiversity hotspots identified by conservation biologists and Figure 10-27 shows major biodiversity hotspots in the United States. In the 34 global areas, a total of 86% of the habitat has been destroyed. They cover only a little more than 2% of the earth's land surface, but they contain an estimated 50% of the world's flowering plant species and 42% of all terrestrial vertebrates (mammals, birds, reptiles, and amphibians). They are also home for a large majority of the world's endangered or critically endangered species. Says Norman Myers, "I can think of no other biodiversity initiative that could achieve so much at a comparatively small cost, as the hotspots strategy."

One drawback of the biodiversity hotspots approach is that some areas rich in plant diversity are not necessarily rich in animal diversity. And when hotspots are protected, local people can be displaced and lose access to important resources. However, the goal of this approach—to protect the unique biodiversity in areas under great stress from human activities—remains urgent. Despite its importance, this approach has not succeeded in capturing sufficient public support and funding.

■ CASE STUDY
A Biodiversity Hotspot in East Africa

The forests covering the flanks of the Eastern Arc Mountains of the African nation of Tanzania contain the highest concentration of endangered animals on earth.

Plants and animals that exist nowhere else (species *endemic* to this area) live in these mountainside forests in considerable numbers. They include 96 species of vertebrates—10 mammal, 19 bird, 29 reptile, and 38 amphibian—43 species of butterflies, and at least 800 endemic species of plants, including most species of African violets.

An international network of scientists, who had extensively surveyed these mountain forests, reported these findings in 2007. They also reported newly discovered species in these forests, including a tree-dwelling monkey called the Kipunji and some surprisingly large reptiles and amphibians.

This area is a major biodiversity hotspot because humans now threaten to do what the ice ages could not do—kill off its forests. Farmers and loggers have cleared 70% of the ancient forests. This loss of habitat, along with hunting, has killed off many species, including elephants and buffalo, and now 71 of the 96 endemic species are threatened, 8 of them critically, with biological extinction.

These species are now forced to survive within 13 patches of forest that total an area about the size of the U.S. state of Rhode Island. Most of these forests are contained within 150 government reserves. New settlements are not allowed, but people still forage in these reserves for fuelwood and building materials, severely degrading some of the forests. Fire is also a threat, because the shrinking, degraded patches of forest are drying out, and this is likely to get worse as global warming takes hold.

CENGAGENOW™ Learn more about biodiversity hotspots around the world, what is at stake there, and how they are threatened at CengageNOW.

RESEARCH FRONTIER

Identifying and preserving all of the world's terrestrial and aquatic biodiversity hotspots. See **academic.cengage.com/ biology/miller.**

Protecting Ecosystem Services Is Also an Urgent Priority

Another way to help sustain the earth's biodiversity and its people is to identify and protect areas where vital *ecosystem services* (orange items in Figure 1-3, p. 8) are being impaired enough to reduce biodiversity or harm local residents. This approach has gotten more attention since the release in 2005 of the U.N. *Millennium Ecosystem Assessment*—a 4-year study by 1,360 experts from 95 countries. It identified key ecosystem services that provide numerous ecological and economic benefits. (Those provided by forests are summarized in Figure 10-4.) The study pointed out that human activities are degrading or overusing about 62% of the earth's natural services in various ecosystems around the world, and it outlined ways to help sustain these vital ecosystem services for human and nonhuman life.

This approach recognizes that most of the world's ecosystems are already dominated or influenced by human activities and that such pressures are increasing as population, urbanization, and resource use increase and the human ecological footprint increases (Figure 1-10, p. 15, and Figure 3, p. S24–S25, in Supplement 4). Proponents of this approach recognize that it is vital to set aside and protect reserves and wilderness areas and to protect highly endangered biodiversity hotspots (Figure 10-26). But they contend that such efforts by themselves will not significantly slow the steady erosion of the earth's biodiversity and ecosystem services.

These analysts argue that we must also identify highly stressed *life raft ecosystems*. In such areas, people live in severe poverty, and a large part of the economy depends on various ecosystem services that are being degraded severely enough to threaten the well-being of people and other forms of life. In these areas, residents, public officials, and conservation scientists are urged to work together to develop strategies to protect both biodiversity and human communities. Instead of emphasizing

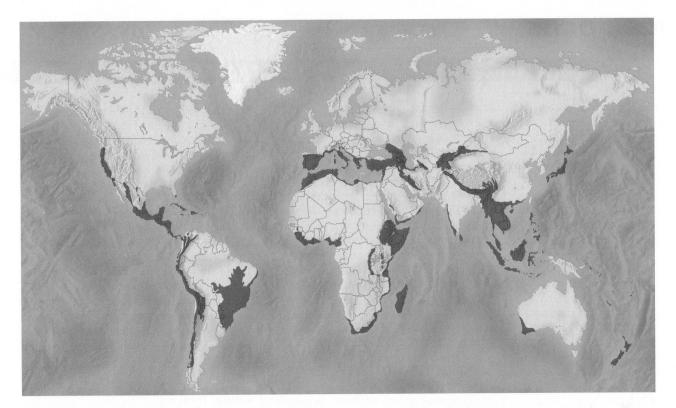

CENGAGENOW™ **Active Figure 10-26 Endangered natural capital:** 34 biodiversity hotspots identified by ecologists as important and endangered centers of terrestrial biodiversity that contain a large number of species found nowhere else. Identifying and saving these critical habitats requires a vital emergency response (**Concept 10-5A**). Compare these areas with those on the map of the human ecological footprint in the world as shown in Figure 3, pp. S24–S25, in Supplement 4. According to the IUCN, the average proportion of biodiversity hotspot areas truly protected with funding and enforcement is only 5%. *See an animation based on this figure at CengageNOW.*
Questions: Are any of these hotspots near where you live? Is there a smaller, localized hotspot in the area where you live? (Data from Center for Applied Biodiversity Science at Conservation International).

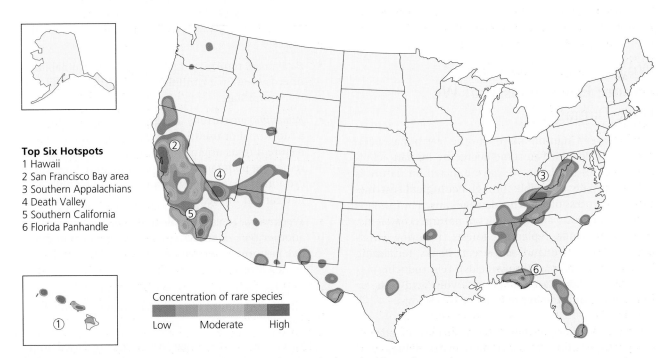

Top Six Hotspots
1 Hawaii
2 San Francisco Bay area
3 Southern Appalachians
4 Death Valley
5 Southern California
6 Florida Panhandle

Concentration of rare species

Low   Moderate   High

Figure 10-27 Endangered natural capital: biodiversity hotspots in the United States that need emergency protection. The shaded areas contain the largest concentrations of rare and potentially endangered species. Compare these areas with those on the map of the human ecological footprint in North America shown in Figure 7, pp. S28–S29, in Supplement 4. **Question:** Do you think that hotspots near urban areas would be harder to protect than those in rural areas? Explain. (Data from State Natural Heritage Programs, The Nature Conservancy, and Association for Biodiversity Information)

SCIENCE FOCUS

Ecological Restoration of a Tropical Dry Forest in Costa Rica

Costa Rica is the site of one of the world's largest *ecological restoration* projects. In the lowlands of its Guanacaste National Park (Figure 10-25), a small tropical dry forest was burned, degraded, and fragmented by large-scale conversion to cattle ranches and farms. Now it is being restored and relinked to the rain forest on adjacent mountain slopes. The goal is to eliminate damaging nonnative grasses and re-establish a tropical dry forest ecosystem over the next 100–300 years.

Daniel Janzen, professor of biology at the University of Pennsylvania and a leader in the field of restoration ecology, helped galvanize international support for this restoration project. He used his own MacArthur grant money to purchase this Costa Rican land to be set aside as a national park. He also raised more than $10 million for restoring the park.

Janzen realized that the original forests had been maintained partly by large native animals that ate the fruit of the Guanacaste tree and spread its seeds in their droppings.

But these animals disappeared about 10,000 years ago. About 500 years ago, horses and cattle introduced by Europeans also spread the seeds, but farming and ranching took their toll on the forest's trees. Janzen decided to speed up restoration of this tropical dry forest by incorporating limited numbers of horses and cattle as seed dispersers in his recovery plan.

Janzen recognizes that ecological restoration and protection of the park will fail unless the people in the surrounding area believe they will benefit from such efforts. His vision is to see the nearly 40,000 people who live near the park play an essential role in the restoration of the degraded forest, a concept he calls *biocultural restoration*.

By actively participating in the project, local residents reap educational, economic, and environmental benefits. Local farmers make money by sowing large areas with tree seeds and planting seedlings started in Janzen's lab. Local grade school, high school, and university students and citizens' groups study the park's ecology during field trips. The park's location near the Pan American Highway makes it an ideal area for ecotourism, which stimulates the local economy.

The project also serves as a training ground in tropical forest restoration for scientists from all over the world. Research scientists working on the project give guest classroom lectures and lead field trips.

In a few decades, today's children will be running the park and the local political system. If they understand the ecological importance of their local environment, they will be more likely to protect and sustain its biological resources. Janzen believes that education, awareness, and involvement—not guards and fences—are the best ways to restore degraded ecosystems and to protect largely intact ecosystems from unsustainable use.

Critical Thinking

Would such an ecological restoration project be possible in the area where you live? Explain.

---

nature-versus-people, this approach focuses on finding win–win ways to protect both people and the ecosystem services that support all life and economies.

We Can Rehabilitate and Restore Ecosystems That We Have Damaged

Almost every natural place on the earth has been affected or degraded to some degree by human activities. Much of the harm we have inflicted on nature is at least partially reversible through **ecological restoration:** the process of repairing damage caused by humans to the biodiversity and dynamics of natural ecosystems. Examples include replanting forests, restoring grasslands, restoring wetlands and stream banks, reclaiming urban industrial areas (brownfields), reintroducing native species (**Core Case Study**), removing invasive species, and freeing river flows by removing dams.

CORE CASE STUDY

Evidence indicates that in order to sustain biodiversity, we must make a global effort to rehabilitate and restore ecosystems we have damaged (**Concept 10-5B**). An important strategy is to mimic nature and natural processes and let nature do most of the work, usually through secondary ecological succession (Figure 5-17, p. 117).

By studying how natural ecosystems recover, scientists are learning how to speed up repair operations using a variety of approaches. They include the following measures:

- *Restoration:* returning a particular degraded habitat or ecosystem to a condition as similar as possible to its natural state.

- *Rehabilitation:* turning a degraded ecosystem into a functional or useful ecosystem without trying to restore it to its original condition. Examples include removing pollutants and replanting to reduce soil erosion in abandoned mining sites and landfills and in clear-cut forests.

- *Replacement:* replacing a degraded ecosystem with another type of ecosystem. For example, a productive pasture or tree plantation may replace a degraded forest.

- *Creating artificial ecosystems:* for example, creating artificial wetlands to help reduce flooding or to treat sewage.

Researchers have suggested a science-based four-point strategy for carrying out most forms of ecological restoration and rehabilitation.

- Identify what caused the degradation (such as pollution, farming, overgrazing, mining, or invasive species).

Figure 10-28 Solutions: *Curtis Prairie*, in the University of Wisconsin's arboretum, Madison, Wisconsin (USA), was restored from abandoned farm fields. It serves as a highly instructive example of successful prairie restoration and is studied by restoration ecology students from around the world. The inset photo, taken in about 1934, shows part of the process of restoration: a controlled burn to prepare the land for establishment of prairie plants. These researchers have just burned a part of the vegetation to simulate a prairie fire, which is an important natural event in the prairie ecosystem. The second person from the left in this photo is the pioneering conservation biologist Aldo Leopold (Individuals Matter, p. 22).

- Stop the abuse by eliminating or sharply reducing these factors. This would include removing toxic soil pollutants, adding nutrients to depleted soil, adding new topsoil, preventing fires, and controlling or eliminating disruptive nonnative species (Science Focus, left).

- If necessary, reintroduce species—especially pioneer, keystone, and foundation species—to help restore natural ecological processes, as was done with wolves in the Yellowstone ecosystem (**Core Case Study**).

- Protect the area from further degradation (Figure 10-21, right).

Most of the tall-grass prairies in the United States have been plowed up and converted to crop fields. However, these prairies are ideal subjects for ecological restoration for three reasons. *First,* many residual or transplanted native plant species can be established within a few years. *Second,* the technology involved is similar to that of gardening and agriculture. *Third,* the process is well suited for volunteer labor needed to plant native species and weed out invading species until the natural species can take over. There are a number of prairie restoration projects in the United States, a prime example of which is Curtis Prairie in the U. S. state of Wisconsin (Figure 10-28).

RESEARCH FRONTIER

Exploring ways to improve ecological restoration efforts. See **academic.cengage.com/biology/miller**.

Will Restoration Encourage Further Destruction?

Some analysts worry that ecological restoration could encourage continuing environmental destruction and degradation by suggesting that any ecological harm we do can be undone. Restoration ecologists disagree with that suggestion. They point out that preventing ecosystem damage in the first place is cheaper and more effective than any form of ecological restoration. But they agree that restoration should not be used as an excuse for environmental destruction.

Restoration ecologists note that so far, we have been able to protect or preserve only about 5% of the earth's land from the effects of human activities, so ecological restoration is badly needed for many of the world's ecosystems. Even if a restored ecosystem differs from the original system, they argue, the result is better than no restoration at all. In time, further experience with ecological restoration will improve its effectiveness. Chapter 12 describes examples of the ecological restoration of aquatic systems such as wetlands and rivers.

┌─ **HOW WOULD YOU VOTE?** ─┐

Should we mount a massive effort to restore the ecosystems
we have degraded, even though this will be quite costly?
Cast your vote online at **academic.cengage.com
/biology/miller**.

We Can Share Areas We Dominate with Other Species

In 2003, ecologist Michael L. Rosenzweig wrote a book entitled *Win–Win Ecology: How Earth's Species Can Survive in the Midst of Human Enterprise.* Rosenzweig strongly supports proposals to help sustain the earth's biodiversity through species protection strategies such as the U.S. Endangered Species Act (Case Study, p. 207).

But Rosenzweig contends that, in the long run, these approaches will fail for two reasons. *First,* fully protected reserves currently are devoted to saving only about 5% of the world's terrestrial area, excluding polar and other uninhabitable areas. To Rosenzweig, the real challenge is to sustain wild species in more of the human-dominated portion of nature that makes up 95% of the planet's terrestrial area (**Concept 10-5C**).

Second, Rosenzweig says, setting aside funds and refuges and passing laws to protect endangered and threatened species are essentially desperate attempts to save species that are in deep trouble. These emergency efforts can help a few species, but it is equally important to learn how to keep more species away from the brink of extinction. This is a prevention approach.

Rosenzweig suggests that we develop a new form of conservation biology, called **reconciliation** or **applied ecology.** This science focuses on inventing, establishing, and maintaining new habitats to conserve species diversity in places where people live, work, or play. In other words, we need to learn how to share with other species some of the spaces we dominate.

Implementing reconciliation ecology will involve the growing practice of *community-based conservation,* in which conservation biologists work with people to help them protect biodiversity in their local communities. With this approach, scientists, citizens, and sometimes national and international conservation organizations seek ways to preserve local biodiversity while allowing people who live in or near protected areas to make sustainable use of some of the resources there (Case Study, right).

For example, people learn how protecting local wildlife and ecosystems can provide economic resources for their communities by encouraging sustainable forms of ecotourism. In the Central American country of Belize, conservation biologist Robert Horwich has helped to establish a local sanctuary for the black howler monkey. He convinced local farmers to set aside strips of forest to serve as habitats and corridors through which these monkeys can travel. The reserve, run by a local women's cooperative, has attracted ecotourists and biologists. The community has built a black howler museum,

and local residents receive income by housing and guiding visiting ecotourists and biological researchers.

In other parts of the world, people are learning how to protect vital insect pollinators, such as native butterflies and bees, which are vulnerable to insecticides and habitat loss. Neighborhoods and municipal governments are doing this by agreeing to reduce or eliminate the use of pesticides on their lawns, fields, golf courses, and parks. Neighbors also work together in planting gardens of flowering plants as a source of food for pollinating insect species. And neighborhoods and farmers build devices using wood and plastic straws, which serve as hives for increasingly threatened pollinating bees.

People have also worked together to help protect bluebirds within human-dominated habitats where most of the bluebirds' nesting trees have been cut and the bluebird populations have declined. Special boxes were designed to accommodate nesting bluebirds, and the North American Bluebird Society has encouraged Canadians and Americans to use these boxes on their properties and to keep house cats away from nesting bluebirds. Now bluebird numbers are growing again.

In Berlin, Germany, people have planted gardens on many large rooftops. These gardens support a variety of wild species by containing varying depths and types of soil and exposures to sunlight. Such roofs also save energy by providing insulation and absorbing less heat than conventional rooftops do, thereby helping to keep cities cooler. They also conserve water by reducing evapotranspiration. Some reconciliation ecology proponents call for a global campaign to use the roofs of the world to help sustain biodiversity. **GREEN CAREER:** Rooftop garden designer

In the U.S. state of California, San Francisco's Golden Gate Park is a large oasis of gardens and trees in the midst of a major city. It is a good example of reconciliation ecology, because it was designed and planted by people who transformed it from a system of sand dunes. There are many other examples of individuals and groups working together on projects to restore grasslands, wetlands, streams, and other degraded areas rain forest Case Study below). **GREEN CAREER:** Reconciliation ecology specialist

■ CASE STUDY

The Blackfoot Challenge— Reconciliation Ecology in Action

The Blackfoot River flows among beautiful mountain ranges in the west central part of the U.S. state of Montana. This large watershed is home to more than 600 species of plants, 21 species of waterfowl, bald eagles, peregrine falcons, grizzly bears, and rare species of trout. Some species, such as the Howell's gumweed and the bull trout, are threatened with extinction.

The Blackfoot River Valley is also home to people who live in seven communities and 2,500 rural households. A book and movie, both entitled *A River Runs*

Through It, tell of how residents of the valley cherish their lifestyles.

In the 1970s, many of these people recognized that their beloved valley was threatened by poor mining, logging, and grazing practices, water and air pollution, and unsustainable commercial and residential development. They also understood that their way of life depended on wildlife and wild ecosystems located on private and public lands. They began meeting informally over kitchen tables to discuss how to maintain their way of life while sustaining the other species living in the valley. These small gatherings spawned community meetings attended by individual and corporate landowners, state and federal land managers, scientists, and local government officials.

Out of these meetings came action. Teams of residents organized weed-pulling parties, built nesting structures for waterfowl, and developed sustainable grazing systems. Landowners agreed to create perpetual conservation easements, setting land aside for only conservation and sustainable uses such as hunting and fishing. They also created corridors between large tracts of undeveloped land. In 1993, these efforts were organized under a charter called the Blackfoot Challenge.

The results were dramatic. Blackfoot Challenge members have restored and enhanced large areas of wetlands, streams, and native grasslands. They have reserved large tracts of private land under perpetual conservation easements.

These pioneers might not have known it, but they were initiating what has become a classic example of *reconciliation ecology.* They worked together, respected each other's views, accepted compromises, and found ways to share their land with the area's plants and animals. They understood that all sustainability is local.

THINKING ABOUT
Wolves and Reconciliation Ecology

What are some ways in which the wolf restoration project in Yellowstone National Park (**Core Case Study**) is similar to some reconciliation ecology examples described above?

RESEARCH FRONTIER

Determining where and how reconciliation ecology can work best. See **academic.cengage.com/biology/miller**.

Figure 10-29 lists some ways in which you can help sustain the earth's terrestrial biodiversity.

WHAT CAN YOU DO?

Sustaining Terrestrial Biodiversity

- Adopt a forest
- Plant trees and take care of them
- Recycle paper and buy recycled paper products
- Buy sustainably produced wood and wood products
- Choose wood substitutes such as bamboo furniture and recycled plastic outdoor furniture, decking, and fencing
- Help to restore a nearby degraded forest or grassland
- Landscape your yard with a diversity of plants natural to the area

Figure 10-29 Individuals Matter: ways to help sustain terrestrial biodiversity. **Questions:** Which two of these actions do you think are the most important? Why? Which of these things do you already do?

REVISITING — Yellowstone Wolves and Sustainability

In this chapter, we looked at how terrestrial biodiversity is being destroyed or degraded. We also saw how we can reduce this destruction and degradation by using forests and grasslands more sustainably, protecting species and ecosystems in parks, wilderness, and other nature reserves, and protecting ecosystem services that support all life and economies. We learned the importance of preserving what remains of richly biodiverse and highly endangered ecosystems (biodiversity hotspots) and identifying and protecting areas where deteriorating ecosystem services threaten people and other forms of life.

We also learned about the value of restoring or rehabilitating some of the ecosystems we have degraded. Reintroducing keystone species such as the gray wolf into ecosystems they once inhabited (**Core Case Study**) is a form of ecological restoration that can result in the reestablishment of certain ecological functions and species interactions in such systems, thereby helping to preserve biodiversity. Finally, we explored ways in which people can share with other species some of the land they occupy (95% of all the earth's land) in order to help sustain biodiversity.

Preserving terrestrial biodiversity involves applying the four **scientific principles of sustainability** (see back cover). First, it means respecting biodiversity by trying to sustain it. If we are successful, we will also be restoring and preserving the flows of energy from the sun through food webs, the cycling of nutrients in ecosystems, and the species interactions in food webs that help prevent excessive population growth of any species, including our own.

We abuse land because we regard it as a commodity belonging to us.
When we see land as a community to which we belong,
we may begin to use it with love and respect.

ALDO LEOPOLD

1. Review the Key Questions and Concepts for this chapter on p. 215. Describe the beneficial effects of reintroducing the keystone gray wolf species (Figure 10-1) into Yellowstone National Park in the United States (**Core Case Study**).

2. Distinguish among an **old-growth forest,** a **second-growth forest,** and a **tree plantation (tree farm or commercial forest).** What major ecological and economic benefits do forests provide? Describe the efforts of scientists and economists to put a price tag on the major ecological services provided by forests and other ecosystems.

3. What harm is caused by building roads into previously inaccessible forests? Distinguish among *selective cutting, clearcutting,* and *strip cutting* in the harvesting of trees. What are the major advantages and disadvantages of clear-cutting forests?

4. What are two types of forest fires? What are some ecological benefits of occasional surface fires? What are four ways to reduce the harmful impacts of diseases and insects on forests? What effects might projected global warming have on forests?

5. What parts of the world are experiencing the greatest forest losses? Define **deforestation** and list some of its major harmful environmental effects. Describe the encouraging news about deforestation in the United States. What are the major basic and secondary causes of tropical deforestation?

6. Describe four ways to manage forests more sustainably. What is certified timber? What are four ways to reduce the harms to forests and to people from forest fires? What are three ways to reduce the need to harvest trees? What is the fuelwood crisis and what are three ways to reduce its severity? Describe the Green Belt Movement. What are five ways to protect tropical forests and use them more sustainably?

7. Distinguish between **rangelands** and **pastures.** Distinguish between the **overgrazing** and **undergrazing** of rangelands. What are three ways to reduce overgrazing and use rangelands more sustainably? Describe the conflict between ranching and urban development in the American West.

8. What major environmental threats affect national parks? How could national parks in the United States be used more sustainably? Describe some of the ecological effects of reintroducing the gray wolf to Yellowstone National Park in the United States (**Core Case Study**). What percentage of the world's land has been set aside and protected as nature reserves, and what percentage do conservation biologists believe should be protected?

9. How should nature reserves be designed and connected? Describe what Costa Rica has done to establish nature reserves. What is **wilderness** and why is it important? Describe the controversy over protecting wilderness in the United States. What is a **biological hotspot** and why is it important to protect such areas? Why is it also important to protect areas where deteriorating ecosystem services threaten people and other forms of life?

10. What is **ecological restoration?** What are the four parts of a prominent strategy for carrying out ecological restoration and rehabilitation? Describe the ecological restoration of a tropical dry forest in Costa Rica. Define and give three examples of **reconciliation (applied) ecology.** Describe the relationship between reestablishing wolves in Yellowstone National park (**Core Case Study**) and the four **scientific principles of sustainability.**

Note: Key Terms are in **bold** type.

1. List three ways in which you could apply **Concept 10-5C** to help sustain terrestrial ecosystems and biodiversity.

2. Do you support the reintroduction of the gray wolf into the Yellowstone ecosystem in the United States (**Core Case Study**)? Explain. Do you think the reintroduction of wolves should be expanded to areas outside of the park? Explain.

3. Some argue that growing oil palm trees in plantations in order to produce biodiesel fuel will help us to lessen our dependence on oil and will cut vehicle CO_2 emis-

sions. Do you think these benefits are important enough to justify burning and clearing some tropical rain forests? Why or why not? Can you think of ways to produce biofuels, other than cutting trees? What are they?

4. In the early 1990s, Miguel Sanchez, a subsistence farmer in Costa Rica, was offered $600,000 by a hotel developer for a piece of land that he and his family had been using sustainably for many years. The land contained an old-growth rain forest and a black sand beach in an area under rapid development. Sanchez refused the offer. What would you have done if you were in Miguel Sanchez's position? Explain your decision.

5. There is controversy over whether Yellowstone National Park in the United States should be accessible by snowmobile during winter. Conservationists and backpackers, who use cross-country skis or snowshoes for excursions in the park during winter, say no. They contend that snowmobiles are noisy, pollute the air, and can destroy vegetation and disrupt some of the park's wildlife. Proponents say that snowmobiles should be allowed so that snowmobilers can enjoy the park during winter when cars are mostly banned. They point out that new snowmobiles are made to cut pollution and noise. A proposed compromise plan would allow no more than 950 of these new machines into the park per day, only on roads, and primarily on guided tours. What is your view on this issue? Explain.

6. In 2007, Lester R. Brown estimated that reforesting the earth and restoring the earth's degraded rangelands would cost about $15 billion a year. Suppose the United States, the world's most affluent country, agreed to put up half this money, at an average annual cost of $25 per American. Would you support doing this? Explain. What other part or parts of the federal budget would you decrease to come up with these funds?

7. Should developed countries provide most of the money needed to help preserve remaining tropical forests in developing countries? Explain.

8. Are you in favor of establishing more wilderness areas in the United States, especially in the lower 48 states (or in the country where you live)? Explain. What might be some drawbacks of doing this?

9. Congratulations! You are in charge of the world. List the three most important features of your policies for using and managing (a) forests, (b) grasslands, (c) nature reserves such as parks and wildlife refuges, (d) biological hotspots, and (e) areas with deteriorating ecosystem services.

10. List two questions that you would like to have answered as a result of reading this chapter.

Note: See Supplement 13 (p. S78) for a list of Projects related to this chapter.

ECOLOGICAL FOOTPRINT ANALYSIS

Study the data below on deforestation in five countries, and answer the questions that follow.

Country	Area of Tropical Rain Forest (square kilometers)	Area of Deforestation per Year (square kilometers)	Annual Rate of Tropical Forest Loss
A	1,800,000	50,000	
B	55,000	3,000	
C	22,000	6,000	
D	530,000	12,000	
E	80,000	700	

1. What is the annual rate of tropical rain forest loss, as a percentage of total forest area, in each of the five countries? Answer by filling in the blank column in the table.

2. What is the annual rate of tropical deforestation collectively in all of the countries represented in the table?

3. According to the table, and assuming the rates of deforestation remain constant, which country's tropical rain forest will be completely destroyed first?

4. Assuming the rate of deforestation in Country C remains constant, how many years will it take for all of its tropical rain forests to be destroyed?

5. Assuming that a hectare (1.0 hectare = 0.01 square kilometer) of tropical rain forest absorbs 0.85 metric tons of carbon dioxide per year, what would be the total annual growth in the carbon footprint (carbon emitted but not absorbed by vegetation because of deforestation) in metric tons of carbon dioxide per year from deforestation for each of the five countries in the table?

LEARNING ONLINE

Log on to the Student Companion Site for this book at **academic.cengage.com/biology/miller**, and choose Chapter 10 for many study aids and ideas for further reading and research. These include flash cards, practice quizzing, Weblinks, information on Green Careers, and InfoTrac® College Edition articles.

Sustaining Aquatic Biodiversity

A Biological Roller Coaster Ride in Lake Victoria

Lake Victoria, a large, shallow lake in East Africa (Figure 11-1), has been in ecological trouble for more than 2 decades.

Until the early 1980s, the lake had 500 species of fish found nowhere else. About 80% of them were small fish known as cichlids (pronounced "SIK-lids"), which feed mostly on detritus, algae, and zooplankton. Since 1980, some 200 of the cichlid species have become extinct, and some of those that remain are in trouble.

Several factors caused this dramatic loss of aquatic biodiversity. First, there was a large increase in the population of the Nile perch (Figure 11-2). This large predatory fish was deliberately introduced into the lake during the 1950s and 1960s to stimulate exports of the fish to several European countries, despite warnings by biologists that this huge fish with a big appetite would reduce or eliminate many defenseless native fish species. The population of this large and prolific fish exploded, devoured the cichlids and by 1986 had wiped out over 200 cichlid species.

Introducing the perch had other social and ecological effects. The new mechanized fishing industry increased poverty and malnutrition by putting most small-scale fishers and fish vendors out of business. And because the oily flesh of the perch are preserved by use of a wood smoker, local forests were depleted for firewood.

Another factor in loss of biodiversity in Lake Victoria was frequent algal blooms. These blooms became more common in the 1980s, due to nutrient runoff from surrounding farms and deforested land, spills of untreated sewage, and declines in the populations of algae-eating cichlids.

Also, the Nile perch population is decreasing because it severely reduced its own food supply of smaller fishes—an example of one of the four **scientific principles of sustainability** (see back cover) in action—and it also shows signs of being overfished. This may allow a gradual increase in the populations of some of the remaining cichlids.

This ecological story about the dynamics of large aquatic systems illustrates that there are unintended consequences when we intrude into a poorly understood ecosystem.

This chapter is devoted to helping us to understand the threats to aquatic biodiversity and what we can do to help sustain this vital part of the earth's natural capital.

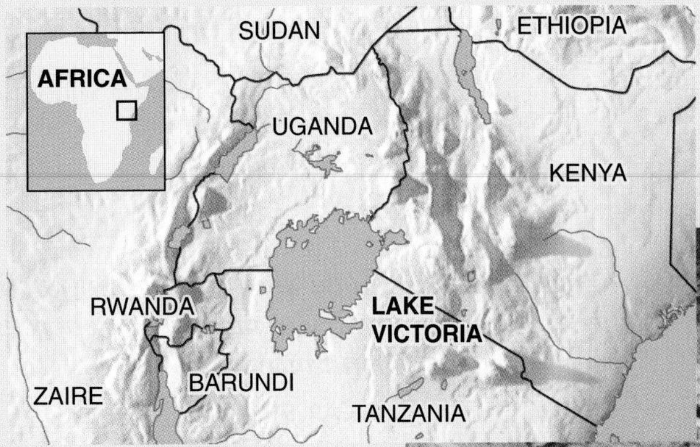

Figure 11-1 Lake Victoria is a large lake in East Africa.

Courtesy of the African Angler

Figure 11-2 Natural capital degradation: the Nile perch is a fine food fish that can weigh more than 91 kilograms (200 pounds). However, this deliberately introduced fish has played a key role in a major loss of biodiversity in East Africa's Lake Victoria (left).

11-1 What are the major threats to aquatic biodiversity?

CONCEPT 11-1 Aquatic species are threatened by habitat loss, invasive species, pollution, climate change, and over-exploitation, all made worse by the growth of the human population.

11-2 How can we protect and sustain marine biodiversity?

CONCEPT 11-2 We can help to sustain marine biodiversity by using laws and economic incentives to protect species, setting aside marine reserves to protect ecosystems, and using community-based integrated coastal management.

11-3 How should we manage and sustain marine fisheries?

CONCEPT 11-3 Sustaining marine fisheries will require improved monitoring of fish populations, cooperative fisheries management among communities and nations, reduction of fishing subsidies, and careful consumer choices in seafood markets.

11-4 How can we protect and sustain wetlands?

CONCEPT 11-4 To maintain the ecological and economic services of wetlands, we must maximize preservation of remaining wetlands and restoration of degraded and destroyed wetlands.

11-5 How can we protect and sustain freshwater lakes, rivers, and fisheries?

CONCEPT 11-5 Freshwater ecosystems are strongly affected by human activities on adjacent lands, and protecting these ecosystems must include protection of their watersheds.

11-6 What should be our priorities for sustaining biodiversity and ecosystem services?

CONCEPT 11-6 Sustaining the world's biodiversity and ecosystem services will require mapping terrestrial and aquatic biodiversity, maximizing protection of undeveloped terrestrial and aquatic areas, and carrying out ecological restoration projects worldwide.

Note: Supplements 2 (p. S4), 8 (p. S47), and 13 (p. S78) can be used with this chapter.

The coastal zone may be the single most important portion of our planet.
The loss of its biodiversity may have repercussions
far beyond our worst fears.

G. CARLETON RAY

11-1 What Are the Major Threats to Aquatic Biodiversity?

▶ **CONCEPT 11-1** Aquatic species are threatened by habitat loss, invasive species, pollution, climate change, and overexploitation, all made worse by the growth of the human population.

We Have Much to Learn about Aquatic Biodiversity

Although we live on a watery planet, we have explored only about 5% of the earth's global ocean (Figure 8-2, p. 163) and know relatively little about its biodiversity and how it works. We also have limited knowledge about freshwater biodiversity.

However, scientists have observed three general patterns of marine biodiversity. *First,* the greatest marine biodiversity occurs in coral reefs (Chapter 8 Core Case Study, p. 162), estuaries, and the deep-ocean floor. *Second,* biodiversity is higher near coasts than in the open sea because of the greater variety of producers and habitats in coastal areas (Figure 8-5, p. 166). *Third,* biodiversity is higher in the bottom region of the ocean than in the surface region because of the greater variety of habitats and food sources on the ocean bottom.

The world's marine systems provide important ecological and economic services (Figure 8-4, p. 165). Thus, scientific investigation of poorly understood marine aquatic systems is a *research frontier* that could lead to immense ecological and economic benefits. Freshwater systems, which occupy only 1% of the earth's surface, also provide important ecological and economic services (Figure 8-14, p. 174).

─── RESEARCH FRONTIERS ───

Exploring marine and freshwater ecosystems, their species, and species interactions. See **academic.cengage.com/ biology/miller**.

Human Activities Are Destroying and Degrading Aquatic Habitats

As with terrestrial biodiversity, the greatest threats to the biodiversity of the world's marine and freshwater ecosystems (**Concept 11-1**) can be remembered with the aid of the acronym HIPPCO, with H standing for *habitat loss and degradation*. Some 90% of fish living in the ocean spawn in coral reefs (Figure 4-10, left, p. 89, and Figure 8-1, p. 162), mangrove forests (Figure 8-8, p. 168), coastal wetlands (Figure 8-7, p. 167), or rivers (Figure 8-17, p. 176). And these areas are under intense pressure from human activities (Figure 8-12, p. 172). Scientists reported in 2006 that these coastal habitats are disappearing at rates 2–10 times higher than the rate of tropical forest loss.

A major threat is loss and degradation of many sea-bottom habitats by dredging operations and trawler fishing boats. Trawlers drag huge nets weighted down with heavy chains and steel plates like giant submerged bulldozers over ocean bottoms to harvest a few species of bottom fish and shellfish (Figure 11-3). Trawling nets reduce coral reef habitats to rubble and kill a variety of creatures on the bottom by crushing them, burying them in sediment, and exposing them to predators. Each year, thousands of trawlers scrape and disturb an area of ocean floor about 150 times larger than the area of forests that are clear-cut annually.

In 2004, some 1,134 scientists signed a statement urging the United Nations to declare a moratorium on bottom trawling on the high seas by 2006 and to eliminate it globally by 2010. Fishing nations led by Iceland, Russia, China, and South Korea blocked such a ban. But in 2007, those countries (except for Iceland) and 18 others agreed to voluntary restrictions on bottom trawling in the South Pacific. The agreement will partially protect about one-quarter of the world's ocean bottom but monitoring and enforcement will be difficult.

Habitat disruption is also a problem in freshwater aquatic zones. Dams and excessive water withdrawal from rivers and lakes (mostly for agriculture) destroy aquatic habitats and water flows and disrupt freshwater biodiversity. As a result of these and other human activities, 51% of freshwater fish species—more than any other major type of species—are threatened with premature extinction (Figure 9-6, p. 189).

Figure 11-3 Natural capital degradation: area of ocean bottom before (left) and after (right) a trawler net scraped it like a gigantic plow. These ocean floor communities could take decades or centuries to recover. According to marine scientist Elliot Norse, "Bottom trawling is probably the largest human-caused disturbance to the biosphere." Trawler fishers claim that ocean bottom life recovers after trawling and that they are helping to satisfy the increasing consumer demand for fish. **Question:** What land activities are comparable to this?

Peter J. Auster/National Undersea Research Center

Invasive Species Are Degrading Aquatic Biodiversity

Another problem is the deliberate or accidental introduction of hundreds of harmful invasive species—the I in HIPPCO—(Figure 9-14, p. 199) into coastal waters, wetlands, and lakes throughout the world (**Concept 11-1**). These bioinvaders can displace or cause the extinction of native species and disrupt ecosystem services. For example, since the late 1980s, Lake Victoria (**Core Case Study**) has been invaded by the water hyacinth (Figure 11-4). This rapidly growing plant has carpeted large areas of the lake, blocked sunlight, deprived fish and plankton of oxygen, and reduced aquatic plant diversity.

> **THINKING ABOUT**
>
> **The Nile Perch and Lake Victoria**
> Would most of the now extinct cichlid fish species in Lake Victoria (**Core Case Study**) still exist today if the Nile perch had not been introduced? Or might other factors come into play? Explain.

According to a 2008 study by The Nature Conservancy, 84% of the world's coastal waters are being colonized by invasive species. Bioinvaders are blamed for about two-thirds of fish extinctions in the United States between 1900 and 2000. They cost the country an average of about $14 million *per hour.* Many of these invaders arrive in the ballast water stored in tanks in large cargo ships to keep them stable. These ships take in ballast water—along with whatever microorganisms and tiny species it contains—in one harbor and dump it in another.

Consumers also introduce invasive species. For example, the *Asian swamp eel* has invaded the waterways of south Florida (USA), probably from the dumping of a home aquarium. This rapidly reproducing eel eats almost anything—including many prized fish species—by sucking them in like a vacuum cleaner. It can elude cold weather, drought, and predators by burrowing into mud banks. It is also resistant to waterborne poisons because it can breathe air, and it can wriggle across dry land to invade new waterways, ditches, canals, and marshes. Eventually, this eel could take over much of the waterways of the southeastern United States.

Another example is the *purple loosestrife,* a perennial plant that grows in wetlands in parts of Europe. Since the 1880s, it has been imported and used in gardens as an ornamental plant in many parts of the world. A single plant can produce more than 2.5 million seeds a year, which are spread by flowing water and by becoming attached to wildlife, livestock, hikers, and vehicle tire treads. It reduces wetland biodiversity by displacing native vegetation and reducing habitat for some forms of wetland wildlife.

Some U.S. states have recently introduced two natural predators of loosestrife from Europe: a weevil species and a leaf-eating beetle. It will take some time to determine the effectiveness of this biological control approach and to be sure the introduced predators themselves do not become pests.

While threatening native species, invasive species can also disrupt and degrade whole ecosystems. This is the focus of study for a growing number of researchers (Science Focus, at right).

Figure 11-4 *Invasive water hyacinths,* supported by nutrient runoff, clogged a ferry terminal on the Kenyan part of Lake Victoria in 1997. By blocking sunlight and consuming oxygen, this invasion has reduced biodiversity in the lake. Scientists reduced the problem at strategic locations by mechanical removal and by introducing two weevils for biological control of the hyacinth.

Population Growth and Pollution Can Reduce Aquatic Biodiversity

The U.N. Environment Programme (UNEP) projects that, by 2020, 80% of the world's people will be living along or near the coasts, mostly in gigantic coastal cities. This coastal population growth—the first P in HIPPCO—will add to the already intense pressure on the world's coastal zones, primarily by destroying more aquatic habitat and increasing pollution (**Concept 11-1**).

A 2008 study by Benjamin S. Halpern and other scientists found that only 4% of the world's oceans are not affected by pollution—the second P in HIPPCO—and 40% are strongly affected. In 2004, the UNEP estimated that 80% of all ocean pollution comes from land-based coastal activities. Humans have doubled the flow of nitrogen, mostly from nitrate fertilizers, into the oceans since 1860, and the 2005 Millennium Ecosystem Assessment estimated that this flow will increase by another two-thirds by 2050. These inputs of nitrogen (and similar inputs of phosphorus) result in eutrophication

SCIENCE FOCUS

How Carp Have Muddied Some Waters

Lake Wingra lies within the city of Madison, Wisconsin (USA), surrounded mostly by a forest preserve. While almost all of its shoreline is undeveloped, the lake receives excessive nutrient inputs from runoff, containing fertilizers from area farms and lawns, and storm water flowing in from city streets and parking lots. Its waters are green and murky throughout the warmer months of the year.

Lake Wingra also contains a number of invasive plant and fish species, including purple loosestrife and common carp. The carp, which were introduced in the late 1800s, now make up about half of the fish biomass in the lake. They devour algae called chara, which would normally cover the lake bottom and stabilize sediments. Consequently, fish movements and currents stir these sediments, which accounts for much of the water's excessive turbidity, or cloudiness.

Knowing this, Dr. Richard Lathrup, a limnologist (lake scientist) who works with Wisconsin's Department of Natural Resources, hypothesized that removing the carp would help to restore the natural ecosystem of Lake Wingra. Lathrop speculated that with the carp gone, the bottom sediments would settle and become stabilized, allowing the water to clear. Clearer water would in turn allow native plants to receive more sunlight and become reestablished on the lake bottom, replacing purple loosestrife and other invasive plants that now dominate the shallow shoreline waters.

Lathrop and his colleagues built a *fish exclosure* by installing a thick, heavy vinyl curtain around a 1-hectare (2.5-acre), square-

Figure 11-A *Lake Wingra* in Madison, Wisconsin (USA) has become eutrophic largely because of the introductions of invasive species, including the common carp, which now represents half of the fish biomass in the lake. Removal of carp in the experimental area shown here resulted in a dramatic improvement in the clarity of the water and subsequent regrowth of native plant species in shallow water.

Mike Kakuska

shaped perimeter extending out from the shore (Figure 11-A). This barrier hangs from buoys on the surface to the bottom of the lake, isolating the volume of water within it. The researchers then removed all of the carp from this study area and began observing results. Within one month, the waters within the exclosure were noticeably clearer, and within a year, the difference in clarity was dramatic, as Figure 11-A shows.

In 2008, the scientists began removing carp from the rest of the lake. Lathrop notes

that keeping carp out of Lake Wingra will be a daunting task, but his controlled scientific experiment clearly shows the effects that an invasive species can have on an aquatic ecosystem. And it reminds us that preventing the introduction of invasive species in the first place is the best way to avoid such effects.

Critical Thinking

What are two other results of this controlled experiment that you might expect? (*Hint:* think food webs.)

---

of marine and freshwater systems, which can lead to algal blooms (Figure 8-16, right, p. 175), fish die-offs, and degradation of ecosystem services.

In Lake Victoria, such eutrophication was a key development in the takeover by invasive Nile perch and the loss of cichlid populations, as described in the **Core Case Study**. With increased runoff generated by the growing human population in nearby towns and farms came algal blooms. Because cichlids feed on algae, their populations rose dramatically. Before Nile perch were introduced, such population explosions would have ended in die-offs of the cichlids. But the Nile perch suddenly had a bigger food supply, and thus their population grew and led to changes in the lake's ecosystem and all the resulting problems.

CORE CASE STUDY

Similar pressures are growing in freshwater systems, as more people seek homes and places for recreation near lakes and streams. The result is massive inputs of sediment and other wastes from land into these aquatic systems.

Toxic pollutants from industrial and urban areas can kill some forms of aquatic life by poisoning them. And each year, plastic items dumped from ships and left as litter on beaches kill up to 1 million seabirds and 100,000 mammals and sea turtles. Such pollutants and debris threaten the lives of millions of marine mammals (Figure 11-5, p. 254) and countless fish that ingest, become entangled in, or are poisoned by them. These forms of pollution lead to an overall reduction in aquatic biodiversity and degradation of ecosystem services.

Doris Alcorn/U.S. National Maritime Fisheries

Figure 11-5 This Hawaiian monk seal was slowly starving to death before this discarded piece of plastic was removed from its snout. Each year, plastic items dumped from ships and left as litter on beaches threaten the lives of millions of marine mammals, turtles, and seabirds that ingest, become entangled in, or are poisoned by such debris.

Climate Change Is a Growing Threat

Climate change—the C in HIPPCO—threatens aquatic biodiversity (**Concept 11-1**) and ecosystem services partly by causing sea levels to rise. During the past 100 years, average sea levels have risen by 10–20 centimeters (4–8 inches), and scientists estimate they will rise another 18–59 centimeters (0.6–1.9 feet) and perhaps as high as 1–1.6 meters (3.2–5.2 feet) between 2050 and 2100 mostly, because of projected global warming. This would destroy more coral reefs, swamp some low-lying islands, drown many highly productive coastal wetlands, and put much of the U.S. state of Louisiana's coast, including New Orleans, under water (Figure 8-18, p. 177). And some Pacific island nations could lose more than half of their protective coastal mangrove forests by 2100, according to a 2006 study by UNEP (Science Focus, at right). See *The Habitable Planet*, Video 5, at **www.learner.org/resources/series209. html** for projected sea level changes in densely populated coastal areas such as Vietnam and New York City.

Overfishing and Extinction: Gone Fishing, Fish Gone

Overfishing—the O in HIPPCO—is not new. Archaeological evidence indicates that for thousands of years, humans living in some coastal areas have overharvested fishes, shellfish, seals, turtles, whales, and other marine mammals (**Concept 11-1**). But today's industrialized fishing fleets can overfish much more of the oceans and deplete marine life at a much faster rate. Today, fish are hunted throughout the world's oceans by a global fleet of millions of fishing boats—some of them longer than a football field. Modern industrial

fishing can cause 80% depletion of a target fish species in only 10–15 years (Case Study, p. 256).

The human demand for seafood is outgrowing the sustainable yield of most ocean fisheries. To keep consuming seafood at the current rate, we will need 2.5 times the area of the earth's oceans, according to the *Fishprint of Nations 2006*, a study based on the concept of the human ecological footprint (**Concept 1-3**, p. 12, and Figure 1-10, p. 15). The **fishprint** is defined as the area of ocean needed to sustain the consumption of an average person, a nation, or the world. The study found that all nations together are overfishing the world's global oceans by an unsustainable 157%.

In most cases, overfishing leads to *commercial extinction*, which occurs when it is no longer profitable to continue fishing the affected species. Overfishing usually results in only a temporary depletion of fish stocks, as long as depleted areas and fisheries are allowed to recover. But as industrialized fishing fleets vacuum up more and more of the world's available fish and shellfish, recovery times for severely depleted populations are increasing and can take 2 decades or more. In 1992, for example, Canada's 500-year-old Atlantic cod fishery off the coast of Newfoundland collapsed and was closed. This put at least 20,000 fishers and fish processors out of work and severely damaged Newfoundland's economy. As Figure 11-6 shows,

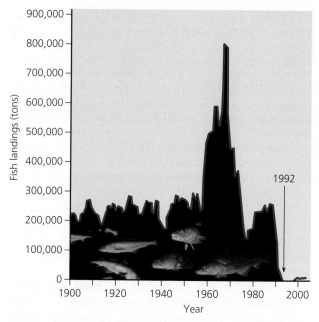

Figure 11-6 Natural capital degradation: this graph illustrates the collapse of the cod fishery in the northwest Atlantic off the Canadian coast. Beginning in the late 1950s, fishers used bottom trawlers to capture more of the stock, reflected in the sharp rise in this graph. This resulted in extreme overexploitation of the fishery, which began a steady fall throughout the 1970s, followed by a slight recovery in the 1980s and total collapse by 1992 when the site was closed to fishing. Canadian attempts to regulate fishing through a quota system had failed to stop the sharp decline. The fishery was reopened on a limited basis in 1998 but then closed indefinitely in 2003. (Data from Millennium Ecosystem Assessment)

Sustaining Ecosystem Services by Protecting and Restoring Mangroves

Mangroves are not among the world's biodiversity hotspots and do not contain a large number of species, as do endangered tropical rain forests. However, they too require protection because of the important ecosystem services they provide for coastal dwellers.

For example, protecting mangroves and restoring them in areas where they have been destroyed are important ways to reduce the impacts of rising sea levels and more intense storm surges. These ecosystem services will become more important in the face of tropical storms, possibly becoming more intense as a result of global warming, and of tsunamis, caused mostly by earthquakes on ocean seafloors. Protecting and restoring these natural coastal barriers is also much cheaper than building concrete sea walls or moving threatened coastal towns and cities inland.

Indonesia, a sprawling nation of about 17,000 islands, is especially vulnerable to rising sea levels and storm surges. But decades of rampant development along its island coastlines have destroyed or degraded about 70% of its mangrove forests. Even so, Indonesia still has the world's largest area of mangroves, amounting to about one-fourth of the world's remaining mangrove forests. In the 1990s, it started a program to protect more of these areas and to restore large areas of degraded mangrove forests.

Expanding mangrove protection and restoration in Indonesia and in other nations will require educating citizens, government officials, and business leaders about the huge economic value of the natural ecosystem services they provide. These economic benefits should be considered during any decision-making process concerning development of these fragile coastal areas.

Critical Thinking

Do you agree that the estimated economic value of ecosystem services provided by mangroves should be considered in making coastal development decisions? If you agree, how would you accomplish this politically?

---

this cod population has not recovered, despite the fishing ban.

Such a collapse can create a domino effect, leading to collapses of other species. After the cod were fished out in the North Atlantic, fishers turned to sharks, which provide important ecosystem services and help to control the populations of other species (Case Study, p. 96). Since then, overfishing of big sharks has cut Atlantic stocks by 99%, according to a 2007 Canadian fisheries study. Scientists reported that with the large sharks essentially gone, the northwest Atlantic populations of rays and skates, which the sharks once fed on, have exploded and have wiped out most of the bay scallops.

One result of the increasingly efficient global hunt for fish is that big individuals in many populations of commercially valuable predatory species—including cod, salmon, mackerel, herring, and dogfish—are becoming scarce. And according to a 2003 study by conservation biologist Boris Worm and his colleagues, 90% or more of the large, predatory, open-ocean fishes such as tuna, swordfish, and marlin have disappeared since 1950 (see *The Habitable Planet*, Video 9, at **www.learner.org/resources/series209. html**). The large bluefin tuna, with a typical weight of 340 kilograms (750 pounds) and a length of 2 meters (6.5 feet), is the premier choice for sushi and sashimi, and, as the world's most desirable fish, can sell for as much as $880 per kilogram ($400 per pound). As a result, it is probably the most endangered of all large fish species.

The fishing industry has begun working its way down marine food webs by shifting from large species to smaller ones. This practice reduces the breeding stock needed for recovery of depleted species, which unravels marine food webs and disrupts marine ecosystems and their ecosystem services.

Most fishing boats hunt and capture one or a small number of commercially valuable species. However, their gigantic nets and incredibly long lines of hooks also catch nontarget species, called *bycatch*. Almost one-third of the world's annual fish catch, by weight, consists of these nontarget species, which are thrown overboard dead or dying. This can deplete the populations of bycatch species that play important roles in marine food webs. Marine mammals such as seals and dolphins can also become part of bycatch.

Fish species are also threatened with *biological extinction*, mostly from overfishing, water pollution, wetlands destruction, and excessive removal of water from rivers and lakes. According to the IUCN, 34% of the world's known marine fish species and 71% of the world's freshwater fish species face extinction within your lifetime. Indeed, *marine and freshwater fishes are threatened with extinction by human activities more than any other group of species.*

> **RESEARCH FRONTIER**
>
> Learning more about how aquatic systems work and how human activities affect aquatic biodiversity and aquatic ecosystem services. See **academic.cengage.com/biology/ miller**.

Industrial Fish Harvesting Methods

Industrial fishing fleets dominate the world's marine fishing industry. They use global satellite positioning equipment, sonar, huge nets and long fishing lines, spotter planes, and gigantic refrigerated factory ships that can process and freeze their catches. These fleets help meet the growing demand for seafood. But critics say that these highly efficient fleets are *vacuuming the seas*, decreasing marine biodiversity, and degrading important marine ecosystem services. Today 75% of the world's commercial fisheries are being fished at or beyond their estimated sustainable yields, according to the U.N. Food and Agricultural Organization.

Figure 11-7 shows the major methods used for the commercial harvesting of various marine fishes and shellfish. Until the mid-1980s, fishing fleets from de-veloped countries dominated the ocean catch. Today, most of these fleets come from developing countries, especially in Asia.

Let us look at a few of these methods. *Trawler fishing* is used to catch fishes and shellfish—especially shrimp, cod, flounder, and scallops—that live on or near the ocean floor. It involves dragging a funnel-shaped net held open at the neck along the ocean bottom. It is weighted down with chains or metal plates and scrapes up almost everything that lies on the ocean floor and often destroys bottom habitats—somewhat like clear-cutting the ocean floor (Figure 11-3). Newer trawling nets are large enough to swallow 12 jumbo jet planes and even larger ones are on the way.

Another method, *purse-seine fishing*, is used to catch surface-dwelling species such as tuna, mackerel, ancho-vies, and herring, which tend to feed in schools near the surface or in shallow areas. After a spotter plane lo-

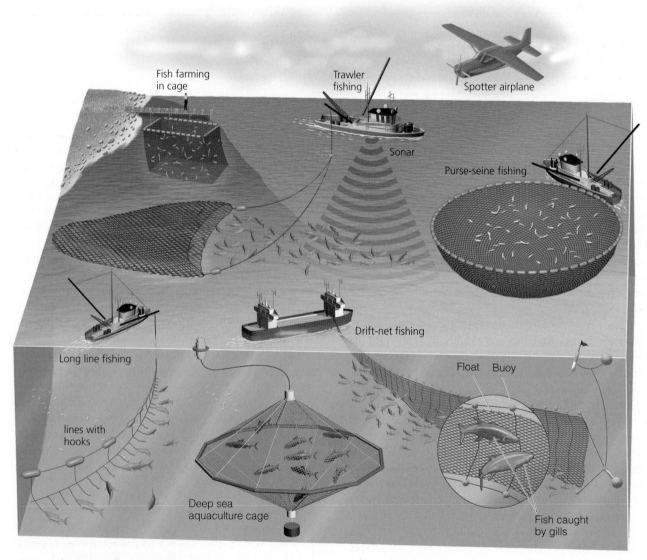

Figure 11-7 Major commercial fishing methods used to harvest various marine species. These methods have become so effective that many fish species have become commercially extinct.

cates a school, the fishing vessel encloses it with a large net called a purse seine. Nets used to capture yellow fin tuna in the eastern tropical Pacific Ocean have killed large numbers of dolphins that swim on the surface above schools of tuna.

Fishing vessels also use *longlining,* which involves putting out lines up to 130 kilometers (80 miles) long, hung with thousands of baited hooks. The depth of the lines can be adjusted to catch open-ocean fish species such as swordfish, tuna, and sharks or bottom fishes such as halibut and cod. Longlines also hook and kill large numbers endangered sea turtles, dolphins, and seabirds each year. Making simple modifications to fishing gear and practices can decrease seabird deaths.

With *drift-net fishing,* fish are caught by huge drifting nets that can hang as deep as 15 meters (50 feet) below the surface and extend to 64 kilometers (40 miles) long. This method can lead to overfishing of the desired species and may trap and kill large quantities of unwanted fish, marine mammals, sea turtles, and seabirds.

Since 1992, a U.N. ban on the use of drift nets longer than 2.5 kilometers (1.6 miles) in international waters has sharply reduced use of this technique. But longer nets continue to be used because compliance is voluntary and it is difficult to monitor fishing fleets over vast ocean areas. Also, the decrease in drift net use has led to increased use of longlines, which often have similar harmful effects on marine wildlife.

11-2 How Can We Protect and Sustain Marine Biodiversity?

▶ **CONCEPT 11-2** We can help to sustain marine biodiversity by using laws and economic incentives to protect species, setting aside marine reserves to protect ecosystems, and using community-based integrated coastal management.

Laws and Treaties Have Protected Some Endangered and Threatened Marine Species

Protecting marine biodiversity is difficult for several reasons. First, the human ecological footprint (Figure 1-10, p. 15) and fishprint are expanding so rapidly into aquatic areas that it is difficult to monitor the impacts (**Concept 1-3**, p. 12). Second, much of the damage to the oceans and other bodies of water is not visible to most people. Third, many people incorrectly view the seas as an inexhaustible resource that can absorb an almost infinite amount of waste and pollution and still produce all the seafood we want. Finally, most of the world's ocean area lies outside the legal jurisdiction of any country. Thus, it is an open-access resource, subject to overexploitation.

Nevertheless, there are ways to protect and sustain marine biodiversity, one of which is the regulatory approach (**Concept 11-2**). National and international laws and treaties to help protect marine species include the 1975 Convention on International Trade in Endangered Species (CITES), the 1979 Global Treaty on Migratory Species, the U.S. Marine Mammal Protection Act of 1972, the U.S. Endangered Species Act of 1973, the U.S. Whale Conservation and Protection Act of 1976, and the 1995 International Convention on Biological Diversity.

The U.S. Endangered Species Act (Case Study, p. 207) and international agreements have been used to identify and protect endangered and threatened marine species such as whales (see the following Case Study), seals, sea lions, and sea turtles.

■ CASE STUDY
Protecting Whales: A Success Story . . . So Far

Cetaceans are an order of mostly marine mammals ranging in size from the 0.9-meter (3-foot) porpoise to the giant 15- to 30-meter (50- to 100-foot) blue whale. They are divided into two major groups: *toothed whales* and *baleen whales* (Figure 11-8, p. 258).

Toothed whales, such as the porpoise, sperm whale, and killer whale (orca), bite and chew their food and feed mostly on squid, octopus, and other marine animals. *Baleen whales,* such as the blue, gray, humpback, minke, and fin, are filter feeders. Attached to their upper jaws are plates made of baleen, or whalebone, which they use to filter plankton, especially tiny shrimp-like krill (Figure 3-14, p. 63), from the seawater.

Whales are fairly easy to kill because of their large size and their need to come to the surface to breathe. Whale hunters became efficient at hunting and killing whales using radar, spotters in airplanes, fast ships, and harpoon guns. Whale harvesting, mostly in international waters, has followed the classic pattern of a tragedy of the commons, with whalers killing an estimated 1.5 million whales between 1925 and 1975. This overharvesting drove 8 of the 11 major species to commercial extinction.

Overharvesting also drove some commercially prized species such as the giant blue whale (Figure 11-8) to the brink of biological extinction. The endangered blue whale is the world's largest animal. Fully grown, it is longer than three train boxcars and weighs more than 25 adult elephants. The adult has a heart the size of a Volkswagen Beetle, and some of its arteries are big enough for a child to swim through.

Blue whales spend about 8 months a year in Antarctic waters. During the winter, they migrate to warmer waters where their young are born. Before commercial whaling began, an estimated 250,000 blue whales roamed the Antarctic Ocean. Today, the species has been hunted to near biological extinction for its oil, meat, and bone. There are probably fewer than 5,000 blue whales left. They take 25 years to mature sexually and have only one offspring every 2–5 years. This low reproductive rate will make it difficult for the species to recover.

Blue whales have not been hunted commercially since 1964 and have been classified as an endangered species since 1975. Despite this protection, some marine biologists fear that too few blue whales remain for the species to recover and avoid premature extinction. Others believe that with continued protection they will make a slow comeback.

In 1946, the International Convention for the Regulation of Whaling established the International Whaling Commission (IWC). Its mission was to regulate the whaling industry by setting annual quotas to prevent overharvesting and commercial extinction. But IWC quotas often were based on inadequate data or were ignored by whaling countries. Without powers of enforcement, the IWC was not able to stop the decline of most commercially hunted whale species.

In 1970, the United States stopped all commercial whaling and banned all imports of whale products. Under pressure from conservationists, the U.S. government, and governments of many nonwhaling countries, the IWC imposed a moratorium on commercial whaling starting in 1986. It worked. The estimated number of whales killed commercially worldwide dropped from 42,480 in 1970 to about 1,300 in 2007. However, despite the ban, more than 26,000 whales were hunted and killed between 1986 and 2007.

Japan hunts and kills at least 1,000 minke and fin whales each year for what it calls "scientific purposes." Critics see this annual whale hunt as poorly disguised commercial whaling because the whale meat is sold to restaurants. Each whale is worth up to $30,000 wholesale. Norway openly defies the international ban on commercial whaling and hunts and kills up to 1,000 minke whales a year (Figure 11-9).

Japan, Norway, Iceland, Russia, and a growing number of small tropical island countries—which Japan brought into the IWC to support its position—hope to overthrow the IWC ban on commercial whaling and reverse the international ban on buying and selling whale products. They argue that commercial whaling should be allowed because it has been a traditional part of their economies and cultures.

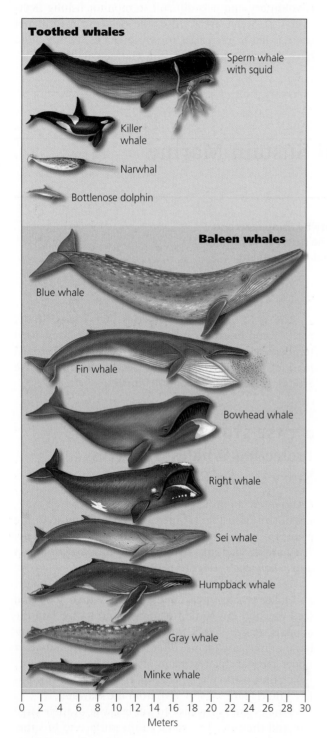

Figure 11-8 *Examples of cetaceans,* which can be classified as either toothed whales or baleen whales.

Figure 11-9 Norwegian whalers harpooning a sperm whale. Norway and Japan kill up to 2,000 whales a year. They also believe that increased but sustainable commercial whaling should be allowed for sperm, minke, and pilot whales whose stocks have built back to large numbers.

Tony Martin/WWI Peter Arnold, Inc.

Proponents of whaling also contend that the ban is emotionally motivated and not supported by current scientific estimates of whale populations. The moratorium on commercial whaling has led to a sharp rebound in these estimates for sperm, pilot, and minke whales.

Most conservationists disagree. Some argue that whales are peaceful, intelligent, sensitive, and highly social mammals that should be protected for ethical reasons. Others question IWC estimates of the allegedly recovered whale populations, noting the inaccuracy of such estimates in the past. And many conservationists fear that opening the door to any commercial whaling may eventually weaken current international disapproval and legal sanctions against commercial whaling and lead to widespread harvests of most whale species.

> **HOW WOULD YOU VOTE?**
>
> Should controlled commercial whaling be resumed for species with populations judged to be stable? Cast your vote online at **academic.cengage.com/biology/miller**.

Some coastal communities have an interest in maintaining the ban on whaling because they can provide jobs and income through increasingly popular whale watching. For example, The Nature Conservancy promoted whale watching in the town of Samaná in the Dominican Republic and has trained fishermen to work as whale-watching guides. The once run-down town has become a tourist hotspot, with spruced up houses, hotels, and inns, largely because of the popularity of whale watching. Local residents now have an economic interest in protecting the whales.

Economic Incentives Can Be Used to Sustain Aquatic Biodiversity

Other ways to protect endangered and threatened aquatic species involve using economic incentives (**Concept 11-2**). For example, according to a 2004 World Wildlife Fund study, sea turtles are worth more to local communities alive than dead. The report estimates that sea turtle tourism brings in almost three times more money than does the sale of turtle products such as meat, leather, and eggs.

The problem is that individuals seeking to make a quick gain take the turtles before their surrounding communities can realize the longer-term economic benefits by protecting them. Educating citizens about this issue could inspire communities to protect the turtles (Case Study, below).

Some individuals find economic rewards in restoring and sustaining aquatic systems. One example is an application of *reconciliation ecology* by a restaurant owner (Individuals Matter, p. 261).

■ CASE STUDY
Holding Out Hope for Marine Turtles

Of the seven species of marine turtles, six are either critically endangered or endangered. Among the latter is the leatherback sea turtle (Figure 11-10, p. 260), a species that has survived 100 million years, but now faces possible extinction. While their population is stable in the Atlantic Ocean, their numbers have declined by 95% in the Pacific.

Naturalist Carl Safina wrote about his studies of the leatherback in his book *Voyage of the Turtle*. He describes the leatherback as "the last living dinosaur." The largest of all sea turtles, and the only warm-blooded one, an adult turtle can weigh as much as 91 kilograms (200 pounds). It swims great distances, migrating across the Atlantic and Pacific Oceans, and it can dive as deep as 1,200 meters (3,900 feet). It is named for its leathery shell.

The leatherback female lays her eggs on sandy ocean beaches in the dark of night and then returns to the sea. The babies hatch simultaneously in large numbers and immediately scamper across the sand to try to survive to adulthood in ocean waters. As Safina describes it, "They start the size of a cookie, and come back the size of a dinosaur."

While leatherbacks survived the impact of the giant asteroid that probably wiped out the dinosaurs, they may not survive the growing human impact on their environment. Bottom trawlers are destroying the coral

Figure 11-10 An endangered *leatherback sea turtle* is entangled in a fishing net and lines and could have starved to death had it not been rescued.

Renatura Congo. www.renatura.osso.eu.org

long-line swordfish fishing off the Pacific coast to help save dwindling sea turtle populations.

On Costa Rica's northwest coast in the community of Playa Junquillal, an important leatherback nesting area, residents learned that tourism can bring in almost three times as much money as selling turtle products can earn. Biologists working with the World Wildlife Fund there directed a community-based program to educate people about the importance of protecting leatherbacks and to create revenue sources for local residents based on tourism instead of on harvesting turtle eggs. Volunteers were enlisted to find and rescue nests before they could be poached and to build hatcheries to protect the eggs.

For the leatherback turtles, this program was a success. In 2004, on the local beaches, all known nests had been poached. The following year, all known nests were protected and none were poached. The leatherback had become an important economic resource for all, not for just a few, of the residents of Playa Junquillal.

THINKING ABOUT
The Leatherback Sea Turtle

Why should we care if the leatherback sea turtle becomes extinct? What are three things you would do to help protect this species from premature extinction?

gardens that serve as their feeding grounds. The turtles are hunted for meat and leather, and their eggs are taken for food. They often drown after becoming entangled in fishing nets and lines (Figure 11-10) and lobster and crab traps. A 2004 study by R. I. Lewison and his colleagues estimated that in 2000 alone, longline fishing operations killed an estimated 50,000 leatherback and 200,000 loggerhead sea turtles. Shrimp trawling fisheries also kill large numbers of leatherback and other sea turtle species.

Pollution is another threat. Sea turtles can mistake discarded plastic bags for jellyfish and choke to death on them. Beachgoers sometimes trample their nests. And artificial lights can disorient hatchlings as they try to find their way to the ocean; going in the wrong direction increases their chances of ending up as food for predators.

Add to this the threat of climate change. Global warming will raise sea levels, which will flood nesting and feeding habitats, and change ocean currents, which could disrupt the turtles' migration routes.

Many people are working to protect the leatherbacks. On some Florida beaches, lights are turned off or blacked out during hatching season. Nesting areas are roped off, and people respect the turtles, according to Safina. Since 1991, the U.S. government has required offshore shrimp trawlers to use turtle excluder devices (TEDs), which help to keep sea turtles out of their nets and to allow caught turtles to escape. TEDs have been adopted in 15 countries that export shrimp to the United States. And, in 2004, the United States banned

Marine Sanctuaries Protect Ecosystems and Species

By international law, a country's offshore fishing zone extends to 370 kilometers (200 statute miles) from its shores. Foreign fishing vessels can take certain quotas of fish within such zones, called *exclusive economic zones*, but only with a government's permission. Ocean areas beyond the legal jurisdiction of any country are known as the *high seas*, and laws and treaties pertaining to them are difficult to monitor and enforce.

Through the Law of the Sea Treaty, the world's coastal nations have jurisdiction over 36% of the ocean surface and 90% of the world's fish stocks. Instead of using this law to protect their fishing grounds, many governments have promoted overfishing, subsidized new fishing fleets, and failed to establish and enforce stricter regulation of fish catches in their coastal waters.

Some countries are attempting to protect marine biodiversity and sustain fisheries by establishing marine sanctuaries. Since 1986, the IUCN has helped to establish a global system of *marine protected areas* (MPAs)—areas of ocean partially protected from human activities. There are more than 4,000 MPAs worldwide, 200 of them in U.S. waters. Despite their name, most MPAs are only partially protected. Nearly all allow dredging, trawler fishing, and other ecologically harmful resource extraction activities. However, the U.S. state

INDIVIDUALS MATTER

Creating an Artificial Coral Reef in Israel

Near the city of Eliat, Israel, at the northern tip of the Red Sea, is a magnificent coral reef, which is a major tourist attraction. To help protect the reef from excessive development and destructive tourism, Israel set aside part of the reef as a nature reserve. But tourism, industrial pollution, and inadequate sewage treatment have destroyed most of the unprotected part of the reef.

Enter Reuven Yosef, a pioneer in coral reef restoration and reconciliation ecology, who has developed an underwater restaurant called the Red Sea Star Restaurant. Patrons take an elevator down two floors and walk into a restaurant surrounded with windows looking out on a beautiful coral reef.

This reef was created from pieces of broken coral. Typically, when coral breaks, the pieces become infected and die. But researchers have learned how to treat the coral fragments with antibiotics and to store them while they are healing in large tanks of fresh seawater. Yosef has such a facility, and when divers find broken pieces of coral in the reserve near Yosef's restaurant, they bring them to his coral hospital. After several months of healing, the fragments are taken to the underwater area outside the Red Sea Star Restaurant's windows where they are wired to panels of iron mesh cloth. The corals grow and cover the iron matrix. Then fish and other creatures show up.

Similarly, other damaged coral reefs are being restored. In 20 different countries, scientists have increased the revival and growth rate of coral on submerged metal structures by exposing them to low-voltage electricity.

Using his creativity and working with nature, Yosef has helped to create a marine ecosystem that people can view and enjoy while they dine at his restaurant. At the same time, he has helped to restore and preserve aquatic biodiversity.

---

of California in 2007 began establishing the nation's most extensive network of MPAs where fishing will be banned or strictly limited. Conservation biologists say this could be a model for other MPAs.

Establishing a Global Network of Marine Reserves: An Ecosystem Approach to Marine Sustainability

Many scientists and policymakers call for adopting an entirely new approach to managing and sustaining marine biodiversity and the important ecological and economic services provided by the seas. The primary objective of this *ecosystem approach* is to protect and sustain whole marine ecosystems for current and future generations instead of focusing primarily on protecting individual species.

The cornerstone of this ecological approach is to establish a global network of fully protected *marines reserves*, some of which already exist. These areas are put off-limits to destructive human activities in order to enable their ecosystems to recover and flourish. This global network would include large reserves on the high seas, especially near highly productive nutrient upwelling areas (Figure 7-2, p. 142), and a mixture of smaller reserves in coastal zones that are adjacent to well-managed, sustainable commercial fishing areas. This would encourage local fishers and coastal communities to support them and participate in determining their locations. Some reserves could be made temporary or moveable to protect migrating species such as turtles. Governments could use satellite technologies to update fishing fleets about the locations of designated reserves.

Such reserves would be closed to extractive activities such as commercial fishing, dredging, and mining, as well as to waste disposal. Most reserves in the proposed global network would permit less harmful activities such as recreational boating, shipping, and in some cases, certain levels of small-scale, nondestructive fishing. However, most reserves would contain core zones where no human activity is allowed. Outside the reserves, commercial fisheries would be managed more sustainably by use of an ecosystem approach instead of the current approach, which focuses on individual species without considering their roles in the marine ecosystems where they live.

Marine reserves work and they work fast (see *The Habitable Planet,* Video 9, at **www.learner.org/resources/series209.html**). Scientific studies show that within fully protected marine reserves, fish populations double, fish size grows by almost a third, reproduction triples, and species diversity increases by almost one-fourth. Furthermore, this improvement occurs within 2–4 years after strict protection begins, and it lasts for decades (**Concept 11-2**). Research also shows that reserves benefit nearby fisheries, because fish move into and out of the reserves, and currents carry fish larvae produced inside reserves to adjacent fishing grounds, thereby bolstering the populations there.

In 2008, the Pacific island nation of Kiribati created the world's largest protected marine reserve. This California-sized area is found about halfway between the Pacific islands of Fiji and Hawaii. In 2006, the United States created the world's second largest protected reserve northwest of the U.S. state of Hawaii. The area is about the size of the U.S. state of Montana and supports more than 7,000 marine species, including the endangered Hawaiian monk seal (Figure 11-5) and the endangered green sea turtle (see photo on the title page of this book).

Still, less than 1% of the world's oceans are closed to fishing and other harmful human activities in marine reserves and only 0.1% is fully protected—compared to 5% of the world's land. Thus, we have reserved essentially 99.9% of the world's oceans to use as we see fit. Furthermore, many current marine reserves are too small to protect most of the species within them and do not provide adequate protection from illegal fishing or from pollution that flows from the land into coastal waters.

In 2006, a statement signed by 161 leading marine scientists called for urgent action to create a global network of fully protected marine reserves. Many marine scientists call for fully protecting at least 30% of the world's oceans as marine reserves, and some call for protecting up to 50%. They also urge connecting the global network of marine reserves, especially those in coastal waters, with protected corridors. This would also help species to move to different habitats in the process of adapting to the effects of ocean warming, acidification, and many forms of ocean pollution.

Establishing and managing a global network of marine reserves would cost an estimated $12–14 billion a year and create more than 1 million jobs, according to a 2004 study by the World Wildlife Fund International and Great Britain's Royal Society for Protection of Birds. This investment in protecting aquatic biodiversity and regenerating fisheries is roughly equal to the amount currently spent by governments on subsidies for the fishing industry, which conservationists say encourage overfishing.

THINKING ABOUT
Marine Reserves

Do you support setting aside at least 30% of the world's oceans as fully protected marine reserves? Explain. How would this affect your life?

Protecting Marine Biodiversity Requires Commitments from Individuals and Communities

There is hope for significant progress in sustaining marine biodiversity, but it will require that we change our ways—and soon. For example, IUCN and The Nature Conservancy scientists reported in 2006 that the world's coral reefs and mangrove forests could survive currently projected global warming if we relieve other stressors such as overfishing and pollution. And while some coral species may be able to adapt to warmer temperatures, they may not have enough time to do this unless we act now to slow down the projected rate of global warming.

Increasing ocean acidity could have a major impact on corals and other marine organisms that build shells and skeletal structures out of calcium carbonate, which can dissolve at certain acidity levels. Increasing ocean acidity is likely to have serious impacts on the biodiversity and functioning of coral reefs. A 2005 report by the United Kingdom's Royal Society concluded that there was no way to reverse the widespread chemical and biological affects of increasing ocean acidification except by sharply reducing human inputs of CO_2 into the atmosphere, without delay.

To deal with these problems, communities must closely monitor and regulate fishing and coastal land development and prevent pollution from land-based activities. More important, each of us can make careful choices in purchasing only sustainably harvested seafood. Coastal residents must also think carefully about the chemicals they put on their lawns, and the kinds of waste they generate and where it ends up. And individuals can reduce their carbon footprints to slow climate change and its numerous harmful effects on marine and other ecosystems, as discussed in more detail in Chapter 19.

One strategy emerging in some coastal communities is *integrated coastal management*—a community-based ef-

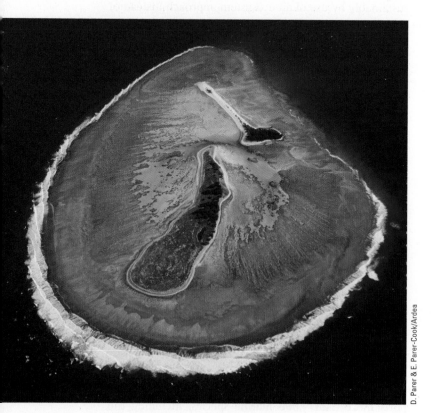

Figure 11-11 An atoll of Australia's Great Barrier Reef.

D. Parer & E. Parer-Cook/Ardea

fort to develop and use coastal resources more sustainably (**Concept 11-2**). Australia manages its huge Great Barrier Reef Marine Park this way, and more than 100 integrated coastal management programs are being developed throughout the world. Figure 11-11 shows an atoll of Australia's Great Barrier Reef, which employs integrated coastal management programs.

The overall aim of such programs is for fishers, business owners, developers, scientists, citizens, and politicians to identify shared problems and goals in their use of marine resources. The idea is to develop workable, cost-effective, and adaptable solutions that help to preserve biodiversity and environmental quality while also meeting various economic and social goals.

This requires all participants to seek reasonable short-term trade-offs that can lead to long-term ecological and economic benefits. For example, fishers might have to give up fishing in certain areas until stocks recover enough to restore biodiversity in those areas, which might then provide fishers with a more sustainable future for their businesses.

11-3 How Should We Manage and Sustain Marine Fisheries?

▶ **CONCEPT 11-3** Sustaining marine fisheries will require improved monitoring of fish populations, cooperative fisheries management among communities and nations, reduction of fishing subsidies, and careful consumer choices in seafood markets.

Estimating and Monitoring Fishery Populations Is the First Step

The first step in protecting and sustaining the world's marine fisheries is to make the best possible estimates of their fish populations (**Concept 11-3**). The traditional approach has used a *maximum sustained yield* (MSY) model to project the maximum number of fish that can be harvested annually from a fish stock without causing a population drop. However, the MSY concept has not worked very well because of the difficulty in estimating the populations and growth rates of fish stocks. Also, harvesting a particular species at its estimated maximum sustainable level can affect the populations of other target and nontarget fish species and other marine organisms.

In recent years, some fishery biologists and managers have begun using the *optimum sustained yield* (OSY) concept. It attempts to take into account interactions among species and to provide more room for error. Similarly, another approach is *multispecies management* of a number of interacting species, which takes into account their competitive and predator–prey interactions. An even more ambitious approach is to develop complex computer models for managing multispecies fisheries in *large marine systems*. However, it is a political challenge to get groups of nations to cooperate in planning and managing such large systems.

There are uncertainties built into any of these approaches because there is much to learn about the biology of fishes and because of changing ocean conditions.

As a result, many fishery and environmental scientists are increasingly interested in using the *precautionary principle* (**Concept 9-4C**, p. 210) for managing fisheries and large marine systems. This means sharply reducing fish harvests and closing some overfished areas until they recover and until we have more information about what levels of fishing can be sustained.

> **RESEARCH FRONTIER**
>
> Studying fish and their habitats to make better estimates of optimum sustained yields for fisheries. See **academic .cengage.com/biology/miller**.

Some Communities Cooperate to Regulate Fish Harvests

An obvious step to take in protecting marine biodiversity—and therefore fisheries—is to regulate fishing. Traditionally, many coastal fishing communities have developed allotment and enforcement systems that have sustained their fisheries, jobs, and communities for hundreds and sometimes thousands of years. An example is Norway's Lofoten fishery, one of the world's largest cod fisheries. For 100 years, it has been self-regulated, with no participation by the Norwegian government. Cooperation can work (**Concept 11-3**).

However, the influx of large modern fishing boats and international fishing fleets has weakened the ability of many coastal communities to regulate and sustain local fisheries. Community management systems have often been replaced by *comanagement*, in which coastal communities and the government work together to manage fisheries.

In comanagement, a central government typically sets quotas for various species and divides the quotas among communities. The government may also limit fishing seasons and regulate the types of fishing gear that can be used to harvest a particular species. Each community then allocates and enforces its quota among its members based on its own rules. Often communities focus on managing inshore fisheries, and the central government manages the offshore fisheries. When it works, community-based comanagement proves that overfishing is not inevitable.

Government Subsidies Can Encourage Overfishing

A 2006 study by fishery experts U. R. Sumaila and Daniel Pauly estimated that governments around the world give a total of about $30–34 billion per year to fishers to help them keep their businesses running. That represents about a third of all revenues earned through commercial fishing. Of that amount, about $20 billion helps fishers to buy ships, fuel, and fishing equipment; the remaining money pays for research and management of fisheries.

Some marine scientists argue that, each year, $10–14 billion of these subsidies are spent to encourage overfishing and expansion of the fishing industry. At a 2007 meeting of the World Trade Organization, the United States proposed a ban on such subsidies. Actions to slash fishing subsidies were supported by a group of 125 marine scientists from 27 countries.

Many marine scientists also call for stronger global efforts to reduce illegal fishing on the high seas and in coastal waters. Actions could include closing ports and markets to such fishers, checking on the authenticity of ship flags, and prosecuting companies that carry out illegal fishing.

> **THINKING ABOUT**
> **Fishing Subsidies**
>
> What are three possible harmful effects of eliminating government fishing subsidies? Do you think they outweigh the benefits of such an action? Explain.

Some Countries Use the Marketplace to Control Overfishing

Some countries use a market-based system called *individual transfer rights* (ITRs) to control access to fisheries. In such a system, the government gives each fishing

vessel owner a specified percentage of the total allowable catch (TAC) for a fishery in a given year. Owners are permitted to buy, sell, or lease their fishing rights as private property.

The ITR market-based system was introduced in New Zealand in 1986 and in Iceland in 1990. In these countries, there has been some reduction in overfishing and in the sizes of their fishing fleets, and the governments have ended fishing subsidies that encourage overfishing. But enforcement has been difficult, some fishers illegally exceed their quotas, and the wasteful bycatch has not been reduced.

In 1995, the United States introduced tradable quotas to regulate Alaska's halibut fishery, which had declined so much that the fishing season had been cut to only 2 days per year. Some fishers sold their quotas and retired, and the number of fishers declined. Halibut prices and fisher income rose, and with less pressure from the fishing industry, the halibut population recovered. By 2005, the season was 258 days long.

Critics have identified three problems with the ITR approach and have made suggestions for its improvement. *First,* in effect, it transfers ownership of fisheries in publicly owned waters to private commercial fishers but still makes the public responsible for the costs of enforcing and managing the system. Critics suggest collecting fees of up to 5% of the value of any catch from quota holders to pay for these costs.

Second, an ITR system can squeeze out small fishing companies that do not have the capital to buy ITRs from others, and it can promote illegal fishing by companies squeezed out of the market. For example, 20 years after the ITR system was implemented in New Zealand, five companies controlled 85% of the ITRs. Critics suggest limiting the number of rights that any one company can obtain.

Third, TACs are often set too high to prevent overfishing. Scientists argue the limit should be set at 50–90% of the estimated *optimal* sustainable yield. Most fishing industry interests oppose setting stricter rules for ITR systems.

> **THINKING ABOUT**
> **Individual Transfer Rights**
>
> Do you support or oppose widespread use of ITR systems to help control access to fisheries? Explain.

Consumer Choices Can Help to Sustain Fisheries and Aquatic Biodiversity

An important component of sustaining aquatic biodiversity and ecosystem services is bottom-up pressure from consumers demanding *sustainable seafood,* which will encourage more responsible fishing practices. In choosing seafood in markets and restaurants, consumers can make choices that will further help to sustain fisheries (**Concept 11-3**).

One way to enable this is through labeling of fresh and frozen seafood to inform consumers about how and where the fish and shellfish were caught. In the United Kingdom, the Waitrose supermarket food chain provides such information for all of the seafood sold at its fresh fish counters. See information on more sustainable seafood choices and download a convenient pocket guide at **www.seafoodwatch.org**.

Another important component is certification of sustainably caught seafood. The London-based Marine Stewardship Council (MSC) was created in 1997 to support sustainable fishing and to certify sustainably produced seafood. It operates in more than 20 nations. Only certified fisheries are allowed to use the MSC's "Fish Forever" eco-label. This certification shows that the fish were caught using environmentally sound and socially responsible practices. By 2007, some 21 wild capture fisheries worldwide were MSC-certified, 18 more were being assessed, and more than 600 seafood products were available with the MSC eco-label. Even so, by 2007 only about 6% of the world's wild capture fisheries were certified.

In 2006, Wal-Mart, the world's largest food retailer, pledged to sell only "MSC-certified" wild-caught fresh and frozen fish in North America within 3–5 years. If implemented, this will have a significant impact on the sustainability of the fresh and frozen seafood market.

Figure 11-12 summarizes actions that individuals, organizations, and governments can take to manage global fisheries more sustainably and to protect marine biodiversity and ecosystem services.

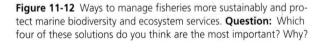

SOLUTIONS

Managing Fisheries

Fishery Regulations

Set catch limits well below the maximum sustainable yield

Improve monitoring and enforcement of regulations

Economic Approaches

Sharply reduce or eliminate fishing subsidies

Charge fees for harvesting fish and shellfish from publicly owned offshore waters

Certify sustainable fisheries

Protect Areas

Establish no-fishing areas

Establish more marine protected areas

Rely more on integrated coastal management

Consumer Information

Label sustainably harvested fish

Publicize overfished and threatened species

Bycatch

Use wide-meshed nets to allow escape of smaller fish

Use net escape devices for seabirds and sea turtles

Ban throwing edible and marketable fish back into the sea

Aquaculture

Restrict coastal locations for fish farms

Control pollution more strictly

Depend more on herbivorous fish species

Nonnative Invasions

Kill organisms in ship ballast water

Filter organisms from ship ballast water

Dump ballast water far at sea and replace with deep-sea water

Figure 11-12 Ways to manage fisheries more sustainably and protect marine biodiversity and ecosystem services. **Question:** Which four of these solutions do you think are the most important? Why?

11-4 How Should We Protect and Sustain Wetlands?

▶ **CONCEPT 11-4** To maintain the ecological and economic services of wetlands, we must maximize preservation of remaining wetlands and restoration of degraded and destroyed wetlands.

Coastal and Inland Wetlands Are Disappearing around the World

Coastal and inland wetlands are important reservoirs of aquatic biodiversity that provide vital ecological and economic services. Despite their ecological value, the United States has lost more than half of its coastal and inland wetlands since 1900, and other countries have lost even more. New Zealand, for example, has lost 92% of its original coastal wetlands, and Italy has lost 95%.

People have drained, filled in, or covered over swamps, marshes, and other wetlands for centuries to create rice fields and to make land available for growing crops, expanding cities, and building roads. Wetlands have also been destroyed in the process of extracting minerals, oil, and natural gas, and in order to reduce diseases such as malaria by eliminating breeding grounds for disease-causing insects.

Wetlands serve as natural filters. Those around Lake Victoria (**Core Case Study**) have historically captured human and animal wastes and kept the lake water clean enough to be used as drinking water for millions of Africans. In 2006, the director of Uganda's wetlands program reported that extensive draining and building on Lake Victoria's coastal wetlands had led to serious water pollution that was killing fish and contaminating drinking water supplies for several countries. He noted that as the waste flow increases, still more wetlands are being destroyed. The Ugandan government is now working to protect its remaining wetlands.

To make matters worse, coastal wetlands in many parts of the world will probably be under water during your lifetime because of rising sea levels caused by global warming. This could seriously degrade aquatic biodiversity supported by coastal wetlands, including commercially important fishes and shellfish and millions of migratory ducks and other birds. It will also diminish the many other ecological and economic services provided by these wetlands.

We Can Preserve and Restore Wetlands

Scientists, land managers, landowners, and environmental groups are involved in intensive efforts to preserve existing wetlands and restore degraded ones (**Concept 11-4**). Laws have been passed to protect existing wetlands. Zoning laws, for example, can be used to steer development away from wetlands.

A U.S. law requires a federal permit to fill in or to deposit dredged material into wetlands occupying more than 1.2 hectares (3 acres). According to the U.S. Fish and Wildlife Service, this law helped cut the average annual wetland loss by 80% since 1969. However, there are continuing attempts by land developers to weaken such wetlands protection. Only about 6% of remaining U.S. inland wetlands are under federal protection, and state and local wetland protection is inconsistent and generally weak.

The stated goal of current U.S. federal policy is zero net loss in the function and value of coastal and inland wetlands. A policy known as *mitigation banking* allows destruction of existing wetlands as long as an equal area of the same type of wetland is created or restored. However, a 2001 study by the National Academy of Sciences found that at least half of the attempts to create new wetlands failed to replace lost ones, and most of the created wetlands did not provide the ecological functions of natural wetlands. The study also found that wetland creation projects often fail to meet the standards set for them and are not adequately monitored.

Creating and restoring wetlands can be profitable. Private investment bankers make money by buying wetland areas and restoring or upgrading them, working with the U.S. Army Corps of Engineers and the EPA. They thus create wetland banks or credits that they can sell to developers. Currently, there are more than 400 wetland banks in the United States with a total of more than $3 billion a year in sales.

It is difficult to restore or create wetlands. Thus, most U.S. wetland banking systems require replacing each hectare of destroyed wetland with 2–3 or more hectares of restored or created wetlands as a built-in ecological insurance policy. **GREEN CAREER:** Wetlands restoration expert

Ecologists argue that mitigation banking should be used only as a last resort. They also call for making sure that new replacement wetlands are created and evaluated *before* existing wetlands are to be destroyed. This example of applying the precautionary principle is often the reverse of what is actually done.

INDIVIDUALS MATTER

Restoring a Wetland

As we learn more about the ecological and economic importance of coastal and inland wetlands, some people have begun to question common practices that damage or destroy these ecosystems. Can we turn back the clock to restore or rehabilitate lost wetlands?

California rancher Jim Callender decided to try. In 1982, he bought 20 hectares (50 acres) of a Sacramento Valley rice field that had been a marsh until the early 1970s. To grow rice, the previous owner had destroyed the marsh by bulldozing, draining, and leveling it, uprooting the native plants, and spraying with chemicals to kill snails.

Callender and his friends set out to restore the marsh. They hollowed out low areas, built up islands, replanted bulrushes and other plants that once were there, reintroduced smartweed and other plants used by migrating and marsh-dwelling birds, and planted fast-growing Peking willows. After years of care, hand planting, and annual seeding with a mixture of watergrass, smartweed, and rice, the land is once again a marsh used by migratory waterfowl.

Jim Callender and others have shown that at least some of the continent's degraded or destroyed wetlands can be reclaimed with scientific knowledge and hard work. Such restoration is useful, but to most ecologists, the real challenge is to protect remaining wetlands from harm in the first place.

Figure 11-13 **Natural capital restoration:** wetland restoration at Fort Erie, Ontario, Canada before (right) and after (left).

Niagara Peninsula Conservation Authority

Niagara Peninsula Conservation Authority

Niagara Peninsula Conservation Authority

THINKING ABOUT

Wetlands Mitigation

Should a new wetland be created and evaluated before anyone is allowed to destroy the wetland it is supposed to replace? Explain.

Wetlands restoration is becoming a big business. While some wetlands restoration projects have failed, others have been very successful (Figure 11-13 and Individuals Matter, at left).

RESEARCH FRONTIER

Evaluating ecological services provided by wetlands, human impacts on wetlands, and how to preserve and restore wetlands. See **academic.cengage.com/biology/miller**.

A good example of an attempt to restore a once vast wetland is that of the Everglades in the U.S. state of Florida, as described in the following case study.

■ CASE STUDY

Can We Restore the Florida Everglades?

South Florida's Everglades (USA) was once a 100-kilometer-wide (60-mile-wide), knee-deep sheet of water flowing slowly south from Lake Okeechobee to Florida Bay (Figure 11-14, p. 268). As this shallow body of water—known as the "River of Grass"—trickled south it created a vast network of wetlands with a variety of wildlife habitats.

Since 1948, a massive water control project has provided south Florida's rapidly growing population with a reliable water supply and flood protection. But is has also contributed to widespread degradation of the original Everglades ecosystem.

Much of the original Everglades has been drained, diverted, paved over, ravaged by nutrient pollution from agriculture, and invaded by a number of plant species. As a result, the Everglades is now less than half its original size. Much of it has also dried out, leaving large areas vulnerable to summer wildfires. And much of its biodiversity has been lost because of reduced water flows, invasive species, and habitat loss and fragmentation from urbanization.

Between 1962 and 1971, the U.S. Army Corps of Engineers transformed the wandering 166-kilometer-long (103-mile-long) Kissimmee River (Figure 11-14, p. 268) into a straight 84-kilometer (56-mile) canal flowing into Lake Okeechobee. The canal provided flood control by speeding the flow of water but it drained large wetlands north of Lake Okeechobee, which farmers then turned into cow pastures.

To help preserve the wilderness in the lower end of the Everglades system, in 1947, the U.S. government established Everglades National Park, which contains about a fifth of the remaining Everglades. But this protection effort did not work—as conservationists had predicted—because the massive water distribution and land development project to the north cut off much of the water flow needed to sustain the park's wildlife.

As a result, 90% of the park's wading birds have vanished, and populations of other vertebrates, from deer to turtles, are down 75–95%. Florida Bay, south of the Everglades is a shallow estuary with many tiny islands, or *keys.* Large volumes of freshwater that once

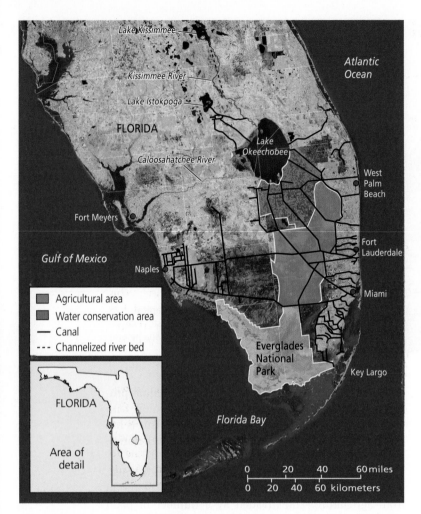

Figure 11-14 The world's largest ecological restoration project is an attempt to undo and redo an engineering project that has been destroying Florida's Everglades (USA) and threatening water supplies for south Florida's rapidly growing population.

flowed through the park into Florida Bay have been diverted for crops and cities, causing the bay to become saltier and warmer. This, along with increased nutrient input from crop fields and cities, has stimulated the growth of large algal blooms that sometimes cover 40% of the bay. This has threatened the coral reefs and the diving, fishing, and tourism industries of the bay and the Florida Keys—another example of harmful unintended consequences.

By the 1970s, state and federal officials recognized that this huge plumbing project was reducing wildlife populations—a major source of tourism income for Florida—and cutting the water supply for the 6 million residents of south Florida. After more than 20 years of political haggling, in 1990, Florida's state government and the federal government agreed on the world's largest ecological restoration project, known as the Comprehensive Everglades Restoration Plan (CERP). The U.S. Army Corps of Engineers is supposed to carry out this joint federal and state plan to partially restore the Everglades.

The project has several ambitious goals. *First,* restore the curving flow of more than half of the Kissimmee

River. *Second,* remove 400 kilometers (250 miles) of canals and levees blocking water flow south of Lake Okeechobee. *Third,* buy 240 square kilometers (93 square miles) of farmland and allow it to be flooded to create artificial marshes that will filter agricultural run-off before it reaches Everglades National Park. *Fourth,* create 18 large reservoirs and underground water storage areas to ensure an adequate water supply for south Florida's current and projected population and for the lower Everglades. *Fifth,* build new canals, reservoirs, and huge pumping systems to capture 80% of the water currently flowing out to sea and return it to the Everglades.

Will this huge ecological restoration project work? It depends not only on the abilities of scientists and engineers but also on prolonged political and economic support from citizens, the state's powerful sugarcane and agricultural industries, and elected state and federal officials.

The carefully negotiated plan has begun to unravel. In 2003, sugarcane growers persuaded the Florida legislature to increase the amount of phosphorus they could discharge and to extend the deadline for reducing such discharges from 2006 to 2016. The project had originally been estimated to cost $7.8 billion and to take 30 years. By 2007, the price tag had risen to $10.5 billion and was expected to go much higher, mostly because of an almost tenfold increase in land prices in South Florida between 2000 and 2007. Overall, funding for the project, especially federal funding, has fallen short of the projected needs, and federal and state agencies are far behind on almost every component of the project. Now the project could take 50 years to complete, or it could be abandoned because of a lack of funding.

According to critics, the main goal of the Everglades restoration plan is to provide water for urban and agricultural development with ecological restoration as a secondary goal. Also, the plan does not specify how much of the water rerouted toward south and central Florida will go to the parched park instead of to increased industrial, agricultural, and urban development. And a National Academy of Sciences panel has found that the plan would probably not clear up Florida Bay's nutrient enrichment problems.

The need to make expensive and politically controversial efforts to undo some of the ecological damage done to the Everglades, caused by 120 years of agricultural and urban development, is another example of failure to heed two fundamental lessons from nature: prevention is the cheapest and best way to go; and when we intervene in nature, unintended and often harmful consequences always occur.

THINKING ABOUT
Everglades Restoration

Do you support carrying out the proposed plan for partially restoring the Florida Everglades, including having the federal government provide half of the funding? Explain.

11-5 How Can We Protect and Sustain Freshwater Lakes, Rivers, and Fisheries?

▶ **CONCEPT 11-5** Freshwater ecosystems are strongly affected by human activities on adjacent lands, and protecting these ecosystems must include protection of their watersheds.

Freshwater Ecosystems Are under Major Threats

The ecological and economic services provided by freshwater lakes, rivers, and fisheries (Figure 8-14, p. 174) are severely threatened by human activities (**Concept 8-5**).

Again, we can use the acronym HIPPCO to summarize these threats. As 40% of the world's rivers have been dammed or otherwise engineered, and as vast portions of the world's freshwater wetlands have been destroyed, aquatic species have been crowded out of at least half of their habitat areas, worldwide. Invasive species, pollution, and climate change threaten the ecosystems of lakes (Case Study, below), rivers, and wetlands. Freshwater fish stocks are overharvested. And increasing human population pressures and global warming make these threats worse.

Sustaining and restoring the biodiversity and ecological services provided by freshwater lakes and rivers is a complex and challenging task, as shown by the story of Lake Victoria (**Core Case Study**) as well as by the following Case Study.

■ CASE STUDY

Can the Great Lakes Survive Repeated Invasions by Alien Species?

Invasions by nonnative species is a major threat to the biodiversity and ecological functioning of lakes, as illustrated by what has happened to the five Great Lakes, located between the United States and Canada.

Collectively, the Great Lakes are the world's largest body of fresh water. Since the 1920s, they have been invaded by at least 162 nonnative species, and the number keeps rising. Many of the alien invaders arrive on the hulls or in bilge water discharges of oceangoing ships that have been entering the Great Lakes through the St. Lawrence Seaway for almost 50 years.

One of the biggest threats, the *sea lamprey*, reached the western lakes through the Welland Canal in Canada as early as 1920. This parasite attaches itself to almost any kind of fish and kills the victim by sucking out its blood (Figure 5-4b, p. 105). Over the years it has depleted populations of many important sport fish species

such as lake trout. The United States and Canada keep the lamprey population down by applying a chemical that kills lamprey larvae in their spawning streams—at a cost of about $15 million a year.

In 1986, larvae of the *zebra mussel* (Figure 9-14, p. 199) arrived in ballast water discharged from a European ship near Detroit, Michigan (USA). This thumbnail-sized mollusk reproduces rapidly and has no known natural enemies in the Great Lakes. As a result, it has displaced other mussel species and depleted the food supply for some other Great Lakes species. The mussels have also clogged irrigation pipes, shut down water intake pipes for power plants and city water supplies, and fouled beaches. They have jammed ship rudders and grown in huge masses on boat hulls, piers, pipes, rocks, and almost any exposed aquatic surface (Figure 11-15). This mussel has also spread to freshwater communities in parts of southern Canada and 18 U.S. states. Currently, the mussels cost the two

Figure 11-15 *Zebra mussels* attached to a water current meter in Lake Michigan. This invader entered the Great Lakes through ballast water dumped from a European ship. It has become a major nuisance and a threat to commerce as well as to biodiversity in the Great Lakes.

countries about $140 million a year—an average of $16,000 per hour.

Sometimes, nature aids us in controlling an invasive alien species. For example, populations of zebra mussels are declining in some parts of the Great Lakes because a native sponge growing on their shells is preventing them from opening up their shells to breathe. However, it is not clear whether the sponges will be effective in controlling the invasive mussels in the long run.

Zebra mussels may not be good for some fish species or for us, but they can benefit a number of aquatic plants. By consuming algae and other microorganisms, the mussels increase water clarity, which permits deeper penetration of sunlight and more photosynthesis. This allows some native plants to thrive and could return the plant composition of Lake Erie (and presumably other lakes) closer to what it was 100 years ago. Because the plants provide food and increase dissolved oxygen, their comeback may benefit certain aquatic animals.

In 1989, a larger and potentially more destructive species, the *quagga mussel,* invaded the Great Lakes, probably discharged in the ballast water of a Russian freighter. It can survive at greater depths and tolerate more extreme temperatures than the zebra mussel can. There is concern that it may spread by river transport and eventually colonize eastern U.S. ecosystems such as Chesapeake Bay and waterways in parts of Florida. In 2007, it was found to have crossed the United States, probably hitching a ride on a boat or trailer being hauled cross-country. It now resides in the Colorado River and reservoir system.

The *Asian carp* may be the next invader. These highly prolific fish, which can quickly grow as long as 1.2 meters (4 feet) and weigh up to 50 kilograms (110 pounds), have no natural predators in the Great Lakes. In less than a decade, this hearty fish with a voracious appetite has dominated sections of the Mississippi River and its tributaries and is spreading toward the Great Lakes. The only barriers are a few kilometers of waterway and a little-tested underwater electric barrier spanning a canal near Chicago, Illinois.

> **THINKING ABOUT**
> **Invasive Species in Lakes**
>
> CORE CASE STUDY
>
> What role did invasive species play in the degradation of Lake Victoria (**Core Case Study**)? What are three ways in which people could avoid introducing more harmful invasive species into lakes?

Managing River Basins Is Complex and Controversial

Rivers and streams provide important ecological and economic services (Figure 11-16). But overfishing, pollution, dams, and water withdrawal for irrigation disrupt these services.

An example of such disruption—one that especially illustrates biodiversity loss—is what happened in the

NATURAL CAPITAL

Ecological Services of Rivers

- Deliver nutrients to sea to help sustain coastal fisheries
- Deposit silt that maintains deltas
- Purify water
- Renew and renourish wetlands
- Provide habitats for wildlife

Figure 11-16 Important ecological services provided by rivers. Currently, these services are given little or no monetary value when the costs and benefits of dam and reservoir projects are assessed. According to environmental economists, attaching even crudely estimated monetary values to these ecosystem services would help to sustain them. **Questions:** Which two of these services do you believe are the most important? Why? Which two of these services do you think we are most likely to decline? Why?

Columbia River, which runs through parts of southwestern Canada and the northwestern United States. It has 119 dams, 19 of which are major generators of inexpensive hydroelectric power. It also supplies water for several major urban areas and for irrigating large areas of agricultural land.

The Columbia River dam system has benefited many people, but it has sharply reduced populations of wild salmon. These migratory fish hatch in the upper reaches of streams and rivers, migrate to the ocean where they spend most of their adult lives, and then swim upstream to return to the places where they were hatched to spawn and die. Dams interrupt their life cycle.

Since the dams were built, the Columbia River's wild Pacific salmon population has dropped by 94% and nine Pacific Northwest salmon species are listed as endangered or threatened. Since 1980, the U.S. federal government has spent more than $3 billion in efforts to save the salmon, but none have been effective.

In another such case—on the lower Snake River in the U.S. state of Washington—conservationists, Native American tribes, and commercial salmon fishers want the government to remove four small hydroelectric dams to restore salmon spawning habitat. Farmers, barge operators, and aluminum workers argue that removing the dams would hurt local economies by reducing irrigation water, eliminating shipping in the affected areas, and reducing the supply of cheap electricity for industries and consumers.

> **HOW WOULD YOU VOTE?**
>
> Should U.S. government efforts to rebuild wild salmon populations in the Columbia River Basin be abandoned? Cast your vote online at **academic.cengage.com/biology/ miller**.

We Can Protect Freshwater Ecosystems by Protecting Watersheds

Sustaining freshwater aquatic systems begins with our realizing that whatever each of us does on land and in the water has some effect on those systems (**Concept 11-5**).

In other words, land and water are always connected in some way. For example, lakes and streams receive many of their nutrients from the ecosystems of bordering land. Such nutrient inputs come from falling leaves, animal feces, and pollutants generated by people, all of which are washed into bodies of water by rainstorms and melting snow. Therefore, to protect a stream or lake from excessive inputs of nutrients and pollutants, we must protect its watershed.

As with marine systems, freshwater ecosystems can be protected through laws, economic incentives, and restoration efforts. For example, restoring and sustaining the ecological and economic services of rivers will probably require taking down some dams and restoring river flows, as may be the case with the Snake River, as mentioned above. And some scientists and politicians have argued for protecting all remaining free-flowing rivers.

With that in mind, in 1968, the U.S. Congress passed the National Wild and Scenic Rivers Act to establish protection of rivers with outstanding scenic, recreational, geological, wildlife, historical, or cultural values. The law classified *wild rivers* as those that are relatively inaccessible (except by trail), and *scenic rivers* as rivers of great scenic value that are free of dams, mostly undeveloped, and accessible in only a few places by roads. These rivers are now protected from widen-ing, straightening, dredging, filling, and damming. But the Wild and Scenic Rivers System keeps only 2% of U.S. rivers free-flowing and protects only 0.2% of the country's total river length.

Sustainable management of freshwater fishes involves supporting populations of commercial and sport fish species, preventing such species from being overfished, and reducing or eliminating populations of harmful invasive species. The traditional way of managing freshwater fish species is to regulate the time and length of fishing seasons and the number and size of fish that can be taken.

Other techniques include building reservoirs and farm ponds and stocking them with fish, fertilizing nutrient-poor lakes and ponds, and protecting and creating fish spawning sites. In addition, fishery managers can protect fish habitats from sediment buildup and other forms of pollution and from excessive growth of aquatic plants due to large inputs of plant nutrients.

Some fishery managers seek to control predators, parasites, and diseases by improving habitats, breeding genetically resistant fish varieties, and using antibiotics and disinfectants. Hatcheries can be used to restock ponds, lakes, and streams with prized species such as trout, and entire river basins can be managed to protect valued species such as salmon. However, all of these practices should be based on on-going studies of their effects on aquatic ecosystems and biodiversity. **GREEN CAREERS:** limnology, fishery management, and wildlife biology

---

RESEARCH FRONTIER

Studying the effects of resource management techniques on aquatic ecosystems. See **academic.cengage.com/biology/miller**.

11-6 What Should Be Our Priorities for Sustaining Biodiversity and Ecosystem Services?

▶ **CONCEPT 11-6** Sustaining the world's biodiversity and ecosystem services will require mapping terrestrial and aquatic biodiversity, maximizing protection of undeveloped terrestrial and aquatic areas, and carrying out ecological restoration projects worldwide.

We Need to Establish Priorities for Protecting Biodiversity and Ecosystem Services

In 2002, Edward O. Wilson, considered to be one of the world's foremost experts on biodiversity, proposed the following priorities for protecting most of the world's remaining ecosystems and species (**Concept 11-6**):

- Complete the mapping of the world's terrestrial and aquatic biodiversity so we know what we have and therefore can make conservation efforts more precise and cost-effective.

- Keep intact the world's remaining old-growth forests and cease all logging of such forests.

- Identify and preserve the world's terrestrial and aquatic biodiversity hotspots and areas where

- deteriorating ecosystem services threaten people and many other forms of life.

- Protect and restore the world's lakes and river systems, which are the most threatened ecosystems of all.

- Carry out ecological restoration projects worldwide to heal some of the damage we have done and to increase the share of the earth's land and water allotted to the rest of nature.

- Find ways to make conservation financially rewarding for people who live in or near terrestrial and aquatic reserves so they can become partners in the protection and sustainable use of the reserves.

There is growing evidence that the current harmful effects of human activities on the earth's terrestrial and aquatic biodiversity and ecosystem services could be reversed over the next 2 decades. Doing this will require implementing an ecosystem approach to protecting and sustaining terrestrial and aquatic ecosystems. According to biologist Edward O. Wilson, such a conservation strategy would cost about $30 billion per year—an amount that could be provided by a tax of one penny per cup of coffee consumed in the world each year.

This strategy for protecting the earth's precious biodiversity will not be implemented without bottom-up political pressure on elected officials from individual citizens and groups. People will also have to vote with their wallets by not buying products and services that destroy or degrade biodiversity. Finally, implementing this strategy will require concerted efforts and cooperation among scientists, engineers, and key people in government and the private sector.

REVISITING — Lake Victoria and Sustainability

This chapter began with a look at how human activities have upset the ecological processes of Africa's Lake Victoria (**Core Case Study**).

Lake Victoria and other cases examined in this chapter illustrate the significant human impacts that have contributed to habitat loss, the spread of invasive species, pollution, climate change, and depletion of commercially valuable fish populations, as well as degradation of aquatic biodiversity in general. We have seen that these threats are growing and are even greater than threats to terrestrial biodiversity.

We also explored ways to manage the world's oceans, fisheries, wetlands, lakes, and rivers more sustainably by applying the four **scientific principles of sustainability**. This means reducing inputs of sediments and excess nutrients, which cloud water, lessen the input of solar energy, and upset the natural cycling of nutrients in aquatic systems. It means placing a high priority on preserving the biodiversity and ecological functioning of aquatic systems and on maintaining natural species interactions that help to prevent excessive population growth of any one species, as happened in Lake Victoria.

By treating the oceans with more respect
and by using them more wisely,
we can obtain more from these life-supporting waters
while also maintaining healthy
and diverse marine ecosystems.

BRIAN HALWEIL

REVIEW

1. Review the Key Questions and Concepts for this chapter on p. 250. Describe how human activities have upset ecological processes in East Africa's Lake Victoria (**Core Case Study**).

2. What are three general patterns of marine biodiversity? Why is marine biodiversity higher **(a)** near coasts than in the open sea and **(b)** on the ocean's bottom than at its surface? Describe the threat to marine biodiversity from bottom trawling. Give two examples of threats to aquatic systems from invasive species. Describe the eco-logical experiment involving carp removal in Wisconsin's Lake Wingra. How does climate change threaten aquatic biodiversity?

3. What is a **fishprint?** Describe the collapse of the cod fishery in the northwest Atlantic and some of its side effects. Describe the effects of trawler fishing, purse-seine fishing, longlining, and drift-net fishing.

4. How have laws and treaties been used to help sustain aquatic species? Describe international efforts to protect

whales from overfishing and premature extinction. Describe threats to sea turtles and efforts to protect them.

5. Describe the use of marine protected areas and marine reserves to help sustain aquatic biodiversity and ecosystem services. What percentage of the world's oceans is fully protected from harmful human activities in marine reserves? Describe the roles of fishing communities and individual consumers in regulating fishing and coastal development. What is integrated coastal management?

6. Describe and discuss the limitations of three ways to estimate the sizes of fish populations. How can the precautionary principle help in managing fisheries and large marine systems? Describe the efforts of local fishing communities in helping to sustain fisheries. How can government subsidies encourage overfishing? Describe the advantages and disadvantages of using individual transfer rights to help manage fisheries.

7. Describe how consumers can help to sustain fisheries, aquatic biodiversity, and ecosystem services by making careful choices in purchasing seafood.

8. What percentage of the U.S. coastal and inland wetlands has been destroyed since 1900? What are three major ecological services provided by wetlands? How does the United States attempt to reduce wetland losses? Describe efforts to restore the Florida Everglades.

9. Describe the major threats to the world's rivers and other freshwater systems. What major ecological services do rivers provide? Describe invasions of the U.S. Great Lakes by nonnative species. Describe ways to help sustain rivers.

10. What are six priorities for protecting terrestrial and aquatic biodiversity? Relate the ecological problems of Lake Victoria (**Core Case Study**) to the four **scientific principles of sustainability**.

Note: Key Terms are in **bold** type.

CRITICAL THINKING

1. Explain how introducing the Nile perch into Lake Victoria (**Core Case Study**) violated all four **scientific principles of sustainability** (see back cover).

2. What difference does it make that the introduction of the Nile perch into Lake Victoria (**Core Case Study**) caused the extinction of more than 200 cichlid fish species? Explain.

3. What do you think are the three greatest threats to aquatic biodiversity and ecosystem services? Why? Why are aquatic species overall more vulnerable to premature extinction resulting from human activities than terrestrial species are? Why is it more difficult to identify and protect endangered marine species than to protect endangered species on land?

4. Why do you think no-fishing marine reserves recover their biodiversity faster and more surely than do areas where fishing is allowed but restricted?

5. Should fishers who harvest fish from a country's publicly owned waters be required to pay the government (taxpayers) fees for the fish they catch? Explain. If your livelihood depended on commercial fishing, would you be for or against such fees?

6. Why do you think that about half of all attempts to create new wetlands fail to replace lost wetlands? Give three reasons why a constructed wetland might not provide the same level of ecological services as a natural wetland. Do you agree with some ecologists' argument that mitiga-

tion wetland banking should be used only as a last resort? Explain.

7. Do you think the plan for restoring Florida's Everglades will succeed? Give three reasons why or why not?

8. Dams on some rivers provide inexpensive hydroelectric power, but they also disrupt aquatic ecosystems. For example, production of hydroelectric power on the Columbia River has resulted in the degradation of the river's Pacific salmon population. Do you think the benefits of these dams justify the ecological damage they cause? Explain. If you see this as a problem, describe a possible solution.

9. Congratulations! You are in charge of protecting the world's aquatic biodiversity and ecosystem services. List the three most important points of your policy to accomplish this goal.

10. List two questions that you would like to have answered as a result of reading this chapter.

Note: See Supplement 13 (p. S78) for a list of Projects related to this chapter.

A *fishprint* provides a measure of a country's fish harvest in terms of area. The unit of area used in fishprint analysis is the global hectare (gha), a unit weighted to reflect the relative ecological productivity of the area fished. When compared with the fishing area's sustainable *biocapacity*, its ability to provide a stable supply of fish year after year in terms of area, its fishprint indicates whether the country's fishing intensity is sustainable. The fishprint and biocapacity are calculated using the following formulae:

$$\text{Fishprint (in gha)} = \frac{\text{metric tons of fish harvested per year}}{\text{productivity in metric tons per hectare} \times \text{weighting factor}}$$

$$\text{Biocapacity (in gha)} = \frac{\text{sustained yield of fish in metric tons per year}}{\text{productivity in metric tons per hectare} \times \text{weighting factor}}$$

The following graph shows the earth's total fishprint and biocapacity. Study it and answer the following questions.

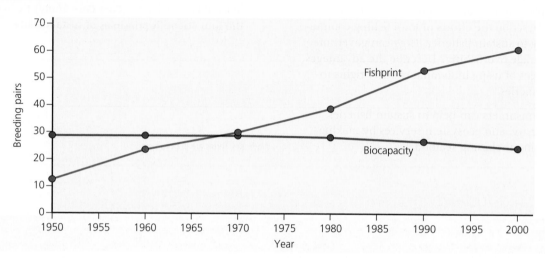

1. Based on the graph,
   a. What is the current status of the global fisheries with respect to sustainability?
   b. In what year did the global fishprint begin exceeding the biological capacity of the world's oceans?
   c. By how much did the global fishprint exceed the biological capacity of the world's oceans in 2000?

2. Assume a country harvests 18 million metric tons of fish annually from an ocean area with an average productivity of 1.3 metric tons per hectare and a weighting factor of 2.68. What is the annual fishprint of that country?

3. If biologists determine that this country's sustained yield of fish is 17 million metric tons per year,
   a. What is the country's sustainable biological capacity?
   b. Is the county's fishing intensity sustainable?
   c. To what extent, as a percentage, is the country under- or overshooting its biological capacity?

Supplements

Measurement Units, Precision, and Accuracy (Chapter 2)

LENGTH

Metric

1 kilometer (km) = 1,000 meters (m)
1 meter (m) = 100 centimeters (cm)
1 meter (m) = 1,000 millimeters (mm)
1 centimeter (cm) = 0.01 meter (m)
1 millimeter (mm) = 0.001 meter (m)

English

1 foot (ft) = 12 inches (in)
1 yard (yd) = 3 feet (ft)
1 mile (mi) = 5,280 feet (ft)
1 nautical mile = 1.15 miles

Metric–English

1 kilometer (km) = 0.621 mile (mi)
1 meter (m) = 39.4 inches (in)
1 inch (in) = 2.54 centimeters (cm)
1 foot (ft) = 0.305 meter (m)
1 yard (yd) = 0.914 meter (m)
1 nautical mile = 1.85 kilometers (km)

AREA

Metric

1 square kilometer (km^2) = 1,000,000 square meters (m^2)
1 square meter (m^2) = 1,000,000 square millimeters (mm^2)
1 hectare (ha) = 10,000 square meters (m^2)
1 hectare (ha) = 0.01 square kilometer (km^2)

English

1 square foot (ft^2) = 144 square inches (in^2)
1 square yard (yd^2) = 9 square feet (ft^2)
1 square mile (mi^2) = 27,880,000 square feet (ft^2)
1 acre (ac) = 43,560 square feet (ft^2)

Metric–English

1 hectare (ha) = 2.471 acres (ac)
1 square kilometer (km^2) = 0.386 square mile (mi^2)
1 square meter (m^2) = 1.196 square yards (yd^2)
1 square meter (m^2) = 10.76 square feet (ft^2)
1 square centimeter (cm^2) = 0.155 square inch (in^2)

VOLUME

Metric

1 cubic kilometer (km^3) = 1,000,000,000 cubic meters (m^3)
1 cubic meter (m^3) = 1,000,000 cubic centimeters (cm^3)
1 liter (L) = 1,000 milliliters (mL) = 1,000 cubic centimeters (cm^3)
1 milliliter (mL) = 0.001 liter (L)
1 milliliter (mL) = 1 cubic centimeter (cm^3)

English

1 gallon (gal) = 4 quarts (qt)
1 quart (qt) = 2 pints (pt)

Metric–English

1 liter (L) = 0.265 gallon (gal)
1 liter (L) = 1.06 quarts (qt)
1 liter (L) = 0.0353 cubic foot (ft^3)
1 cubic meter (m^3) = 35.3 cubic feet (ft^3)
1 cubic meter (m^3) = 1.30 cubic yards (yd^3)
1 cubic kilometer (km^3) = 0.24 cubic mile (mi^3)
1 barrel (bbl) = 159 liters (L)
1 barrel (bbl) = 42 U.S. gallons (gal)

MASS

Metric

1 kilogram (kg) = 1,000 grams (g)
1 gram (g) = 1,000 milligrams (μg)
1 gram (g) = 1,000,000 micrograms ((g)
1 milligram (mg) = 0.001 gram (g)
1 microgram (μg) = 0.000001 gram (g)
1 metric ton (mt) = 1,000 kilograms (kg)

English

1 ton (t) = 2,000 pounds (lb)
1 pound (lb) = 16 ounces (oz)

Metric–English

1 metric ton (mt) = 2,200 pounds (lb) = 1.1 tons (t)
1 kilogram (kg) = 2.20 pounds (lb)
1 pound (lb) = 454 grams (g)
1 gram (g) = 0.035 ounce (oz)

ENERGY AND POWER

Metric

1 kilojoule (kJ) = 1,000 joules (J)
1 kilocalorie (kcal) = 1,000 calories (cal)
1 calorie (cal) = 4.184 joules (J)

Metric–English

1 kilojoule (kJ) = 0.949 British thermal unit (Btu)
1 kilojoule (kJ) = 0.000278 kilowatt-hour (kW-h)
1 kilocalorie (kcal) = 3.97 British thermal units (Btu)
1 kilocalorie (kcal) = 0.00116 kilowatt-hour (kW-h)
1 kilowatt-hour (kW-h) = 860 kilocalories (kcal)
1 kilowatt-hour (kW-h) = 3,400 British thermal units (Btu)
1 quad (Q) = 1,050,000,000,000,000 kilojoules (kJ)
1 quad (Q) = 293,000,000,000 kilowatt-hours (kW-h)

Temperature Conversions

Fahrenheit (°F) to Celsius °C):
    °C (°F − 32.0) ÷ 1.80
Celsius (°C) to Fahrenheit (°F):
    °F = (°C × 1.80) + 32.0

Uncertainty, Accuracy, and Precision in Scientific Measurements

How do we know whether a scientific measurement is correct? All scientific observations and measurements have some degree of uncertainty because people and measuring devices are not perfect.

However, scientists take great pains to reduce the errors in observations and measurements by using standard procedures and testing (calibrating) measuring devices. They also repeat their measurements several times, and then find the average value of these measurements.

It is important to distinguish between accuracy and precision when determining the uncertainty involved in a measurement. *Accuracy* is how well a measurement conforms to the accepted correct value for the measured quantity, based on many careful measurements made over a long time. *Precision* is a measure of reproducibility, or how closely a series of measurements of the same quantity agree with one another.

Good accuracy and good precision

Poor accuracy and poor precision

Poor accuracy and good precision

Figure 1 The distinction between accuracy and precision. In scientific measurements, a measuring device that has not been calibrated to determine its accuracy may give precise or reproducible results that are not accurate.

A dartboard analogy (Figure 1) shows the difference between precision and accuracy. Accuracy depends on how close the darts are to the bull's-eye. Precision depends on how close the darts are to each other. Note that good precision is necessary for accuracy but does not guarantee it. Three closely spaced darts may be far from the bull's-eye.

Graphs and Maps are Important Visual Tools

A *graph* is a tool for conveying information that can be summarized numerically by illustrating the information in a visual format. This information, called *data*, is collected in experiments, surveys, historical studies, and other information-gathering activities. Graphing can be a powerful tool for summarizing and conveying complex information, especially in the ever-expanding fields of environmental science.

In this textbook and accompanying web-based Active Graphing exercises, we use three major types of graphs: line graphs, bar graphs, and pie graphs. Here you will explore each of these types of graphs and learn how to read them. In the web-based Active Graphing exercises, you can try your hand at creating some graphs.

Another important visual tool that can serve the same purpose of communicating complex information is a *map*. Maps can be used to summarize data that vary over small or large areas—from ecosystems to the biosphere, and from backyards to continents. We discuss some aspects of reading maps relating to environmental science at the end of this supplement.

Line Graphs

Line graphs usually represent data that fall in some sort of sequence, such as a series of measurements over time or distance. In most such cases, units of time or distance lie on the horizontal *x-axis*. The possible measurements of some quantity or variable, such as temperature that changes over time or distance, usually lie on the vertical *y-axis*. In Figure 1, the x-axis shows the years between 1950 and 2010, and the y-axis displays the possible values for the annual amounts of oil consumed worldwide in millions of tons, ranging from 0 to 4,000 million (or 4 billion) tons. Usually, the y-axis appears on the left end of the x-axis, although y-axes can appear on the right end, in the middle, or on both ends of the x-axis.

The line on a line graph, sometimes referred to as the *curve*, represents the measurements taken at certain time or distance intervals. In Figure 1, the line represents changes in oil consumption between 1950 and 2007. To find the oil consumption for any year, find that year on the x-axis (a point called the *abscissa*) and run a vertical line from the axis to the consumption line. At the point where your line intersects the consumption line, run a horizontal line to the y-axis. The value at that point on the y-axis, called the *ordinate*, is the amount you are seeking. You can

go through the same process in reverse to find a year in which oil consumption was at a certain point.

Questions

1. What was the total amount of oil consumed in the world in 1990?
2. In about what year between 1950 and 2000 did oil consumption first start declining?
3. About how much oil was consumed in 2007? Roughly how many times more oil was consumed in 2007 than in 1970? How many times more oil was consumed in 2007 than in 1950?

Line graphs have several important uses. One of the most common applications is to compare two or more variables. For example, while Figure 1 shows worldwide oil consumption over a certain time period, Figure 2 enables

us to compare consumption of coal and natural gas during the same period.

Questions

1. In what year was the gap between coal use and natural gas use the widest? In what year was it narrowest? Describe the trend in coal use since 1995.
2. Compare Figures 1 and 2. Among trends in oil, coal, and natural gas use, which one grew the most sharply between 1950 and 1980?

Likewise, we can compare many variables on the same scales. Figure 3 depicts U.S. energy consumption for six different energy resources between 1980 and 2007 and projections to 2030. It uses scales to measure time on the x-axis and energy consumption in quadrillions of British thermal units (a standard unit of energy required to produce a certain amount of heat) on the y-axis.

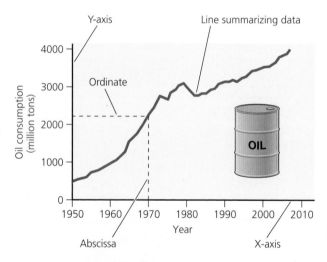

Figure 1 World oil consumption, 1950–2007. (Data from U.S. Energy Information Administration, British Petroleum, International Energy Agency, and United Nations)

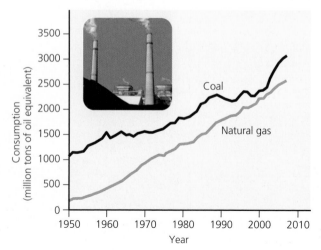

Figure 2 World coal and natural gas consumption, 1950–2007. (Data from U.S. Energy Information Administration, British Petroleum, International Energy Agency, and United Nations)

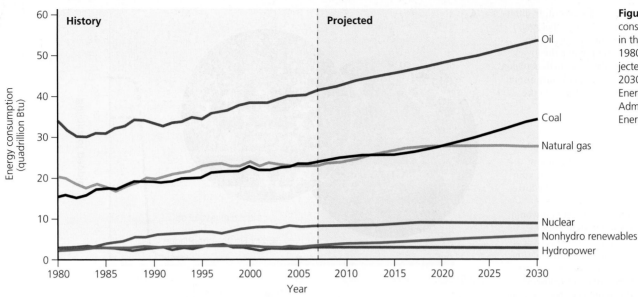

Figure 3 Energy consumption by fuel in the United States, 1980–2007, with projected consumption to 2030. (Data from U.S. Energy Information Administration/Annual Energy Outlook 2007)

Questions

1. Which of the energy sources has seen the least growth in consumption through 2007? Which two sources are projected to have the sharpest growth after 2020? Coal and hydropower are both used mostly to generate electricity. The U.S. consumed about how many times as much coal as hydropower in 2007?

2. Compare Figures 2 and 3. Has the U.S. increased its use of coal and natural gas more sharply or less sharply than the world as a whole?

Figure 4 compares two variables—monthly temperature and precipitation (rain and snowfall) during a typical year in a temperate deciduous forest. However, in this case, the variables are measured on two different scales, so there are two y-axes. The y-axis on the left end of the graph shows a Centigrade temperature scale, while the y-axis on the right shows the range of precipitation measurements in millimeters. The x-axis displays the first letters of each of the 12 month names.

Questions

1. In what month does most precipitation fall? What is the driest month of the year? What is the hottest month?

2. If the temperature curve were almost flat, running throughout the year at about its highest point of about 30 °C, how do you think this forest would be different from what it is? (See Figure 7-15, p. 154.) If the annual precipitation suddenly dropped and remained under 25 centimeters all year, what do you think would eventually happen to this forest?

Line graphs can also show dramatic differences between two types of the same phenomenon, such as growth. *Linear growth* is growth by a given amount in each interval. *Exponential growth* is growth by a fixed percentage of a growing amount in each interval. For example, if you save $1,000 per year under your mattress for 70 years, you will end up with $70,000. But if you invest $1,000 per year and earn 10% interest on your total amount invested every year, and if you keep it all invested for 70 years, you

will end up with $1,024,000. Figure 5 shows the difference between these two types of growth.

Questions

1. How do you think the upper curve would differ if the exponential growth rate were 1% instead of 10%? How would it differ if the rate were 50%? Explain why any exponential growth curve eventually becomes nearly vertical.

2. Which one of the curves in Figure 2 most closely resembles the upper curve in Figure 5? What, if anything, might stop that curve from becoming nearly vertical?

It is important to know that line graphs can give different impressions of data, depending on how they are designed. Changing the measurement ranges on either of the x- or y-axes can change the shape of the curve, and two curves

Temperate deciduous forest

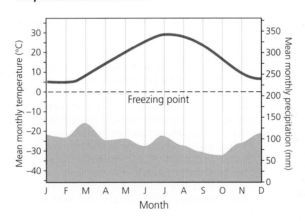

Figure 4 Climate graph showing typical variations in annual temperature (red) and precipitation (blue) in a temperate deciduous forest.

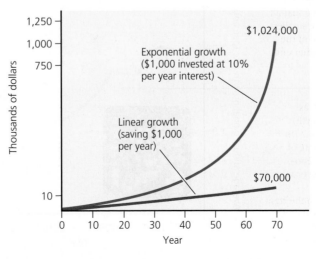

Figure 5 Linear and exponential growth. If resource use, economic growth, or money invested grows exponentially for 70 years at 10%, it will increase 1,024-fold.

representing the same data can look quite different. For example, in Figure 6, the upper graph shows the growth of the human population from 8000 B.C. to the present, while the lower graph shows human population growth between 1950 and 2010. The latter would be only a small segment on the right end of the curve in the upper graph.

Questions

1. What would be your overall impression of human population growth if you saw only the upper graph? What would be your overall impression if you saw only the lower graph?

2. On the upper graph, mark what you would estimate to be the left and right ends of the segment of the curve that fall between 1950 and 2007. Why do you think the slope (steepness) of this segment varies so much from the slope of the curve in the lower graph? Describe the differences between the two graphs that might explain this difference in slopes.

3. You can see from these graphs that by adjusting a graph's time span and the height of its y-axis, you can change the slope of the curve. This can give the reader a different first impression of the data. Does this make the changed graph inaccurate or somehow wrong? Explain. What does this tell you about what you need to look for when reading a graph?

It is also important to consider what aspect of a data set is being displayed on a graph. The creator of a graph can take two different aspects of one data set and create two very different looking graphs that would give two different impressions of the same phenomenon. For example, we must be careful when talking about any type of growth to distinguish the question of whether something is growing from the question of how fast it is growing. While a quantity can keep growing continuously, its rate of growth can go up and down.

One of many important examples of growth used in this book is human population growth. Look again at Figure 6. The two graphs in this figure give you the impression that human population growth has been continuous and uninterrupted, for the most part. However consider Figure 7, which plots the rate of growth of the human population since 1950. Note that all of the numbers on the y-axis, even the smallest ones, represent growth. The lower end of the scale represents slower growth, the higher end, faster growth.

Questions

1. If this graph were presented to you as a picture of human population growth, what would be your first impression? Do you think that reaching a growth rate of 0.5% would relieve those who are concerned about overpopulation? Why or why not?

2. As the curve in Figure 7 proceeds to the right and downward, what do you think will happen to the curve in the lower graph in Figure 6? Explain.

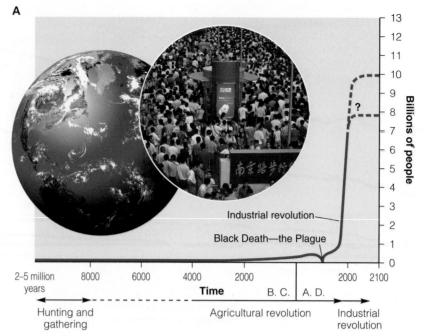

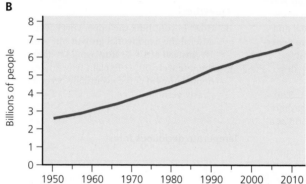

Figure 6 Two graphs showing human population growth. The upper graph spans the time between 8000 B.C. and 2100 A.D. The lower graph runs from 1950 A.D. to 2010.

Finally, but no less important, a common scientific use of the line graph is to show experimental results. Usually, such graphs represent *variables,* which are factors or values that can change. Experimenters measure changes to a *dependent variable*—a variable that changes in response to changes in another variable called the *independent variable.* The latter may be manipulated by experimenters in order to cause changes in the dependent variable. For example, in the Core Case Study of Chapter 2 (p. 28), we report on the Hubbard Brook experiment in which scientists measured changes over time in the presence of soil nutrients in a forest (the dependent variable) in response to removal of trees from the forest (the independent variable).

On a line graph, the range of values for the independent variable is usually placed on the x-axis, while range of values for the independent variable usually appears on the y-axis (for example, see Figure 17-14, p. 455). However, another way to represent changes

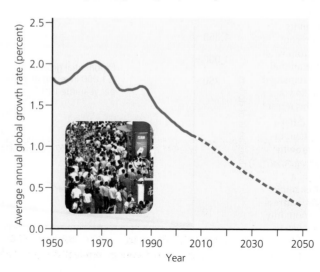

Figure 7 Annual growth rate in world population, 1950–2007, with projections to 2050. (Data from U.N. Population Division and U.S. Census Bureau)

to the independent and dependent variables is to show them on two separate curves. This is useful when an experiment takes place over a long period of time, as did the Hubbard Brook experiment. In Figure 8, the years in which this experiment was conducted appear on the x-axis. The range of values for presence of a soil nutrient called nitrate appears on the y-axis. And two curves were plotted: one showing the values of the dependent variable in the uncut forest (the *control* site), and the other showing the values of the dependent variable in the clear-cut forest (the *experimental* site).

Questions

1. Approximately what was the maximum amount of nitrate lost from the undisturbed (control) forest? Approximately what was the maximum amount of nitrate lost from the clear-cut (experimental) forest? At the point of maximum nitrate loss from the experimental forest, about how many times more nitrate was lost there than in the control forest?

2. In what year do you think the experimental forest was cut? How long did it take, once nutrient loss started there, for the losses to reach their maximum? How long did it take for the forest to regain its pre-experimental level of nutrients?

Bar Graphs

The *bar graph* is used to compare measurements for one or more variables across categories. Unlike the line graph, a bar graph typically does not involve a sequence of measurements over time or distance. The measurements compared on a bar graph usually represent data collected at some point in time or during a well-defined period. For instance, we can compare the

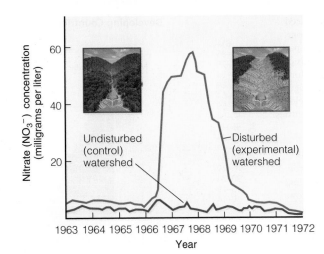

Figure 8 Loss of nitrate ions from a deforested watershed (upper curve), mostly due to precipitation washing away the nutrients, compared with loss of nitrate ions from an undisturbed forest (lower curve). (Data from F. H. Bormann and Gene Likens)

net primary productivity (or NPP, a measure of chemical energy produced by plants in an ecosystem) for different ecosystems, as represented in Figure 9.

In most bar graphs, the categories to be compared are laid out on the x-axis, while the range of measurements for the variable under consideration lies along the y-axis. In our example in Figure 9, the categories (ecosystems) are on the y-axis, and the variable range (NPP) lies on the x-axis. In either case, reading the graph is straightforward. Simply run a line perpendicular to the bar you are reading from the top of that bar (or the right or left end, if it lies horizontally) to the variable value axis. In Figure 9, you can see that the NPP for continental shelf, for example, is close to 1,600 kcal/m²/yr.

Questions

1. About how many times greater is the NPP in a tropical rain forest than the NPP in a savannah?

2. What is the most productive of aquatic ecosystems shown here? What is the least productive?

An important application of the bar graph used in this book is the *age structure diagram* (Figure 10, p. S8), which describes a population by showing the numbers of males and females in certain age groups (see pp. 130–132). Environmental scientists are concerned about human population growth, and one of the key factors determining a particular population's growth rate is the relative numbers of people in various age categories.

In particular, the number of women of childbearing age and younger gives an important clue to whether the population might grow rapidly. If most of the women fall into that category, the population will grow much more quickly than it would if most of the women are beyond their child-bearing years. Likewise, a population with most of its women beyond childbearing age will likely shrink in future years.

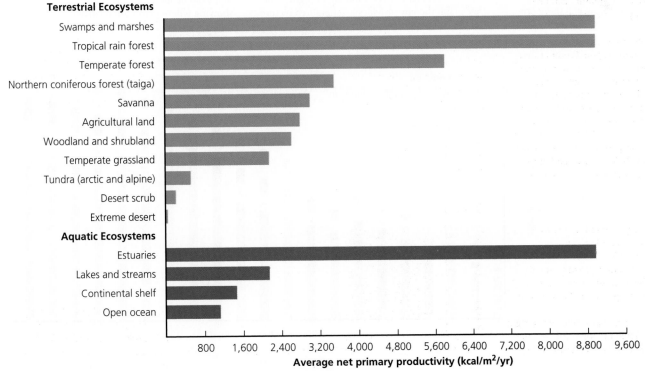

Figure 9 Estimated annual average net primary productivity (NPP) in major life zones and ecosystems, expressed as kilocalories of energy produced per square meter per year (kcal/m²/yr). (Data from R. H. Whittaker, *Communities and Ecosystems*, 2nd ed., New York: Macmillan, 1975)

Note that in Figure 10, the bars are placed horizontally and run both left and right from the center. This allows comparison of two parts of the population across age groups. In this case, values for males lie to the left and values for females lie to the right of the y-axis. The figure contains two separate graphs, one for developing countries and the other for developed countries.

Questions

1. What are the three largest age groups in developing countries? What are the three largest age groups in developed countries? Which group of countries will likely grow more rapidly in coming years?
2. If all girls under the age of 15 had only one child during their lifetimes, how do you think these structures would change over time?

Another interesting application of the bar graph is called a *stacked bar graph*. In such a graph, each bar is actually a combination of several smaller bars representing multiple data groups, stacked up to make one bar divided by different colors or shades to depict the subgroups. Figure 11 is a good example of this, combining population data for six different regions of the world. Each bar contains six different sets of data, all collected in, or projected for, a particular year. This bar graph is a powerful illustration of comparative growth rates of the human populations in the various regions, and it effectively shows the cumulative effects of growth.

Questions

1. Which region will likely grow the fastest between 2010 and 2050? Which region's population will likely stay about the same size?
2. About what proportion of the world's population lived in Asia, Oceania, Africa, and Europe (the eastern hemisphere) and what proportion lived in North America, Latin America, and the Caribbean (the western hemisphere) in 2005? How will these projected proportions change, if at all, by 2050?

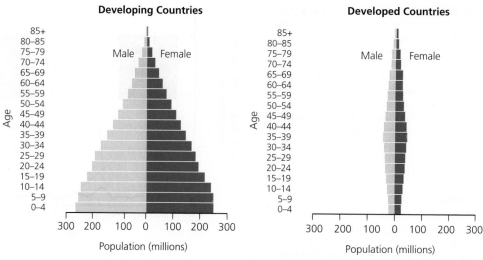

Figure 10 Population structure by age and sex in developing countries and developed countries, 2006. (Data from United Nations Population Division and Population Reference Bureau)

Pie Graphs

Like bar graphs, *pie graphs* illustrate numerical values for two or more categories. But in addition to that, they can also show each category's proportion of the total of all measurements. Usually, the categories are ordered on the graph from largest to smallest, for ease of comparison, although this is not always the case. Also, as with bar graphs, pie graphs are generally snapshots of a set of data at a point in time or during a defined time period. Unlike line graphs, one pie graph cannot show changes over time.

For example, Figure 12 shows how much each major energy source contributes to the world's total amount of energy used in 2006. This graph includes the numerical data used to construct it—the percentages of the total taken up by each part of the pie. But pie graphs can be used without the numerical data included, and such percentages can be estimated roughly. The pie graph thereby provides a generalized picture of the composition of a set of data.

Figure 15-3 (p. 373) shows this and other data in more detail, and illustrates how pie graphs can be used to compare different groups of categories and different data sets. Also, see the Active Graphing exercise for Chapter 15 on the website for this book.

Questions

1. Do you think that the fact that use of natural gas, coal, and oil have all grown over the years (see Figures 1 and 2, p. S4) means that other categories on this graph have shrunk in that time? Explain.
2. Use the projected data in Figure 3 (p. S5) to estimate how the relative sizes of the pie slices for each energy resource might change by 2030.

Reading Maps

Maps can be used for considerably more than showing where places are relative to one another. They can also show comparisons among

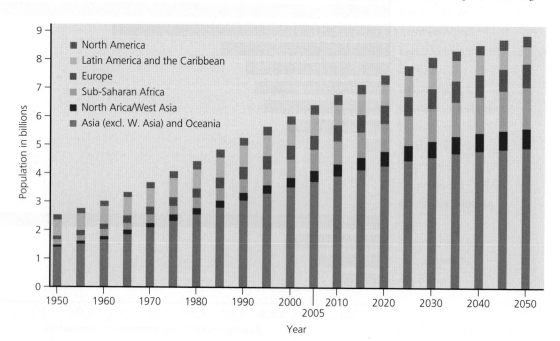

Figure 11 Projected world population growth, by region. (Data from United Nations)

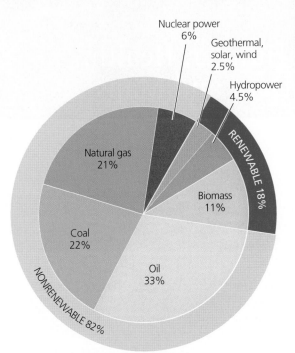

Nuclear power
6%

Geothermal,
solar, wind
2.5%

Hydropower
4.5%

Natural gas
21%

RENEWABLE 18%

Biomass
11%

Coal
22%

Oil
33%

NONRENEWABLE 82%

Figure 12 World energy use by source in 2006. (Data from U.S. Department of Energy, British Petroleum, Worldwatch Institute, and Inernational Energy Institute)

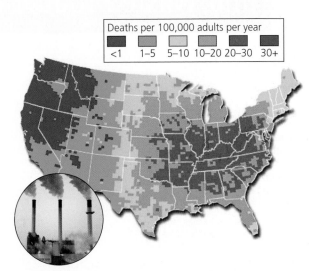

Deaths per 100,000 adults per year
| <1 | 1–5 | 5–10 | 10–20 | 20–30 | 30+ |

Figure 13 Premature deaths from air pollution in the United States, mostly from very small particles added to the atmosphere by coal-burning power plants. (Data from U.S. Environmental Protection Agency)

different areas with regard to any number of factors or conditions. This is the basis for the field of geographical information systems (GIS, see Figure 3-23, p. 73). Using powerful GIS tools, scientists can create detailed maps that show where natural resources are concentrated, for example, within a given region or in the world.

In environmental science, such maps can be very helpful. For example, maps can be used to compare how people or different areas are affected by environmental problems such as air pollution and acid deposition (a form of air pollution). Figure 13 is a map of the United States showing the relative numbers of premature deaths due to

air pollution in the various regions of the country. Figure 14 compares various regions of the country in terms of levels of acidity in precipitation. Study these figures and their captions.

Questions

1. Generally, what part of the country has the lowest level of premature deaths due to air pollution? What part of the country has the highest level? What is the level in the area where you live or go to school?

2. Generally, what part of the country has the highest levels of acidity in its precipitation? What area has the lowest? Do you see any

similarities between the maps in Figures 13 and 14? If so, what are they?

The Active Graphing exercises available for various chapters on the website for this textbook will help you to apply this information. Register and log on to CengageNOW™ using the access code card in the front of your book. Choose a chapter with an Active Graphing exercise, click on the exercise, and begin learning more about graphing. There is also a data analysis exercise at the end of each chapter in this book. Some of these exercises involve analysis of various types of graphs and maps.

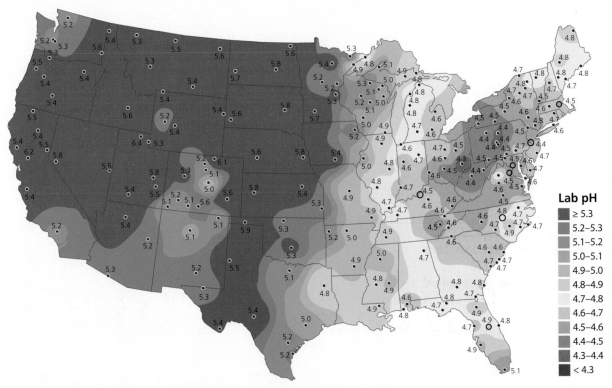

Lab pH

	≥ 5.3
	5.2–5.3
	5.1–5.2
	5.0–5.1
	4.9–5.0
	4.8–4.9
	4.7–4.8
	4.6–4.7
	4.5–4.6
	4.4–4.5
	4.3–4.4
	< 4.3

Figure 14 Measurements of the pH of precipitation at various sites in the lower 48 states in 2005 as a result of acid deposition, mostly from a combination of motor vehicles and coal burning power plants (red dots). The pH is a measure of acidity; the lower the pH, the higher the acidity. For more details see Figure 5, p. S41, in Supplement 6. (Data from National Atmospheric Deposition Program/National Trends Network, 2006)

Economic, Population, Hunger, Health, and Waste Production Data and Maps

Figure 1 Countries of the world.

Map Analysis:

1. What is the largest country (in area) in **(a)** North America, **(b)** Central America, **(c)** South America, **(d)** Europe, and **(e)** Asia?
2. What countries surround **(a)** China, **(b)** Mexico, **(c)** Germany, and **(d)** Sudan?

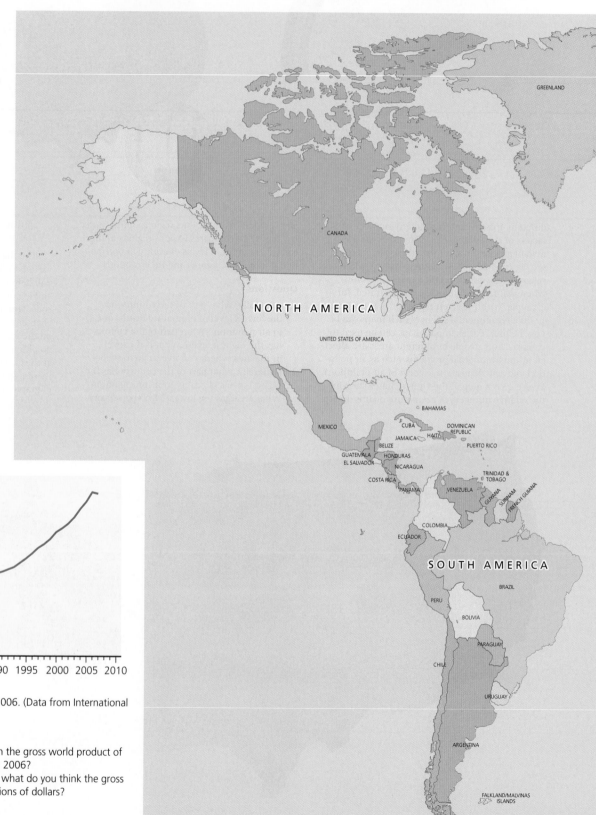

Figure 2 Gross world product, 1950–2006. (Data from International Monetary Fund and the World Bank)

Data and Graph Analysis

1. Roughly how many times bigger than the gross world product of 1985 was the gross world product in 2006?
2. If the current trend continues, about what do you think the gross world product will be in 2010, in trillions of dollars?

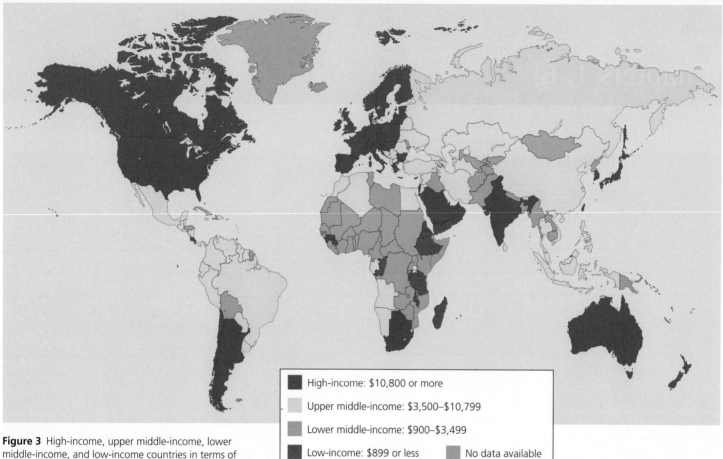

Figure 3 High-income, upper middle-income, lower middle-income, and low-income countries in terms of gross national income (GNI) PPP per capita (U.S. dollars) in 2006. (Data from World Bank and International Monetary Fund)

Legend:
- High-income: $10,800 or more
- Upper middle-income: $3,500–$10,799
- Lower middle-income: $900–$3,499
- Low-income: $899 or less
- No data available

Data and Map Analysis

1. In how many countries is the per capita average income $899 or less? Look at Figure 1 and find the names of three of these countries.
2. In how many instances does a lower-middle- or low-income country share a border with a high-income country? Look at Figure 1 and find the names of the countries involved in three of these instances.

Year	Event	Human population
50,000 B.C.	Hunter-gatherer societies	1.2 million
10,000 B.C.	End of last Ice Age	4 million
8,000 B.C.	Agricultural Revolution	5 million
2,000 B.C.	Contraceptives in use in Egypt	
500 B.C.		100 million
1,000 A.D.		250 million
1347–1351	Black Death (Plague); 75 million people die	
1500		450 million
1750	Industrial Revolution begins in Europe	791 million
1800	Industrial Revolution begins in the United States	
1804		1 billion
1845–1849	Irish potato famine: 1 million people die	
1927		2 billion
1943	Penicillin used against infection	
1952	Contraceptive pill introduced	
1957	Great famine in China; 20 million die	
1961		3 billion
1974		4 billion
1984		5 billion
1987		6 billion
2011	Projected human population:	7 billion
2024	Projected human population:	8 billion
2042	Projected human population:	9 billion

Figure 4 Population timeline, 10,000 B.C.–2008.

Data Analysis

1. About how many years did it take the human population to reach 1 billion? How long after that did it take to reach 2 billion?
2. In about what year was the population half of what it is projected to be in 2011?

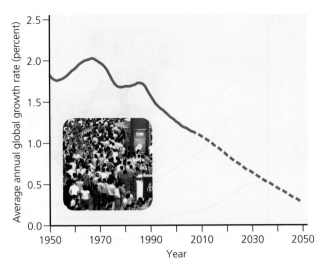

Figure 5 Annual growth rate in world population, 1950–2007 with projections to 2050. (Data from U.N. Population Division and U.S. Census Bureau

Data and Graph Analysis

1. In about what year since 1950 did the growth rate first start increasing? In about what year did it first start decreasing?
2. In about what year will the growth rate reach half of what it was at its peak since 1950, according to projections?

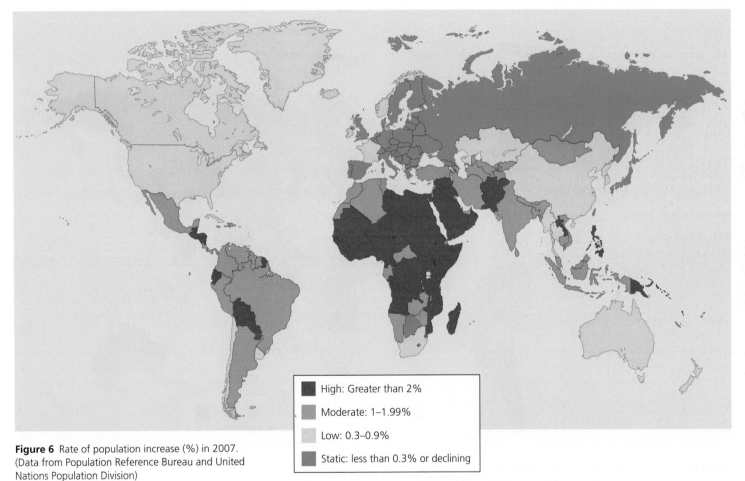

■	High: Greater than 2%
■	Moderate: 1–1.99%
■	Low: 0.3–0.9%
■	Static: less than 0.3% or declining

Figure 6 Rate of population increase (%) in 2007. (Data from Population Reference Bureau and United Nations Population Division)

Data and Map Analysis

1. What continent holds the highest number of countries with high rates of population increase? What continent has the highest number of countries with static rates? (See Figure 1 on pp. S10–S11 for country and continent names.)
2. For each category on this map, name the two countries that you think are largest in terms of total area?

Figure 7 Total fertility rate for the world, developed regions, and less developed regions, 1950–2007, with projection to 2050 (based on medium population projections). (Data from U.N. Population Division)

Data and Graph Analysis

1. What are two conclusions you can draw from comparing these curves?
2. In 2010, about how many more children will be born to each woman in the developing regions than will be born to each woman in the developed regions?

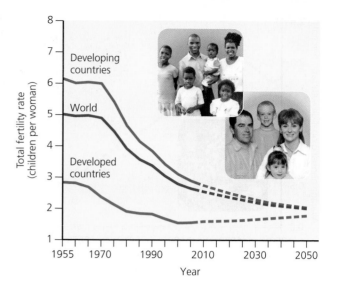

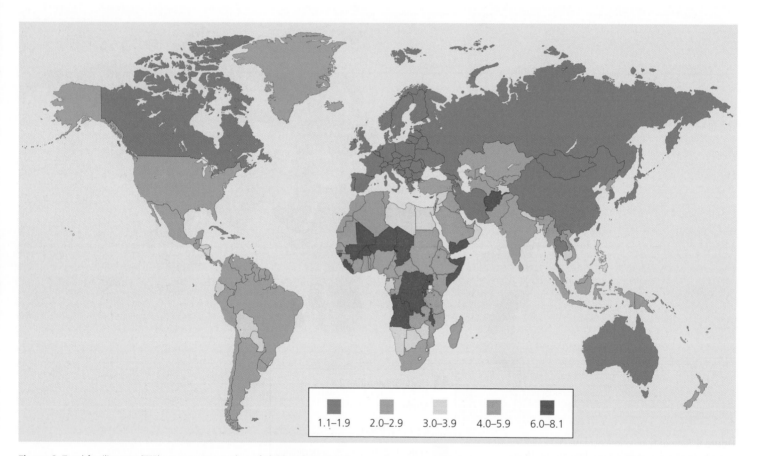

Figure 8 Total fertility rate (TFR), or average number of children born to the world's women throughout their lifetimes, as measured in 2007. (Data from Population Reference Bureau and United Nations Population Division)

Data and Map Analysis

1. Which country in the highest TFR category borders two countries in the lowest TFR category? What are those two countries? (See Figure 1 on pp. S10–S11.)
2. Describe two geographic patterns that you see on this map.

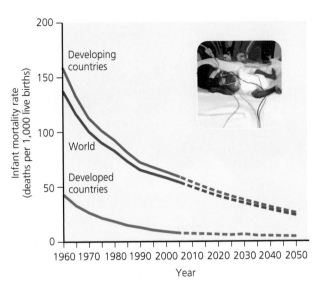

Figure 9 Infant mortality rate for the world, developed regions, and less developed regions, 1950–2007, with projection to 2050 (based on medium population projections). (Data from United Nations Population Division)

Data and Graph Analysis

1. What are two conclusions you can draw from comparing these curves?
2. When the world infant mortality was 100 deaths per 1,000 live births, what were the approximate infant mortality rates for developed and developing regions?

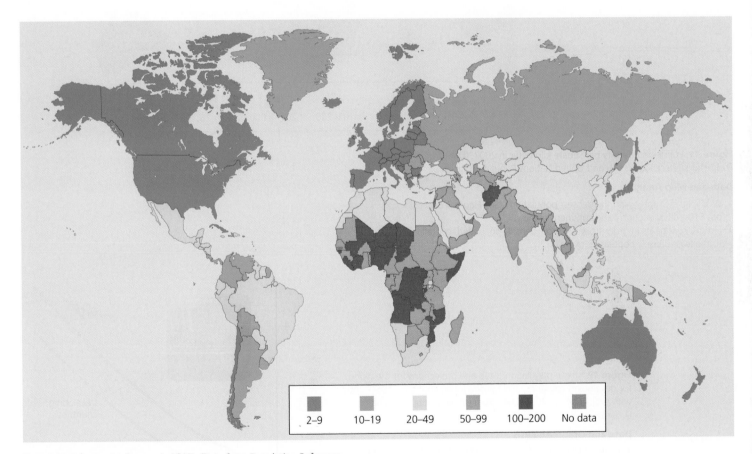

Figure 10 Infant mortality rate in 2007. (Data from Population Reference Bureau and United Nations Population Division)

Data and Map Analysis

1. Describe a geographic pattern that you can see, related to infant mortality rates as reflected on this map.
2. Describe any similarities that you see in geographic patterns between this map and the one in Figure 8.

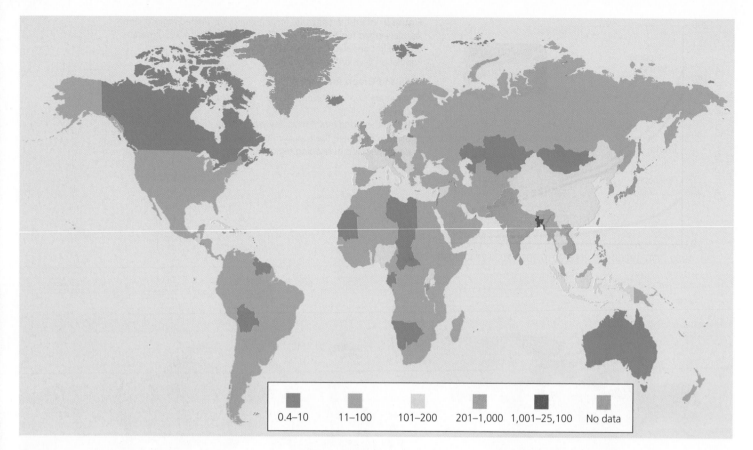

Figure 11 Population density per square kilometer in 2007. (Data from Population Reference Bureau and United Nations Population Division)

| 0.4–10 | 11–100 | 101–200 | 201–1,000 | 1,001–25,100 | No data |

Data and Map Analysis

1. What is the country with the densest population? (See Figure 1 on pp. S10–S11 for country and continent names.)
2. List the continents in order from the most densely populated, overall, to the least densely populated, overall.

Figure 12 Urban population totals and projections for the world, for developing countries, and for developed countries, 1950–2007 with projections to 2030 (Data from United Nations Population Division)

Data and Graph Analysis

1. In about what year did the urban population in developing countries surpass the urban population in developed countries? In about what year was the former twice that of the latter?
2. About how many people will be living in urban areas in developing countries in 2030? In 2030, about how many people will be living in a developing country urban area for every person living in a developed country urban area?

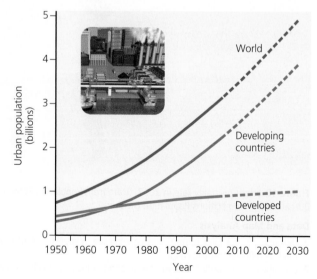

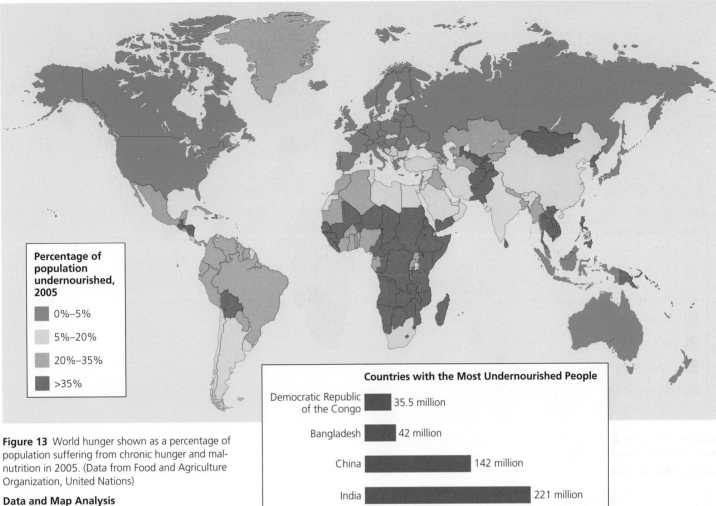

Countries with the Most Undernourished People

Democratic Republic of the Congo — 35.5 million
Bangladesh — 42 million
China — 142 million
India — 221 million

Figure 13 World hunger shown as a percentage of population suffering from chronic hunger and malnutrition in 2005. (Data from Food and Agriculture Organization, United Nations)

Percentage of population undernourished, 2005

- 0%–5%
- 5%–20%
- 20%–35%
- >35%

Data and Map Analysis

1. List the continents in order from the highest percentage of undernourished people to the lowest such percentage. (See Figure 1 on pp. S10–S11 for country and continent names.)
2. On which continent is the largest block of countries that suffer the highest levels of undernourishment? List five of these countries.

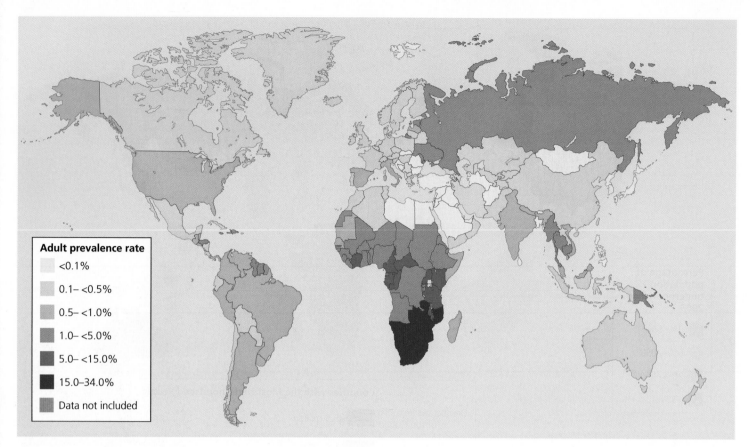

Figure 14 Percentage of adults infected with HIV in 2005. (Data from the World Health Organization)

Data and Map Analysis

1. How many countries have an adult HIV prevalence rate of 5% or higher? List ten of these countries. (See Figure 1 on pp. S10–S11 for country and continent names.)
2. Find an instance where a country with a high rate of HIV prevalence borders a country or countries with a low rate. List the countries involved.

Figure 15 Total and per capita production of municipal solid waste in the United States, 1960–2005. (Data from the U.S. Environmental Protection Agency)

Data and Graph Analysis

1. How much more municipal solid waste was generated in 2005 than in 1960, in millions of tons?
2. In what year did per capita solid waste generation reach a level four times as high as it was in 1960?

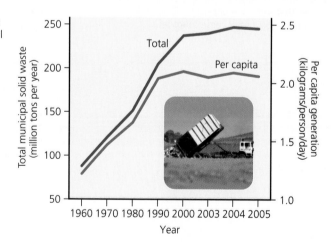

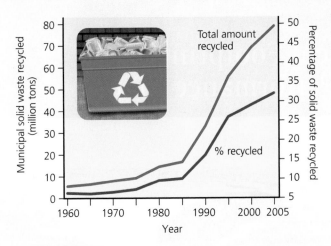

Figure 16 Total amount and percent of municipal solid waste recycled in the United States, 1960–2005. (Data from the U.S. Environmental Protection Agency)

Data and Graph Analysis

1. After 1980, how long did it take for the United States to triple the total amount of materials recycled in 1980?
2. In what 10-year period was the sharpest increase in the percentage of solid waste recycled in the United States?

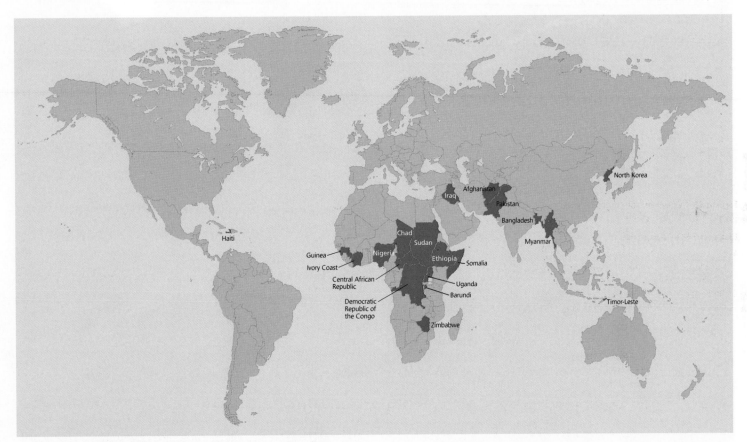

Figure 17 *Top 20 failing states in 2006.* According to a 2007 study by International Alert, 56 countries could become failing or failed states during this century as a result of conflict and instability caused by projected climate change. (Data from the U.S. Central Intelligence Agency, Carnegie Endowment for International Peace, and Fund for Peace)

Biodiversity, Ecological Footprints, and Environmental Performance Maps

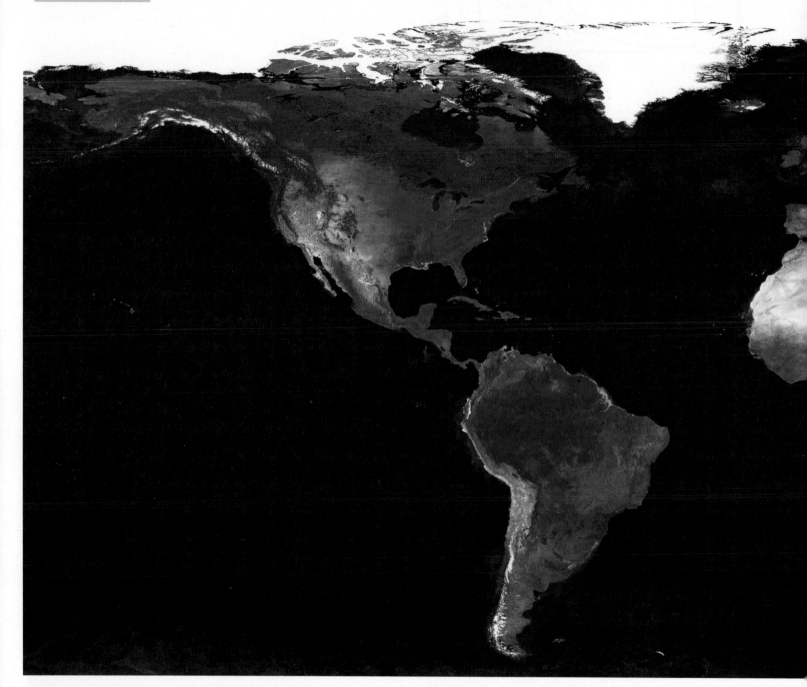

Figure 1 Composite satellite view of the earth showing its major terrestrial and aquatic features.

Data and Map Analysis

1. On what continent does desert make up the largest percentage of the continent's total land area?
2. Which two continents contain large areas of polar ice?

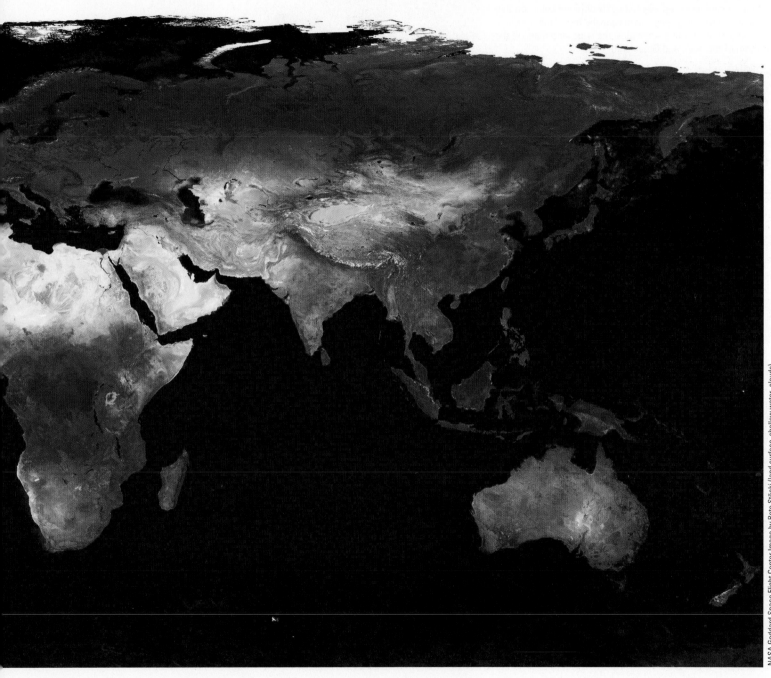

NASA Goddard Space Flight Center Image by Reto Stöcki (land surface, shallow water, clouds). Enhancements by Robert Simmon (ocean color, compositing, 3D globes, animation)

Figure 2 Global map of plant biodiversity. (Used by permission from Kier, et al. 2005. "Global Patterns of Plant Diversity and Floristic Knowledge." *Journal of Biogeography,* Vol. 32, Issue 6, pp. 921–1106, and Blackwell Publishing)

Data and Map Analysis

1. What continent holds the largest continuous area of land that hosts more than 5,000 species per eco-region? On what continent is the second largest area of such land? (See Figure 1, pp. S10–S11, in Supplement 3 for the names of countries and continents.)

2. Of the six categories represented by six different colors on this map, which category seems to occupy the most land area in the world (not counting Antarctica, the large land mass on the bottom of the map, and Greenland)?

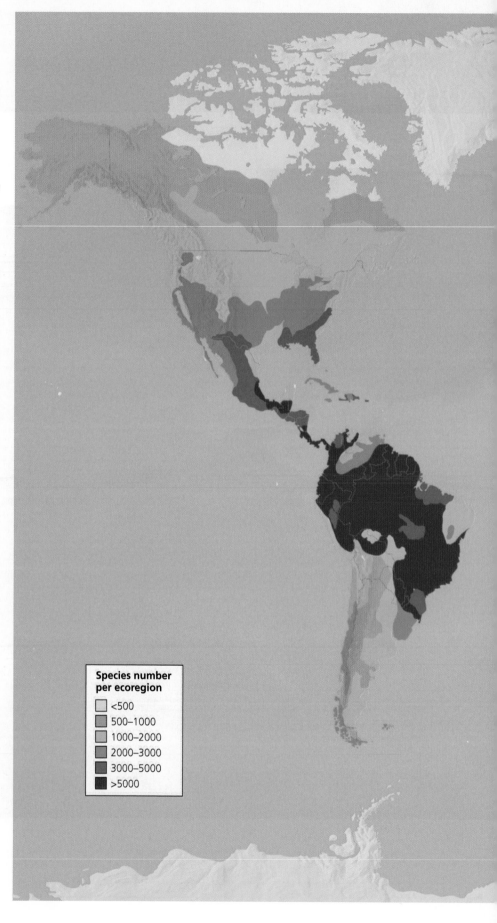

Species number per ecoregion

☐ <500
■ 500–1000
☐ 1000–2000
■ 2000–3000
■ 3000–5000
■ >5000

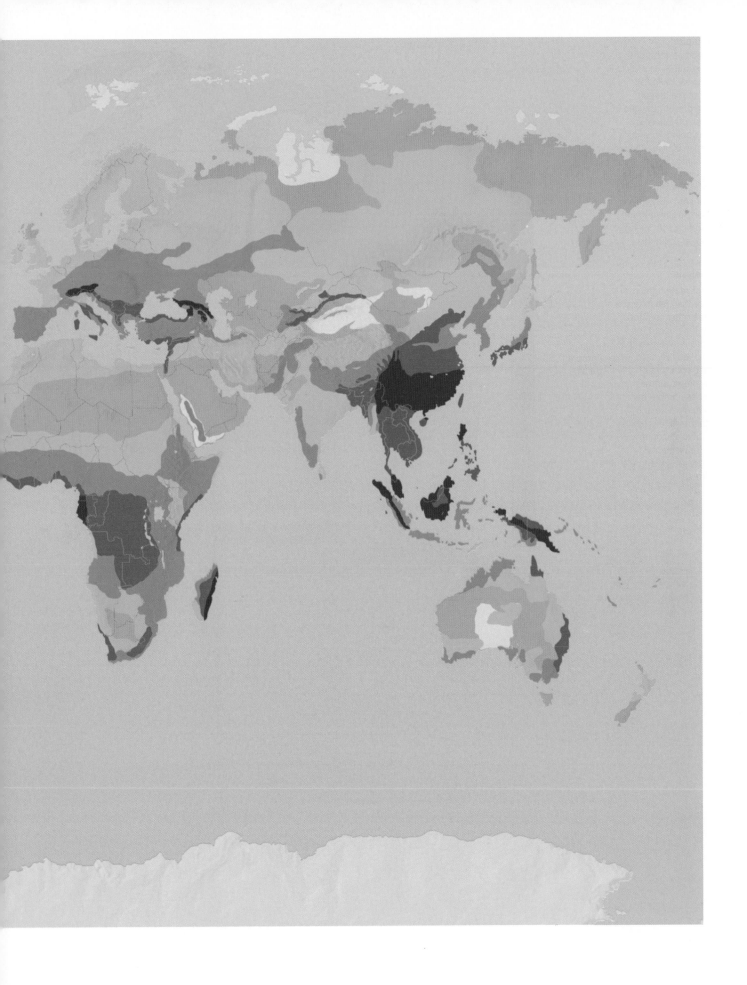

Figure 3 Natural capital degradation: the human footprint on the earth's land surface—in effect the sum of all ecological footprints (Figure 1-10, p. 15) of the human population. Colors represent the percentage of each area influenced by human activities. Excluding Antarctica and Greenland, human activities have directly affected to some degree about 83% of the earth's land surface and 98% of the area where it is possible to grow rice, wheat, or corn. (Data from Wildlife Conservation Society and the Center for International Earth Science Information Network at Columbia University)

Data and Map Analysis

1. What is the approximate human footprint value for the area in which you live? List three other countries in the world that have about the same human footprint value as that of the area where you live.

2. Compare this map with that of Figure 2 and list three countries in which the species number per eco-region is 2000 or more and the human footprint value is higher than 40 in parts of each country. (See Figure 1 in Supplement 3, pp. S10–S11, for names of countries and continents.)

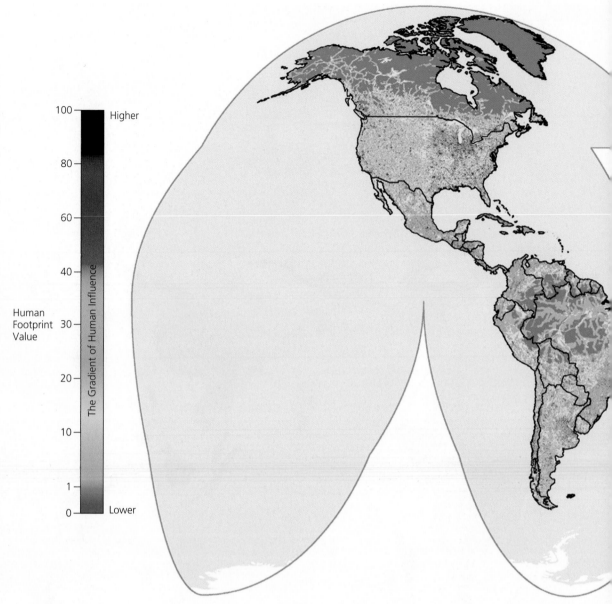

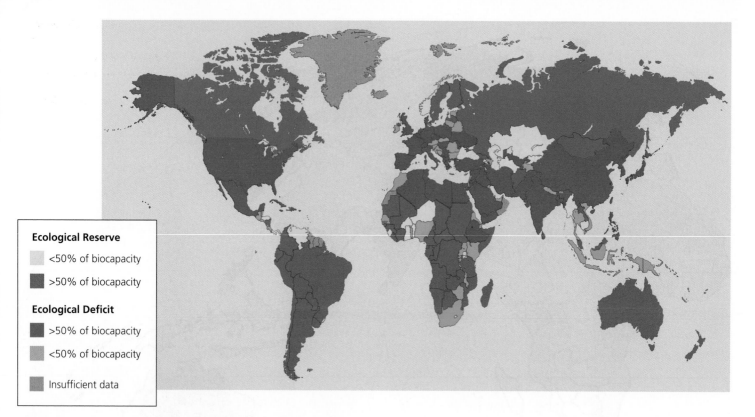

Ecological Reserve

<50% of biocapacity

>50% of biocapacity

Ecological Deficit

>50% of biocapacity

<50% of biocapacity

Insufficient data

Figure 4 *Ecological Debtors and Creditors.* The ecological footprints of some countries exceed their biocapacity, while others still have ecological reserves. (Data from Global Footprint Network)

Data and Map Analysis

1. List five countries, including the three largest, in which the ecological deficit is greater than 50% of biocapacity. (See Figure 1, pp. S10–S11, in Supplement 3 for names of countries and continents.)

2. On which two continents does land with ecological reserves of more than 50% of biocapacity occupy the largest percentage of total land area? Look at Figure 3 and for each of these two continents, list the highest human footprint value that you see on the map.

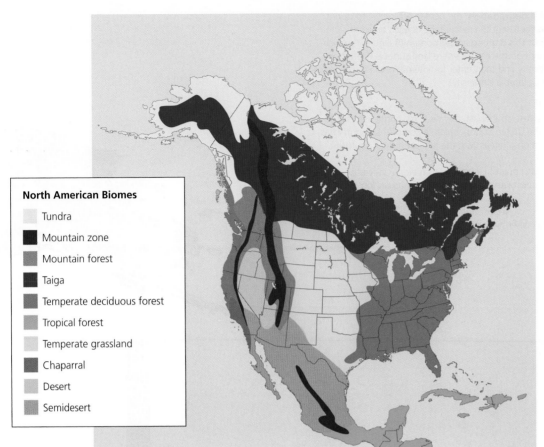

Figure 5 Natural capital: biomes of North America.

Data and Map Analysis

1. What type of biome occupies the most coastal area?
2. Which biome is the rarest in North America?

North American Biomes

- Tundra
- Mountain zone
- Mountain forest
- Taiga
- Temperate deciduous forest
- Tropical forest
- Temperate grassland
- Chaparral
- Desert
- Semidesert

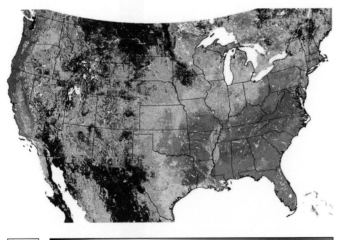

0 0.1 4.5 8.6 12.6 16.7 20.8 24.9 28.9 32.9

Gross primary productivity
(grams of carbon per square meter)

Figure 6 Gross primary productivity across the continental United States, based on remote satellite data. The differences roughly correlate with variations in moisture and soil types. (NASA's Earth Observatory)

Data and Map Analysis

1. Comparing the five northwestern-most states with the five southeastern-most states, which of these regions has the greater variety of levels of gross primary productivity? Which of the regions has the highest levels, overall?
2. Compare this map with that of Figure 5. Which biome in the United States is associated with the highest level of gross primary productivity?

Figure 7 Natural capital degradation: the human ecological footprint in North America. Colors represent the percentage of each area influenced by human activities. This is an expanded portion of Figure 2, which shows the human footprint on the earth's entire land surface. (Data from Wildlife Conservation Society and the Center for International Earth Science Information Network at Columbia University)

Data and Map Analysis

1. Considering the northwest, southwest, middle third, northeast, and southeast regions of the United States, which region of the country has the greatest concentration of areas with the highest human footprint values?

2. Comparing this map with that of Figure 5, which biome in North America is located in areas with the highest overall human footprint values?

3. If you live in North America, what is the approximate value of the human ecological footprint **(a)** where you live and **(b)** where you go to school?

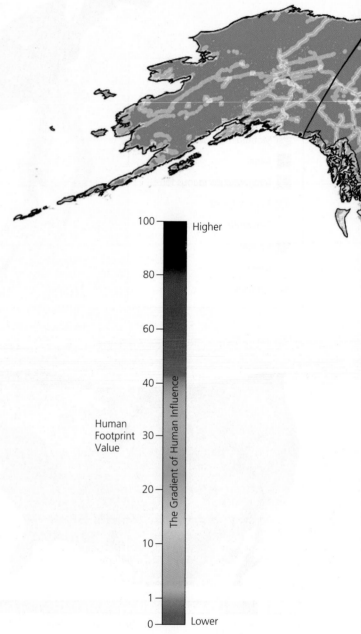

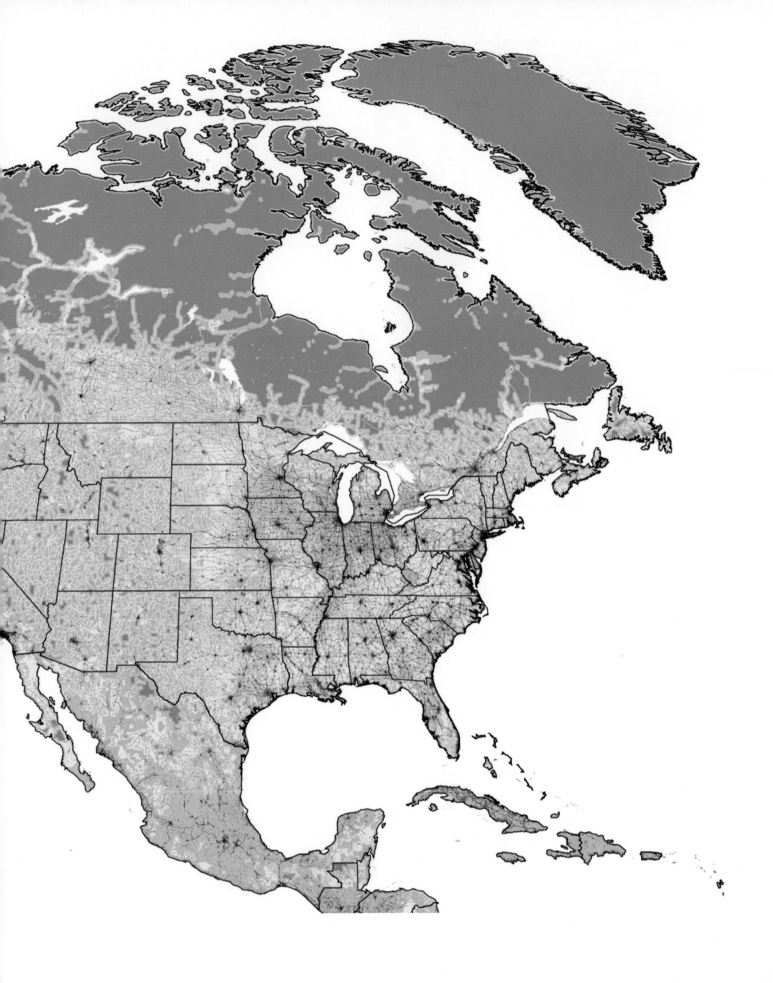

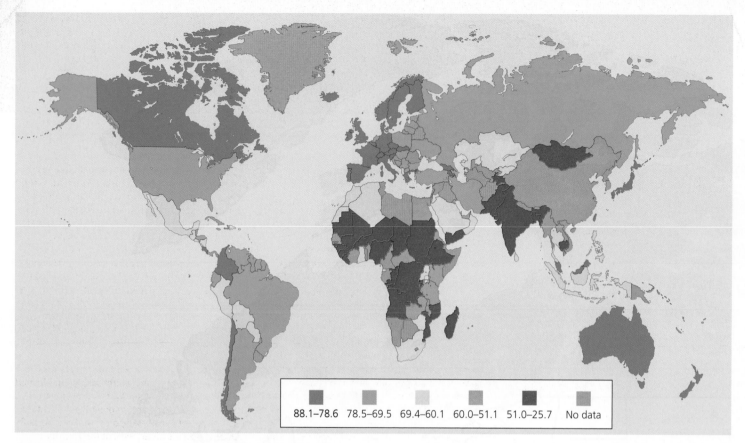

Figure 8 2006 Environmental Performance Index (EPI) uses 16 indicators of environmental health, air quality, water resources, biodiversity and habitat quality, renewable natural resources, and sustainable energy resources to evaluate countries in terms of their ecosystem health and environmental stresses on human health. This map shows the scores by quintiles. (Data from Yale Center for Environmental Law and Policy and the Center for International Earth Science Information Network)

Data and Map Analysis

1. List six countries that lie within the highest EPI range. List six countries that lie within the lowest EPI range. (See Figure 1, pp. S10–S11, in Supplement 3 for country names.)

2. Name a country in the 78.5–69.5 quintile that is completely surrounded by countries in lower quintiles?

Environmental History
(Chapters 1, 2, 5, 7, 8)

S5-1 A Look at Some Past Civilizations

The Norse Greenland Civilization Destroyed Its Resource Base

Greenland is a vast and mostly ice-covered island about three times the size of the U.S. state of Texas. During the 10th century, Viking explorers settled a small, flat portion of this island that was covered with vegetation and located near the water.

In his 2005 book, *Collapse: How Societies Choose to Fail or Succeed*, biogeographer Jared Diamond describes how this 450-year-old Norse settlement in Greenland collapsed in the 1400s from a combination of colder weather in the 1300s and abuse of its soil resources.

Diamond suggests the Norse made three major errors. First, they cut most of the trees and shrubs to clear fields, make lumber, and gather firewood. Without that vegetation, cold winds dried and eroded the already thin soil. The second error was overgrazing, which meant the depletion of remaining vegetation and trampling of the fragile soil.

Finally, when wood used for lumber was depleted, the Norse removed chunks of their turf and used it to build thick walls in their houses to keep out cold winds. Because they removed the turf faster than it could be regenerated, there was less land for grazing so livestock numbers fell. As a result, their food supply and civilization collapsed. Archeological evidence suggests the last residents starved or froze to death.

After about 500 years, nature healed the ecological wounds that the Norse had caused, and Greenland's meadows recovered. In the 20th century, Danes who settled in Greenland reintroduced livestock. Today, more than 56,000 people make their living there by mining, fishing, growing crops, and grazing livestock.

But there is evidence that Greenland's green areas—about 1% of its total land area—are again being overused and strained to their limits. Now Greenlanders have the scientific knowledge to avoid the tragedy of the commons by reducing livestock numbers to a sustainable level, cutting trees no faster than they can be replenished, and practicing soil conservation.

Because of global warming, some of the ice that covers most of Greenland is melting (Figure 19-C, p. 508). This may increase agricultural activity and extraction of mineral and energy resources. Time will tell what beneficial and harmful environmental effects such changes will bring.

The Sumerian Civilization Collapsed Because of Unsustainable Farming

By around 4000 B.C., a highly advanced urban and literate Sumerian civilization had begun emerging on the flood plains of the lower reaches of the Tigris and Euphrates Rivers in parts of what is now Iraq. This civilization developed science and mathematics and built a well-engineered crop irrigation system, which used dams to divert water from the Euphrates River through a network of gravity-fed canals.

The irrigated cropland produced a food surplus and allowed Sumerians to develop the world's first cities and written language (the cunneiform script). But the Sumerians also learned the painful lesson that long-term irrigation can lead to salt buildup in soils and sharp declines in food production.

Poor underground drainage slowly raised the water table to the surface, and evaporation of the water left behind salts that sharply reduced crop productivity—a form of environmental degradation we now call *soil salinization* (Figure 12-14, p. 289). As wheat yields declined, the Sumerians slowed the salinization by shifting to more salt-tolerant barley. But as salt concentrations continued to increase, barley yields declined and food production was undermined.

Around 2000 B.C., this once-great civilization disappeared as a result of such environmental degradation, along with economic decline and invasion by Semitic peoples.

Iceland Has Had Environmental Struggles and Triumphs

Iceland is a Northern European island country slightly smaller than the U.S. state of Kentucky. This volcanic island is located in the North Atlantic Ocean just south of the Arctic Circle between Greenland and Norway, Ireland, and Scotland. Glaciers cover about 10% of the country, and it is subject to earthquakes and volcanic activity.

Immigrants from Scandinavia, Ireland, and Scotland began settling the country during the late 9th and 10th centuries A.D. Since these settlements began, most of the country's trees and other vegetation have been destroyed and about half of its original soils have eroded into the sea. As a result, Iceland suffers more ecological degradation than any other European country.

The early settlers saw what appeared to be a country with deep and fertile soils, dense forests, and highland grasslands similar to those in their native countries. They did not realize that the soils built up by ash from volcanic eruptions were

replenished very slowly and were highly susceptible to water and wind erosion when protective vegetation was removed for growing crops and grazing livestock. Within a few decades, the settlers degraded much of this natural capital that had taken thousands of years to build up.

When the settlers realized what was happening, they took corrective action to save the remaining trees, and they stopped raising ecologically destructive pigs and goats. The farmers joined together to slow soil erosion and preserve their grasslands. They estimated how many sheep the communal highland grasslands could sustain and divided the allotted quotas among themselves.

Icelanders also learned how to fish, how to tap into an abundance of hot springs and heated rock formations for geothermal power, and how to use hydroelectric power from the many rivers. Renewable hydropower and geothermal energy provide about 95% of the country's electricity, and geothermal energy is used to heat 80% of its buildings and to grow most of its fruits and vegetables in greenhouses.

In terms of per capita income, Iceland is one of the world's ten richest countries, and in 2006 it had the world's fifth highest Environmental Sustainability Index. By 2050, Iceland plans to become the world's first country to run its entire economy on renewable hydropower, geothermal energy, and wind, and to use these resources to produce hydrogen for running all of its motor vehicles and ships (see Chapter 16 Core Case Study, p. 399).

> **THINKING ABOUT**
> **Past Civilizations**
>
> What are two ecological lessons that we could learn from these three stories?

S5-2 An Overview of U.S. Environmental History

There Have Been Four Major Eras of U.S. Environmental History

The environmental history of the United States can be divided into four eras. During the *tribal era*, 5–10 million tribal people (now called Native Americans) occupied North America for at least 10,000 years before European settlers began arriving in the early 1600s. These hunter-gatherers generally had sustainable, low-impact ways of life because of their low numbers and low resource use per person.

Next was the *frontier era* (1607–1890) when European colonists began settling North America. Faced with a continent offering seemingly inexhaustible resources, the early colonists developed a **frontier environmental worldview.** They saw a wilderness to be conquered and managed for human use.

Next came the *early conservation era* (1832–1870), which overlapped the end of the frontier era. During this period some people became alarmed at the scope of resource depletion and degradation in the United States. They argued that part of the unspoiled wilderness on public lands should be protected as a legacy to future generations. Most of these warnings and ideas were not taken seriously.

This period was followed by an era—lasting from 1870 to the present—featuring an increased role of the federal government and private citizens in resource conservation, public health, and environmental protection.

The Frontier Era (1607–1890)

During the frontier era, European settlers spread across the land, cleared forests for cropland and settlements, and displaced the Native Americans who generally had lived on the land sustainably for thousands of years.

The U.S. government accelerated this settling of the continent and use of its resources by transferring vast areas of public land to private interests. Between 1850 and 1890, more than half of the country's public land was given away or sold cheaply by the government to railroad, timber, and mining companies, land developers, states, schools, universities, and homesteaders to encourage settlement. This era came to an end when the government declared the frontier officially closed in 1890.

Early Conservationists (1832–1870)

Between 1832 and 1870, some people became alarmed at the scope of resource depletion and degradation in the United States. They urged the government to preserve part of the unspoiled wilderness on public lands owned jointly by all people (but managed by the government) and to protect it as a legacy to future generations.

Two of these early conservationists were Henry David Thoreau (1817–1862) and George Perkins Marsh (1801–1882). Thoreau (Figure 1) was alarmed at the loss of numerous wild species from his native eastern Massachusetts. To gain a better understanding of nature, he built a cabin in the woods on Walden Pond near Concord, Massachusetts, lived there alone for 2 years, and wrote *Life in the Woods,* an environmental classic.*

In 1864, George Perkins Marsh, a scientist and member of Congress from Vermont, published *Man and Nature,* which helped legislators and citizens see the need for resource conservation. Marsh questioned the idea that the country's resources were inexhaustible. He also used scientific studies and case studies to show how the rise and fall of past civilizations were linked to the use and misuse of their soils, water supplies, and other resources. Some of his resource conservation principles are still used today.

What Happened between 1870 and 1930?

Between 1870 and 1930, a number of actions increased the role of the federal government and private citizens in resource conservation and public health (Figure 2). The *Forest Reserve Act of 1891* was a turning point in establishing the responsibility of the federal government for protecting public lands from resource exploitation.

In 1892, nature preservationist and activist John Muir (1838–1914)(Figure 3) founded the Sierra Club. He became the leader of the *preservationist movement,* which called for protecting large areas of wilderness on public lands from human exploitation, except for low-impact recreational activities such as hiking and camping. This idea was not enacted into law until 1964. Muir also proposed and lobbied for creation of a national park system on public lands.

Mostly because of political opposition, effective protection of forests and wildlife did not begin until Theodore Roosevelt (Figure 4, p. S34), an ardent conservationist, became president. His term of office, 1901–1909, has been called the country's *Golden Age of Conservation.*

While in office he persuaded Congress to give the president power to designate public land as federal wildlife refuges. In 1903, Roosevelt established the first federal refuge at Pelican Island off the east coast of Florida for preservation of the endangered brown pelican (Photo 1 in the Detailed Contents), and he added 35 more reserves by 1904. He also more than tripled the size of the national forest reserves.

In 1905, Congress created the U.S. Forest Service to manage and protect the forest reserves. Roosevelt appointed Gifford Pinchot (1865–1946) as its first chief. Pinchot pioneered scientific management of forest resources on public lands. In 1906, Congress passed the *Antiquities Act,* which allows the president to protect areas of scientific or historical interest on federal lands as national monuments. Roosevelt used this act to protect the Grand Canyon and other areas that would later become national parks.

Congress became upset with Roosevelt in 1907, because by then he had added vast tracts to the forest reserves. Congress passed a law banning further executive withdrawals of public forests. On the day before the bill became law, Roosevelt defiantly reserved another large block of land. Most environmental historians view Roosevelt (a Republican) as the country's best environmental president.

Figure 1 Henry David Thoreau (1817–1862) was an American writer and naturalist who kept journals about his excursions into wild areas in parts of the northeastern United States and Canada and at Walden Pond in Massachusetts. He sought self-sufficiency, a simple lifestyle, and a harmonious coexistence with nature.

Early in the 20th century, the U.S. conservation movement split into two factions over how public lands should be used. The *wise-use,* or *conservationist,* school, led by Roosevelt and Pinchot, believed all public lands should be managed wisely and scientifically to provide needed resources. The *preservationist* school, led by Muir wanted wilderness areas on public lands to be left untouched. This controversy over use of public lands continues today.

In 1916, Congress passed the *National Park Service Act.* It declared that parks are to be maintained in a manner that leaves them unimpaired for future generations. The act also established the National Park Service (within the Department of the Interior) to manage the system. Under its first head, Stephen T. Mather (1867–1930), the dominant park policy was to encourage tourist visits by allowing private concessionaires to operate facilities within the parks.

After World War I, the country entered a new era of economic growth and expansion. During the Harding, Coolidge, and Hoover administrations, the federal government promoted increased sales of timber, energy, mineral and other resources found on public lands at low prices to stimulate economic growth.

President Herbert Hoover (a Republican) went even further and proposed that the federal government return all remaining federal lands to the states or sell them to private interests for economic development. But the Great Depression (1929–1941) made owning such lands unattractive to state governments and private investors. The depression was bad news for the country. But some say that without it we might have little if any of the public lands that make up about one-third of the country's land today (Figure 24-5, p. 641).

1891–97 Timber cutting banned on large tracts of public land.

1920–27 Public health boards established in most cities.

| 1870s | 1880s | 1890s | 1900s | 1910s | 1920s |

1918 Migratory Bird Act restricts hunting of migratory birds.

1880 Killer fog in London kills 700 people.

1893 Few remaining American bison given refuge in Yellowstone National Park.

1916 National Park Service Act creates National Park System and National Park Service.

1872 Yellowstone National Park established. American Forestry Association organized by private citizens to protect forests. American Public Health Association formed.

1892 Sierra Club founded by John Muir to promote increased preservation of public land. Killer fog in London kills 1,000 people.

1915 Ecologists form Ecological Society of America.

1912 Public Health Service Act authorizes government investigation of water pollution.

1870 First official wildlife refuge established at Lake Merritt, California.

1891 Forest Reserve Act authorized the president to set aside forest reserves.

1911 Weeks Act allows Forest Service to purchase land at headwaters of navigable streams as part of National Forest System.

1908 Swedish chemist Svante Arrhenius argues that increased emissions from burning fossil fuels will lead to global warming.

1906 Antiquities Act allows president to set aside areas on federal lands as national monuments. Pure Food and Drug Act enacted.

1890 Government declares the country's frontier closed. Yosemite National Park established.

1905 U.S. Forest Service created. Audubon Society founded by private citizens to preserve country's bird species.

1904 Child lead poisoning linked to lead-based paints.

1903 First National Wildlife Refuge at Pelican Island, Florida, established by President Theodore Roosevelt.

1902 Reclamation Act promotes irrigation and water development projects in arid West.

1900 Lacey Act bans interstate shipment of birds killed in violation of state laws.

1870–1930

Figure 2 Examples of the increased role of the federal government in resource conservation and public health and the establishment of key private environmental groups, 1870–1930.
Question: Which two of these events do you think were the most important?

Figure 3 John Muir (1838–1914) was a geologist, explorer, and naturalist. He spent 6 years studying, writing journals, and making sketches in the wilderness of California's Yosemite Valley and then went on to explore wilderness areas in Utah, Nevada, the Northwest, and Alaska. He was largely responsible for establishing Yosemite National Park in 1890. He also founded the Sierra Club and spent 22 years lobbying actively for conservation laws.

Figure 4 Theodore (Teddy) Roosevelt (1858–1919) was a writer, explorer, naturalist, avid birdwatcher, and twenty-sixth president of the United States. He was the first national political figure to bring the issues of conservation to the attention of the American public. According to many historians, he has contributed more than any other president to natural resource conservation in the United States.

What Happened between 1930 and 1960?

A second wave of national resource conservation and improvements in public health began in the early 1930s (Figure 5) as President Franklin D. Roosevelt (1882–1945) strove to bring the country out of the Great Depression. He persuaded Congress to enact federal government programs to provide jobs and to help restore the country's degraded environment.

During this period, the government purchased large tracts of land from cash-poor landowners, and established the *Civilian Conservation Corps* (CCC) in 1933. It put 2 million unemployed people to work planting trees and developing and maintaining parks and recreation areas. The CCC also restored silted waterways and built levees and dams for flood control.

The government built and operated many large dams in the Tennessee Valley and in the arid western states, including Hoover Dam on the Colorado River (Figure 13-14, p. 327). The goals were to provide jobs, flood control, cheap irrigation water, and cheap electricity for industry.

Congress enacted the Soil Conservation Act in 1935. It established the *Soil Erosion Service* as part of the Department of Agriculture to correct the enormous erosion problems that had ruined many farms in the Great Plains states during the depression, as discussed on pp. 303–305. Its name was later changed to the *Soil Conservation Service,* now called the *Natural Resources Conservation Service.* Many environmental historians praise Roosevelt (a Democrat) for his efforts to get the country out of a major economic depression and to help restore environmentally degraded areas.

Federal resource conservation and public health policy during the 1940s and 1950s changed little, mostly because of preoccupation with World War II (1941–1945) and economic recovery after the war.

Between 1930 and 1960, improvements in public health included establishment of public health boards and agencies at the municipal, state, and federal levels; increased public education about health issues; introduction of vaccination programs; and a sharp reduction in the incidence of waterborne infectious diseases, mostly because of improved sanitation and garbage collection.

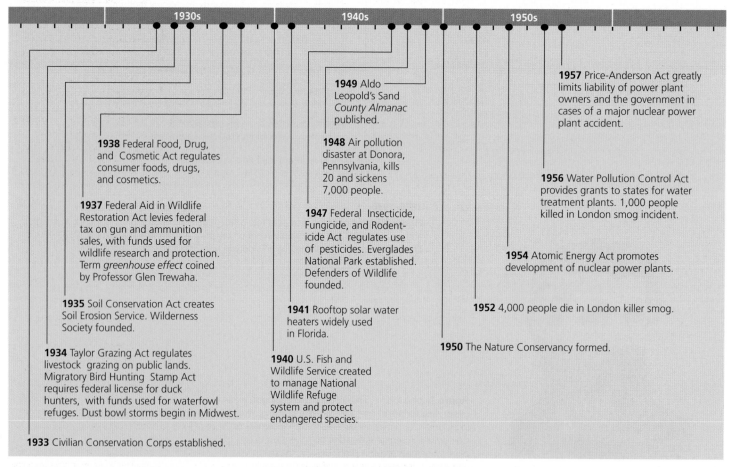

1957 Price-Anderson Act greatly limits liability of power plant owners and the government in cases of a major nuclear power plant accident.

1956 Water Pollution Control Act provides grants to states for water treatment plants. 1,000 people killed in London smog incident.

1954 Atomic Energy Act promotes development of nuclear power plants.

1952 4,000 people die in London killer smog.

1950 The Nature Conservancy formed.

1949 Aldo Leopold's Sand *County Almanac* published.

1948 Air pollution disaster at Donora, Pennsylvania, kills 20 and sickens 7,000 people.

1947 Federal Insecticide, Fungicide, and Rodenticide Act regulates use of pesticides. Everglades National Park established. Defenders of Wildlife founded.

1941 Rooftop solar water heaters widely used in Florida.

1940 U.S. Fish and Wildlife Service created to manage National Wildlife Refuge system and protect endangered species.

1938 Federal Food, Drug, and Cosmetic Act regulates consumer foods, drugs, and cosmetics.

1937 Federal Aid in Wildlife Restoration Act levies federal tax on gun and ammunition sales, with funds used for wildlife research and protection. Term *greenhouse effect* coined by Professor Glen Trewaha.

1935 Soil Conservation Act creates Soil Erosion Service. Wilderness Society founded.

1934 Taylor Grazing Act regulates livestock grazing on public lands. Migratory Bird Hunting Stamp Act requires federal license for duck hunters, with funds used for waterfowl refuges. Dust bowl storms begin in Midwest.

1933 Civilian Conservation Corps established.

1930–1960

Figure 5 Some important conservation and environmental events, 1930–1960. **Question:** Which two of these events do you think were the most important?

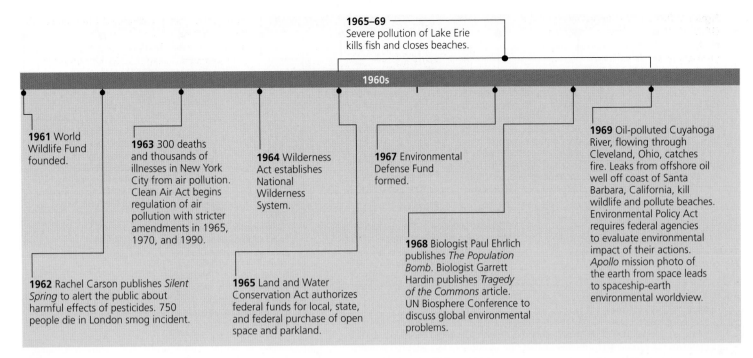

1965–69
Severe pollution of Lake Erie kills fish and closes beaches.

1960s

1961 World Wildlife Fund founded.

1962 Rachel Carson publishes *Silent Spring* to alert the public about harmful effects of pesticides. 750 people die in London smog incident.

1963 300 deaths and thousands of illnesses in New York City from air pollution. Clean Air Act begins regulation of air pollution with stricter amendments in 1965, 1970, and 1990.

1964 Wilderness Act establishes National Wilderness System.

1965 Land and Water Conservation Act authorizes federal funds for local, state, and federal purchase of open space and parkland.

1967 Environmental Defense Fund formed.

1968 Biologist Paul Ehrlich publishes *The Population Bomb*. Biologist Garrett Hardin publishes *Tragedy of the Commons* article. UN Biosphere Conference to discuss global environmental problems.

1969 Oil-polluted Cuyahoga River, flowing through Cleveland, Ohio, catches fire. Leaks from offshore oil well off coast of Santa Barbara, California, kill wildlife and pollute beaches. Environmental Policy Act requires federal agencies to evaluate environmental impact of their actions. *Apollo* mission photo of the earth from space leads to spaceship-earth environmental worldview.

1960s

Figure 6 Some important environmental events during the 1960s. **Question:** Which two of these events do you think were the most important?

What Happened during the 1960s?

A number of milestones in American environmental history occurred during the 1960s (Figure 6). In 1962, biologist Rachel Carson (1907–1964) published *Silent Spring*, which documented the pollution of air, water, and wildlife from use of pesticides such as DDT (see Individuals Matter, p. 295). This influential book helped to broaden the concept of resource conservation to include preservation of the *quality* of the air, water, soil, and wildlife.

Many environmental historians mark Carson's wake-up call as the beginning of the modern **environmental movement** in the United States. It flourished when a growing number of citizens organized to demand that political leaders enact laws and develop policies to curtail pollution, clean up polluted environments, and protect unspoiled areas from environmental degradation.

In 1964, Congress passed the *Wilderness Act*, inspired by the vision of John Muir more than 80 years earlier. It authorized the government to protect undeveloped tracts of public land as part of the National Wilderness System, unless Congress later decides they are needed for the national good. Land in this system is to be used only for nondestructive forms of recreation such as hiking and camping.

Between 1965 and 1970, the emerging science of *ecology* received widespread media attention. At the same time, the popular writings of biologists such as Paul Ehrlich, Barry Commoner, and Garrett Hardin awakened people to the interlocking relationships among population growth, resource use, and pollution.

During that period, a number of events increased public awareness of pollution (Figure 6).

The public also became aware that pollution and loss of habitat were endangering well-known wildlife species such as the North American bald eagle, grizzly bear, whooping crane, and peregrine falcon.

During the 1968 U.S. Apollo 8 mission to the moon, astronauts photographed the earth for the first time from lunar orbit. This allowed people to see the earth as a tiny blue and white planet in the black void of space (Figure 1-1, p. 5), and it led to the development of the *spaceship-earth environmental worldview*. It reminded us that we live on a planetary spaceship that we should not harm because it is the only home we have.

What Happened during the 1970s? The Environmental Decade

During the 1970s, media attention, public concern about environmental problems, scientific research, and action to address environmental concerns grew rapidly. This period is sometimes called the *environmental decade*, or the *first decade of the environment* (Figure 7).

The first annual *Earth Day* was held on April 20, 1970. During this event, proposed by Senator Gaylord Nelson (1916–2005), some 20 million people in more than 2,000 communities took to the streets to heighten awareness and to demand improvements in environmental quality.

Republican President Richard Nixon (1913–1994) responded to the rapidly growing environmental movement. He established the *Environmental Protection Agency* (EPA) in 1970 and supported passage of the *Endangered Species Act of 1973*. This greatly strengthened the role of the federal government in protecting endangered species and their habitats.

In 1978, the *Federal Land Policy and Management Act* gave the *Bureau of Land Management* (BLM) its first real authority to manage the public land under its control, 85% of which is in 12 western states. This law angered a number of western interests whose use of these public lands was restricted for the first time.

In response, a coalition of ranchers, miners, loggers, developers, farmers, some elected officials, and others launched a political campaign known as the *sagebrush rebellion*. It had two major goals. *First*, sharply reduce government regulation of the use of public lands. *Second*, remove most public lands in the western United States from federal ownership and management and turn them over to the states. Then the plan was to persuade state legislatures to sell or lease the resource-rich lands at low prices to ranching, mining, timber, land development, and other private interests. This represented a return to President Hoover's plan to get rid of all public land, which had been thwarted by the Great Depression.

Jimmy Carter (a Democrat), president between 1977 and 1981, was very responsive to environmental concerns. He persuaded Congress to create the Department of Energy in order to develop a long-range energy strategy to reduce the country's heavy dependence on imported oil. He appointed respected environmental leaders to key positions in environmental and resource agencies and consulted with environmental interests on environmental and resource policy matters.

In 1980, Carter helped to create a *Superfund* as part of the *Comprehensive Environment Response, Compensation, and Liability Act* (pp. 582–583) to clean up abandoned hazardous waste sites, including the Love Canal housing development in

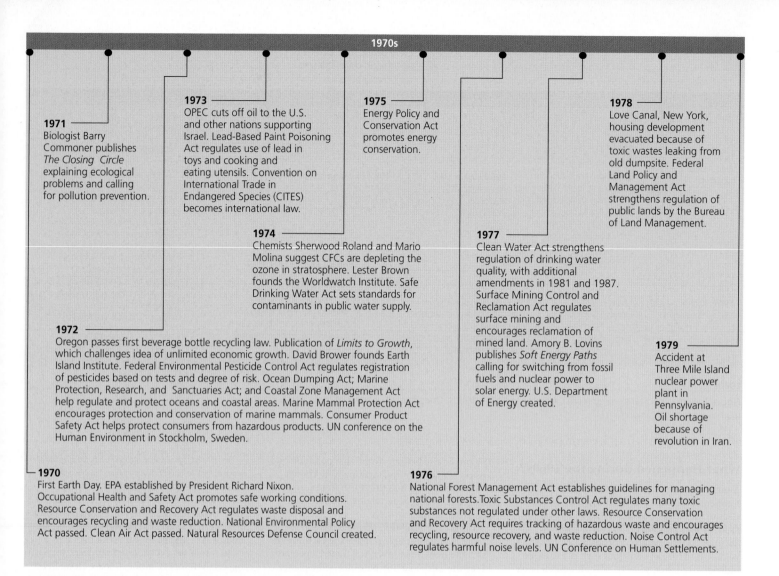

Figure 7 Some important environmental events during the 1970s, sometimes called the *environmental decade*. **Question:** Which two of these events do you think were the most important?

Niagara Falls, New York, which was abandoned when hazardous wastes began leaking into yards, school grounds, and basements.

Carter also used the *Antiquities Act of 1906* to triple the amount of land in the National Wilderness System and double the area in the National Park System (primarily by adding vast tracts in Alaska). He used the Antiquities Act to protect more public land, in all 50 states, than any president before him had done.

What Happened during the 1980s? Environmental Backlash

Figure 8 summarizes some key environmental events during the 1980s that shaped U.S. environmental policy. During this decade, farmers and ranchers and leaders of the oil, coal, automobile, mining, and timber industries strongly opposed many of the environmental laws and regulations developed in the 1960s and 1970s. They organized and funded multiple efforts to

defeat environmental laws and regulations—efforts that persist today.

In 1981, Ronald Reagan (a Republican, 1911–2004), a self-declared *sagebrush rebel* and advocate of less federal control, became president. During his 8 years in office, he angered environmentalists by appointing to key federal positions people who opposed most existing environmental and public land-use laws and policies.

Reagan greatly increased private energy and mineral development and timber cutting on public lands. He also drastically cut federal funding for research on energy conservation and renewable energy resources and eliminated tax incentives for residential solar energy and energy conservation enacted during the Carter administration. In addition, he lowered automobile gas mileage standards and relaxed federal air and water quality pollution standards.

Although Reagan was immensely popular, many people strongly opposed his environmental and resource policies. This resulted in strong

opposition in Congress, public outrage, and legal challenges by environmental and conservation organizations, whose memberships soared during this period.

In 1988, an industry-backed, coalition called the *wise-use movement* was formed. Its major goals were to weaken or repeal most of the country's environmental laws and regulations and destroy the effectiveness of the environmental movement in the United States. Politically powerful coal, oil, mining, automobile, timber, and ranching interests helped back this movement.

Upon his election in 1989, George H. W. Bush (a Republican) promised to be "the environmental president." But he received criticism from environmentalists for not providing leadership on such key environmental issues as population growth, global warming, and loss of biodiversity. He also continued support of exploitation of valuable resources on public lands at giveaway prices. In addition, he allowed some environmental laws to be undercut by the politi-

1980–89 Rise of a strong anti-environmental movement.

1980s

1985 Scientists discover annual seasonal thinning of the ozone layer above Antarctica.

1984 Toxic fumes leaking from pesticide plant in Bhopal, India, kill at least 6,000 people and injure 50,000–60,000. Lester R. Brown publishes first annual *State of the World* report.

1983 U.S. EPA and National Academy of Sciences publish reports finding that buildup of carbon dioxide and other greenhouse gases will lead to global warming.

1980 Superfund law passed to clean up abandoned toxic waste dumps. Alaska National Interest Lands Conservation Act protects 42 million hectares (104 million acres) of land in Alaska.

1986 Explosion of Chernobyl nuclear power plant in Ukraine. Times Beach, Missouri, evacuated and bought by EPA because of dioxin contamination.

1987 Montreal Protocol to halve emissions of ozone-depleting CFCs signed by 24 countries. International Basel Convention controls movement of hazardous wastes from one country to another.

1989 *Exxon Valdez* oil tanker accident in Alaska's Prince William Sound.

1988 Industry-backed wise-use movement established to weaken and destroy U.S. environmental movement. Biologist E. O. Wilson publishes *Biodiversity*, detailing how human activities are affecting the earth's diversity of species.

1980s

Figure 8 Some important environmental events during the 1980s. **Question:** Which two of these events do you think were the most important?

cal influence of industry, mining, ranching, and real estate development interests. They argued that environmental laws had gone too far and were hindering economic growth.

What Happened from 1990 to 2008?

Between 1990 and 2008, opposition to environmental laws and regulations gained strength. This occurred because of continuing political and economic support from corporate backers, who argued that environmental laws were hindering economic growth, and because federal elections gave Republicans (many of whom were generally unsympathetic to environmental concerns) a majority in Congress.

Consequently, during this period, leaders and supporters of the environmental movement have had to spend much of their time and funds fighting efforts to discredit the movement and weaken or eliminate most environmental laws passed during the 1960s and 1970s. They also have had to counter claims by anti-environmental groups that problems such as global warming and ozone depletion are hoaxes or are not very serious and that environmental laws and regulations have hindered economic growth.

In 1993, Bill Clinton (a Democrat) became president and promised to provide national and global environmental leadership. During his 8 years in office, he appointed respected environmental leaders to key positions in environmental and resource agencies and consulted with

environmental interests about environmental policy, as Carter had done.

He also vetoed most of the anti-environmental bills (or other bills passed with anti-environmental riders attached) passed by a Republican-dominated Congress between 1995 and 2000. He announced regulations requiring sport utility vehicles (SUVs) to meet the same air pollution emission standards as cars. Clinton also used executive orders to make forest health the primary priority in managing national forests and to declare many roadless areas in national forests off limits to the building of roads and to logging.

In addition, he used the Antiquities Act of 1906 to protect various parcels of public land in the West from development and resource exploitation by declaring them national monuments. He protected more public land as national monuments in the lower 48 states than any other president, including Teddy Roosevelt and Jimmy Carter. However, environmental leaders criticized Clinton for failing to push hard enough on key environmental issues such as global warming energy policy and global and national biodiversity protection.

During the 1990s, many small and mostly local grassroots environmental organizations sprang up to deal with environmental threats in their local communities. Interest in environmental issues increased on many college campuses and environmental studies programs at colleges and universities expanded. In addition, awareness of important, complex environmental issues, such as sustainability, population growth,

biodiversity protection, and threats from global warming, increased.

In 2001, George W. Bush (a Republican) became president. Like Reagan in the 1980s, he appointed to key federal positions people who opposed or wanted to weaken many existing environmental and public land-use laws and policies because they were alleged to threaten economic growth. Also like Reagan, he did not consult with environmental groups and leaders in developing environmental policies, and he greatly increased private energy and mineral development and timber cutting on public lands.

Bush also opposed increasing automobile gas mileage standards as a way to save energy and reduce dependence on oil imports, and he supported relaxation of various federal air and water quality standards. Like Reagan, he developed an energy policy that emphasized use of fossil fuels and nuclear power with much less support during his first term for reducing energy waste and relying more on renewable energy resources.

In addition, he withdrew the United States from participation in the international Kyoto treaty, designed to help reduce carbon dioxide emissions that can promote global warming and lead to long-lasting climate change. He also repealed or tried to weaken most of the pro-environmental measures established by Bill Clinton. On the other hand, in 2006, he created the world's second largest marine reserve in waters around some of the Hawaiian Islands.

According to leaders of a dozen major environmental organizations Bush, backed by a

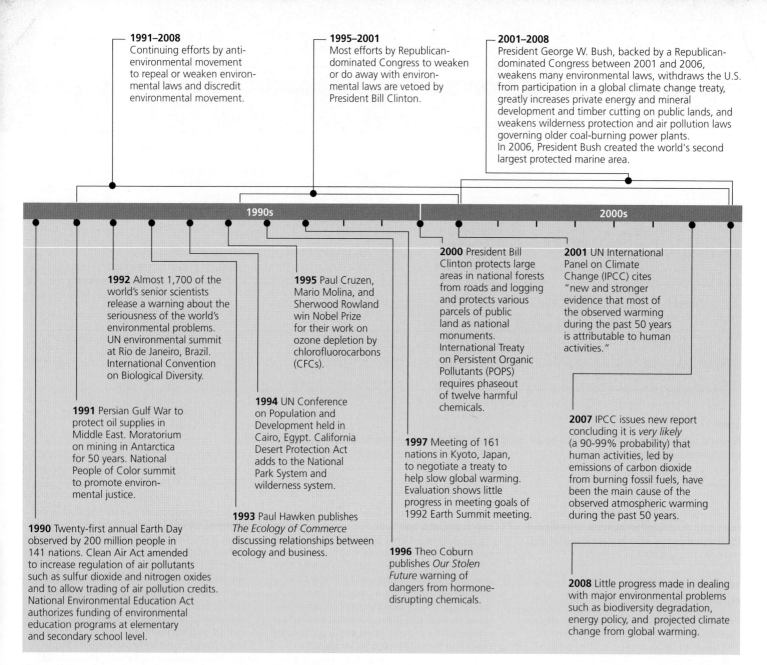

1991–2008 Continuing efforts by anti-environmental movement to repeal or weaken environmental laws and discredit environmental movement.

1995–2001 Most efforts by Republican-dominated Congress to weaken or do away with environmental laws are vetoed by President Bill Clinton.

2001–2008 President George W. Bush, backed by a Republican-dominated Congress between 2001 and 2006, weakens many environmental laws, withdraws the U.S. from participation in a global climate change treaty, greatly increases private energy and mineral development and timber cutting on public lands, and weakens wilderness protection and air pollution laws governing older coal-burning power plants.
In 2006, President Bush created the world's second largest protected marine area.

1990s   2000s

1992 Almost 1,700 of the world's senior scientists release a warning about the seriousness of the world's environmental problems. UN environmental summit at Rio de Janeiro, Brazil. International Convention on Biological Diversity.

1995 Paul Cruzen, Mario Molina, and Sherwood Rowland win Nobel Prize for their work on ozone depletion by chlorofluorocarbons (CFCs).

2000 President Bill Clinton protects large areas in national forests from roads and logging and protects various parcels of public land as national monuments. International Treaty on Persistent Organic Pollutants (POPS) requires phaseout of twelve harmful chemicals.

2001 UN International Panel on Climate Change (IPCC) cites "new and stronger evidence that most of the observed warming during the past 50 years is attributable to human activities."

1991 Persian Gulf War to protect oil supplies in Middle East. Moratorium on mining in Antarctica for 50 years. National People of Color summit to promote environmental justice.

1994 UN Conference on Population and Development held in Cairo, Egypt. California Desert Protection Act adds to the National Park System and wilderness system.

1997 Meeting of 161 nations in Kyoto, Japan, to negotiate a treaty to help slow global warming. Evaluation shows little progress in meeting goals of 1992 Earth Summit meeting.

2007 IPCC issues new report concluding it is *very likely* (a 90–99% probability) that human activities, led by emissions of carbon dioxide from burning fossil fuels, have been the main cause of the observed atmospheric warming during the past 50 years.

1990 Twenty-first annual Earth Day observed by 200 million people in 141 nations. Clean Air Act amended to increase regulation of air pollutants such as sulfur dioxide and nitrogen oxides and to allow trading of air pollution credits. National Environmental Education Act authorizes funding of environmental education programs at elementary and secondary school level.

1993 Paul Hawken publishes *The Ecology of Commerce* discussing relationships between ecology and business.

1996 Theo Coburn publishes *Our Stolen Future* warning of dangers from hormone-disrupting chemicals.

2008 Little progress made in dealing with major environmental problems such as biodiversity degradation, energy policy, and projected climate change from global warming.

Figure 9 Some important environmental events, 1990–2008 **Question:** Which two of these events do you think were the most important?

Republican-dominated Congress during his first term, compiled the worst environmental record of any president in the history of the country during his two terms in office.

A few moderate Republican members of Congress have urged their party to return to its environmental roots, which were put down during Teddy Roosevelt's presidency, and to shed its anti-environmental approach to legislation. Most Democrats agree and assert that the environmental problems we face are much too serious to be held hostage by political squabbling. They call for cooperation, not confronta-

tion. These Democrats and Republicans urge elected officials, regardless of party, to enter into a new pact in which the United States becomes the world leader in making this the *environmental century*. This would help to sustain the country's rich heritage of natural capital and provide economic development, jobs, and profits in rapidly growing businesses such as solar and wind energy, energy-efficient vehicles and buildings, ecological restoration, and pollution prevention.

In 2007, the Intergovernmental Panel on Climate Change (IPCC), which includes more

than 2,500 of the world's climate experts, issued its third report on global climate change. According to this overwhelming consensus among the world's climate scientists, global warming is occurring and it is very likely (a 90–99% probability) that human activities, led by the burning of fossil fuels, have been the main cause of the observed atmospheric warming during the past 50 years. By 2008, the U.S. government had made little progress in facing up to the threat of climate change from global warming.

Some Basic Chemistry (Chapters 1–5)

Chemists Use the Periodic Table to Classify Elements on the Basis of Their Chemical Properties

Chemists have developed a way to classify the elements according to their chemical behavior, in what is called the *periodic table of elements* (Figure 1). Each horizontal row in the table is called a *period*. Each vertical column lists elements with similar chemical properties and is called a *group*.

The partial periodic table in Figure 1 shows how the elements can be classified as *metals, nonmetals,* and *metalloids*. Most of the elements found to the left and at the bottom of the table are *metals,* which usually conduct electricity and heat, and are shiny. Examples are sodium (Na), calcium (Ca), aluminum (Al), iron (Fe), lead (Pb), silver (Ag), and mercury (Hg).

Atoms of metals tend to lose one or more of their electrons to form positively charged ions such as Na^+, Ca^{2+}, and Al^{3+}. For example, an atom of the metallic element sodium (Na, with

atomic number 11) with 11 positively charged protons and 11 negatively charged electrons can lose one of its electrons. It then becomes a sodium ion with a positive charge of 1 (Na^+) because it now has 11 positive charges (protons) but only 10 negative charges (electrons).

Nonmetals, found in the upper right of the table, do not conduct electricity very well. Examples are hydrogen (H), carbon (C), nitrogen (N), oxygen (O), phosphorus (P), sulfur (S), chlorine (Cl), and fluorine (F).

Atoms of some nonmetals such as chlorine, oxygen, and sulfur tend to gain one or more electrons lost by metallic atoms to form negatively charged ions such as O^{2-}, S^{2-}, and Cl^-. For example, an atom of the nonmetallic element chlorine (Cl, with atomic number 17) can gain an electron and become a chlorine ion. The ion has a negative charge of 1 (Cl^-) because it has 17 positively charged protons and 18 negatively charged electrons. Atoms of nonmetals can also combine with one another to form molecules in which they share one or more pairs of their electrons. Hydrogen, a nonmetal,

is placed by itself above the center of the table because it does not fit very well into any of the groups

The elements arranged in a diagonal staircase pattern between the metals and nonmetals have a mixture of metallic and nonmetallic properties and are called *metalloids*.

Figure 1 also identifies the elements required as *nutrients* (black squares) for all or some forms of life and elements that are moderately or highly toxic (red squares) to all or most forms of life. Six nonmetallic elements—carbon (C), oxygen (O), hydrogen (H), nitrogen (N), sulfur (S), and phosphorus (P)—make up about 99% of the atoms of all living things.

THINKING ABOUT
The Periodic Table

Use the periodic table to identify by name and symbol two elements that should have chemical properties similar to those of **(a)** Ca, **(b)** potassium, **(c)** S, **(d)** lead.

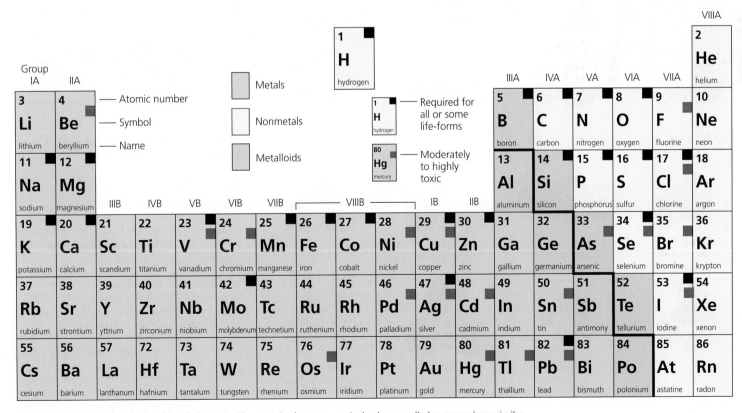

Figure 1 Abbreviated periodic table of elements. Elements in the same vertical column, called a *group,* have similar chemical properties. To simplify matters at this introductory level, only 72 of the 118 known elements are shown.

Figure 2 A solid crystal of an ionic compound such as sodium chloride consists of a three–dimensional array of oppositely charged ions held together by *ionic bonds* resulting from the strong forces of attraction between opposite electrical charges. They are formed when an electron is transferred from a metallic atom such as sodium (Na) to a nonmetallic element such as chlorine (Cl).

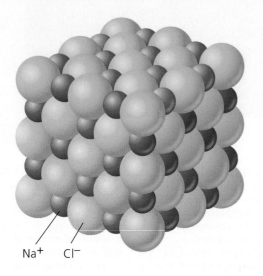

Na⁺   Cl⁻

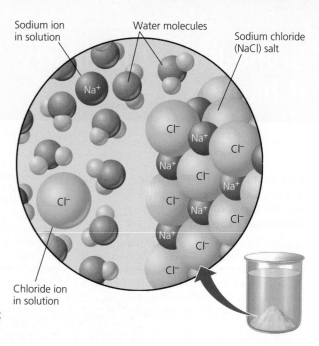

Sodium ion in solution

Water molecules

Sodium chloride (NaCl) salt

Chloride ion in solution

Figure 3 How a salt dissolves in water.

Ionic and Covalent Bonds Hold Compounds Together

Sodium chloride (NaCl) consists of a three-dimensional network of oppositely charged *ions* (Na⁺ and Cl⁻) held together by the forces of attraction between opposite charges (Fig-

ure 2). The strong forces of attraction between such oppositely charged ions are called *ionic bonds*. Because ionic compounds consist of ions formed from atoms of metallic (positive ions) and nonmetallic (negative ions) elements (Figure 1), they can be described as *metal–nonmetal compounds*.

Sodium chloride and many other ionic compounds tend to dissolve in water and break apart into their individual ions (Figure 3).

NaCl → Na⁺ + Cl⁻
sodium chloride   sodium ion + chloride ion
(in water)

Water, a *covalent compound*, consists of molecules made up of uncharged atoms of hydrogen (H) and oxygen (O). Each water molecule consists of two hydrogen atoms chemically bonded to an oxygen atom, yielding H_2O molecules. The bonds between the atoms in such molecules are called *covalent bonds* and form when the atoms in the molecule share one or more pairs of their electrons. Because they are formed from atoms of nonmetallic elements (Figure 1), covalent compounds can be described as *nonmetal–nonmetal compounds*. Figure 4 shows the chemical formulas and shapes of the molecules that are the building blocks for several common *covalent compounds*.

What Makes Solutions Acidic? Hydrogen Ions and pH

The *concentration*, or number of hydrogen ions (H⁺) in a specified volume of a solution (typically a liter), is a measure of its acidity. Pure water (not tap water or rainwater) has an equal number of hydrogen (H⁺) and hydroxide (OH⁻) ions. It is called a **neutral solution.** An **acidic solution** has more hydrogen ions than hydroxide ions per liter. A **basic solution** has more hydroxide ions than hydrogen ions per liter.

Scientists use **pH** as a measure of the acidity of a solution based on its concentration of hydrogen ions (H⁺). By definition, a neutral solution has a pH of 7, an acidic solution has a pH of less than 7, and a basic solution has a pH greater than 7.

Each single unit change in pH represents a 10-fold increase or decrease in the concentration of hydrogen ions per liter. For example, an acidic solution with a pH of 3 is 10 times more acidic than a solution with a pH of 4. Figure 5

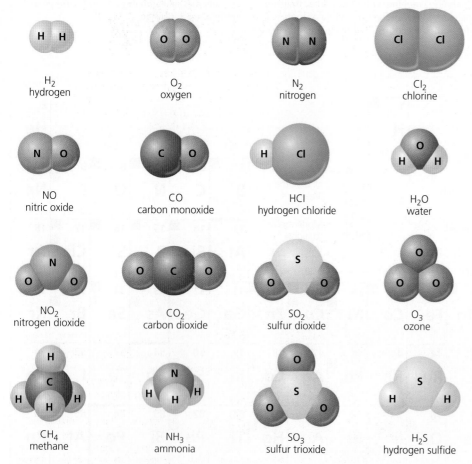

Figure 4 Chemical formulas and shapes for some *covalent compounds* formed when atoms of one or more nonmetallic elements combine with one another and share one or more pairs of their electrons. The bonds between the atoms in such molecules are called *covalent bonds*.

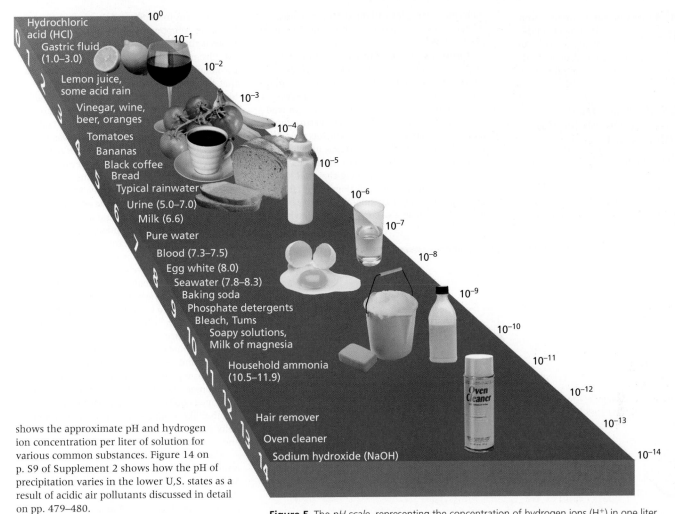

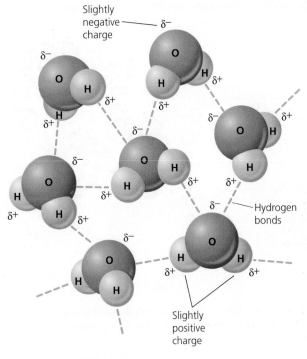

shows the approximate pH and hydrogen ion concentration per liter of solution for various common substances. Figure 14 on p. S9 of Supplement 2 shows how the pH of precipitation varies in the lower U.S. states as a result of acidic air pollutants discussed in detail on pp. 479–480.

**THINKING ABOUT
pH**

A solution has a pH of 2. How many times more acidic is this solution than one with a pH of 6?

Figure 5 The *pH scale*, representing the concentration of hydrogen ions (H^+) in one liter of solution is shown on the righthand side. On the left side are the approximate pH values for solutions of some common substances. A solution with a pH less than 7 is *acidic,* one with a pH of 7 is *neutral,* and one with a pH greater than 7 is *basic.* A change of 1 on the pH scale means a tenfold increase or decrease in H^+ concentration. (Modified from Cecie Starr, *Biology: Today and Tomorrow,* Pacific Grove, Calif.: Brooks/Cole, © 2005)

There Are Weak Forces of Attraction between Some Molecules

Ionic and covalent bonds form between the ions or atoms *within* a compound. There are also weaker forces of attraction *between* the molecules of covalent compounds (such as water) resulting from an unequal sharing of electrons by two atoms.

For example, an oxygen atom has a much greater attraction for electrons than does a hydrogen atom. Thus, in a water molecule the electrons shared between the oxygen atom and its two hydrogen atoms are pulled closer to the oxygen atom, but not actually transferred to the oxygen atom. As a result, the oxygen atom in a water molecule has a slightly negative partial charge and its two hydrogen atoms have a slightly positive partial charge (Figure 6).

The slightly positive hydrogen atoms in one water molecule are then attracted to the slightly negative oxygen atoms in another water molecule. These forces of attraction *between* water molecules are called *hydrogen bonds* (Figure 6).

Figure 6 *Hydrogen bond:* slightly unequal sharing of electrons in the water molecule creates a molecule with a slightly negatively charged end and a slightly positively charged end. Because of this electrical polarity, the hydrogen atoms of one water molecule are attracted to the oxygen atoms in other water molecules. These fairly weak forces of attraction *between* molecules (represented by the dashed lines) are called *hydrogen bonds.*

They account for many of water's unique properties (Science Focus, p. 67). Hydrogen bonds also form between other covalent molecules or portions of such molecules containing hydrogen and nonmetallic atoms with a strong ability to attract electrons.

Four Types of Large Organic Compounds Are the Molecular Building Blocks of Life

Larger and more complex organic compounds, called *polymers*, consist of a number of basic structural or molecular units (*monomers*) linked by chemical bonds, somewhat like rail cars linked in a freight train. Four types of macromolecules—complex carbohydrates, proteins, nucleic acids, and lipids—are molecular building blocks of life.

Complex carbohydrates consist of two or more monomers of *simple sugars* (such as glucose, Figure 7) linked together. One example is the starches that plants use to store energy and also to provide energy for animals that feed on plants. Another is cellulose, the earth's most abundant organic compound, which is found in the cell walls of bark, leaves, stems, and roots.

Proteins, are large polymer molecules formed by linking together long chains of monomers called *amino acids* (Figure 8). Living organisms use about 20 different amino acid molecules to build a variety of proteins, which play different roles. Some help to store energy. Some are components of the *immune system* that protects the body against diseases and harmful substances by forming antibodies that make invading agents harmless. Others are *hormones* that are used as chemical messengers in the bloodstreams of animals to turn various bodily functions on or off. In animals, proteins are also components of hair, skin, muscle, and tendons. In addition, some proteins act as *enzymes* that catalyze or speed up certain chemical reactions.

Nucleic acids are large polymer molecules made by linking hundreds to thousands of four types of monomers called *nucleotides*. Two nucleic acids—DNA (**d**eoxyribo**n**ucleic **a**cid) and RNA (**ri**bo**n**ucleic **a**cid)—participate in the building of proteins and carry hereditary information used to pass traits from parent to offspring. Each nucleotide consists of a *phosphate group*, a *sugar molecule* containing five carbon atoms (deoxyribose in DNA molecules and ribose in RNA molecules), and one of four different *nucleotide bases* (represented by A, G, C, and T, the first letter in each of their names, or A, G, C, and U in RNA; see Figure 9). In the cells of living organisms, these nucleotide units combine in different numbers and sequences to form *nucleic acids* such as various types of RNA and DNA (Figure 10).

Hydrogen bonds formed between parts of the four nucleotides in DNA hold two DNA strands together like a spiral staircase, forming a double helix (Figure 10). DNA molecules can unwind and replicate themselves.

The total weight of the DNA needed to reproduce all of the world's people is only about 50 milligrams—the weight of a small match. If the DNA coiled in your body were unwound, it would stretch about 960 million kilometers (600 million miles)—more than six times the distance between the sun and the earth.

The different molecules of DNA that make up the millions of species found on the earth are like a vast and diverse genetic library. Each species is a unique book in that library. The *genome* of a species is made up of the entire sequence of DNA "letters" or base pairs that combine to "spell out" the chromosomes in typical members of each species. In 2002, scientists were able to map out the genome for the human species by

General structure ➤ Chain of glucose units ➤ Starch

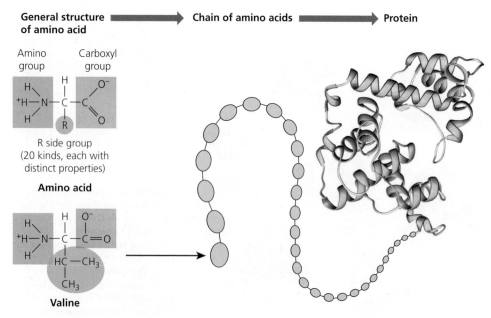

Glucose ($C_6H_{12}O_6$)

Figure 7 Straight-chain and ring structural formulas of glucose, a simple sugar that can be used to build long chains of complex carbohydrates such as starch and cellulose.

General structure of amino acid ➤ Chain of amino acids ➤ Protein

Amino group   Carboxyl group

R side group (20 kinds, each with distinct properties)

Amino acid

Valine

Figure 8 General structural formula of amino acids and a specific structural formula of one of the 20 different amino acid molecules that can be linked together in chains to form proteins that fold up into more complex shapes.

Deoxyribose in DNA
Ribose in RNA

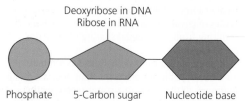

Phosphate   5-Carbon sugar   Nucleotide base

Figure 9 Generalized structures of the nucleotide molecules linked in various numbers and sequences to form large nucleic acid molecules such as various types of DNA (deoxyribonucleic acid) and RNA (ribonucleic acid). In DNA, the 5-carbon sugar in each nucleotide is deoxyribose; in RNA it is ribose. The four basic nucleotides used to make various forms of DNA molecules differ in the types of nucleotide bases they contain—guanine (G), cytosine (C), adenine (A), and thymine (T). (Uracil, labeled U, occurs instead of thymine in RNA.)

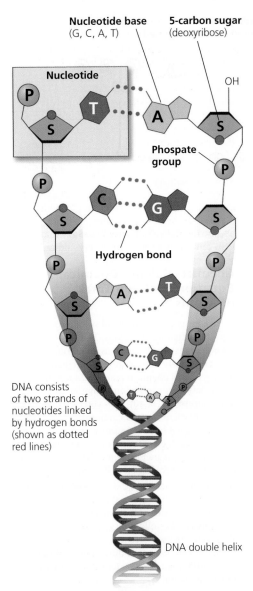

Nucleotide base
(G, C, A, T)

5-carbon sugar
(deoxyribose)

Nucleotide

Phospate group

Hydrogen bond

DNA consists of two strands of nucleotides linked by hydrogen bonds (shown as dotted red lines)

DNA double helix

Figure 10 Portion of the double helix of a DNA molecule. The double helix is composed of two spiral (helical) strands of nucleotides. Each nucleotide contains a unit of phosphate (P), deoxyribose (S), and one of four nucleotide bases: guanine (G), cytosine (C), adenine (A), and thymine (T). The two strands are held together by hydrogen bonds formed between various pairs of the nucleotide bases. Guanine (G) bonds with cytosine (C), and adenine (A) with thymine (T).

analyzing the 3.1 billion base sequences in human DNA.

Lipids, a fourth building block of life, are a chemically diverse group of large organic compounds that do not dissolve in water. Examples are *fats and oils* for storing energy (Figure 11), *waxes* for structure, and *steroids* for producing hormones.

Figure 12 shows the relative sizes of simple and complex molecules, cells, and multicelled organisms.

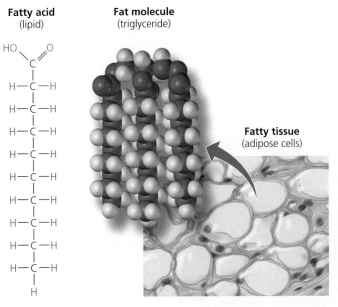

Fatty acid
(lipid)

Fat molecule
(triglyceride)

Fatty tissue
(adipose cells)

Figure 11 Structural formula of fatty acid that is one form of lipid (left). Fatty acids are converted into more complex fat molecules that are stored in adipose cells (right).

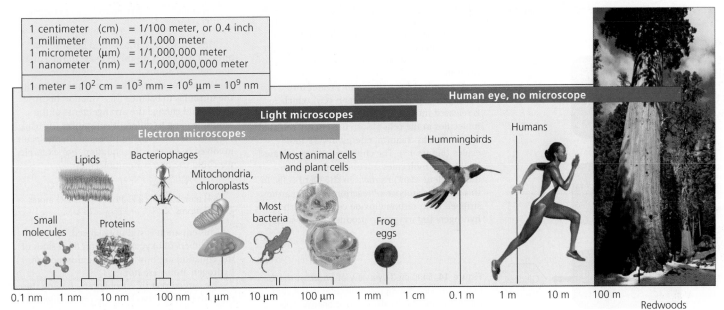

1 centimeter (cm) = 1/100 meter, or 0.4 inch
1 millimeter (mm) = 1/1,000 meter
1 micrometer (µm) = 1/1,000,000 meter
1 nanometer (nm) = 1/1,000,000,000 meter

1 meter = 10^2 cm = 10^3 mm = 10^6 µm = 10^9 nm

Human eye, no microscope

Light microscopes

Electron microscopes

Lipids

Bacteriophages

Most animal cells and plant cells

Humans

Hummingbirds

Mitochondria, chloroplasts

Small molecules

Proteins

Most bacteria

Frog eggs

0.1 nm 1 nm 10 nm 100 nm 1 µm 10 µm 100 µm 1 mm 1 cm 0.1 m 1 m 10 m 100 m

Redwoods

Figure 12 Relative size of simple molecules, complex molecules, cells, and multicellular organisms. This scale is exponential, not linear. Each unit of measure is 10 times larger than the unit preceding it. (Used by permission from Cecie Starr and Ralph Taggart, *Biology*, 11th ed, Belmont, Calif.: Thomson Brooks/Cole, © 2006)

Figure 13
Energy storage
and release in
cells.

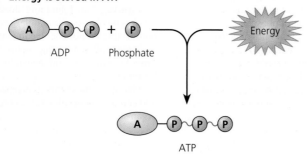

**ATP synthesis:
Energy is stored in ATP**

ADP   Phosphate

Energy

ATP

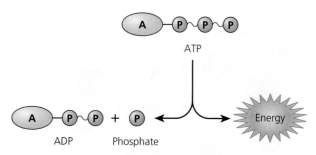

**ATP breakdown:
Energy stored in ATP is released**

ATP

ADP   Phosphate

Energy

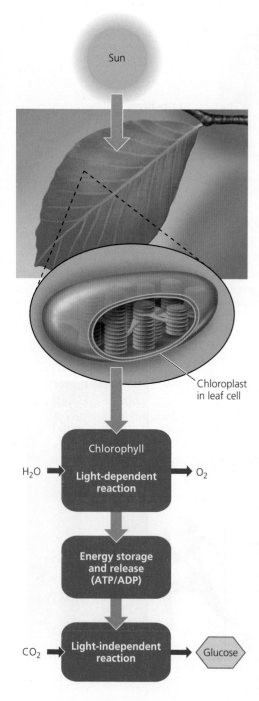

Sun

Chloroplast
in leaf cell

Chlorophyll

H_2O → **Light-dependent reaction** → O_2

Energy storage and release (ATP/ADP)

CO_2 → **Light-independent reaction** → Glucose

Sunlight

$6CO_2 + 6H_2O \longrightarrow C_6H_{12}O_6 + 6O_2$

Certain Molecules Store and Release Energy in Cells

Chemical reactions occurring in photosynthesis (pp. 58–59) release energy that is absorbed by adenosine diphosphate (ADP) molecules and stored as chemical energy in adenosine triphosphate (ATP) molecules (Figure 13, left). When cellular processes require energy, ATP molecules release it to form ADP molecules (Figure 13, right).

A Closer Look at Photosynthesis

In photosynthesis, sunlight powers a complex series of chemical reactions that combine water taken up by plant roots and carbon dioxide from the air to produce sugars such as glucose. This process converts solar energy into chemical energy in sugars for use by plant cells, with the solar energy captured, stored, and released as chemical energy in ATP and ADP molecules (Figure 13). Figure 14 is a greatly simplified summary of the photosynthesis process.

Photosynthesis takes place within tiny enclosed structures called *chloroplasts* found within plant cells. Chlorophyll, a special compound in chloroplasts, absorbs incoming visible light mostly in the violet and red wavelengths. The green light that is not absorbed is reflected back, which is why photosynthetic plants look green. The absorbed wavelengths of solar energy initiate a sequence of chemical reactions with other molecules in what are called *light-dependent reactions*.

This series of reactions splits water into hydrogen ions (H^+) and oxygen (O_2) which is released into the atmosphere. Small ADP molecules in the cells absorb the energy released and store it as chemical energy in ATP molecules (Figure 13). The chemical energy released by the ATP molecules drives a series of *light-independent (dark) reactions* in the plant cells. In this second sequence of reactions, carbon atoms stripped from carbon dioxide combine with hydrogen and oxygen to produce sugars such

Figure 14 Simplified overview of *photosynthesis*. In this process, chlorophyll molecules in the chloroplasts of plant cells absorb solar energy. This initiates a complex series of chemical reactions in which carbon dioxide and water are converted to sugars such as glucose and oxygen.

as glucose ($C_6H_{12}O_6$, Figure 7, p. S42) that plant cells can use as a source of energy and carbon.

Chemists Balance Chemical Equations to Keep Track of Atoms

Chemists use a shorthand system to represent chemical reactions. These chemical equations are also used as an accounting system to verify that no atoms are created or destroyed in a chemical reaction as required by the law of conservation of matter (p. 39 and **Concept 2-3**, p. 40). As a consequence, each side of a chemical equation must have the same number of atoms or ions of each element involved. Ensuring that this condition is met leads to what chemists call a *balanced chemical equation*. The equation for the burning of carbon ($C + O_2 \longrightarrow CO_2$) is balanced because one atom of carbon and two atoms of oxygen are on both sides of the equation.

Consider the following chemical reaction: When electricity passes through water (H_2O), the latter can be broken down into hydrogen (H_2) and oxygen (O_2), as represented by the following equation:

$$H_2O \longrightarrow H_2 + O_2$$

2 H atoms   2 H atoms   2 O atoms
1 O atom

This equation is unbalanced because one atom of oxygen is on the left side of the equation but two atoms are on the right side.

We cannot change the subscripts of any of the formulas to balance this equation because that would change the arrangements of the atoms, leading to different substances. Instead, we must use different numbers of the molecules involved to balance the equation. For example, we could use two water molecules:

$$2 H_2O \longrightarrow H_2 + O_2$$

4 H atoms   2 H atoms   2 O atoms
2 O atoms

This equation is still unbalanced. Although the numbers of oxygen atoms on both sides of the equation are now equal, the numbers of hydrogen atoms are not.

We can correct this problem by having the reaction produce two hydrogen molecules:

$$2 H_2O \longrightarrow 2 H_2 + O_2$$

4 H atoms   4 H atoms   2 O atoms
2 O atoms

Now the equation is balanced, and the law of conservation of matter has been observed. For every two molecules of water through which we pass electricity, two hydrogen molecules and one oxygen molecule are produced.

THINKING ABOUT
Chemical Equations

Try to balance the chemical equation for the reaction of nitrogen gas (N_2) with hydrogen gas (H_2) to form ammonia gas (NH_3).

Scientists Are Learning How to Build Materials from the Bottom Up: Nanotechnology

Nanotechnology (Science Focus, p. 362) uses atoms and molecules to build materials from the bottom up using atoms of the elements in the periodic table as its raw materials. A *nanometer* (nm) is one-billionth of a meter—equal to the length of about 10 hydrogen atoms lined up side by side. A DNA molecule (Figure 10) is about 2.5 nanometers wide. A human hair has a width of 50,000 to 100,000 nanometers.

For objects smaller than about 100 nanometers, the properties of materials change dramatically. At this *nanoscale* level, materials can exhibit new properties, such as extraordinary strength or increased chemical activity, that they do not exhibit at the much larger *macroscale* level that we are all familiar with.

For example, scientists have learned how to make tiny tubes of carbon atoms linked together in hexagons. Experiments have shown that these carbon nanotubes are the strongest material ever made—60 times stronger than high-grade steel. Such nanotubes have been linked together to form a rope so thin that it is invisible, but strong enough to suspend a pickup truck.

At the macroscale, zinc oxide (ZnO) can be rubbed on the skin as a white paste to protect against the sun's harmful UV rays; at the nanoscale it becomes transparent and is being used as invisible coatings to protect the skin and fabrics from UV damage. Because silver (Ag) can kill harmful bacteria, silver nanocrystals are being incorporated into bandages for wounds.

Researchers hope to incorporate nanoparticles of hydroxyapatite, with the same chemical structure as tooth enamel, into toothpaste to put coatings on teeth that prevent bacteria from penetrating. Nanotech coatings now being used on cotton fabrics form an impenetrable barrier that causes liquids to bead and roll off. Such stain-resistant fabrics used to make clothing, rugs, and furniture upholstery could eliminate the need to use harmful chemicals for removing stains.

Self-cleaning window glass coated with a layer of nanoscale titanium dioxide (TiO_2) particles is now available. As the particles interact with UV rays from the sun, dirt on the surface of the glass loosens and washes off when it rains. Similar products can be used for self-cleaning sinks and toilet bowls.

Scientists are working on ways to replace the silicon in computer chips with carbon-based nanomaterials that greatly increase the processing power of computers. Biological engineers are working on nanoscale devices that could deliver drugs. Such devices could penetrate cancer cells and deliver nanomolecules that could kill the cancer cells from the inside. Researchers also hope to develop nanoscale crystals that could change color when they detect tiny amounts (measured in parts per trillion) of harmful substances such chemical and biological warfare agents and food pathogens. For example, a color change in food packaging could alert a consumer when a food is contaminated or has begun to spoil. The list of possibilities could go on.

By 2008, more than 1,000 products containing nanoscale particles were commercially available and thousands more were in the pipeline. Examples are found in cosmetics, sunscreens, fabrics, pesticides, and food additives.

So far, these products are unregulated and unlabeled. This concerns many health and environmental scientists because the tiny size of nanoparticles can allow them to penetrate the natural defenses of the body against invasions by foreign and potentially harmful chemicals and pathogens. Nanoparticles of a chemical tend to be much more chemically reactive than macroparticles of the chemical, largely because the tiny nanoparticles have relatively large surface areas for their small mass. This means that a chemical that is harmless at the macroscale may be hazardous at the nanoscale when they are inhaled, ingested, or absorbed through the skin.

We know little about such effects and risks at a time when the use of untested and unregulated nanoparticles is increasing exponentially. A few toxicological studies are sending up red flags:

- In 2004, Eva Olberdorster, an environmental toxicologist at Southern Methodist University, found that fish swimming in water loaded with a certain type of carbon nanomolecule called buckyballs experienced brain damage within 48 hours.
- In 2005, NASA researchers found that injecting commercially available carbon nanotubes into rats caused significant lung damage.
- A 2005 study by researchers at the U.S. National Institute of Occupational Safety and Health found substantial damage to the heart and aortic arteries of mice exposed to carbon nanotubes.
- In 2005, researchers at New York's University of Rochester found increased blood clotting in rabbits inhaling carbon buckyballs.

In 2004, the British Royal Society and Royal Academy of Engineering recommended that we avoid the environmental release of nanoparticles and nanotubes as much as possible until more is known about their potential harmful impacts. They recommended as a precautionary measure that factories and research laboratories treat manufactured nanoparticles and nanotubes as if they were hazardous to their workers and to the general public. **GREEN CAREER:** Nanotechnology

---

THINKING ABOUT
Nanotechnology

Do you think that the benefits of nanotechnology outweigh its potentially harmful effects? Explain. What are three things you would do to reduce its potentially harmful effects?

RESEARCH FRONTIER

Learning more about nanotechnology and how to reduce its potentially harmful effects. See **academic.cengage.com/biology/miller**.

Classifying and Naming Species (Chapters 3, 4, 8)

All organisms on the earth today are descendants of single-cell organisms that lived almost 4 billion years ago. As a result of biological evolution through natural selection, life has evolved into six major groups of species, called *kingdoms: eubacteria, archaebacteria, protists, fungi, plants,* and *animals* (Figure 4-3, p. 81).

Eubacteria are prokaryotes with single cells that lack a nucleus and other internal compartments (Figure 3-2b, p. 52) found in the cells of species from other kingdoms. Examples include various cyanobacteria and bacteria such as *staphylococcus* and *streptococcus.*

Archaebacteria are single-celled bacteria that are closer to eukaryotic cells (Figure 3-2a, p. 52) than to eubacteria. Examples include methanogens, which live in oxygen-free sediments of lakes and swamps and in animal guts; halophiles, which live in extremely salty water; and thermophiles, which live in hot springs, hydro-thermal vents, and acidic soil. These organisms live in extreme environments.

The remaining four kingdoms—protists, fungi, plants, and animals (Figure 4-3, p. 81) are eukarotes with one or more cells that have a nucleus and complex internal compartments (Figure 3-2a, p. 52). *Protists* are mostly single-celled eukaryotic organisms, such as diatoms, dinoflagellates, amoebas, golden brown and yellow-green algae, and protozoans. Some protists cause human diseases such as malaria (pp. 444–447) and sleeping sickness.

Fungi are mostly many-celled, sometimes microscopic, eukaryotic organisms such as mushrooms, molds, mildews, and yeasts. Many fungi are decomposers (Figure 3-11, p. 60). Other fungi kill various plants and animals and cause huge losses of crops and valuable trees.

Plants are mostly many-celled eukaryotic organisms such as red, brown, and green algae and mosses, ferns, and flowering plants (whose flowers produce seeds that perpetuate the species). Some plants such as corn and marigolds are *annuals,* meaning that they complete their life cycles in one growing season. Others are *perennials,* which can live for more than 2 years, such as roses, grapes, elms, and magnolias.

Animals are also many-celled, eukaryotic organisms. Most have no backbones and hence are called *invertebrates.* Invertebrates include sponges, jellyfish, worms, arthropods (e.g., insects, shrimp, and spiders), mollusks (e.g., snails, clams, and octopuses), and echinoderms (e.g., sea urchins and sea stars). *Vertebrates* (animals with backbones and a brain protected by skull bones) include fishes (e.g., sharks and tuna), amphibians (e.g., frogs and salamanders), reptiles (e.g., crocodiles and snakes), birds (e.g., eagles and robins), and mammals (e.g., bats, elephants, whales, and humans).

Within each kingdom, biologists have created subcategories based on anatomical, physiological, and behavioral characteristics. Kingdoms are divided into *phyla,* which are divided into subgroups called *classes.* Classes are subdivided into *orders,* which are further divided into *families.* Families consist of *genera* (singular, *genus*), and each genus contains one or more *species.* Note that the word *species* is both singular and plural. Figure 1 shows this detailed taxonomic classification for the current human species.

Most people call a species by its common name, such as robin or grizzly bear. Biologists use scientific names (derived from Latin) consisting of two parts (printed in italics, or underlined) to describe a species. The first word is the capitalized name (or abbreviation) for the genus to which the organism belongs. It is followed by a lowercase name that distinguishes the species from other members of the same genus. For example, the scientific name of the robin is *Turdus migratorius* (Latin for "migratory thrush") and the grizzly bear goes by the scientific name *Ursus horribilis* (Latin for "horrible bear").

Kingdom

Animalia Many-celled eukaryotic organisms

Phylum

Chordata Animals with notochord (a long rod of stiffened tissue), nerve cord, and a pharynx (a muscular tube used in feeding, respiration, or both)

Subphylum

Vertebrata Spinal cord enclosed in a backbone of cartilage or bone; and skull bones that protect the brain

Class

Mammalia Animals whose young are nourished by milk produced by mammary glands of females, and that have hair or fur and warm blood

Order

Primates Animals that live in trees or are descended from tree dwellers

Family

Hominidae Upright animals with two-legged locomotion and binocular vision

Genus

Homo Upright animals with large brain, language, and extended parental care of young

Species

sapiens Animals with sparse body hair, high forehead, and large brain

Species

sapiens sapiens Animals capable of sophisticated cultural evolution

Figure 1 Taxonomic classification of the latest human species, *Homo sapiens sapiens.*

Weather Basics: El Niño, Tornadoes, and Tropical Cyclones (Chapters 4, 7, 11)

Weather Is Affected by Moving Masses of Warm and Cold Air

Weather is the set of short-term atmospheric conditions—typically those occurring over hours or days—for a particular area. Examples of atmospheric conditions include temperature, pressure, moisture content, precipitation, sunshine, cloud cover, and wind direction and speed.

Meteorologists use equipment mounted on weather balloons, aircraft, ships, and satellites, as well as radar and stationary sensors, to obtain data on weather variables. They then feed these data into computer models to draw weather maps. Other computer models project the weather for a period of several days by calculating the probabilities that air masses, winds, and other factors will change in certain ways.

Much of the weather we experience results from interactions between the leading edges of moving masses of warm or cold air. Weather changes as one air mass replaces or meets another. The most dramatic changes in weather occur along a **front,** the boundary between two air masses with different temperatures and densities.

A **warm front** is the boundary between an advancing warm air mass and the cooler one it is replacing (Figure 1, left). Because warm air is less dense (weighs less per unit of volume) than cool air, an advancing warm front rises up over a mass of cool air. As the warm front rises, its moisture begins condensing into droplets, forming layers of clouds at different altitudes. Gradually the clouds thicken, descend to a lower altitude, and often release their moisture as rainfall. A moist warm front can bring days of cloudy skies and drizzle.

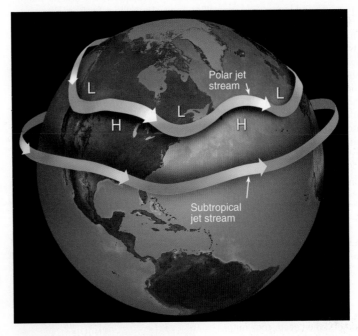

A **cold front** (Figure 1, right) is the leading edge of an advancing mass of cold air. Because cold air is denser than warm air, an advancing cold front stays close to the ground and wedges underneath less dense warmer air. An approaching cold front produces rapidly moving, towering clouds called *thunderheads.*

As a cold front passes through, we may experience high surface winds and thunderstorms. After it leaves the area, we usually have cooler temperatures and a clear sky.

Near the top of the troposphere, hurricane-force winds circle the earth. These powerful winds, called *jet streams,* follow rising and falling paths that have a strong influence on weather patterns (Figure 2).

Weather Is Affected by Changes in Atmospheric Pressure

Changes in atmospheric pressure also affect weather. *Atmospheric pressure* results from molecules of gases (mostly nitrogen and oxygen) in the atmosphere zipping around at very high speeds and hitting and bouncing off everything they encounter.

Figure 2 A *jet stream* is a rapidly flowing air current that moves west to east in a wavy pattern. This figure shows a polar jet stream and a subtropical jet stream in winter. In reality, jet streams are discontinuous and their positions vary from day to day. (Used by permission from C. Donald Ahrens, *Meteorology Today*, 8th ed. Belmont, Calif.: Brooks/ Cole, 2006)

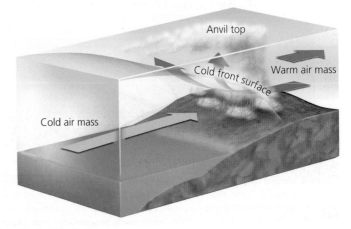

Figure 1 *Weather fronts:* a *warm front* (left) arises when an advancing mass of warm air meets and rises up over a mass of denser cool air. A *cold front* (right) forms when a moving mass of cold air wedges beneath a mass of less dense warm air.

Atmospheric pressure is greater near the earth's surface because the molecules in the atmosphere are squeezed together under the weight of the air above them. An air mass with high pressure, called a **high,** contains cool, dense air that descends toward the earth's surface and becomes warmer. Fair weather follows as long as this high-pressure air mass remains over the area.

In contrast, a low-pressure air mass, called a **low,** produces cloudy and sometimes stormy weather. Because of its low pressure and low density, the center of a low rises, and its warm air expands and cools. When the temperature drops below a certain level where condensation takes place, called the *dew point,* moisture in the air condenses and forms clouds.

If the droplets in the clouds coalesce into larger drops or snowflakes heavy enough to fall from the sky, then precipitation occurs. The condensation of water vapor into water drops usually requires that the air contain suspended tiny particles of material such as dust, smoke, sea salts, or volcanic ash. These so-called *condensation nuclei* provide surfaces on which the droplets of water can form and coalesce.

Every Few Years Major Wind Shifts in the Pacific Ocean Affect Global Weather Patterns

An **upwelling,** or upward movement of ocean water, can mix the water, bringing cool and nutrient-rich water from the bottom of the ocean to the surface where it supports large populations of phytoplankton, zooplankton, fish, and fish-eating seabirds.

Figure 7-2 (p. 142) shows the oceans' major upwelling zones. Upwellings far from shore occur when surface currents move apart and draw water up from deeper layers. Strong upwellings are also found along the steep western coasts of some continents when winds blowing along the coasts push surface water away from the land and draw water up from the ocean bottom (Figure 3).

Every few years in the Pacific Ocean, normal shore upwellings (Figure 4, left) are affected by changes in weather patterns called the *El Niño–Southern Oscillation,* or *ENSO* (Figure 4, right). In an ENSO, often called simply *El Niño,* prevailing tropical trade winds blowing east to west weaken or reverse direction. This allows

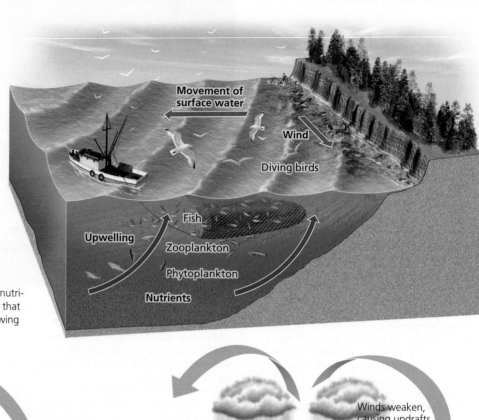

Figure 3 *A shore upwelling* occurs when deep, cool, nutrient-rich waters are drawn up to replace surface water that has been moved away from a steep coast by wind flowing along the coast toward the equator.

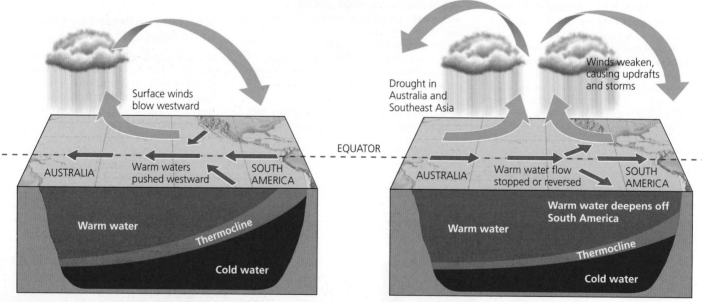

Figure 4 Normal trade winds blowing east to west cause shore upwellings of cold, nutrient-rich bottom water in the tropical Pacific Ocean near the coast of Peru (left). A zone of gradual temperature change called the *thermocline* separates the warm and cold water. Every few years a shift in trade winds known as the *El Niño–Southern Oscillation* (*ENSO*) disrupts this pattern. Trade winds blowing from east to west weaken or reverse direction, which depresses the coastal upwellings and warms the surface waters off South America (right). When an ENSO lasts 12 months or longer, it severely disrupts populations of plankton, fish, and seabirds in upwelling areas and can alter weather conditions over much of the globe (Figure 5).

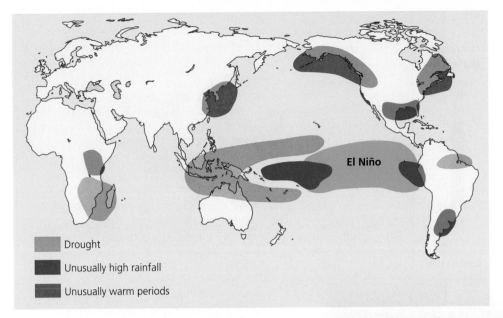

Drought

Unusually high rainfall

Unusually warm periods

El Niño

Figure 5 Typical global weather effects of an El Niño–Southern Oscillation. During the 1996–1998 ENSO, huge waves battered the coast in the U.S. state of California and torrential rains caused widespread flooding and mudslides. In Peru, floods and mudslides killed hundreds of people, left about 250,000 people homeless, and ruined harvests. Drought in Brazil, Indonesia, and Australia led to massive wildfires in tinder-dry forests. India and parts of Africa also experienced severe drought. A catastrophic ice storm hit Canada and the northeastern United States, but the southeastern United States had fewer hurricanes. (Data from United Nations Food and Agriculture Organization)

Data and Map Analysis

1. How might an ENSO affect the weather where you live or go to school?
2. Why do you think the area to the west of El Niño suffers drought?

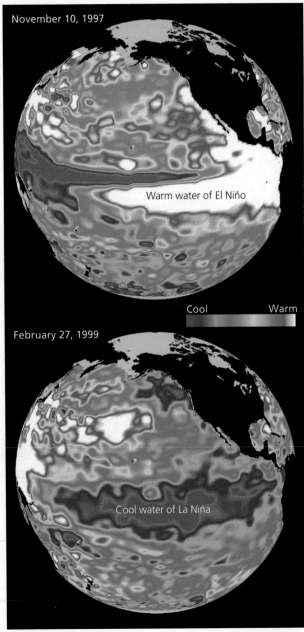

November 10, 1997

Warm water of El Niño

Cool   Warm

February 27, 1999

Cool water of La Niña

Figure 6 Locations of flowing masses of warm and cold water in the Pacific Ocean during an *El Niño* (top) and a *La Niña* (bottom). (Data from Jet Propulsion Lab, NASA)

the warmer waters of the western Pacific to move toward the coast of South America, which suppresses the normal upwellings of cold, nutrient-rich water (Figure 4, right). The decrease in nutrients reduces primary productivity and causes a sharp decline in the populations of some fish species.

A strong ENSO can alter the weather of at least two-thirds of the globe (Figure 5)—especially in lands along the Pacific and Indian Oceans. Scientists do not know for sure the causes of an ENSO, but they know how to detect its formation and track its progress.

La Niña, the reverse of El Niño, cools some coastal surface waters, and brings back upwellings. Typically, La Niña means more Atlantic Ocean hurricanes, colder winters in Canada and the northeastern United States, and warmer and drier winters in the southeastern and southwestern United States. It also usually leads to wetter winters in the Pacific Northwest, torrential rains in Southeast Asia, lower wheat yields in Argentina, and more wildfires in Florida. Figure 6 uses satellite data to show changes in the locations of masses of warm and cold water in the Pacific Ocean during an *El Niño* (top) and a *La Niña* (bottom).

Tornadoes and Tropical Cyclones Are Violent Weather Extremes

Sometimes we experience *weather extremes*. Two examples are violent storms called *tornadoes* (which form over land) and *tropical cyclones* (which form over warm ocean waters and sometimes pass over coastal land).

Tornadoes or *twisters* are swirling funnel-shaped clouds that form over land. They can destroy houses, cause other serious damage, and kill people in areas where they touch down on the earth's surface. The United States is the world's most tornado-prone country, followed by Australia.

Tornadoes in the plains of the midwestern United States usually occur when a large, dry, cold-air front moving southward from Canada runs into a large mass of humid air moving northward from the Gulf of Mexico. Most tornadoes occur in the spring and summer when fronts of cold air from the north penetrate deeply into the midwestern plains.

As the large warm-air mass moves rapidly over the more dense cold-air mass, it rises swiftly and forms strong vertical convection currents that suck air upward, as shown in Figure 7. Scientists hypothesize that the rising vortex of air starts spinning because the air near the ground in the funnel is moving more slowly than the air above. This difference causes the air ahead of the advancing front to roll or spin in a vertically rising air mass or vortex.

Figure 8 shows the areas of greatest risk from tornadoes in the continental United States.

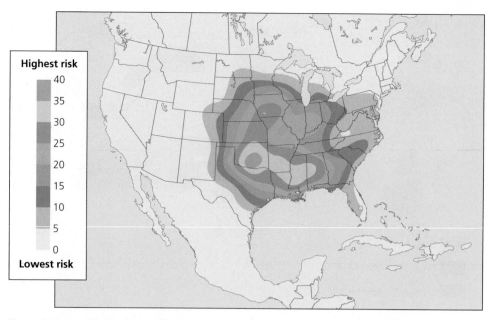

Figure 8 States with *very high* and *high* tornado risk in the continental United States. (Data from NOAA)

Data and Map Analysis

1. How many states have areas with a risk factor of 25 or higher? How many have areas with a risk factor of 20 or higher?
2. What is the level of risk where you live? If you live in a far western state, does this mean you are guaranteed never to see a tornado in your area?

Tropical cyclones are spawned by the formation of low-pressure cells of air over warm tropical seas. Figure 9 shows the formation and structure of a tropical cyclone. *Hurricanes* are tropical cyclones that form in the Atlantic Ocean; those forming in the Pacific Ocean usually are called *typhoons*. Tropical cyclones take a long time to form and gain strength. As a result, meteorologists can track their paths and wind speeds and warn people in areas likely to be hit by these violent storms.

For a tropical cyclone to form, the temperature of ocean water has to be at least 27 °C (80 °F) to a depth of 46 meters (150 feet). A tropical cyclone forms when areas of low pressure over the warm ocean draw in air from surrounding higher-pressure areas. The earth's rotation makes these winds spiral counterclockwise in the northern hemisphere and clockwise in the southern hemisphere (Figure 7-3, p. 142). Moist air warmed by the heat of the ocean rises in a vortex through the center of the storm until it becomes a tropical cyclone (Figure 9).

The intensities of tropical cyclones are rated in different categories based on their sustained wind speeds: *Category 1:* 119–153 kilometers per hour (74–95 miles per hour); *Category 2:* 154–177 kilometers per hour (96–110 miles per hour); *Category 3:* 178–209 kilometers per hour (111–130 miles per hour); *Category 4:* 210–249 kilometers per hour (131–155 miles per hour); and *Category 5:* greater than 249 kilometers per hour (155 miles per hour). The longer a tropical cyclone stays over warm waters, the stronger it gets. Significant hurricane-force winds can extend 64–161 kilometers (40–100 miles) from the center, or eye, of a tropical cyclone.

Figure 10 shows the change in the average surface temperature of the global ocean between 1871 and 2000. Note the rise in this temperature since 1980. These higher temperatures, especially in tropical waters, may explain why

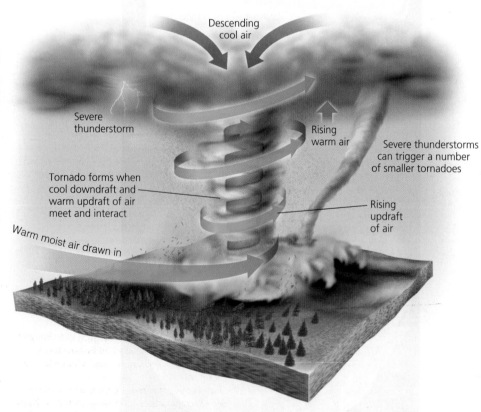

Figure 7 Formation of a *tornado* or *twister*. Although twisters can form at any time of the year, the most active tornado season in the United States is usually March through August. Meteorologists cannot tell us with great accuracy when and where most tornadoes will form.

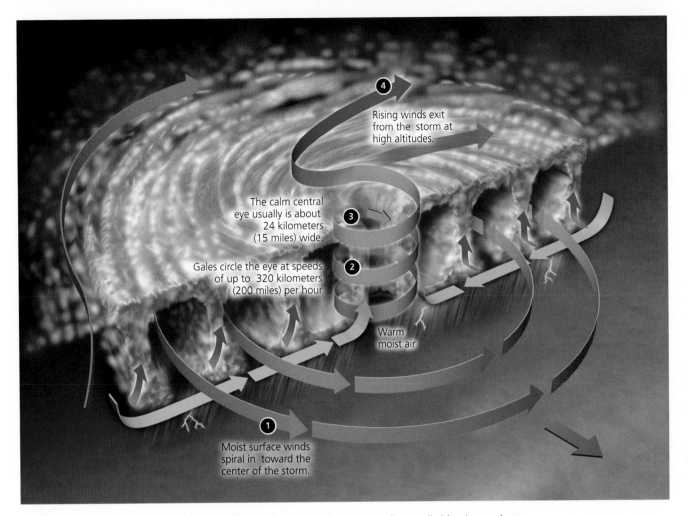

Figure 9 Formation of a *tropical cyclone*. Those forming in the Atlantic Ocean usually are called *hurricanes;* those forming in the Pacific Ocean usually are called *typhoons*.

the average intensity of tropical cyclones has increased since 1990. With the number of people living along the world's coasts increasing, the danger to lives and property has risen dramatically. The greatest risk from hurricanes in the continental United States is along the gulf and eastern coasts, as shown in Figure 11 (p. S52).

Hurricanes and typhoons kill and injure people and damage property (Figure 8-18, p. 177) and agricultural production. Sometimes, however, the long-term ecological and economic benefits of a tropical cyclone exceed its short-term harmful effects.

For example, in parts of the U.S. state of Texas along the Gulf of Mexico, coastal bays and marshes normally are closed off from freshwater and saltwater inflows. In August 1999, Hurricane Brett struck this coastal area. According to marine biologists, it flushed out excess nutrients from land runoff and swept dead sea grasses and rotting vegetation from the coastal bays and marshes. It also carved out 12 channels through the barrier islands along the coast, allowing huge quantities of seawater to flood the bays and marshes.

This flushing of the bays and marshes reduced brown tides consisting of explosive growths of algae that had fed on excess nutrients. It also increased growth of sea grasses, which serve as nurseries for shrimp, crabs, and fish and provide food for millions of ducks wintering in Texas bays. Production of commercially important species of shellfish and fish also increased.

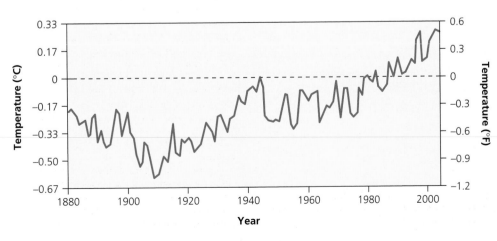

Figure 10 Change in global ocean temperature from its average baseline temperature from 1880 to 2000. (Data from National Oceanic and Atmospheric Administration)

Data and Graph Analysis

1. In 1900, the global ocean temperature dropped from its baseline temperature by how many degrees? (Give answer in both Centigrade and Fahrenheit.)
2. Since about what year after 1980 have global ocean temperatures consistently increased (with all changes being positive)? What has been the highest temperature increase since then, and in about what year did it happen?

Figure 11 The number of hurricanes expected to occur during a 100-year period in the continental United States (based on historical data from U.S. Geological Survey)

Data and Map Analysis

1. What is the degree of risk where you live or go to school?
2. How many states include an area with some risk of hurricanes?

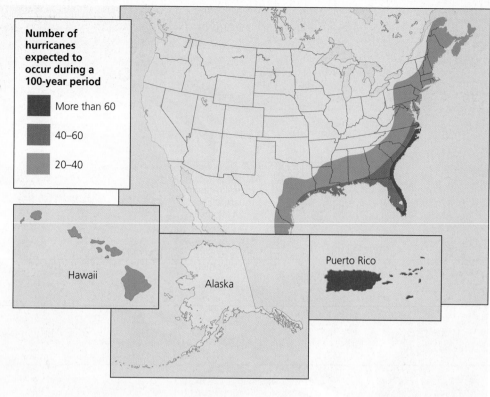

Number of hurricanes expected to occur during a 100-year period

More than 60

40–60

20–40

Hawaii

Alaska

Puerto Rico

Components and Interactions in Major Biomes (Chapters 8, 9, 10)

Figure 1 Some components and interactions in a *temperate desert ecosystem.* When these organisms die, decomposers break down their organic matter into minerals that plants use. Colored arrows indicate transfers of matter and energy among producers, primary consumers (herbivores), secondary or higher-level consumers (carnivores), and decomposers. Organisms are not drawn to scale. **Question:** What species might undergo population growth and what species might suffer a population decline if the diamondback rattlesnake were eliminated from this ecosystem?

Red-tailed hawk

Gambel's quail

Agave

Yucca

Jack rabbit

Collared lizard

Prickly pear cactus

Roadrunner

Darkling beetle

Bacteria

Fungi

Diamondback rattlesnake

Kangaroo rat

Producer to primary consumer

Primary to secondary consumer

Secondary to higher-level consumer

All producers and consumers to decomposers

CENGAGENOW Active Figure 2 Some components and interactions in a *temperate tall-grass prairie ecosystem* in North America. When these organisms die, decomposers break down their organic matter into minerals that plants can use. Colored arrows indicate transfers of matter and energy among producers, primary consumers (herbivores), secondary or higher-level consumers (carnivores), and decomposers. Organisms are not drawn to scale. *See an animation based on this figure at* CengageNOW™. **Question:** What species might increase and what species might decrease in population size if the threatened prairie dog were eliminated from this ecosystem?

Golden eagle

Pronghorn antelope

Coyote

Grasshopper sparrow

Grasshopper

Blue stem grass

Prairie dog

Bacteria

Fungi

Prairie coneflower

→ Producer to primary consumer

→ Primary to secondary consumer

→ Secondary to higher-level consumer

→ All producers and consumers to decomposers

Figure 3 Some components and interactions in an *arctic tundra (cold grassland) ecosystem.* When these organisms die, decomposers break down their organic matter into minerals that plants use. Colored arrows indicate transfers of matter and energy among producers, primary consumers (herbivores), secondary or higher-level consumers (carnivores), and decomposers. Organisms are not drawn to scale. **Question:** What species might increase and what species might decrease in population size if the arctic fox were eliminated from this ecosystem?

Long-tailed jaeger

Grizzly bear

Caribou

Mosquito

Snowy owl

Horned lark

Arctic fox

Willow ptarmigan

Dwarf willow

Bacteria

Lemming

Fungi

Mountain cranberry

Moss campion

| ➤ Producer to primary consumer | ➤ Primary to secondary consumer | ➤ Secondary to higher-level consumer | ➤ All producers and consumers to decomposers |

Figure 4 Some components and interactions in a *temperate deciduous forest ecosystem.* When these organisms die, decomposers break down their organic matter into minerals that plants use. Colored arrows indicate transfers of matter and energy among producers, primary consumers (herbivores), secondary or higher-level consumers (carnivores), and decomposers. Organisms are not drawn to scale. **Question:** What species might increase and what species might decrease in population size if the broad-winged hawk were eliminated from this ecosystem?

Figure 5 Some components and interactions in an *evergreen coniferous* (*boreal* or *taiga*) *forest ecosystem*. When these organisms die, decomposers break down their organic matter into minerals that plants use. Colored arrows indicate transfers of matter and energy among producers, primary consumers (herbivores), secondary or higher-level consumers (carnivores), and decomposers. Organisms are not drawn to scale. **Question:** What species might increase and what species might decrease in population size if the great horned owl were eliminated from this ecosystem?

Blue jay

Great horned owl

Balsam fir

Marten

Moose

White spruce

Wolf

Bebb willow

Snowshoe hare

Pine sawyer beetle and larvae

Fungi

Starflower

Bacteria

Bunchberry

Producer to primary consumer

Primary to secondary consumer

Secondary to higher-level consumer

All producers and consumers to decomposers

Figure 6 Components and interactions in a *coral reef ecosystem*. When these organisms die, decomposers break down their organic matter into minerals used by plants. Colored arrows indicate transfers of matter and energy among producers, primary consumers (herbivores), secondary or higher-level consumers (carnivores), and decomposers. Organisms are not drawn to scale. See the photo of a coral reef in Figure 8-1, left, p. 162. **Question:** How would the species in this ecosystem be affected if phytoplankton populations suffered a sharp drop?

Chapter Projects (Chapters 3–11)

CHAPTER 3　Ecosystems: What Are They and How Do They Work?

1. Visit a nearby terrestrial ecosystem or aquatic life zone and try to identify major producers, primary and secondary consumers, detritus feeders, and decomposers.

2. Write a brief scenario describing the major consequences for us and other species if each of the following biogeochemical cycles were to stop functioning: **(a)** water; **(b)** carbon; **(c)** nitrogen; **(d)** phosphorus; and **(e)** sulfur.

CHAPTER 4　Biodiversity and Evolution

1. Visit a nearby terrestrial or aquatic ecosystem and try to identify a native, nonnative, indicator, keystone, and foundation species.

2. Visit a nearby terrestrial or aquatic ecosystem, try to identify a generalist species and a specialist species, and try to estimate the area's species richness and species evenness.

3. Use the library or the Internet to learn about the emerging field of synthetic biology, which combines biology, genetics, and engineering. How might synthetic biology get around some of the problems raised by genetic engineering? What problems does it pose?

CHAPTER 5　Biodiversity, Species Interactions, and Population Control

1. Use the library or Internet to find and describe two species not discussed in this textbook that are engaged in a **(a)** commensalistic interaction, **(b)** mutualistic interaction, and **(c)** parasite–host relationship.

2. Visit a nearby natural area and try to identify examples of **(a)** mutualism and **(b)** resource partitioning.

3. Do some research to identify the parasites likely to be found in your body.

4. Choose one wild plant species and one wild animal species and use the library or the Internet to help you analyze the factors that are likely to limit the population of each species.

5. Visit a nearby land area such as a partially cleared or burned forest or grassland or an abandoned crop field and record signs of secondary ecological succession. Was the disturbance that led to this succession natural or caused by humans? Study the area carefully to see whether you can find patches that are at different stages of succession because of various disturbances.

CHAPTER 6　The Human Population and Its Impact

1. Assume your entire class (or each of a number of groups from your class) is charged with coming up with a plan for cutting the world's population growth rate by half within the next 20 years. Develop a detailed plan that would achieve this goal, including any differences between policies in developing countries and those in developed countries. Justify each part of your plan. Try to anticipate what problems you might face in implementing the plan, and devise strategies for dealing with these problems.

2. Prepare an age structure diagram for your community. Use the diagram to project future population growth and economic and social problems.

CHAPTER 7　Climate and Terrestrial Biodiversity

1. How has the climate changed in the area where you live during the past 50 years? Investigate the beneficial and harmful effects of these changes. How have these changes benefited or harmed you personally?

2. How have human activities over the past 50 years affected the characteristic vegetation and animal life normally found where you live?

CHAPTER 8　Aquatic Biodiversity

1. Develop three guidelines for preserving the earth's aquatic biodiversity based on the four **scientific principles of sustainability** (see back cover).

2. Visit a nearby lake or reservoir. Would you classify it as oligotrophic, mesotrophic, eutrophic, or hypereutrophic? What are the primary factors contributing to its nutrient enrichment? Which of these factors are related to human activities? Try to determine the specific activities in your area that may be affecting this body of water.

3. Developers want to drain a large area of inland wetland in your community and build a large housing development. List **(a)** the main arguments the developers would use to support this project and **(b)** the main arguments ecologists would use in opposing it. If you were an elected city official, would you vote for or against this project? Can you come up with a compromise plan?

CHAPTER 9　Sustaining Biodiversity: The Species Approach

1. Make a record of your own consumption of all products for a single day. Relate your level and types of consumption to the decline of wildlife species and the increased destruction, degradation, and fragmentation of wildlife habitats in **(a)** the country where you live and **(b)** tropical forests. Compare your results with those of your classmates.

2. Identify examples of habitat destruction or degradation in your community that have had harmful effects on the populations of various wild plant and animal species. Develop a management plan for rehabilitating these habitats and species.

3. Choose a particular animal or plant species that interests you and use the library or the Internet to find out (a) its numbers and distribution, (b) whether it is threatened with extinction, (c) the major future threats to its survival, (d) actions that are being taken to help sustain this species, and (e) a type of reconciliation ecology that might be useful in sustaining this species.

CHAPTER 10 Sustaining Terrestrial Biodiversity: The Ecosystem Approach

1. If possible, try to visit (a) a diverse old-growth forest, (b) an area that has been recently clear-cut, and (c) an area that was clear-cut 5–10 years ago. Compare the biodiversity, soil erosion, and signs of rapid water runoff in each of these areas.

2. For many decades, New Zealand has had a policy of meeting all its demand for wood and wood products by growing timber on intensively managed tree plantations. Use the library or Internet to evaluate the effectiveness of this approach and its major advantages and disadvantages.

3. Try to find an area near where you live that includes a degraded ecosystem. Assume you are the conservation biologist in charge of restoring the ecosystem, and write a five-step plan for completing the project. Would your plan involve use of reconciliation ecology? Explain.

4. Use the library or the Internet to find one example of a successful ecological restoration project not discussed in this chapter and an example of one that failed. For each example, describe the strategy used, the ecological principles involved, and why the project succeeded or failed.

CHAPTER 11 Sustaining Aquatic Biodiversity

1. Survey the condition of a nearby wetland, coastal area, river, or stream and research its history. Has its condition improved or deteriorated during the last 10 years? What local, state, or national efforts are being used to protect this aquatic system? Develop a plan for protecting it.

2. Pick a major seafood species and describe its life cycle, including its reproductive cycle. Find out if the species has been overfished, and if so, where in the world it has been depleted, to what degree, and by what methods. Describe your findings. Write a brief script for telling a friend or relative why you would or would not recommend choosing that species from a seafood restaurant menu.

3. Work with your classmates to develop an experiment in aquatic reconciliation ecology for your campus or local community.

Key Concepts (by Chapter)

CHAPTER 1   Environmental Problems, Their Causes, and Sustainability

Concept 1-1A Our lives and economies depend on energy from the sun (*solar capital*) and on natural resources and natural services (*natural capital*) provided by the earth.

Concept 1-1B Living sustainably means living off the earth's natural income without depleting or degrading the natural capital that supplies it.

Concept 1-2 Societies can become more environmentally sustainable through economic development dedicated to improving the quality of life for everyone without degrading the earth's life support systems.

Concept 1-3 As our ecological footprints grow, we are depleting and degrading more of the earth's natural capital.

Concept 1-4 Preventing pollution is more effective and less costly than cleaning up pollution.

Concept 1-5A Major causes of environmental problems are population growth, wasteful and unsustainable resource use, poverty, exclusion of environmental costs of resource use from the market prices of goods and services, and attempts to manage nature with insufficient knowledge.

Concept 1-5B People with different environmental worldviews often disagree about the seriousness of environmental problems and what we should do about them.

Concept 1-6 Nature has sustained itself for billions of years by using solar energy, biodiversity, population control, and nutrient cycling—lessons from nature that we can apply to our lifestyles and economies.

CHAPTER 2   Science, Matter, Energy, and Systems

Concept 2-1 Scientists collect data and develop theories, models, and laws about how nature works.

Concept 2-2 Matter consists of elements and compounds, which are in turn made up of atoms, ions, or molecules.

Concept 2-3 When matter undergoes a physical or chemical change, no atoms are created or destroyed (the law of conservation of matter).

Concept 2-4A When energy is converted from one form to another in a physical or chemical change, no energy is created or destroyed (first law of thermodynamics).

Concept 2-4B Whenever energy is changed from one form to another, we end up with lower-quality or less usable energy than we started with (second law of thermodynamics).

Concept 2-5A Systems have inputs, flows, and outputs of matter and energy, and their behavior can be affected by feedback.

Concept 2-5B Life, human systems, and the earth's lifesupport systems must conform to the law of conservation of matter and the two laws of thermodynamics.

CHAPTER 3   Ecosystems: What Are They and How Do They Work?

Concept 3-1 Ecology is the study of how organisms interact with one another and with their physical environment of matter and energy.

Concept 3-2 Life is sustained by the flow of energy from the sun through the biosphere, the cycling of nutrients within the biosphere, and gravity.

Concept 3-3A Ecosystems contain living (biotic) and nonliving (abiotic) components.

Concept 3-3B Some organisms produce the nutrients they need, others get their nutrients by consuming other organisms, and some recycle nutrients back to producers by decomposing the wastes and remains of organisms.

Concept 3-4A Energy flows through ecosystems in food chains and webs.

Concept 3-4B As energy flows through ecosystems in food chains and webs, the amount of chemical energy available to organisms at each succeeding feeding level decreases.

Concept 3-5 Matter, in the form of nutrients, cycles within and among ecosystems and the biosphere, and human activities are altering these chemical cycles.

Concept 3-6 Scientists use field research, laboratory research, and mathematical and other models to learn about ecosystems.

CHAPTER 4   Biodiversity and Evolution

Concept 4-1 The biodiversity found in genes, species, ecosystems, and ecosystem processes is vital to sustaining life on earth.

Concept 4-2A The scientific theory of evolution explains how life on earth changes over time through changes in the genes of populations.

Concept 4-2B Populations evolve when genes mutate and give some individuals genetic traits that enhance their abilities to survive and to produce offspring with these traits (natural selection).

Concept 4-3 Tectonic plate movements, volcanic eruptions, earthquakes, and climate change have shifted wildlife habitats, wiped out large numbers of species, and created opportunities for the evolution of new species.

Concept 4-4A As environmental conditions change, the balance between formation of new species and extinction of existing species determines the earth's biodiversity.

Concept 4-4B Human activities can decrease biodiversity by causing the premature extinction of species and by destroying or degrading habitats needed for the development of new species.

Concept 4-5 Species diversity is a major component of biodiversity and tends to increase the sustainability of ecosystems.

Concept 4-6A Each species plays a specific ecological role called its niche. Any given species may play one or more of five important roles—native, nonnative, indicator, keystone, or foundation-roles—in a particular ecosystem.

Concept 10-4 Sustaining biodiversity will require protecting much more of the earth's remaining undisturbed land area as parks and nature reserves.

Concept 10-5A We can help to sustain biodiversity by identifying severely threatened areas and protecting those with high plant diversity (biodiversity hotspots) and those where ecosystem services are being impaired.

Concept 10-5B Sustaining biodiversity will require a global effort to rehabilitate and restore damaged ecosystems.

Concept 10-5C Humans dominate most of the earth's land, and preserving biodiversity will require sharing as much of it as possible with other species.

CHAPTER 11   Sustaining Aquatic Biodiversity

Concept 11-1 Aquatic species are threatened by habitat loss, invasive species, pollution, climate change, and overexploitation, all made worse by the growth of the human population.

Concept 11-2 We can help to sustain marine biodiversity by using laws and economic incentives to protect species, setting aside marine reserves to protect ecosystems, and using community-based integrated coastal management.

Concept 11-3 Sustaining marine fisheries will require improved monitoring of fish populations, cooperative fisheries management among communities and nations, reduction of fishing subsidies, and careful consumer choices in seafood markets.

Concept 11-4 To maintain the ecological and economic services of wetlands, we must maximize preservation of remaining wetlands and restoration of degraded and destroyed wetlands.

Concept 11-5 Freshwater ecosystems are strongly affected by human activities on adjacent lands, and protecting these ecosystems must include protection of their watersheds.

Concept 11-6 Sustaining the world's biodiversity and ecosystem services will require mapping terrestrial and aquatic biodiversity, maximizing protection of undeveloped terrestrial and aquatic areas, and carrying out ecological restoration projects worldwide.

Glossary

abiotic   Nonliving. Compare *biotic*.

acid   See *acid solution*.

acid deposition   The falling of acids and acid-forming compounds from the atmosphere to the earth's surface. Acid deposition is commonly known as *acid rain*, a term that refers to the wet deposition of droplets of acids and acid-forming compounds.

adaptation   Any genetically controlled structural, physiological, or behavioral characteristic that helps an organism survive and reproduce under a given set of environmental conditions. It usually results from a beneficial mutation. See *biological evolution, differential reproduction, mutation, natural selection*.

adaptive radiation   Process in which numerous new species evolve to fill vacant and new ecological niches in changed environments, usually after a mass extinction. Typically, this process takes millions of years.

adaptive trait   See *adaptation*.

aerobic respiration   Complex process that occurs in the cells of most living organisms, in which nutrient organic molecules such as glucose ($C_6H_{12}O_6$) combine with oxygen (O_2) to produce carbon dioxide (CO_2), water (H_2O), and energy. Compare *photosynthesis*.

age structure   Percentage of the population (or number of people of each sex) at each age level in a population.

air pollution   One or more chemicals in high enough concentrations in the air to harm humans, other animals, vegetation, or materials. Excess heat is also considered a form of air pollution. Such chemicals or physical conditions are called air pollutants.

alien species   See *nonnative species*.

alpha particle   Positively charged matter, consisting of two neutrons and two protons, which is emitted as radioactivity from the nuclei of some radioisotopes. See also *beta particle, gamma rays*.

altitude   Height above sea level. Compare *latitude*.

anaerobic respiration   Form of cellular respiration in which some decomposers get the energy they need through the breakdown of glucose (or other nutrients) in the absence of oxygen. Compare *aerobic respiration*.

ancient forest   See *old-growth forest*.

annual   Plant that grows, sets seed, and dies in one growing season. Compare *perennial*.

anthropocentric   Human-centered.

applied ecology   See *reconciliation ecology*.

aquatic   Pertaining to water. Compare *terrestrial*.

aquatic life zone   Marine and freshwater portions of the biosphere. Examples include freshwater life zones (such as lakes and streams) and ocean or marine life zones (such as estuaries, coastlines, coral reefs, and the open ocean).

aquifer   Porous, water-saturated layers of sand, gravel, or bedrock that can yield an economically significant amount of water.

arid   Dry. A desert or other area with an arid climate has little precipitation.

artificial selection   Process by which humans select one or more desirable genetic traits in the population of a plant or animal species and then use *selective breeding* to produce populations containing many individuals with the desired traits. Compare *genetic engineering, natural selection*.

asexual reproduction   Reproduction in which a mother cell divides to produce two identical daughter cells that are clones of the mother cell. This type of reproduction is common in single-celled organisms. Compare *sexual reproduction*.

atmosphere   Whole mass of air surrounding the earth. See *stratosphere, troposphere*. Compare *biosphere, geosphere, hydrosphere*.

atmospheric pressure   Force or mass per unit area of air, caused by the bombardment of a surface by the molecules in air.

atom   Minute unit made of subatomic particles that is the basic building block of all chemical elements and thus all matter; the smallest unit of an element that can exist and still have the unique characteristics of that element. Compare *ion, molecule*.

atomic number   Number of protons in the nucleus of an atom. Compare *mass number*.

atomic theory   Idea that all elements are made up of atoms; the most widely accepted scientific theory in chemistry.

autotroph   See *producer*.

background extinction   Normal extinction of various species as a result of changes in local environmental conditions. Compare *mass extinction*.

bacteria   Prokaryotic, one-celled organisms. Some transmit diseases. Most act as decomposers and get the nutrients they need by breaking down complex organic compounds in the tissues of living or dead organisms into simpler inorganic nutrient compounds.

barrier islands   Long, thin, low offshore islands of sediment that generally run parallel to the shore along some coasts.

benthos   Bottom-dwelling organisms. Compare *decomposer, nekton, plankton*.

beta particle   Swiftly moving electron emitted by the nucleus of a radioactive isotope. See also *alpha particle, gamma ray*.

biocentric   Life-centered. Compare *anthropocentric*.

biodegradable   Capable of being broken down by decomposers.

biodegradable pollutant   Material that can be broken down into simpler substances (elements and compounds) by bacteria or other decomposers. Paper and most organic wastes such as animal manure are biodegradable but can take decades to biodegrade in modern landfills. Compare *nondegradable pollutant*.

biodiversity   Variety of different species (*species diversity*), genetic variability among individuals within each species (*genetic diversity*), variety of ecosystems (*ecological diversity*), and functions such as energy flow and matter cycling needed for the survival of species and biological communities (*functional diversity*).

biodiversity hot spots   Areas especially rich in plant species that are found nowhere else and are in great danger of extinction. These areas suffer serious ecological disruption, mostly because of rapid human population growth and the resulting pressure on natural resources.

biogeochemical cycle   Natural processes that recycle nutrients in various chemical forms from the nonliving environment to living organisms and then back to the nonliving environment. Examples include the carbon, oxygen, nitrogen, phosphorus, sulfur, and hydrologic cycles.

biological community   See *community*.

biological diversity   See *biodiversity*.

biological evolution   Change in the genetic makeup of a population of a species in successive generations. If continued long enough, it can lead to the formation of a new species. Note that populations, not individuals, evolve. See also *adaptation, differential reproduction, natural selection, theory of evolution*.

biological pest control   Control of pest populations by natural predators, parasites, or disease-causing bacteria and viruses (pathogens).

biomass Organic matter produced by plants and other photosynthetic producers; total dry weight of all living organisms that can be supported at each trophic level in a food chain or web; dry weight of all organic matter in plants and animals in an ecosystem; plant materials and animal wastes used as fuel.

biome Terrestrial regions inhabited by certain types of life, especially vegetation. Examples include various types of deserts, grasslands, and forests.

biosphere Zone of the earth where life is found. It consists of parts of the atmosphere (the troposphere), hydrosphere (mostly surface water and groundwater), and lithosphere (mostly soil and surface rocks and sediments on the bottoms of oceans and other bodies of water) where life is found. Compare *atmosphere, geosphere, hydrosphere.*

biotic Living organisms. Compare *abiotic.*

biotic pollution The effect of invasive species that can reduce or wipe out populations of many native species and trigger ecological disruptions.

biotic potential Maximum rate at which the population of a given species can increase when there are no limits on its rate of growth. See *environmental resistance.*

birth rate See *crude birth rate.*

broadleaf deciduous plants Plants such as oak and maple trees that survive drought and cold by shedding their leaves and becoming dormant. Compare *broadleaf evergreen plants, coniferous evergreen plants.*

broadleaf evergreen plants Plants that keep most of their broad leaves year-round. An example is the trees found in the canopies of tropical rain forests. Compare *broadleaf deciduous plants, coniferous evergreen plants.*

calorie Unit of energy; amount of energy needed to raise the temperature of 1 gram of water by 1 C° (unit on Celsius temperature scale). See also *kilocalorie.*

carbon cycle Cyclic movement of carbon in different chemical forms from the environment to organisms and then back to the environment.

carnivore Animal that feeds on other animals. Compare *herbivore, omnivore.*

carrying capacity (K) Maximum population of a particular species that a given habitat can support over a given period. Compare *cultural carrying capacity.*

cell Smallest living unit of an organism. Each cell is encased in an outer membrane or wall and contains genetic material (DNA) and other parts to perform its life function. Organisms such as bacteria consist of only one cell, but most organisms contain many cells.

cell theory The idea that all living things are composed of cells; the most widely accepted scientific theory in biology.

chain reaction Multiple nuclear fissions, taking place within a certain mass of a fissionable

isotope, which release an enormous amount of energy in a short time.

chemical One of the millions of different elements and compounds found naturally and synthesized by humans. See *compound, element.*

chemical change Interaction between chemicals in which the chemical composition of the elements or compounds involved changes. Compare *nuclear change, physical change.*

chemical formula Shorthand way to show the number of atoms (or ions) in the basic structural unit of a compound. Examples include H_2O, NaCl, and $C_6H_{12}O_6$.

chemical reaction See *chemical change.*

chemosynthesis Process in which certain organisms (mostly specialized bacteria) extract inorganic compounds from their environment and convert them into organic nutrient compounds without the presence of sunlight. Compare *photosynthesis.*

chlorinated hydrocarbon Organic compound made up of atoms of carbon, hydrogen, and chlorine. Examples include DDT and PCBs.

chlorofluorocarbons (CFCs) Organic compounds made up of atoms of carbon, chlorine, and fluorine. An example is Freon-12 (CCl_2F_2), which is used as a refrigerant in refrigerators and air conditioners and in making plastics such as Styrofoam. Gaseous CFCs can deplete the ozone layer when they slowly rise into the stratosphere and their chlorine atoms react with ozone molecules. Their use is being phased out.

chromosome A grouping of genes and associated proteins in plant and animal cells that carry certain types of genetic information. See *genes.*

chronic undernutrition Condition suffered by people who cannot grow or buy enough food to meet their basic energy needs. Most chronically undernourished children live in developing countries and are likely to suffer from mental retardation and stunted growth and to die from infectious diseases. Compare *malnutrition, overnutrition.*

clear-cutting Method of timber harvesting in which all trees in a forested area are removed in a single cutting. Compare *selective cutting, strip cutting.*

climate Physical properties of the troposphere of an area based on analysis of its weather records over a long period (at least 30 years). The two main factors determining an area's climate are its average *temperature,* with its seasonal variations, and the average amount and distribution of *precipitation.* Compare *weather.*

climax community See *mature community.*

coal Solid, combustible mixture of organic compounds with 30–98% carbon by weight, mixed with various amounts of water and small amounts of sulfur and nitrogen compounds. It forms in several stages as the remains of plants are subjected to heat and pressure over millions of years.

coastal wetland Land along a coastline, extending inland from an estuary that is covered with salt water all or part of the year. Examples include marshes, bays, lagoons, tidal flats, and mangrove swamps. Compare *inland wetland.*

coastal zone Warm, nutrient-rich, shallow part of the ocean that extends from the high-tide mark on land to the edge of a shelf-like extension of continental land masses known as the continental shelf. Compare *open sea.*

coevolution Evolution in which two or more species interact and exert selective pressures on each other that can lead each species to undergo adaptations. See *evolution, natural selection.*

cold front Leading edge of an advancing mass of cold air. Compare *warm front.*

commensalism An interaction between organisms of different species in which one type of organism benefits and the other type is neither helped nor harmed to any great degree. Compare *mutualism.*

commercial extinction Depletion of the population of a wild species used as a resource to a level at which it is no longer profitable to harvest the species.

commercial forest See *tree plantation.*

common-property resource Resource that is owned jointly by a large group of individuals. One example is the roughly one-third of the land in the United States that is owned jointly by all U.S. citizens and held and managed for them by the government. Another example is an area of land that belongs to a whole village and that can be used by anyone for grazing cows or sheep. Compare *open access renewable resource* and *private property resource.* See *tragedy of the commons.* Compare *open access renewable resource, private property resource.*

community Populations of all species living and interacting in an area at a particular time.

competition Two or more individual organisms of a single species (*intraspecific competition*) or two or more individuals of different species (*interspecific competition*) attempting to use the same scarce resources in the same ecosystem.

compound Combination of atoms, or oppositely charged ions, of two or more elements held together by attractive forces called chemical bonds. Examples are NaCl, CO_2, and $C_6H_{12}O_6$. Compare *element.*

concentration Amount of a chemical in a particular volume or weight of air, water, soil, or other medium.

condensation nuclei Tiny particles on which droplets of water vapor can collect.

coniferous evergreen plants Cone-bearing plants (such as spruces, pines, and firs) that keep some of their narrow, pointed leaves (needles) all year. Compare *broadleaf deciduous plants, broadleaf evergreen plants.*

coniferous trees Cone-bearing trees, mostly evergreens, that have needle-shaped or scale-like leaves. They produce wood known commercially as softwood. Compare *deciduous plants*.

consensus science   See *reliable science*.

conservation   Sensible and careful use of natural resources by humans. People with this view are called *conservationists*.

conservation biology   Multidisciplinary science created to deal with the crisis of maintaining the genes, species, communities, and ecosystems that make up earth's biological diversity. Its goals are to investigate human impacts on biodiversity and to develop practical approaches to preserving biodiversity.

conservationist   Person concerned with using natural areas and wildlife in ways that sustain them for current and future generations of humans and other forms of life.

constancy   Ability of a living system, such as a population, to maintain a certain size. Compare *inertia, resilience*.

consumer   Organism that cannot synthesize the organic nutrients it needs and gets its organic nutrients by feeding on the tissues of producers or of other consumers; generally divided into *primary consumers* (herbivores), *secondary consumers* (carnivores), *tertiary (higher-level) consumers*, *omnivores*, and *detritivores* (decomposers and detritus feeders). In economics, one who uses economic goods. Compare *producer*.

controlled burning   Deliberately set, carefully controlled surface fires that reduce flammable litter and decrease the chances of damaging crown fires. See *ground fire, surface fire*.

coral reef   Formation produced by massive colonies containing billions of tiny coral animals, called polyps, that secrete a stony substance (calcium carbonate) around themselves for protection. When the corals die, their empty outer skeletons form layers and cause the reef to grow. Coral reefs are found in the coastal zones of warm tropical and subtropical oceans.

core   Inner zone of the earth. It consists of a solid inner core and a liquid outer core. Compare *crust, mantle*.

corrective feedback loop   See *negative feedback loop*.

crown fire   Extremely hot forest fire that burns ground vegetation and treetops. Compare *controlled burning, ground fire, surface fire*.

crude birth rate   Annual number of live births per 1,000 people in the population of a geographic area at the midpoint of a given year. Compare *crude death rate*.

crude death rate   Annual number of deaths per 1,000 people in the population of a geographic area at the midpoint of a given year. Compare *crude birth rate*.

crude oil   Gooey liquid consisting mostly of hydrocarbon compounds and small amounts of compounds containing oxygen, sulfur, and nitrogen. Extracted from underground accumulations, it is sent to oil refineries, where it is converted to heating oil, diesel fuel, gasoline, tar, and other materials.

crust   Solid outer zone of the earth. It consists of oceanic crust and continental crust. Compare *core, mantle*.

cultural carrying capacity   The limit on population growth that would allow most people in an area or the world to live in reasonable comfort and freedom without impairing the ability of the planet to sustain future generations. Compare *carrying capacity*.

cultural eutrophication   Overnourishment of aquatic ecosystems with plant nutrients (mostly nitrates and phosphates) because of human activities such as agriculture, urbanization, and discharges from industrial plants and sewage treatment plants. See *eutrophication*.

culture   Whole of a society's knowledge, beliefs, technology, and practices.

currents   Mass movements of surface water produced by prevailing winds blowing over the oceans.

dam   A structure built across a river to control the river's flow or to create a reservoir. See *reservoir*.

data   Factual information collected by scientists.

DDT   Dichlorodiphenyltrichloroethane, a chlorinated hydrocarbon that has been widely used as an insecticide but is now banned in some countries.

death rate   See *crude death rate*.

deciduous plants   Trees, such as oaks and maples, and other plants that survive during dry seasons or cold seasons by shedding their leaves. Compare *coniferous trees, succulent plants*.

decomposer   Organism that digests parts of dead organisms and cast-off fragments and wastes of living organisms by breaking down the complex organic molecules in those materials into simpler inorganic compounds and then absorbing the soluble nutrients. Producers return most of these chemicals to the soil and water for reuse. Decomposers consist of various bacteria and fungi. Compare *consumer, detritivore, producer*.

deductive reasoning   Use of logic to arrive at a specific conclusion based on a generalization or premise. Compare *inductive reasoning*.

deep ecology worldview   Worldview holding that each form of life has inherent value, that the fundamental interdependence and diversity of life forms helps all life to thrive, that humans have no right to reduce this interdependence and diversity except to satisfy vital needs, and that present human interference with the nonhuman world is excessive, and the situation is worsening rapidly. Compare *environmental wisdom worldview, frontier worldview, planetary management worldview, stewardship worldview*.

deforestation   Removal of trees from a forested area.

demographic transition   Hypothesis that countries, as they become industrialized, have declines in death rates followed by declines in birth rates.

density   Mass per unit volume.

desert   Biome in which evaporation exceeds precipitation and the average amount of precipitation is less than 25 centimeters (10 inches) per year. Such areas have little vegetation or have widely spaced, mostly low vegetation. Compare *forest, grassland*.

desertification   Conversion of rangeland, rain-fed cropland, or irrigated cropland to desert-like land, with a drop in agricultural productivity of 10% or more. It usually is caused by a combination of overgrazing, soil erosion, prolonged drought, and climate change.

detritivore   Consumer organism that feeds on detritus, parts of dead organisms, and cast-off fragments and wastes of living organisms. Examples include earthworms, termites, and crabs. Compare *decomposer*.

detritus   Parts of dead organisms and cast-off fragments and wastes of living organisms.

detritus feeder   See *detritivore*.

developed country   Country that is highly industrialized and has a high per capita GDP. Compare *developing country*.

developing country   Country that has low to moderate industrialization and low to moderate per capita GDP. Most are located in Africa, Asia, and Latin America. Compare *developed country*.

dieback   Sharp reduction in the population of a species when its numbers exceed the carrying capacity of its habitat. See *carrying capacity*.

differential reproduction   Phenomenon in which individuals with adaptive genetic traits produce more living offspring than do individuals without such traits. See *natural selection*.

dissolved oxygen (DO) content   Amount of oxygen gas (O_2) dissolved in a given volume of water at a particular temperature and pressure, often expressed as a concentration in parts of oxygen per million parts of water.

disturbance   An event that disrupts an ecosystem or community. Examples of *natural disturbances* include fires, hurricanes, tornadoes, droughts, and floods. Examples of *human-caused disturbances* include deforestation, overgrazing, and plowing.

DNA (deoxyribonucleic acid)   Large molecules in the cells of organisms that carry genetic information in living organisms.

domesticated species   Wild species tamed or genetically altered by crossbreeding for use by humans for food (cattle, sheep, and food crops), pets (dogs and cats), or enjoyment (animals in zoos and plants in botanical gardens). Compare *wild species*.

doubling time   Time it takes (usually in years) for the quantity of something growing exponentially to double. It can be calculated by dividing the annual percentage growth rate into 70.

drainage basin   See *watershed*.

drought   Condition in which an area does not get enough water because of lower-than-normal precipitation or higher-than-normal temperatures that increase evaporation.

ecological diversity   The variety of forests, deserts, grasslands, oceans, streams, lakes, and other biological communities interacting with one another and with their nonliving environment. See *biodiversity*. Compare *functional diversity, genetic diversity, species diversity*.

ecological efficiency   Percentage of energy transferred from one trophic level to another in a food chain or web.

ecological footprint   Amount of biologically productive land and water needed to supply a population with the renewable resources it uses and to absorb or dispose of the wastes from such resource use. It is a measure of the average environmental impact of populations in different countries and areas. See *per capita ecological footprint*.

ecological niche   Total way of life or role of a species in an ecosystem. It includes all physical, chemical, and biological conditions that a species needs to live and reproduce in an ecosystem. See *fundamental niche, realized niche*.

ecological restoration   Deliberate alteration of a degraded habitat or ecosystem to restore as much of its ecological structure and function as possible.

ecological succession   Process in which communities of plant and animal species in a particular area are replaced over time by a series of different and often more complex communities. See *primary succession, secondary succession*.

ecologist   Biological scientist who studies relationships between living organisms and their environment.

ecology   Biological science that studies the relationships between living organisms and their environment; study of the structure and functions of nature.

economic development   Improvement of human living standards by economic growth. Compare *economic growth, environmentally sustainable economic development*.

economic growth   Increase in the capacity to provide people with goods and services; an increase in gross domestic product (GDP). Compare *economic development, environmentally sustainable economic development*. See *gross domestic product*.

economic system   Method that a group of people uses to choose which goods and services to produce, how to produce them, how much to produce, and how to distribute them to people.

economy   System of production, distribution, and consumption of economic goods.

ecosphere   See *biosphere*.

ecosystem   One or more communities of different species interacting with one another and with the chemical and physical factors making up their nonliving environment.

ecosystem services   Natural services or natural capital that support life on the earth and are essential to the quality of human life and the functioning of the world's economies. Examples are the chemical cycles, natural pest control, and natural purification of air and water. See *natural resources*.

electromagnetic radiation   Forms of kinetic energy traveling as electromagnetic waves. Examples include radio waves, TV waves, microwaves, infrared radiation, visible light, ultraviolet radiation, X rays, and gamma rays.

electron (e)   Tiny particle moving around outside the nucleus of an atom. Each electron has one unit of negative charge and almost no mass. Compare *neutron, proton*.

element   Chemical, such as hydrogen (H), iron (Fe), sodium (Na), carbon (C), nitrogen (N), or oxygen (O), whose distinctly different atoms serve as the basic building blocks of all matter. Two or more elements combine to form the compounds that make up most of the world's matter. Compare *compound*.

elevation   Distance above sea level.

endangered species   Wild species with so few individual survivors that the species could soon become extinct in all or most of its natural range. Compare *threatened species*.

endemic species   Species that is found in only one area. Such species are especially vulnerable to extinction.

energy   Capacity to do work by performing mechanical, physical, chemical, or electrical tasks or to cause a heat transfer between two objects at different temperatures.

energy conservation   Reducing or eliminating the unnecessary waste of energy.

energy efficiency   Percentage of the total energy input that does useful work and is not converted into low-quality, generally useless heat in an energy conversion system or process. See *energy quality, net energy*. Compare *material efficiency*.

energy productivity   See *energy efficiency*.

energy quality   Ability of a form of energy to do useful work. High-temperature heat and the chemical energy in fossil fuels and nuclear fuels are concentrated high-quality energy. Low-quality energy such as low-temperature heat is dispersed or diluted and cannot do much useful work. See *high-quality energy, low-quality energy*.

enhanced greenhouse effect   See *global warming, greenhouse effect*.

environment   All external conditions, factors, matter, and energy, living and nonliving, that affect any living organism or other specified system.

environmental degradation   Depletion or destruction of a potentially renewable resource such as soil, grassland, forest, or wildlife that is used faster than it is naturally replenished. If such use continues, the resource becomes non-

renewable (on a human time scale) or nonexistent (extinct). See also *sustainable yield*.

environmental ethics   Human beliefs about what is right or wrong with how we treat the environment.

environmentalism   Social movement dedicated to protecting the earth's life support systems for us and other species.

environmentalist   Person who is concerned about the impacts of human activities on the environment.

environmental justice   Fair treatment and meaningful involvement of all people regardless of race, color, sex, national origin, or income with respect to the development, implementation, and enforcement of environmental laws, regulations, and policies.

environmentally sustainable economic development   Development that meets the basic needs of the current generations of humans and other species without preventing future generations of humans and other species from meeting their basic needs. It is the economic component of an *environmentally sustainable society*. Compare *economic development, economic growth*.

environmentally sustainable society   Society that meets the current and future needs of its people for basic resources in a just and equitable manner without compromising the ability of future generations of humans and other species from meeting their basic needs.

environmental movement   Citizens organized to demand that political leaders enact laws and develop policies to curtail pollution, clean up polluted environments, and protect unspoiled areas from environmental degradation.

environmental resistance   All of the limiting factors that act together to limit the growth of a population. See *biotic potential, limiting factor*.

environmental revolution   Cultural change that includes halting population growth and altering lifestyles, political and economic systems, and the way we treat the environment with the goal of living more sustainably. It requires working with the rest of nature by learning more about how nature sustains itself.

environmental science   Interdisciplinary study that uses information and ideas from the physical sciences (such as biology, chemistry, and geology) with those from the social sciences and humanities (such as economics, politics, and ethics) to learn how nature works, how we interact with the environment, and how we can to help deal with environmental problems.

environmental scientist   Scientist who uses information from the physical sciences and social sciences to understand how the earth works, learn how humans interact with the earth, and develop solutions to environmental problems. See *environmental science*.

environmental wisdom worldview   Worldview holding that humans are part of and totally dependent on nature and that nature exists for all species, not just for us. Our success depends on learning how the earth sustains itself and integrating such environmental wisdom into the ways we think and act. Compare *deep ecology worldview, frontier worldview, planetary management worldview, stewardship worldview.*

environmental worldview Set of assumptions and beliefs about how people think the world works, what they think their role in the world should be, and what they believe is right and wrong environmental behavior (environmental ethics). See *deep ecology worldview, environmental wisdom worldview, frontier worldview, planetary management worldview, stewardship worldview.*

EPA U.S. Environmental Protection Agency; responsible for managing federal efforts to control air and water pollution, radiation and pesticide hazards, environmental research, hazardous waste, and solid waste disposal.

epiphyte Plant that uses its roots to attach itself to branches high in trees, especially in tropical forests.

erosion Process or group of processes by which loose or consolidated earth materials are dissolved, loosened, or worn away and removed from one place and deposited in another. See *weathering.*

estuary Partially enclosed coastal area at the mouth of a river where its fresh water, carrying fertile silt and runoff from the land, mixes with salty seawater.

eukaryotic cell Cell that is surrounded by a membrane and has a distinct nucleus. Compare *prokaryotic cell.*

euphotic zone Upper layer of a body of water through which sunlight can penetrate and support photosynthesis.

eutrophication Physical, chemical, and biological changes that take place after a lake, estuary, or slow-flowing stream receives inputs of plant nutrients—mostly nitrates and phosphates—from natural erosion and runoff from the surrounding land basin. See *cultural eutrophication.*

eutrophic lake Lake with a large or excessive supply of plant nutrients, mostly nitrates and phosphates. Compare *mesotrophic lake, oligotrophic lake.*

evaporation Conversion of a liquid into a gas.

evergreen plants Plants that keep some of their leaves or needles throughout the year. Examples include cone-bearing trees (conifers) such as firs, spruces, pines, redwoods, and sequoias. Compare *deciduous plants, succulent plants.*

evolution See *biological evolution.*

exhaustible resource See *nonrenewable resource.*

exotic species See *nonnative species.*

experiment Procedure a scientist uses to study some phenomenon under known conditions. Scientists conduct some experiments in the laboratory and others in nature.

The resulting scientific data or facts must be verified or confirmed by repeated observations and measurements, ideally by several different investigators.

exponential growth Growth in which some quantity, such as population size or economic output, increases at a constant rate per unit of time. An example is the growth sequence 2, 4, 8, 16, 32, 64, and so on, which increases by 100% at each interval. When the increase in quantity over time is plotted, this type of growth yields a curve shaped like the letter J. Compare *linear growth.*

extinction Complete disappearance of a species from the earth. It happens when a species cannot adapt and successfully reproduce under new environmental conditions or when a species evolves into one or more new species. Compare *speciation.* See also *endangered species, mass extinction, threatened species.*

extinction rate Percentage or number of species that go extinct within a certain time such as a year.

family planning Providing information, clinical services, and contraceptives to help people choose the number and spacing of children they want to have.

famine Widespread malnutrition and starvation in a particular area because of a shortage of food, usually caused by drought, war, flood, earthquake, or other catastrophic events that disrupt food production and distribution.

feedback Any process that increases (positive feedback) or decreases (negative feedback) a change to a system.

feedback loop Occurs when an output of matter, energy, or information is fed back into the system as an input and leads to changes in that system. See *positive feedback loop* and *negative feedback loop.*

fermentation See *anaerobic respiration.*

fertility rate Number of children born to an average woman in a population during her lifetime. Compare *replacement-level fertility.*

first law of thermodynamics In any physical or chemical change, no detectable amount of energy is created or destroyed, but energy can be changed from one form to another; you cannot get more energy out of something than you put in; in terms of energy quantity, you cannot get something for nothing. This law does not apply to nuclear changes, in which energy can be produced from small amounts of matter. See *second law of thermodynamics.*

fishery Concentration of particular aquatic species suitable for commercial harvesting in a given ocean area or inland body of water.

fishprint Area of ocean needed to sustain the consumption of an average person, a nation, or the world. Compare *ecological footprint.*

floodplain Flat valley floor next to a stream channel. For legal purposes, the term often applies to any low area that has the potential for flooding, including certain coastal areas.

flows See *throughputs.*

flyway Generally fixed route along which waterfowl migrate from one area to another at certain seasons of the year.

food chain Series of organisms in which each eats or decomposes the preceding one. Compare *food web.*

food web Complex network of many interconnected food chains and feeding relationships. Compare *food chain.*

forest Biome with enough average annual precipitation to support the growth of tree species and smaller forms of vegetation. Compare *desert, grassland.*

fossil fuel Products of partial or complete decomposition of plants and animals; occurs as crude oil, coal, natural gas, or heavy oils as a result of exposure to heat and pressure in he earth's crust over millions of years. See *coal, crude oil, natural gas.*

fossils Skeletons, bones, shells, body parts, leaves, seeds, or impressions of such items that provide recognizable evidence of organisms that lived long ago.

foundation species Species that plays a major role in shaping a community by creating and enhancing a habitat that benefits other species. Compare *indicator species, keystone species, native species, nonnative species.*

free-access resource See *open access renewable resource.*

freons See *chlorofluorocarbons.*

freshwater life zones Aquatic systems where water with a dissolved salt concentration of less than 1% by volume accumulates on or flows through the surfaces of terrestrial biomes. Examples include *standing* (lentic) bodies of fresh water such as lakes, ponds, and inland wetlands and *flowing* (lotic) systems such as streams and rivers. Compare *biome.*

front The boundary between two air masses with different temperatures and densities. See *cold front, warm front.*

frontier science See *tentative science.*

frontier worldview View held by European colonists settling North America in the 1600s that the continent had vast resources and was a wilderness to be conquered by settlers clearing and planting land.

functional diversity Biological and chemical processes or functions such as energy flow and matter cycling needed for the survival of species and biological communities. See *biodiversity, ecological diversity, genetic diversity, species diversity.*

fundamental niche Full potential range of the physical, chemical, and biological factors a species can use if it does not face any competition from other species. See *ecological niche.* Compare *realized niche.*

game species Type of wild animal that people hunt or fish for, for sport and recreation and sometimes for food.

gamma ray Form of electromagnetic radiation with a high energy content emitted by some radioisotopes. It readily penetrates body tissues. See also *alpha particle, beta particle.*

GDP See *gross domestic product.*

gene mutation See *mutation.*

gene pool Sum total of all genes found in the individuals of the population of a particular species.

generalist species Species with a broad ecological niche. They can live in many different places, eat a variety of foods, and tolerate a wide range of environmental conditions. Examples include flies, cockroaches, mice, rats, and humans. Compare *specialist species.*

genes Coded units of information about specific traits that are passed from parents to offspring during reproduction. They consist of segments of DNA molecules found in chromosomes.

genetic adaptation Changes in the genetic makeup of organisms of a species that allow the species to reproduce and gain a competitive advantage under changed environmental conditions. See *differential reproduction, evolution, mutation, natural selection.*

genetically modified organism (GMO) Organism whose genetic makeup has been altered by genetic engineering.

genetic diversity Variability in the genetic makeup among individuals within a single species. See *biodiversity.* Compare *ecological diversity, functional diversity, species diversity.*

genetic engineering Insertion of an alien gene into an organism to give it a beneficial genetic trait. Compare *artificial selection, natural selection.*

geographic isolation Separation of populations of a species for long times into different areas.

geosphere Earth's intensely hot core, thick mantle composed mostly of rock, and thin outer crust that contains most of the earth's rock, soil, and sediment. Compare *atmosphere, biosphere, hydrosphere.*

global climate change Broad term referring to changes in any aspects of the earth's climate, including temperature, precipitation, and storm activity. Compare *weather.*

global warming Warming of the earth's lower atmosphere (troposphere) because of increases in the concentrations of one or more greenhouse gases. It can result in climate change that can last for decades to thousands of years. See *greenhouse effect, greenhouse gases, natural greenhouse effect.*

GMO See genetically modified organism.

GPP See gross primary productivity.

grassland Biome found in regions where enough annual average precipitation to support the growth of grass and small plants but not enough to support large stands of trees. Compare *desert, forest.*

greenhouse effect Natural effect that releases heat in the atmosphere (troposphere) near the earth's surface. Water vapor, carbon dioxide, ozone, and other gases in the lower atmosphere (troposphere) absorb some of the infrared radiation (heat) radiated by the earth's surface. Their molecules vibrate and transform the absorbed energy into longer-wavelength infrared radiation (heat) in the troposphere. If the atmospheric concentrations of these greenhouse gases increase and other natural processes do not remove them, the average temperature of the lower atmosphere will increase gradually. Compare *global warming.* See also *natural greenhouse effect.*

greenhouse gases Gases in the earth's lower atmosphere (troposphere) that cause the greenhouse effect. Examples include carbon dioxide, chlorofluorocarbons, ozone, methane, water vapor, and nitrous oxide.

gross domestic product (GDP) Annual market value of all goods and services produced by all firms and organizations, foreign and domestic, operating within a country. See *per capita GDP.*

gross primary productivity (GPP) Rate at which an ecosystem's producers capture and store a given amount of chemical energy as biomass in a given length of time. Compare *net primary productivity.*

ground fire Fire that burns decayed leaves or peat deep below the ground surface. Compare *crown fire, surface fire.*

groundwater Water that sinks into the soil and is stored in slowly flowing and slowly renewed underground reservoirs called aquifers; underground water in the zone of saturation, below the water table. Compare *runoff, surface water.*

habitat Place or type of place where an organism or population of organisms lives. Compare *ecological niche.*

habitat fragmentation Breakup of a habitat into smaller pieces, usually as a result of human activities.

heat Total kinetic energy of all randomly moving atoms, ions, or molecules within a given substance, excluding the overall motion of the whole object. Heat always flows spontaneously from a warmer sample of matter to a colder sample of matter. This is one way to state the second law of thermodynamics. Compare *temperature.*

herbivore Plant-eating organism. Examples include deer, sheep, grasshoppers, and zooplankton. Compare *carnivore, omnivore.*

heterotroph See *consumer.*

high Air mass with a high pressure. Compare *low.*

high-quality energy Energy that is concentrated and has great ability to perform useful work. Examples include high-temperature heat and the energy in electricity, coal, oil, gasoline, sunlight, and nuclei of uranium-235. Compare *low-quality energy.*

high-quality matter Matter that is concentrated and contains a high concentration of a useful resource. Compare *low-quality matter.*

high-throughput economy Economic system in most advanced industrialized countries, in which ever-increasing economic growth is sustained by maximizing the rate at which matter and energy resources are used, with little emphasis on pollution prevention, recycling, reuse, reduction of unnecessary waste, and other forms of resource conservation. Compare *low-throughput economy, matter-recycling economy.*

high-waste economy See *high-throughput economy.*

HIPPCO Acronym used by conservation biologists for the six most important secondary causes of premature extinction: **H**abitat destruction, degradation, and fragmentation; **I**nvasive (nonnative) species; **P**opulation growth (too many people consuming too many resources); **P**ollution; **C**limate change; and **O**verexploitation.

host Plant or animal on which a parasite feeds.

hydrocarbon Organic compound made of hydrogen and carbon atoms. The simplest hydrocarbon is methane (CH_4), the major component of natural gas.

hydrologic cycle Biogeochemical cycle that collects, purifies, and distributes the earth's fixed supply of water from the environment to living organisms and then back to the environment.

hydrosphere Earth's *liquid water* (oceans, lakes, other bodies of surface water, and underground water), *frozen water* (polar ice caps, floating ice caps, and ice in soil, known as permafrost), and *water vapor* in the atmosphere. See also *hydrologic cycle.* Compare *atmosphere, biosphere, geosphere.*

hypereutrophic Result of excessive inputs of nutrients in a lake. See *cultural eutrophication.*

immature community Community at an early stage of ecological succession. It usually has a low number of species and ecological niches and cannot capture and use energy and cycle critical nutrients as efficiently as more complex, mature communities. Compare *mature community.*

immigrant species See *nonnative species.*

immigration Migration of people into a country or area to take up permanent residence.

indicator species Species that serve as early warnings that a community or ecosystem is being degraded. Compare *foundation species, keystone species, native species, nonnative species.*

inductive reasoning Using specific observations and measurements to arrive at a general conclusion or hypothesis. Compare *deductive reasoning.*

inertia Ability of a living system, such as a grassland or a forest, to survive moderate disturbances. Compare *constancy, resilience.*

infant mortality rate Number of babies out of every 1,000 born each year who die before their first birthday.

infiltration Downward movement of water through soil.

inherent value See *intrinsic value*.

inland wetland Land away from the coast, such as a swamp, marsh, or bog, that is covered all or part of the time with fresh water. Compare *coastal wetland*.

inorganic compounds All compounds not classified as organic compounds. See *organic compounds*.

input Matter, energy, or information entering a system. Compare *output, throughput*.

input pollution control See *pollution prevention*.

instrumental value Value of an organism, species, ecosystem, or the earth's biodiversity based on its usefulness to humans. Compare *intrinsic value*.

interspecific competition Attempts by members of two or more species to use the same limited resources in an ecosystem. See *competition, intraspecific competition*.

intertidal zone The area of shoreline between low and high tides.

intraspecific competition Attempts by two or more organisms of a single species to use the same limited resources in an ecosystem. See *competition, interspecific competition*.

intrinsic rate of increase (r) Rate at which a population could grow if it had unlimited resources. Compare *environmental resistance*.

intrinsic value Value of an organism, species, ecosystem, or the earth's biodiversity based on its existence, regardless of whether it has any usefulness to humans. Compare *instrumental value*.

invasive species See *nonnative species*.

invertebrates Animals that have no backbones. Compare *vertebrates*.

ion Atom or group of atoms with one or more positive (+) or negative (−) electrical charges. Examples are Na^+ and Cl^-. Compare *atom, molecule*.

isotopes Two or more forms of a chemical element that have the same number of protons but different mass numbers because they have different numbers of neutrons in their nuclei.

J-shaped curve Curve with a shape similar to that of the letter J; can represent prolonged exponential growth. See *exponential growth*.

junk science See *unreliable science*.

keystone species Species that play roles affecting many other organisms in an ecosystem. Compare *foundation species, indicator species, native species, nonnative species*.

kilocalorie (kcal) Unit of energy equal to 1,000 calories. See *calorie*.

kilowatt (kW) Unit of electrical power equal to 1,000 watts. See *watt*.

kinetic energy Energy that matter has because of its mass and speed, or velocity. Compare *potential energy*.

K-selected species Species that produce a few, often fairly large offspring but invest a great deal of time and energy to ensure that most of those offspring reach reproductive age. Compare *r-selected species*.

K-strategists See *K-selected species*.

lake Large natural body of standing fresh water formed when water from precipitation, land runoff, or groundwater flow fills a depression in the earth created by glaciation, earth movement, volcanic activity, or a giant meteorite. See *eutrophic lake, mesotrophic lake, oligotrophic lake*.

land degradation Decrease in the ability of land to support crops, livestock, or wild species in the future as a result of natural or human-induced processes.

latitude Distance from the equator. Compare *altitude*.

law of conservation of energy See *first law of thermodynamics*.

law of conservation of matter In any physical or chemical change, matter is neither created nor destroyed but merely changed from one form to another; in physical and chemical changes, existing atoms are rearranged into different spatial patterns (physical changes) or different combinations (chemical changes).

law of nature See *scientific law*.

law of tolerance Existence, abundance, and distribution of a species in an ecosystem are determined by whether the levels of one or more physical or chemical factors fall within the range tolerated by the species. See *threshold effect*.

LDC See *developing country*.

less developed country (LDC) See *developing country*.

life expectancy Average number of years a newborn infant can be expected to live.

limiting factor Single factor that limits the growth, abundance, or distribution of the population of a species in an ecosystem. See *limiting factor principle*.

limiting factor principle Too much or too little of any abiotic factor can limit or prevent growth of a population of a species in an ecosystem, even if all other factors are at or near the optimal range of tolerance for the species.

linear growth Growth in which a quantity increases by some fixed amount during each unit of time. An example is growth that increases by 2 units in the sequence 2, 4, 6, 8, 10, and so on. Compare *exponential growth*.

logistic growth Pattern in which exponential population growth occurs when the population is small, and population growth decreases steadily with time as the population approaches the carrying capacity. See *S-shaped curve*.

low Air mass with a low pressure. Compare *high*.

low-quality energy Energy that is dispersed and has little ability to do useful work. An example is low-temperature heat. Compare *high-quality energy*.

low-quality matter Matter that is dilute or dispersed or contains a low concentration of a useful resource. Compare *high-quality matter*.

low-throughput economy Economy based on working with nature by recycling and reusing discarded matter, preventing pollution, conserving matter and energy resources by reducing unnecessary waste and use, not degrading renewable resources, building things that are easy to recycle, reuse, and repair, not allowing population size to exceed the carrying capacity of the environment, and preserving biodiversity. Compare *high-throughput economy, matter-recycling economy*.

low-waste economy See *low-throughput economy*.

malnutrition Faulty nutrition, caused by a diet that does not supply an individual with enough protein, essential fats, vitamins, minerals, and other nutrients needed for good health.

mangrove swamps Swamps found on the coastlines in warm tropical climates. They are dominated by mangrove trees, any of about 55 species of trees and shrubs that can live partly submerged in the salty environment of coastal swamps.

mantle Zone of the earth's interior between its core and its crust. Compare *core, crust*. See *geosphere, lithosphere*.

mass Amount of material in an object.

mass extinction Catastrophic, widespread, often global event in which major groups of species are wiped out over a short time compared with normal (background) extinctions. Compare *background extinction*.

mass number Sum of the number of neutrons (n) and the number of protons (p) in the nucleus of an atom. It gives the approximate mass of that atom. Compare *atomic number*.

material efficiency Total amount of material needed to produce each unit of goods or services. Also called *resource productivity*. Compare *energy efficiency*.

matter Anything that has mass (the amount of material in an object) and takes up space. On the earth, where gravity is present, we weigh an object to determine its mass.

matter quality Measure of how useful a matter resource is, based on its availability and concentration. See *high-quality matter, low-quality matter*.

matter-recycling-and-reuse economy Economy that emphasizes recycling the maximum amount of all resources that can be recycled and reused. The goal is to allow economic growth to continue without depleting matter resources and without producing excessive pollution and environmental degradation. Compare *high-throughput economy, low-throughput economy*.

mature community   Fairly stable, self-sustaining community in an advanced stage of ecological succession; usually has a diverse array of species and ecological niches; captures and uses energy and cycles critical chemicals more efficiently than simpler, immature communities. Compare *immature community.*

maximum sustainable yield   See *sustainable yield.*

MDC   See *developed country.*

mesotrophic lake   Lake with a moderate supply of plant nutrients. Compare *eutrophic lake, oligotrophic lake.*

metabolism   Ability of a living cell or organism to capture and transform matter and energy from its environment to supply its needs for survival, growth, and reproduction.

microorganisms   Organisms such as bacteria that are so small that it takes a microscope to see them.

migration   Movement of people into and out of specific geographic areas. Compare *emigration* and *immigration.*

mineral resource   Concentration of naturally occurring solid, liquid, or gaseous material in or on the earth's crust in a form and amount such that extracting and converting it into useful materials or items is currently or potentially profitable. Mineral resources are classified as *metallic* (such as iron and tin ores) or *nonmetallic* (such as fossil fuels, sand, and salt).

model   Approximate representation or simulation of a system being studied.

molecule   Combination of two or more atoms of the same chemical element (such as O_2) or different chemical elements (such as H_2O) held together by chemical bonds. Compare *atom, ion.*

more developed country (MDC)   See *developed country.*

mutation   Random change in DNA molecules making up genes that can alter anatomy, physiology, or behavior in offspring. See *mutagen.*

mutualism   Type of species interaction in which both participating species generally benefit. Compare *commensalism.*

native species   Species that normally live and thrive in a particular ecosystem. Compare *foundation species, indicator species, keystone species, nonnative species.*

natural capital   Natural resources and natural services that keep us and other species alive and support our economies. See *natural resources, natural services.*

natural greenhouse effect   Heat buildup in the troposphere caused by the presence of certain gases, called greenhouse gases. Without this effect, the earth would be nearly as cold as Mars, and life as we know it could not exist. See *global warming.*

natural income   Renewable resources such as plants, animals, and soil provided by natural capital.

natural law   See *scientific law.*

natural radioactive decay   Nuclear change in which unstable nuclei of atoms spontaneously shoot out particles (usually alpha or beta particles) or energy (gamma rays) at a fixed rate.

natural rate of extinction   See *background extinction.*

natural resources   Materials such as air, water, and soil and energy in nature that are essential or useful to humans. See *natural capital.*

natural selection   Process by which a particular beneficial gene (or set of genes) is reproduced in succeeding generations more than other genes. The result of natural selection is a population that contains a greater proportion of organisms better adapted to certain environmental conditions. See *adaptation, biological evolution, differential reproduction, mutation.*

natural services   Processes of nature, such as purification of air and water and pest control, which support life and human economies. See *natural capital.*

negative feedback loop   Feedback loop that causes a system to change in the opposite direction from which is it moving. Compare *positive feedback loop.*

nekton   Strongly swimming organisms found in aquatic systems. Compare *benthos, plankton.*

net energy   Total amount of useful energy available from an energy resource or energy system over its lifetime, minus the amount of energy *used* (the first energy law), *automatically wasted* (the second energy law), and *unnecessarily wasted* in finding, processing, concentrating, and transporting it to users.

net primary productivity (NPP)   Rate at which all the plants in an ecosystem produce net useful chemical energy; equal to the difference between the rate at which the plants in an ecosystem produce useful chemical energy (gross primary productivity) and the rate at which they use some of that energy through cellular respiration. Compare *gross primary productivity.*

neutron (n)   Elementary particle in the nuclei of all atoms (except hydrogen-1). It has a relative mass of 1 and no electric charge. Compare *electron, proton.*

niche   See *ecological niche.*

nitrogen cycle   Cyclic movement of nitrogen in different chemical forms from the environment to organisms and then back to the environment.

nitrogen fixation   Conversion of atmospheric nitrogen gas into forms useful to plants by lightning, bacteria, and cyanobacteria; it is part of the nitrogen cycle.

nondegradable pollutant   Material that is not broken down by natural processes. Examples include the toxic elements lead and mercury. Compare *biodegradable pollutant.*

nonnative species   Species that migrate into an ecosystem or are deliberately or accidentally introduced into an ecosystem by humans. Compare *native species.*

nonpoint sources   Broad and diffuse areas, rather than points, from which pollutants enter bodies of surface water or air. Examples include runoff of chemicals and sediments from cropland, livestock feedlots, logged forests, urban streets, parking lots, lawns, and golf courses. Compare *point source.*

nonrenewable resource   Resource that exists in a fixed amount (stock) in the earth's crust and has the potential for renewal by geological, physical, and chemical processes taking place over hundreds of millions to billions of years. Examples include copper, aluminum, coal, and oil. We classify these resources as exhaustible because we are extracting and using them at a much faster rate than they are formed. Compare *renewable resource.*

NPP   See *net primary productivity.*

nuclear change   Process in which nuclei of certain isotopes spontaneously change, or are forced to change, into one or more different isotopes. The three principal types of nuclear change are natural radioactivity, nuclear fission, and nuclear fusion. Compare *chemical change, physical change.*

nuclear energy   Energy released when atomic nuclei undergo a nuclear reaction such as the spontaneous emission of radioactivity, nuclear fission, or nuclear fusion.

nuclear fission   Nuclear change in which the nuclei of certain isotopes with large mass numbers (such as uranium-235 and plutonium-239) are split apart into lighter nuclei when struck by a neutron. This process releases more neutrons and a large amount of energy. Compare *nuclear fusion.*

nuclear fusion   Nuclear change in which two nuclei of isotopes of elements with a low mass number (such as hydrogen-2 and hydrogen-3) are forced together at extremely high temperatures until they fuse to form a heavier nucleus (such as helium-4). This process releases a large amount of energy. Compare *nuclear fission.*

nucleus   Extremely tiny center of an atom, making up most of the atom's mass. It contains one or more positively charged protons and one or more neutrons with no electrical charge (except for a hydrogen-1 atom, which has one proton and no neutrons in its nucleus).

nutrient   Any chemical element or compound an organism must take in to live, grow, or reproduce.

nutrient cycle   See *biogeochemical cycle.*

nutrient cycling   The circulation of chemicals necessary for life, from the environment (mostly from soil and water) through organisms and back to the environment

oil   See *crude oil.*

old-growth forest   Virgin and old, second-growth forests containing trees that are often hundreds—sometimes thousands—of years old. Examples include forests of Douglas fir, western hemlock, giant sequoia, and coastal redwoods in the western United States. Compare *second-growth forest, tree plantation.*

oligotrophic lake   Lake with a low supply of plant nutrients. Compare *eutrophic lake, mesotrophic lake*.

omnivore   Animal that can use both plants and other animals as food sources. Examples include pigs, rats, cockroaches, and humans. Compare *carnivore, herbivore*.

open access renewable resource   Renewable resource owned by no one and available for use by anyone at little or no charge. Examples include clean air, underground water supplies, the open ocean and its fish, and the ozone layer. Compare *common property resource, private property resource*.

open sea   Part of an ocean that lies beyond the continental shelf. Compare *coastal zone*.

organic compounds   Compounds containing carbon atoms combined with each other and with atoms of one or more other elements such as hydrogen, oxygen, nitrogen, sulfur, phosphorus, chlorine, and fluorine. All other compounds are called *inorganic compounds*.

organism   Any form of life.

output   Matter, energy, or information leaving a system. Compare *input, throughput*.

output pollution control   See *pollution cleanup*.

overfishing   Harvesting so many fish of a species, especially immature individuals, that not enough breeding stock is left to replenish the species and it becomes unprofitable to harvest them.

overgrazing   Destruction of vegetation when too many grazing animals feed too long and exceed the carrying capacity of a rangeland or pasture area.

ozone (O₃)   Colorless and highly reactive gas and a major component of photochemical smog. Also found in the ozone layer in the stratosphere. See *photochemical smog*.

ozone layer   Layer of gaseous ozone (O₃) in the stratosphere that protects life on earth by filtering out most harmful ultraviolet radiation from the sun.

paradigm shift   Shift in thinking that occurs when the majority of scientists in a field or related fields agree that a new explanation or theory is better than the old one

parasite   Consumer organism that lives on or in, and feeds on, a living plant or animal, known as the host, over an extended period. The parasite draws nourishment from and gradually weakens its host; it may or may not kill the host. See *parasitism*.

parasitism   Interaction between species in which one organism, called the parasite, preys on another organism, called the host, by living on or in the host. See *host, parasite*.

pathogen   Living organism that can cause disease in another organism. Examples include bacteria, viruses, and parasites.

peer review   Process of scientists reporting details of the methods and models they used, the results of their experiments, and the reasoning behind their hypotheses for other scientists working in the same field (their peers) to examine and criticize.

per capita ecological footprint   Amount of biologically productive land and water needed to supply each person or population with the renewable resources they use and to absorb or dispose of the wastes from such resource use. It measures the average environmental impact of individuals or populations in different countries and areas. Compare *ecological footprint*.

per capita GDP   Annual gross domestic product (GDP) of a country divided by its total population at midyear. It gives the average slice of the economic pie per person. Used to be called per capita gross national product (GNP). See *gross domestic product*.

per capita GDP PPP (Purchasing Power Parity)   Measure of the amount of goods and services that a country's average citizen could buy in the United States.

perennial   Plant that can live for more than 2 years. Compare *annual*.

permafrost   Perennially frozen layer of the soil that forms when the water there freezes. It is found in arctic tundra.

perpetual resource   Essentially inexhaustible resource on a human time scale because it is renewed continuously. Solar energy is an example. Compare *nonrenewable resource, renewable resource*.

persistence   (1) the ability of a living system, such as a grassland or a forest, to survive moderate disturbances (2) the tendency for a pollutant to stay in the air, water, soil, or body. Compare *constancy, resilience*.

pest   Unwanted organism that directly or indirectly interferes with human activities.

petroleum   See *crude oil*.

phosphorus cycle   Cyclic movement of phosphorus in different chemical forms from the environment to organisms and then back to the environment.

photosynthesis   Complex process that takes place in cells of green plants. Radiant energy from the sun is used to combine carbon dioxide (CO₂) and water (H₂O) to produce oxygen (O₂), carbohydrates (such as glucose, C₆H₁₂O₆), and other nutrient molecules. Compare *aerobic respiration, chemosynthesis*.

physical change   Process that alters one or more physical properties of an element or a compound without changing its chemical composition. Examples include changing the size and shape of a sample of matter (crushing ice and cutting aluminum foil) and changing a sample of matter from one physical state to another (boiling and freezing water). Compare *chemical change, nuclear change*.

phytoplankton   Small, drifting plants, mostly algae and bacteria, found in aquatic ecosystems. Compare *plankton, zooplankton*.

pioneer community   First integrated set of plants, animals, and decomposers found in an area undergoing primary ecological succession. See *immature community, mature community*.

pioneer species   First hardy species—often microbes, mosses, and lichens—that begin colonizing a site as the first stage of ecological succession. See *ecological succession, pioneer community*.

planetary management worldview   Worldview holding that humans are separate from nature, that nature exists mainly to meet our needs and increasing wants, and that we can use our ingenuity and technology to manage the earth's life-support systems, mostly for our benefit. It assumes that economic growth is unlimited. Compare *deep ecology worldview, environmental wisdom worldview, stewardship worldview*.

plankton   Small plant organisms (phytoplankton) and animal organisms (zooplankton) that float in aquatic ecosystems.

point source   Single identifiable source that discharges pollutants into the environment. Examples include the smokestack of a power plant or an industrial plant, drainpipe of a meatpacking plant, chimney of a house, or exhaust pipe of an automobile. Compare *nonpoint source*.

pollutant   Particular chemical or form of energy that can adversely affect the health, survival, or activities of humans or other living organisms. See *pollution*.

pollution   Undesirable change in the physical, chemical, or biological characteristics of air, water, soil, or food that can adversely affect the health, survival, or activities of humans or other living organisms.

pollution cleanup   Device or process that removes or reduces the level of a pollutant after it has been produced or has entered the environment. Examples include automobile emission control devices and sewage treatment plants. Compare *pollution prevention*.

pollution prevention   Device, process, or strategy used to prevent a potential pollutant from forming or entering the environment or to sharply reduce the amount entering the environment. Compare *pollution cleanup*.

population   Group of individual organisms of the same species living in a particular area.

population change   Increase or decrease in the size of a population. It is equal to (Births + Immigration) − (Deaths + Emigration).

population density   Number of organisms in a particular population found in a specified area or volume.

population dispersion   General pattern in which the members of a population are arranged throughout its habitat.

population distribution   Variation of population density over a particular geographic area or volume. For example, a country has a high population density in its urban areas and a much lower population density in its rural areas.

population dynamics Major abiotic and biotic factors that tend to increase or decrease the population size and affect the age and sex composition of a species.

population size Number of individuals making up a population's gene pool.

positive feedback loop Feedback loop that causes a system to change further in the same direction. Compare *negative feedback loop*.

potential energy Energy stored in an object because of its position or the position of its parts. Compare *kinetic energy*.

poverty Inability to meet basic needs for food, clothing, and shelter.

prairie See *grassland*.

precautionary principle When there is significant scientific uncertainty about potentially serious harm from chemicals or technologies, decision makers should act to prevent harm to humans and the environment. See *pollution prevention*.

precipitation Water in the form of rain, sleet, hail, and snow that falls from the atmosphere onto land and bodies of water.

predation Interaction in which an organism of one species (the predator) captures and feeds on parts or all of an organism of another species (the prey).

predator Organism that captures and feeds on parts or all of an organism of another species (the prey).

predator–prey relationship Relationship that has evolved between two organisms, in which one organism has become the prey for the other, the latter called the predator. See *predator, prey*.

prey Organism that is captured and serves as a source of food for an organism of another species (the predator).

primary consumer Organism that feeds on all or part of plants (herbivore) or on other producers. Compare *detritivore, omnivore, secondary consumer*.

primary productivity See *gross primary productivity, net primary productivity*.

primary succession Ecological succession in a bare area that has never been occupied by a community of organisms. See *ecological succession*. Compare *secondary succession*.

principles of sustainability Principles by which nature has sustained itself for billions of years by relying on solar energy, biodiversity, population regulation, and nutrient recycling.

private property resource Land, mineral, or other resource owned by individuals or by a firm. Compare *common property resource, open access renewable resource*.

probability Mathematical statement about how likely it is that something will happen.

producer Organism that uses solar energy (green plants) or chemical energy (some bac-

teria) to manufacture the organic compounds it needs as nutrients from simple inorganic compounds obtained from its environment. Compare *consumer, decomposer*.

prokaryotic cell Cell containing no distinct nucleus or organelles. Compare *eukaryotic cell*.

proton (p) Positively charged particle in the nuclei of all atoms. Each proton has a relative mass of 1 and a single positive charge. Compare *electron, neutron*.

pyramid of energy flow Diagram representing the flow of energy through each trophic level in a food chain or food web. With each energy transfer, only a small part (typically 10%) of the usable energy entering one trophic level is transferred to the organisms at the next trophic level.

radioactive decay Change of a radioisotope to a different isotope by the emission of radioactivity.

radioactive isotope See *radioisotope*.

radioactivity Nuclear change in which unstable nuclei of atoms spontaneously shoot out "chunks" of mass, energy, or both at a fixed rate. The three principal types of radioactivity are gamma rays and fast-moving alpha particles and beta particles.

radioisotope Isotope of an atom that spontaneously emits one or more types of radioactivity (alpha particles, beta particles, gamma rays).

rain shadow effect Low precipitation on the leeward side of a mountain when prevailing winds flow up and over a high mountain or range of high mountains, creating semiarid and arid conditions on the leeward side of a high mountain range.

rangeland Land that supplies forage or vegetation (grasses, grass-like plants, and shrubs) for grazing and browsing animals and is not intensively managed. Compare *feedlot, pasture*.

range of tolerance Range of chemical and physical conditions that must be maintained for populations of a particular species to stay alive and grow, develop, and function normally. See *law of tolerance*.

rare species Species that has naturally small numbers of individuals (often because of limited geographic ranges or low population densities) or that has been locally depleted by human activities.

realized niche Parts of the fundamental niche of a species that are actually used by that species. See *ecological niche, fundamental niche*.

reconciliation ecology Science of inventing, establishing, and maintaining habitats to conserve species diversity in places where people live, work, or play.

recycling Collecting and reprocessing a resource so that it can be made into new products. An example is collecting aluminum cans, melting them down, and using the aluminum to make new cans or other aluminum products. Compare *reuse*.

reforestation Renewal of trees and other types of vegetation on land where trees have been removed; can be done naturally by seeds from nearby trees or artificially by planting seeds or seedlings.

reliable science Concepts and ideas that are widely accepted by experts in a particular field of the natural or social sciences. Compare *tentative science, unreliable science*.

renewable resource Resource that can be replenished rapidly (hours to several decades) through natural processes as long as it is not used up faster than it is replaced. Examples include trees in forests, grasses in grasslands, wild animals, fresh surface water in lakes and streams, most groundwater, fresh air, and fertile soil. If such a resource is used faster than it is replenished, it can be depleted and converted into a nonrenewable resource. Compare *nonrenewable resource* and *perpetual resource*. See also *environmental degradation*.

replacement-level fertility Average number of children a couple must bear to replace themselves. The average for a country or the world usually is slightly higher than two children per couple (2.1 in the United States and 2.5 in some developing countries) mostly because some children die before reaching their reproductive years. See also *total fertility rate*.

reproduction Production of offspring by one or more parents.

reproductive isolation Long-term geographic separation of members of a particular sexually reproducing species.

reproductive potential See *biotic potential*.

resilience Ability of a living system to be restored through secondary succession after a moderate disturbance.

resource Anything obtained from the environment to meet human needs and wants. It can also be applied to other species.

resource partitioning Process of dividing up resources in an ecosystem so that species with similar needs (overlapping ecological niches) use the same scarce resources at different times, in different ways, or in different places. See *ecological niche, fundamental niche, realized niche*.

resource productivity See *material efficiency*.

respiration See *aerobic respiration*.

restoration ecology Research and scientific study devoted to restoring, repairing, and reconstructing damaged ecosystems.

reuse Using a product over and over again in the same form. An example is collecting, washing, and refilling glass beverage bottles. Compare *recycling*.

riparian zones Thin strips and patches of vegetation that surround streams. They are very important habitats and resources for wildlife.

r-selected species Species that reproduce early in their life span and produce large numbers of usually small and short-lived offspring in a short period. Compare *K-selected species*.

r-strategists See *r-selected species*.

rule of 70 Doubling time (in years) = 70/(percentage growth rate). See *doubling time, exponential growth*.

salinity Amount of various salts dissolved in a given volume of water.

salinization Accumulation of salts in soil that can eventually make the soil unable to support plant growth.

scavenger Organism that feeds on dead organisms that were killed by other organisms or died naturally. Examples include vultures, flies, and crows. Compare *detritivore*.

science Attempts to discover order in nature and use that knowledge to make predictions about what is likely to happen in nature. See *reliable science, scientific data, scientific hypothesis, scientific law, scientific methods, scientific model, scientific theory, tentative science, unreliable science*.

scientific data Facts obtained by making observations and measurements. Compare *scientific hypothesis, scientific law, scientific methods, scientific model, scientific theory*.

scientific hypothesis An educated guess that attempts to explain a scientific law or certain scientific observations. Compare *scientific data, scientific law, scientific methods, scientific model, scientific theory*.

scientific law Description of what scientists find happening in nature repeatedly in the same way, without known exception. See *first law of thermodynamics, law of conservation of matter, second law of thermodynamics*. Compare *scientific data, scientific hypothesis, scientific methods, scientific model, scientific theory*.

scientific methods The ways scientists gather data and formulate and test scientific hypotheses, models, theories, and laws. See *scientific data, scientific hypothesis, scientific law, scientific model, scientific theory*.

scientific model A simulation of complex processes and systems. Many are mathematical models that are run and tested using computers.

scientific theory A well-tested and widely accepted scientific hypothesis. Compare *scientific data, scientific hypothesis, scientific law, scientific methods, scientific model*.

secondary consumer Organism that feeds only on primary consumers. Compare *detritivore, omnivore, primary consumer*.

secondary succession Ecological succession in an area in which natural vegetation has been removed or destroyed but the soil or bottom sediment has not been destroyed. See *ecological succession*. Compare *primary succession*.

second-growth forest Stands of trees resulting from secondary ecological succession. Compare *old-growth forest, tree farm*.

second law of energy See *second law of thermodynamics*.

second law of thermodynamics In any conversion of heat energy to useful work, some of the initial energy input is always degraded to lower-quality, more dispersed, less useful energy—usually low-temperature heat that flows into the environment; you cannot break even in terms of energy quality. See *first law of thermodynamics*.

selective cutting Cutting of intermediate-aged, mature, or diseased trees in an uneven-aged forest stand, either singly or in small groups. This encourages the growth of younger trees and maintains an uneven-aged stand. Compare *clear-cutting, strip cutting*.

sexual reproduction Reproduction in organisms that produce offspring by combining sex cells or *gametes* (such as ovum and sperm) from both parents. It produces offspring that have combinations of traits from their parents. Compare *asexual reproduction*.

social capital Result of getting people with different views and values to talk and listen to one another, find common ground based on understanding and trust, and work together to solve environmental and other problems.

soil Complex mixture of inorganic minerals (clay, silt, pebbles, and sand), decaying organic matter, water, air, and living organisms.

solar capital Solar energy that warms the planet and supports photosynthesis, the process that plants use to provide food for themselves and for us and other animals. This direct input of solar energy also produces indirect forms of renewable solar energy such as wind and flowing water. Compare *natural capital*.

solar energy Direct radiant energy from the sun and a number of indirect forms of energy produced by the direct input of such radiant energy. Principal indirect forms of solar energy include wind, falling and flowing water (hydropower), and biomass (solar energy converted into chemical energy stored in the chemical bonds of organic compounds in trees and other plants)—none of which would exist without direct solar energy.

sound science See *reliable science*.

specialist species Species with a narrow ecological niche. They may be able to live in only one type of habitat, tolerate only a narrow range of climatic and other environmental conditions, or use only one type or a few types of food. Compare *generalist species*.

speciation Formation of two species from one species because of divergent natural selection in response to changes in environmental conditions; usually takes thousands of years. Compare *extinction*.

species Group of similar organisms, and for sexually reproducing organisms, they are a set of individuals that can mate and produce fertile offspring. Every organism is a member of a certain species.

species diversity Number of different species (species richness) combined with the relative abundance of individuals within each of those species (species evenness) in a given area. See *biodiversity, species evenness, species richness*. Compare *ecological diversity, genetic diversity*.

species equilibrium model See *theory of island biogeography*.

species evenness Relative abundance of individuals within each of the species in a community. See *species diversity*. Compare *species richness*.

species richness Number of different species contained in a community. See *species diversity*. Compare *species evenness*.

S-shaped curve Leveling off of an exponential, J-shaped curve when a rapidly growing population reaches or exceeds the carrying capacity of its environment and ceases to grow.

statistics Mathematical tools used to collect, organize, and interpret numerical data.

stewardship worldview Worldview holding that we can manage the earth for our benefit but that we have an ethical responsibility to be caring and responsible managers, or *stewards*, of the earth. It calls for encouraging environmentally beneficial forms of economic growth and discouraging environmentally harmful forms. Compare *deep ecology worldview, environmental wisdom worldview, planetary management worldview*.

stratosphere Second layer of the atmosphere, extending about 17–48 kilometers (11–30 miles) above the earth's surface. It contains small amounts of gaseous ozone (O_3), which filters out about 95% of the incoming harmful ultraviolet radiation emitted by the sun. Compare *troposphere*.

stream Flowing body of surface water. Examples are creeks and rivers.

subatomic particles Extremely small particles—electrons, protons, and neutrons—that make up the internal structure of atoms.

succession See *ecological succession, primary succession, secondary succession*.

succulent plants Plants, such as desert cacti, that survive in dry climates by having no leaves, thus reducing the loss of scarce water. They store water and use sunlight to produce the food they need in the thick, fleshy tissue of their green stems and branches. Compare *deciduous plants, evergreen plants*.

sulfur cycle Cyclic movement of sulfur in various chemical forms from the environment to organisms and then back to the environment.

sulfur dioxide (SO_2) Colorless gas with an irritating odor. About one-third of the SO_2 in the atmosphere comes from natural sources as part of the sulfur cycle. The other two-thirds come from human sources, mostly combustion of sulfur-containing coal in electric power and industrial plants and from oil refining and smelting of sulfide ores.

surface fire Forest fire that burns only undergrowth and leaf litter on the forest floor. Compare *crown fire, ground fire*. See *controlled burning*.

surface water Precipitation that does not infiltrate the ground or return to the atmosphere by evaporation or transpiration. See *runoff*. Compare *groundwater*.

survivorship curve   Graph showing the number of survivors in different age groups for a particular species.

sustainability   Ability of earth's various systems, including human cultural systems and economies, to survive and adapt to changing environmental conditions indefinitely.

sustainable development   See *environmentally sustainable economic development.*

sustainable living   Taking no more potentially renewable resources from the natural world than can be replenished naturally and not overloading the capacity of the environment to cleanse and renew itself by natural processes.

sustainable society   Society that manages its economy and population size without doing irreparable environmental harm by overloading the planet's ability to absorb environmental insults, replenish its resources, and sustain human and other forms of life over a specified period, usually hundreds to thousands of years. During this period, the society satisfies the needs of its people without depleting natural resources and thereby jeopardizing the prospects of current and future generations of humans and other species.

sustainable yield (sustained yield)   Highest rate at which a potentially renewable resource can be used indefinitely without reducing its available supply. See also *environmental degradation.*

synergistic interaction   Interaction of two or more factors or processes so that the combined effect is greater than the sum of their separate effects.

system   Set of components that function and interact in some regular and theoretically predictable manner.

temperature   Measure of the average speed of motion of the atoms, ions, or molecules in a substance or combination of substances at a given moment. Compare *heat.*

tentative science   Preliminary scientific data, hypotheses, and models that have not been widely tested and accepted. Compare *reliable science, unreliable science.*

terrestrial   Pertaining to land. Compare *aquatic.*

tertiary (higher-level) consumers   Animals that feed on animal-eating animals. They feed at high trophic levels in food chains and webs. Examples include hawks, lions, bass, and sharks. Compare *detritivore, primary consumer, secondary consumer.*

theory of evolution   Widely accepted scientific idea that all life forms developed from earlier life forms. It is the way most biologists explain how life has changed over the past 3.6–3.8 billion years and why it is so diverse today.

theory of island biogeography   Widely accepted scientific theory holding that the number of different species (species richness) found on an island is determined by the interactions of two factors: the rate at which new species immi-grate to the island and the rate at which species become *extinct,* or cease to exist, on the island. See *species richness.*

third and higher-level consumers   Carnivores such as tigers and wolves that feed on the flesh of carnivores.

threatened species   Wild species that is still abundant in its natural range but is likely to become endangered because of a decline in numbers. Compare *endangered species.*

threshold effect   Harmful or fatal effect of a small change in environmental conditions that exceeds the limit of tolerance of an organism or population of a species. See *law of tolerance.*

throughput   Rate of flow of matter, energy, or information through a system. Compare *input, output.*

throwaway society   See *high-throughput economy.*

tipping point   Threshold level at which an environmental problem causes a fundamental and irreversible shift in the behavior of a system.

tolerance limits   Minimum and maximum limits for physical conditions (such as temperature) and concentrations of chemical substances beyond which no members of a particular species can survive. See *law of tolerance.*

total fertility rate (TFR)   Estimate of the average number of children who will be born alive to a woman during her lifetime if she passes through all her childbearing years (ages 15–44) conforming to age-specific fertility rates of a given year. More simply, it is an estimate of the average number of children that women in a given population will have during their childbearing years.

tragedy of the commons   Depletion or degradation of a potentially renewable resource to which people have free and unmanaged access. An example is the depletion of commercially desirable fish species in the open ocean beyond areas controlled by coastal countries. See *common-property resource, open access renewable resource.*

trait   Characteristic passed on from parents to offspring during reproduction in an animal or plant.

transpiration   Process in which water is absorbed by the root systems of plants, moves up through the plants, passes through pores (stomata) in their leaves or other parts, and evaporates into the atmosphere as water vapor.

tree farm   See *tree plantation.*

tree plantation   Site planted with one or only a few tree species in an even-aged stand. When the stand matures it is usually harvested by clear-cutting and then replanted. These farms normally raise rapidly growing tree species for fuelwood, timber, or pulpwood. Compare *old-growth forest, second-growth forest.*

trophic level   All organisms that are the same number of energy transfers away from the original source of energy (for example, sunlight) that enters an ecosystem. For example, all producers belong to the first trophic level, and all herbivores belong to the second trophic level in a food chain or a food web.

troposphere   Innermost layer of the atmosphere. It contains about 75% of the mass of earth's air and extends about 17 kilometers (11 miles) above sea level. Compare *stratosphere.*

unreliable science   Scientific results or hypotheses presented as reliable science but not having undergone the rigors of the peer review process. Compare *reliable science, tentative science.*

upwelling   Movement of nutrient-rich bottom water to the ocean's surface. It can occur far from shore but usually takes place along certain steep coastal areas where the surface layer of ocean water is pushed away from shore and replaced by cold, nutrient-rich bottom water.

utilitarian value   See *instrumental value.*

warm front   Boundary between an advancing warm air mass and the cooler one it is replacing. Because warm air is less dense than cool air, an advancing warm front rises over a mass of cool air. Compare *cold front.*

water cycle   See *hydrologic cycle.*

watershed   Land area that delivers water, sediment, and dissolved substances via small streams to a major stream (river).

weather   Short-term changes in the temperature, barometric pressure, humidity, precipitation, sunshine, cloud cover, wind direction and speed, and other conditions in the troposphere at a given place and time. Compare *climate.*

wetland   Land that is covered all or part of the time with salt water or fresh water, excluding streams, lakes, and the open ocean. See *coastal wetland, inland wetland.*

wilderness   Area where the earth and its ecosystems have not been seriously disturbed by humans and where humans are only temporary visitors.

wildlife   All free, undomesticated species. Sometimes the term is used to describe animals only.

wildlife resources   Wildlife species that have actual or potential economic value to people.

wild species   Species found in the natural environment. Compare *domesticated species.*

worldview   How people think the world works and what they think their role in the world should be. See *environmental wisdom worldview, planetary management worldview, stewardship worldview.*

zooplankton   Animal plankton; small floating herbivores that feed on plant plankton (phytoplankton). Compare *phytoplankton.*

Index